A Course in Linear Algebra

David B. Damiano
Department of Mathematics and Computer Science
College of the Holy Cross

John B. Little
Department of Mathematics and Computer Science
College of the Holy Cross

Dover Publications, Inc.
Mineola, New York

Bibliographical Note

This Dover edition, first published in 2011, is an unabridged republication of the work originally published in 1988 by Harcourt Brace Jovanovich, Inc., New York. The authors have made corrections to the text and added an Errata List for this edition.

Library of Congress Cataloging-in-Publication Data

Damiano, David B.
A course in linear algebra / David B. Damiano, John B. Little.
p. cm.
Originally published: Mineola, N.Y. : Harcourt Brace Jovanovich, 1988.
Includes index.
Summary: "Suitable for advanced undergraduates and graduate students, this text introduces basic concepts of linear algebra. Each chapter contains an introduction, definitions, and propositions, in addition to multiple examples, lemmas, theorems, corollaries, and proofs. Each chapter features numerous supplemental exercises, and solutions to selected problems appear at the end. 1988 edition" — Provided by publisher.
ISBN-13: 978-0-486-46908-9 (pbk.)
ISBN-10: 0-486-46908-5 (pbk.)
1. Algebras, Linear. I. Little, John B. II. Title.

QA184.D35 2011
512'.5—dc22

2011005199

Manufactured in the United States by Courier Corporation
46908501
www.doverpublications.com

Preface

A Course in Linear Algebra was shaped by three major aims: (1) to present a reasonably complete, mathematically honest introduction to the basic concepts of linear algebra; (2) to raise our students' level of mathematical maturity significantly in the crucial sophomore year of their undergraduate careers; and, (3) to present the material in an interesting, inspiring fashion, to provide motivation for the students and give them a feeling both for the overall logical structure of the subject and for the way linear algebra helps to explain phenomena and solve problems in many areas of mathematics and its applications.

We firmly believe that, even today, the traditional theoretical approach to the teaching of linear algebra is appropriate and desirable for students who intend to continue their studies in the mathematical sciences.

With such goals in mind, we have chosen to present linear algebra as the study of vector spaces and linear mappings between vector spaces rather than as the study of matrix algebra or the study of numerical techniques for solving systems of linear equations. We define general vector spaces and linear mappings at the outset, and we base all of the subsequent developments on these ideas.

We feel that this approach has several major benefits for our intended audience of mathematics majors (and for others as well). First, it highlights the way seemingly unrelated sets of phenomena, such as the algebraic properties of vectors in the plane and the algebraic properties of functions $f:\mathbf{R} \rightarrow \mathbf{R}$, or the geometric behavior of projections and rotations in the plane and the differentiation rules for sums and scalar multiples of functions, may be unified and understood as different instances of more fundamental patterns. Furthermore, once these essential similarities are recognized, they may be exploited to solve other problems.

Second, our approach provides a ready-made *context, motivation,* and *geometric interpretation* for each new computational technique that we present. For example, the Gauss-Jordan elimination procedure for solving systems of linear equations is introduced first in order to allow us to answer questions about parametrizations (spanning sets) for subspaces, linear independence and dependence of sets of vectors, and the like.

Finally, our approach offers the opportunity to introduce proofs and abstract problem-solving into the course from the beginning. We believe that all students of mathematics at this level must begin to practice applying what they have learned in new situations, and not merely master routine calculations. In addition, they must begin to learn how to construct correct and convincing proofs of their assertions—that is the way they will be working and communicating with their colleagues as long as they stay in the mathematical sciences. Since the subject matter of linear algebra is relatively uncomplicated, this is the ideal place to start.

We have included some important mathematical applications of the topics that we cover, such as the application of the spectral theorem for symmetric real matrices to the geometry of conic sections and quadric surfaces, and the application of diagonalization and Jordan canonical form to the theory of systems of ordinary differential equations. We have not included many applications of linear algebra to problems in other disciplines, however, both because of the difficulty of presenting convincing, realistic applied problems at this level, and because of the needs of our audience. We prefer to give students a deep understanding of the mathematics that will be useful and to leave the discussion of the applications themselves to other courses.

A WORD ON PREREQUISITES

Since students taking the sophomore linear algebra course have typically had at least one year of one-variable calculus, we have felt free to use various facts from calculus (mainly properties of and formulas for derivatives and integrals) in many of the examples and exercises in the first six chapters. These examples may be omitted if necessary. The seventh chapter, which covers some topics in differential equations, uses substantially more calculus, through derivatives of vector-valued functions of one variable and partial derivatives.

In the text and in proofs, we have also freely used some ideas such as the technique of proof by mathematical induction, the division algorithm for polynomials in one variable and its consequences about roots and factorizations of polynomials, and various notions about sets and functions. Ideally, these topics should be familiar to students from high school mathematics; they are also reviewed briefly in the text or in the appendices for easy reference.

SOME COMMENTS ON THE EXERCISES

In keeping with our goals for the course, we have tried to structure the book so that, as they progress through the course, students will start to become active participants in the theoretical development of linear algebra, rather than remaining passive bystanders. Thus, in the exercises for each section, in addition to computational problems illustrating the main points of the section, we have included proofs of parts of propositions stated in the text and other problems dealing with related topics and extensions of the results of the text. We have sometimes used the exercises to introduce new ideas that will be used later, when those ideas are straightforward enough to make it possible to ask reasonable questions with only a minimum of background. In addition to the exercises at the end of each section, there are supplementary exercises at the end of each chapter. These provide a review of the foregoing material and extend concepts developed in the chapter. Included in the supplementary exercises are true-false questions designed to test the student's mastery of definitions and statements of theorems.

Following the appendices, we provide solutions to selected exercises. In particular, we give solutions to alternative parts of exercises requiring numerical solutions, solutions to exercises that are proofs of propositions in the text, and solutions to exercises that are used subsequently in the text.

Finally, as the course progresses, we have included numerous extended sequences of exercises that develop other important topics in, or applications of, linear algebra. We strongly recommend that instructors using this book assign some of these exercises from time to time. Though some of them are rather difficult, we have tried to structure the questions to lead students to the right approach to the solution. In addition, many hints are provided. These exercise sequences can also serve as the basis for individual or group papers or in-class presentations, if the instructor desires. We have found assignments of this type to be very worthwhile and enjoyable, even for students at the sophomore level. A list of these exercises appears after the acknowledgments.

ACKNOWLEDGMENTS

We would first like to thank our students at Holy Cross for their participation and interest in the development of this book. Their comments and suggestions have been most valuable at all stages of the project. We thank our colleagues Mel Tews, Pat Shanahan, and Margaret Freije, who have also taught from our material and have made numerous recommendations for improvements that have been incorporated in the book. The following reviewers have read the manuscript and contributed many useful comments: Susan C. Geller, Texas A&M University; Harvey J. Schmidt, Jr., Lewis and Clark College; Joseph Neggers, University of Alabama; Harvey C. Greenwald, California Polytechnic State University; and Samuel G. Councilman, California State University, Long Beach.

Finally, we thank Mrs. Joy Bousquet for her excellent typing, and the entire production staff at Harcourt Brace Jovanovich.

David B. Damiano
John B. Little

COURSES OF STUDY

To allow for flexibility in constructing courses using this book, we have included more material than can be covered in a typical one-semester course. In fact, the book as a whole contains almost exactly enough material for a full-year course in linear algebra. The basic material that should be covered in any one-semester course is contained in the first three chapters and the first two sections of the fourth chapter. We have structured the book so that several different types of coherent one-semester courses may be constructed by choosing some additional topics and omitting others.

Option I

A one-semester course culminating in the geometry of the inner product space $\mathbf{R}^n$ and the spectral theorem for symmetric mappings would consist of the core material plus the remaining sections of Chapter 4. We have covered all of this material in a 13-week semester at Holy Cross.

Option II

A one-semester course culminating in an introduction to differential equations would consist of the core material plus Sections 1, 2, 4, and 5 of Chapter 7. (The third section of Chapter 7 presupposes the Jordan canonical form, and the description of the form of the solutions to higher-order, constant-coefficient equations in Section 4 is deduced by reducing to a system of first-order equations and applying the results of Section 3. Nevertheless, the main result of Section 4, Theorem 7.4.5, could be justified in other ways, if desired. The Jordan form could also be introduced without proof in order to deduce these results, since the final method presented for explicitly solving the equations makes no use of matrices.)

Option III

A one-semester course incorporating complex arithmetic and the study of vector spaces over **C** and other fields would consist of the core material plus Sections 1 and 2 of Chapter 5. (If time permits, additional sections from Chapter 4 or Chapter 6 could also be included.)

Errata

• Pages 145-146, from the first displayed equation on page 145 to the end of the third line on page 146 should be replaced by the following:

$$\begin{aligned}
\det(A) &= \sum_{j=1}^{n}(-a)^{1+j}a_{1j}\left[\sum_{q=1}^{j-1}(-1)^{1+q}a_{2q}\det((A_{1j})_{1q})+\right.\\
&\qquad \left.\sum_{q=j+1}^{n}(-1)^{1+(q-1)}a_{2q}\det((A_{1j})_{1(q-1)})\right]\\
&= \sum_{j=1}^{n}\left[\sum_{q=1}^{j-1}(-1)^{1+j+1+q}a_{1j}a_{2q}\det((A_{1j})_{1q}) + \right.\\
&\qquad \left.\sum_{q=j+1}^{n}(-1)^{1+j+1+(q-1)}a_{1j}a_{2q}\det((A_{1j})_{1(q-1)})\right].
\end{aligned}$$

Together j and q range over all possible pairs of integers between 1 and n inclusive, except for the pairs when $j = q$. We can switch the order of summation as follows:

$$\begin{aligned}
\det(A) &= \sum_{q=1}^{n}\left[\sum_{j=1}^{q-1}(-1)^{1+j+1+(q-1)}a_{2q}a_{1j}\det((A_{1q})_{1j})+\right.\\
&\qquad \left.\sum_{j=q+1}^{n}(-1)^{1+j+1+q}a_{2q}a_{1j}\det((A_{1q})_{1(j-1)})\right].
\end{aligned}$$

Note we used the facts that if $j < q$, $(A_{1j})_{1(q-1)} = (A_{1q})_{1j}$ and if $j > q$, $(A_{1j})_{1q} = (A_{1q})_{1(j-1)}$. Multiplying by -1 we recover

$$\begin{aligned}
-\det(A) &= \sum_{q=1}^{n}\left[\sum_{j=1}^{q-1}(-1)^{1+j+1+q}a_{2q}a_{1j}\det((A_{1q})_{1j})+\right.\\
&\qquad \left.\sum_{j=q+1}^{n}(-1)^{1+q+1+(j-1)}a_{2q}a_{1j}\det((A_{1q})_{1(j-1)})\right].
\end{aligned}$$

Reversing the above steps, we see the right side is the determinant of the matrix obtained from A by switching the first two rows. Since we multiplied by -1 to achieve this identity, we conclude that the determinant is alternating.

- Page 246, Proposition (5.3.11) should read:

If $\mathbf{u}$ *and* $\mathbf{v}$ *are eigenvectors, respectively, for distinct eigenvalues* λ *and* μ *of a self adjoint transformation* $T : V \longrightarrow V$*, then* $\langle \mathbf{u}, \mathbf{v} \rangle = 0$*, so* $\mathbf{u}$ *and* $\mathbf{v}$ *are orthogonal.*

- Page 265, equation (6.2) should appear as follows:

$$[N|_{C(\mathbf{x})}]_\alpha^\alpha = \begin{pmatrix} 0 & 1 & 0 & \cdots & 0 \\ 0 & 0 & 1 & \cdots & 0 \\ \vdots & & \ddots & \ddots & \vdots \\ 0 & & & 0 & 1 \\ 0 & 0 & \cdots & 0 & 0 \end{pmatrix}$$

(The entries in row i and column j with $j - i = 1$ are all ones. All other entries are zeroes.)

- Page 274, matrix on line 13 should appear as follows:

$$\begin{pmatrix} \lambda & & & * \\ 0 & \lambda & & \\ \vdots & \ddots & \ddots & \\ 0 & \cdots & 0 & \lambda \end{pmatrix}$$

(All the diagonal entries are λ and all entries below the main diagonal are zeroes.)

- Page 276, should be replaced by the following:

Now, inductively again, assume we have constructed vectors $\mathbf{y}_{\ell,1}, \ldots, \mathbf{y}_{\ell,j-1}$ such that $T(\mathbf{y}_{\ell,n}) = \lambda_\ell \mathbf{y}_{\ell,n} + \mathbf{u}_n$, where $\mathbf{u}_n \in \mathrm{Span}(\{\mathbf{y}_{\ell,1}, \ldots, \mathbf{y}_{\ell,n-1}\})$ for each n $(1 \leq n \leq j-1)$ and

$$\mathrm{Span}(\{\mathbf{y}_{\ell,1}, \ldots, \mathbf{y}_{\ell,j-1}\}) + W_{\ell-1} = \mathrm{Span}(\{\mathbf{x}_{\ell,1}, \ldots, \mathbf{x}_{\ell,j-1}\}) + W_{\ell-1},$$

where $W_{\ell-1} = V_1' + \cdots + V_{\ell-1}'$. We know that $T(\mathbf{x}_{\ell,j}) = \lambda_\ell \mathbf{x}_{\ell,j} + \mathbf{u} + \mathbf{z}$ where $\mathbf{u} \in \mathrm{Span}(\{\mathbf{y}_{\ell,1}, \ldots, \mathbf{y}_{\ell,j-1}\})$ and $\mathbf{z} \in W_{\ell-1}$ by the construction of the triangular basis α and induction. As before, we can rewrite this as $(T - \lambda_\ell I)(\mathbf{x}_{\ell,j}) = \mathbf{u} + \mathbf{z}$. Since the only eigenvalues of $T|_{W_{\ell-1}}$ are $\lambda_1, \ldots, \lambda_{\ell-1}$, as before, we can find some vector $\mathbf{w} \in W_{\ell-1}$ such that $(T - \lambda_\ell I)(\mathbf{w}) = \mathbf{z}$. As a result, $(T - \lambda_\ell I)(\mathbf{x}_{\ell,j} - \mathbf{w}) = \mathbf{u} \in \mathrm{Span}(\{\mathbf{y}_{\ell,1}, \ldots, \mathbf{y}_{\ell,j-1}\})$. Therefore, we let $\mathbf{y}_{\ell,j} = \mathbf{x}_{\ell,j} - \mathbf{w}$. It can be seen that

$$\mathrm{Span}(\{\mathbf{y}_{\ell,1}, \ldots, \mathbf{y}_{\ell,j}\}) + W_{\ell-1} = \mathrm{Span}(\{\mathbf{x}_{\ell,1}, \ldots, \mathbf{x}_{\ell,j}\}) + W_{\ell-1},$$

so that the induction can continue, and we find the subspace

$$V'_\ell = \mathrm{Span}(\{\mathbf{y}_{\ell,1},\ldots,\mathbf{y}_{\ell,m_\ell}\})$$

after m_ℓ steps in all. By construction V'_ℓ is invariant under T, since it is invariant under $T-\lambda_\ell I$. Thus, we find the desired basis for V.

- Page 278, line -6. This matrix should be identical to the matrix of Equation 6.2, page 265. See the correction above for page 265.

- Page 278, Equation 6.7 should appear as follows:

$$\begin{pmatrix} \lambda_i & 1 & 0 & \cdots & 0 \\ 0 & \lambda_i & 1 & \cdots & 0 \\ \vdots & \ddots & \ddots & \ddots & \vdots \\ 0 & & \ddots & \lambda_i & 1 \\ 0 & 0 & \cdots & 0 & \lambda_i \end{pmatrix}$$

- Page 391, solution for Chapter 4, Section 4, Exercise 7 a should say:

Let $\mathbf{x},\mathbf{y}\in W^\perp$ and $c\in\mathbf{R}$. Then for all $\mathbf{z}\in W$, $\langle c\mathbf{x}+\mathbf{y},\mathbf{z}\rangle = c\langle\mathbf{x},\mathbf{z}\rangle+\langle\mathbf{y},\mathbf{z}\rangle = c\cdot 0+0=0$. Hence $c\mathbf{x}+\mathbf{y}\in W^\perp$, so $W^\perp$ is a subspace.

- Page 394, solution for Chapter 4, section 6, Exercise 1 e. The matrices of the orthogonal projections onto the subspaces $E_{(7\pm\sqrt{73})/2}$ should be:

$$\frac{1}{146\mp 10\sqrt{73}}\begin{pmatrix} 16 & 16 & 16 & -4(5\mp\sqrt{73}) \\ 16 & 16 & 16 & -4(5\mp\sqrt{73}) \\ 16 & 16 & 16 & -4(5\mp\sqrt{73}) \\ -4(5\mp\sqrt{73}) & -4(5\mp\sqrt{73}) & -4(5\mp\sqrt{73}) & (5\mp\sqrt{73})^2 \end{pmatrix}$$

- Page 398, solution for Chapter 4, Supplementary Exercises, Exercise 5 a. The first line of the solution should be:

$$\lambda = 1\pm\sqrt{29}. \qquad E_{1\pm\sqrt{29}} = \mathrm{Span}(\{(\pm 5,-2\pm\sqrt{29})/\|(\pm 5,-2\pm\sqrt{29})\|\}).$$

- Page 401, solution for Chapter 5, section 2, Exercise 12 b should be:

$$\lambda = \pm ia. \qquad E_{\pm ia} = \mathrm{Span}(\{(1,\mp i)\}).$$

- Page 419, solution for Chapter 7, section 4, Exercise 5 c should be:

$$u(t) = \frac{2}{3}e^t - \frac{1}{3}e^{-t/2}\cos\left(\frac{\sqrt{3}}{2}t\right) + \frac{\sqrt{3}}{3}e^{-t/2}\sin\left(\frac{\sqrt{3}}{2}t\right).$$

- Page 420, solution for Chapter 6. The third sentence should read:

If $\lambda_+ > 0$ and $c_1 \neq 0$, $\lambda_- > 0$ and $c_2 \neq 0$, or $\lambda > 0$ and either $c_1 \neq 0$ or $c_2 \neq 0$, then $\lim_{t\to\infty} |u(t)| = \infty$.

- Page 424, solution for Chapter 7, Supplementary Exercise 10 c should be:

$$a_n = \begin{cases} 0 & \text{if } n = 2k+1 \\ \dfrac{(-1)^{k+1}\binom{2k}{k-1}}{(2k)(2k-1)2^{2k-1}} & \text{if } n = 2k. \end{cases}$$

A Guide to the Exercises

Below is a listing by topic of the exercises that either introduce a new topic used later in the text or develop important extensions of the material covered in the text. The topics are listed in order of their first occurrence.

Topic	*Section (Exercise)*
Even and odd functions	1.2 (2)
Vector spaces of matrices	1.2 (11–16)
	1.3 (11–12)
	2.2 (10–15)
Direct sums of subspaces	1.3 (8–9)
Alternate version of elimination	1.5 (7–8)
Lagrange interpolation	1.6 (16)
Vector space of linear maps	2.1 (12)
	2.2 (10–15)
Restriction of a linear transformation	2.1 (14–15)
	2.6 (9)
Reflections	2.2 (4)
	2.7 (14)
	4.4 (14)
Transposes, symmetric, and skew-symmetric matrices	2.2 (12–13)
	2.3 (9)
Elimination in matrix form	2.3 (10–11)
Traces of matrix	2.3 (12)
Transpose as a linear transformation	2.3 (13)
Direct sum of linear transformations	2.4 (11)
	6.1 (6)

Topic	*Section (Exercise)*
Upper and lower triangular matrices	2.5 (9,10)
	2.6 (15)
	4.1 (11)
Elementary matrices	2.5 (14–18)
	2.6 (12–14)
	3.3 (12)
Cross product	3.2 (13–16)
Vandermonde determinants	3, Supplementary Exercises (9)
Involutions	4.1 (15–16)
	4.2 (8)
Companion matrices	4.1 (18)
Simultaneous diagonalizability	4.2 (14)
Polarization identity	4.3 (6)
Inner product on a real vector space	4.3 (10–13)
Bilinear mappings	4.3 (14–17)
Normal vector to a hyperplane	4.4 (9–11)
Bessel's inequality	4.4 (18)
Dual space	4.4 (19)
Normal matrices	4.5 (7)
Symmetric map with respect to a general inner product	4.5 (8)
Orthogonal matrices and transformations (orthogonal group)	4.6 (4,8)
Definite matrices	4.6 (11–13)
Special linear matrices (special linear group)	4, Supplementary Exercises (8)
Hessian matrix	4, Supplementary Exercises (12)
Quadratic forms (Sylvester's theorem)	4, Supplementary Exercises (18–20)
Rational numbers and related fields	5.1 (8)
Finite fields	5.1 (10–15)
$\mathbf{C}^n$ as a real vector space	5.1 (6)
	5.3 (6–10)
Vector space over a finite field	5.2 (11,13,14)
R as a vector space over Q	5.2 (15)
Skew-Hermitian matrices	5.3 (7)
Unitary matrices	5.3 (8,9)
Normal matrices	5.3 (10–12)
Hermitian inner products on $\mathbf{C}^n$	5.3 (13–14)
General projection mappings	6.1 (12)
Unipotent transformations	6.2 (8)
	6.3 (6)
Segre characteristics	6.3 (11)

Topic	*Section (Exercise)*
Invariant subspace of set of mappings	6.4 (7)
Irreducible set of mappings	6.4 (8–9)
Markov chains	6.4 (12)
Finite difference equations	7.2 (8–12)
	7.3 (15–18)
	7.4 (14–15)
Inhomogeneous first order ordinary differential equations	7.4 (8)
Alternate method of solving linear differential equations	7.4 (9,10)
Ordinary differential equations as operators	7.4 (11)
Non-constant coefficient ordinary differential equations	7.4 (12)
Heat equation	7.5 (12)
Legendre polynomials	7, Supplementary Exercises (7–12)

Contents

CHAPTER 1

Vector Spaces

Introduction

In this first chapter of *A Course in Linear Algebra,* we begin by introducing the fundamental concept of a vector space, a mathematical structure that has proved to be very useful in describing some of the common features of important mathematical objects such as the set of vectors in the plane and the set of all functions from the real line to itself.

Our first goal will be to write down a list of properties that hold for the algebraic sum and scalar multiplication operations in the two aforementioned examples. We will then take this list of properties as our definition of what a general vector space should be. This is a typical example of the idea of defining an object by specifying what properties it should have, a commonly used notion in mathematics.

We will then develop a repertoire of examples of vector spaces, drawing on ideas from geometry and calculus. Following this, we will explore the inner structure of vector spaces by studying subspaces and spanning sets and bases (special subsets from which the whole vector space can be built up). Along the way, we will find that most of the calculations that we need to perform involve solving simultaneous systems of linear equations, so we will also discuss a general method for doing this.

The concept of a vector space provides a way to organize, explain, and build on many topics you have seen before in geometry, algebra, and calculus. At the same time, as you begin to study linear algebra, you may find that the way everything is presented seems very general and abstract. Of course, to the extent that this is true, it is a reflection of the fact that mathematicians have seen a very basic general pattern that holds in many different situations. They have exploited this information by inventing the ideas discussed in this chapter in order to understand all these situations and treat them all without resorting to dealing with each case separately. With time and practice, working in general vector spaces should become natural to you, just as the equally abstract concept of number (as opposed to specific collections of some number of objects) has become second nature.

---

§1.1. VECTOR SPACES

The basic geometric objects that are studied in linear algebra are called vector spaces. Since you have probably seen vectors before in your mathematical experience, we begin by recalling some basic facts about vectors in the plane to help motivate the discussion of general vector spaces that follows.

In the geometry of the Euclidean plane, a *vector* is usually defined as a directed line segment or "arrow," that is, as a line segment with one endpoint distinguished as the "head" or final point, and the other distinguished as the "tail" or initial point. See Figure 1.1. Vectors are useful for describing quantities with both a magnitude and a direction. Geometrically, the *length* of the directed line segment may be taken to represent the magnitude of the quantity; the direction is given by the direction that the arrow is pointing. Important examples of quantities of this kind are the instantaneous velocity and the instantaneous acceleration at each time of an object moving along a path in the plane, the momentum of the moving object, forces, and so on. In physics these quantities are treated mathematically by using vectors as just described.

In linear algebra one of our major concerns will be the *algebraic properties* of vectors. By this we mean, for example, the operations by which vectors may be combined to produce new vectors and the properties of those operations. For instance, if we consider the set of *all* vectors in the plane with a tail at some fixed point O, then it is possible to combine vectors to produce new vectors in two ways.

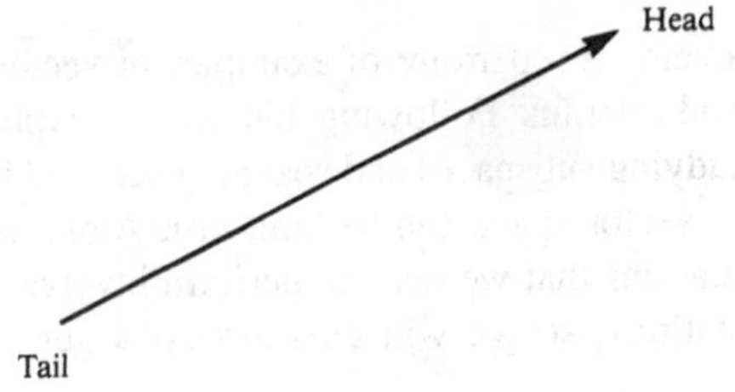

Figure 1.1

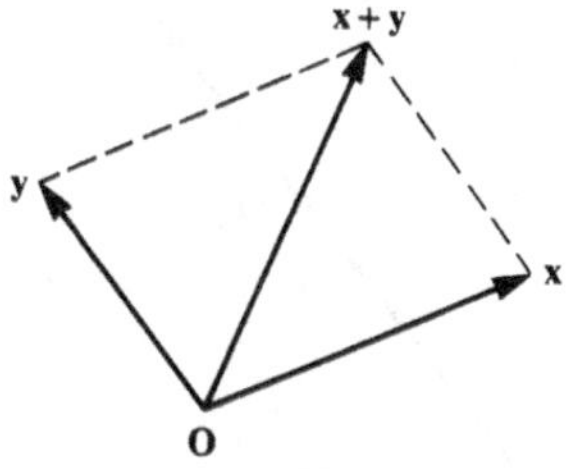

Figure 1.2

First, if we take two vectors **x** and **y**, then we can define their *vector sum* **x** + **y** to be the vector whose tail is at the point O and whose head is at the fourth corner of the parallelogram with sides **x** and **y**. See Figure 1.2. One physical interpretation of this sum operation is as follows. If two forces, represented by vectors **x** and **y**, act on an object located at the point O then the resulting force will be given by the vector sum **x** + **y**.

Second, if we take a vector **x** and a positive real number c (called a *scalar* in this context), then we can define the *product* of the vector **x** and the scalar c to be the vector in the same direction as **x** but with a magnitude or length that is equal to c times the magnitude of **x**. If $c > 1$, this has the effect of magnifying **x**, whereas if $c < 1$, this shrinks **x**. The case $c > 1$ is pictured in Figure 1.3. Physically, a positive scalar multiple of a vector may be thought of in the following way. For example, in the case $c = 2$, if the vector **x** represents a force, then the vector 2**x** represents a force that is "twice as strong" and that acts in the same direction. Similarly, the vector (1/2)**x** represents a force that is "one-half as strong." The product c**x** may also be defined if $c < 0$. In this case the vector c**x** will point along the same line through the origin as **x** but in the opposite direction from **x**. The magnitude of c**x** in this case will be equal to $|c|$ times the magnitude of **x**. See Figure 1.4.

Further properties of these two operations on vectors may be derived directly from these geometric definitions. However, to bring their algebraic nature into clearer focus, we will now consider an alternate way to understand these operations. If we introduce the familiar Cartesian coordinate system in the plane and place the origin at the point O = (0, 0), then a vector whose tail is at O is uniquely specified

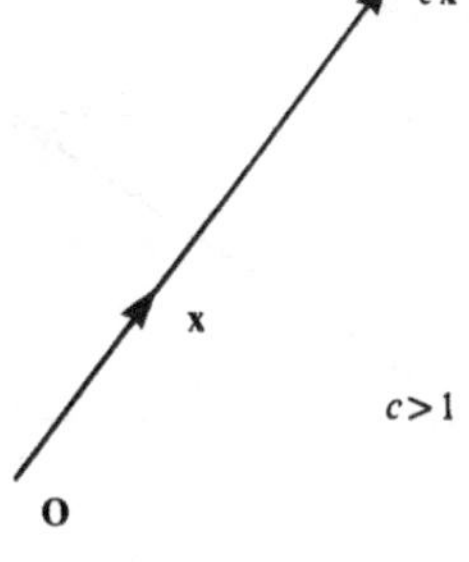

Figure 1.3

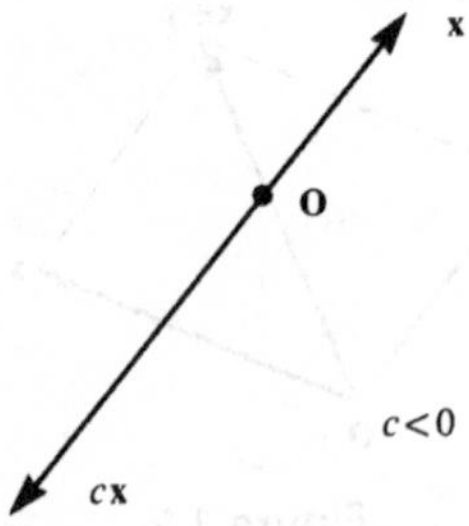

Figure 1.4

by the coordinates of its head. That is, vectors may be described as ordered pairs of real numbers. See Figure 1.5.

In this way we obtain a one-to-one correspondence between the set of vectors with a tail at the origin and the set $\mathbf{R}^2$ (the set of ordered pairs of real numbers) and we write $\mathbf{x} = (x_1, x_2)$ to indicate the vector whose head is at the point (x_1, x_2).

Our two operations on vectors may be described using coordinates. First, from the parallelogram law for the vector sum, we see that if $\mathbf{x} = (x_1, x_2)$ and $\mathbf{y} = (y_1, y_2)$, then $\mathbf{x} + \mathbf{y} = (x_1 + y_1, x_2 + y_2)$. See Figure 1.6. That is, to find the vector sum, we simply add "component-wise." For example the vector sum $(2, 5) + (4, -3)$ is equal to $(6, 2)$. Second, if c is a scalar and $\mathbf{x} = (x_1, x_2)$ is a vector, then $c\mathbf{x} = (cx_1, cx_2)$. The scalar multiple $4(-1, 2)$ is equal to $(-4, 8)$.

With this description, the familiar properties of addition and multiplication of real numbers may be used to show that our two operations on vectors have the following algebraic properties:

1. The vector sum is *associative:* For all $\mathbf{x}$, $\mathbf{y}$, and $\mathbf{z} \in \mathbf{R}^2$ we have

$$(\mathbf{x} + \mathbf{y}) + \mathbf{z} = \mathbf{x} + (\mathbf{y} + \mathbf{z})$$

2. The vector sum is *commutative:* For all $\mathbf{x}$ and $\mathbf{y} \in \mathbf{R}^2$ we have

$$\mathbf{x} + \mathbf{y} = \mathbf{y} + \mathbf{x}$$

3. There is an *additive identity element* $\mathbf{0} = (0, 0) \in \mathbf{R}^2$ with the property that for all $\mathbf{x} \in \mathbf{R}^2$, $\mathbf{x} + \mathbf{0} = \mathbf{x}$.

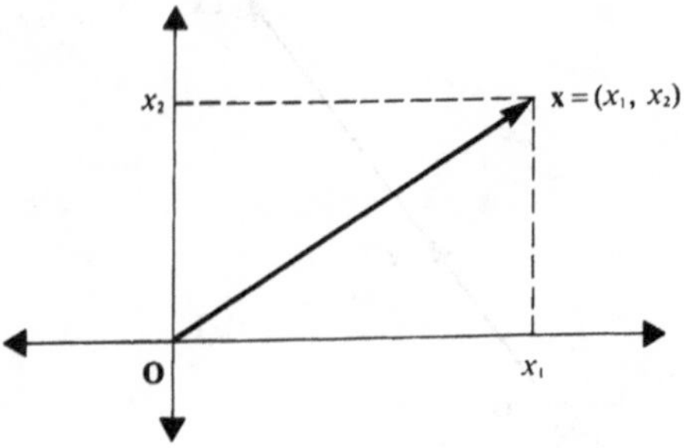

Figure 1.5

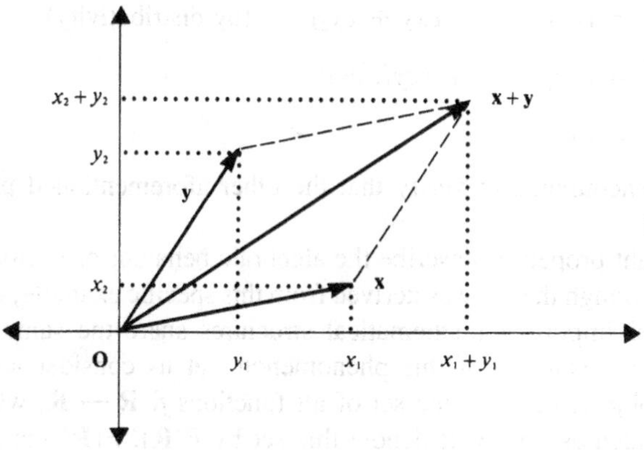

Figure 1.6

4. For each $\mathbf{x} = (x_1, x_2) \in \mathbf{R}^2$ there is an *additive inverse* $-\mathbf{x} = (-x_1, -x_2)$ with the property that $\mathbf{x} + (-\mathbf{x}) = \mathbf{0}$.
5. Multiplication by a scalar is *distributive* over vector sums: For all $c \in \mathbf{R}$ and all $\mathbf{x}, \mathbf{y} \in \mathbf{R}^2$ we have
$$c(\mathbf{x} + \mathbf{y}) = c\mathbf{x} + c\mathbf{y}$$
6. Multiplication of a vector by a sum of scalars is also *distributive:* For all $c, d \in \mathbf{R}$ and all $\mathbf{x} \in \mathbf{R}^2$ we have
$$(c + d)\mathbf{x} = c\mathbf{x} + d\mathbf{x}$$
7. For all $c, d \in \mathbf{R}$ and all $\mathbf{x} \in \mathbf{R}^2$ we have
$$c(d\mathbf{x}) = (cd)\mathbf{x}$$
8. For all $\mathbf{x} \in \mathbf{R}^2$ we have $1\mathbf{x} = \mathbf{x}$.

For example, property 1 follows from the fact that addition of real numbers is also associative. If $\mathbf{x} = (x_1, x_2)$, $\mathbf{y} = (y_1, y_2)$, and $\mathbf{z} = (z_1, z_2)$, then we have

$$\begin{aligned}(\mathbf{x} + \mathbf{y}) + \mathbf{z} &= ((x_1, x_2) + (y_1, y_2)) + (z_1, z_2)\\ &= ((x_1 + y_1) + z_1, (x_2 + y_2) + z_2)\\ &= (x_1 + (y_1 + z_1), x_2 + (y_2 + z_2))\\ &= \mathbf{x} + (\mathbf{y} + \mathbf{z})\end{aligned}$$

Property 5 follows from the fact that multiplication is distributive over addition in $\mathbf{R}$. If $\mathbf{x} = (x_1, x_2)$, $\mathbf{y} = (y_1, y_2)$, and c is a scalar, then

$$\begin{aligned}c(\mathbf{x} + \mathbf{y}) &= c((x_1, x_2) + (y_1, y_2))\\ &= c(x_1 + y_1, x_2 + y_2)\\ &= (c(x_1 + y_1), c(x_2 + y_2))\end{aligned}$$

$$= (cx_1 + cy_1, cx_2 + cy_2) \qquad \text{(by distributivity)}$$

$$= (cx_1, cx_2) + (cy_1, cy_2)$$

$$= c\mathbf{x} + c\mathbf{y}$$

The reader is encouraged to verify that the other aforementioned properties also hold in general.

These eight properties describe the algebraic behavior of vectors in the plane completely. Although the list was derived from this specific example, it is interesting that many other important mathematical structures share the same characteristic properties. As an example of this phenomenon, let us consider another type of mathematical object, namely, the set of all functions $f\colon \mathbf{R} \to \mathbf{R}$, which you have dealt with in calculus. We will denote this set by $F(\mathbf{R})$. $F(\mathbf{R})$ contains elements such as the functions defined by $f(x) = x^3 + 4x - 7$, $g(x) = \sin(4e^x)$, $h(x) = (|x - \cos(x)|)^{1/2}$, and so on.

We recall that if $f, g \in F(\mathbf{R})$, we can produce a new function denoted $f + g \in F(\mathbf{R})$ by defining $(f + g)(x) = f(x) + g(x)$ for all $x \in \mathbf{R}$. For instance, if $f(x) = e^x$ and $g(x) = 2e^x + x^2$, then $(f + g)(x) = 3e^x + x^2$. If $c \in \mathbf{R}$, we may multiply any $f \in F(\mathbf{R})$ by c to produce a new function denoted $cf \in F(\mathbf{R})$, and defined by $(cf)(x) = cf(x)$ for all $x \in \mathbf{R}$. Hence there are sum and scalar multiplication operations that are defined on the elements of $F(\mathbf{R})$ as well.

The analogy between algebraic operations on the vectors in $\mathbf{R}^2$ and the functions in $F(\mathbf{R})$ that is becoming apparent actually goes even deeper. We see that the following properties hold:

1. The sum operation on functions is also *associative:* For all $f, g, h \in F(\mathbf{R})$, $(f + g) + h = f + (g + h)$ as functions. This is so because for all $x \in \mathbf{R}$ we have $((f + g) + h)(x) = (f(x) + g(x)) + h(x) = f(x) + (g(x) + h(x)) = (f + (g + h))(x)$ by the associativity of addition in $\mathbf{R}$.

2. The sum operation on functions is also *commutative:* For all $f, g \in F(\mathbf{R})$, $f + g = g + f$ as functions. This follows since $f(x) + g(x) = g(x) + f(x)$ for all $x \in \mathbf{R}$. (Addition of real numbers is commutative.)

3. There is an *additive identity element* in $F(\mathbf{R})$—the constant function $z(x) = 0$. For all functions f, $f + z = f$ since $f(x) + z(x) = f(x) + 0 = f(x)$ for all $x \in \mathbf{R}$.

4. For each function f there is an *additive inverse* $-f \in F(\mathbf{R})$ (defined by $(-f)(x) = -f(x)$) with the property that $f + (-f) = z$ (the zero function from (3)).

5. For all functions f and g and all $c \in \mathbf{R}$, $c(f + g) = cf + cg$. This follows since $c(f + g)(x) = c(f(x) + g(x)) = cf(x) + cg(x) = (cf)(x) + cg(x) = (cf + cg)(x)$ by the distributivity of multiplication over addition in $\mathbf{R}$.

6. For all functions f and all $c, d \in \mathbf{R}$, $(c + d)f = cf + df$. This also follows from ordinary distributivity.

7. For all functions f and all $c, d \in \mathbf{R}$, $(cd)f = c(df)$. That is, the scalar multiple of f by the product cd is the same function as is obtained by multiplying f by d, then multiplying the result by c.

8. For all functions f, $1f = f$.

Thus, with regard to the sum and scalar multiplication operations we have defined, the elements of our two sets $\mathbf{R}^2$ and $F(\mathbf{R})$ actually behave in exactly the same way. Because of the importance and usefulness of these vectors and functions, the fact that they are so similar from an algebraic point of view is a very fortunate occurrence. Both $\mathbf{R}^2$ and $F(\mathbf{R})$ are examples of a more general mathematical structure called a vector space, which is defined as follows.

(1.1.1) Definition. A (real) *vector space* is a set V (whose elements are called *vectors* by analogy with the first example we considered) together with

a) an operation called *vector addition*, which for each pair of vectors $\mathbf{x}, \mathbf{y} \in V$ produces another vector in V denoted $\mathbf{x} + \mathbf{y}$, and

b) an operation called *multiplication by a scalar* (a real number), which for each vector $\mathbf{x} \in V$, and each scalar $c \in \mathbf{R}$ produces another vector in V denoted $c\mathbf{x}$.

Furthermore, the two operations must satisfy the following *axioms:*

1. For all vectors $\mathbf{x}$, $\mathbf{y}$, and $\mathbf{z} \in V$, $(\mathbf{x} + \mathbf{y}) + \mathbf{z} = \mathbf{x} + (\mathbf{y} + \mathbf{z})$.
2. For all vectors $\mathbf{x}$ and $\mathbf{y} \in V$, $\mathbf{x} + \mathbf{y} = \mathbf{y} + \mathbf{x}$.
3. There exists a vector $\mathbf{0} \in V$ with the property that $\mathbf{x} + \mathbf{0} = \mathbf{x}$ for all vectors $\mathbf{x} \in V$.
4. For each vector $\mathbf{x} \in V$, there exists a vector denoted $-x$ with the property that $\mathbf{x} + -\mathbf{x} = \mathbf{0}$.
5. For all vectors $\mathbf{x}$ and $\mathbf{y} \in V$ and all scalars $c \in \mathbf{R}$, $c(\mathbf{x} + \mathbf{y}) = c\mathbf{x} + c\mathbf{y}$.
6. For all vectors $\mathbf{x} \in V$, and all scalars c and $d \in \mathbf{R}$, $(c + d)\mathbf{x} = c\mathbf{x} + d\mathbf{x}$.
7. For all vectors $\mathbf{x} \in V$, and all scalars c and $d \in \mathbf{R}$, $(cd)\mathbf{x} = c(d\mathbf{x})$.
8. For all vectors $\mathbf{x} \in V$, $1\mathbf{x} = \mathbf{x}$.

The reason we introduce abstract definitions such as this one is that they allow us to focus on the *similarities* between different mathematical objects and treat objects with the same properties in a unified way. We will now consider several other examples of vector spaces.

(1.1.2) Example. As a first example, let us consider the set $V = \mathbf{R}^n = \{(x_1, \ldots, x_n) | x_i \in \mathbf{R}$ for all $i\}$. We define operations called vector addition and multiplication by a scalar by rules that are similar to the ones we saw before in $\mathbf{R}^2$.

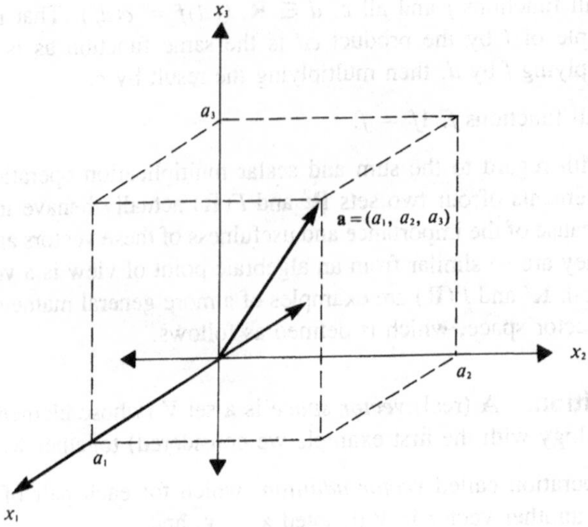

Figure 1.7

Namely, if $\mathbf{x} = (x_1, \ldots, x_n)$ and $\mathbf{y} = (y_1, \ldots, y_n)$, then we define $\mathbf{x} + \mathbf{y} = (x_1 + y_1, \ldots, x_n + y_n)$ in $\mathbf{R}^n$. Next, if $\mathbf{x} = (x_1, \ldots, x_n)$ and $c \in \mathbf{R}$, we define $c\mathbf{x} = (cx_1, \ldots, cx_n)$. With these operations V is a vector space, since all eight axioms in the definition hold. (This may be seen by arguments that are exactly analogous to the ones given before for $\mathbf{R}^2$.) Thus we have a whole infinite collection of new vector spaces $\mathbf{R}^n$ for different integers $n \geq 1$.

When $n = 1$ we obtain $\mathbf{R}$ itself, which we may picture geometrically as a line. As we have seen, vectors in $\mathbf{R}^2$ correspond to points in the plane. Similarly, vectors in $\mathbf{R}^3$ correspond to points in a three-dimensional space such as the physical space we live in, described by Cartesian coordinates. See Figure 1.7. The vector spaces $\mathbf{R}^4$, $\mathbf{R}^5$, and so on may be thought of geometrically in a similar fashion. Due to the fact that our sense organs are adapted to an apparently three-dimensional world, it is certainly true that they are harder to visualize. Nevertheless, the reader encountering these spaces for the first time should *not* make the mistake of viewing them only as meaningless generalizations of the mathematics underlying the two- and three-dimensional physical spaces in which our geometric intuition is more at home. They are useful because frequently more than three coordinates are needed to specify a configuration or situation arising in applications of mathematics. For instance, to describe the position *and* velocity of a moving object in ordinary physical space at a given time, we actually need *six* coordinates in all (three for the position and three for the velocity, which is also a vector). Thus our "position-velocity space" may be seen as $\mathbf{R}^6$. To describe the state of the U.S. economy at a given time in a realistic economic model, hundreds or thousands of different variables or coordinates might be specified. In each of these cases, the space of all possible configurations or situations may be thought of as one of our spaces $\mathbf{R}^n$ (or a subset of one of these spaces).

(1.1.3) Example. A different kind of example of vector spaces comes from considering certain subsets of our vector space of functions. In particular, let n be a fixed nonnegative integer, and let $P_n(\mathbf{R}) = \{p: \mathbf{R} \to \mathbf{R} | \, p(x) = a_nx^n + a_{n-1}x^{n-1} + \cdots + a_0$, where the $a_i \in \mathbf{R}\}$, the set of all polynomial functions of a degree no larger than n. We can define the sum of two polynomial functions as we did before in the larger space $F(\mathbf{R})$. If $p(x) = a_nx^n + \cdots + a_0$ and $q(x) = b_nx^n + \cdots + b_0$, then we define

$$(p + q)(x) = p(x) + q(x) = (a_n + b_n)x^n + \cdots + (a_0 + b_0) \in P_n(\mathbf{R})$$

Similarly, if $c \in \mathbf{R}$, we define:

$$(cp)(x) = cp(x) = ca_nx^n + \cdots + ca_0 \in P_n(\mathbf{R})$$

To show that $P_n(\mathbf{R})$ is a vector space, we may verify the eight axioms given in Definition (1.1.1). These are all straightforward, and may be verified by computations that are similar to those we did before in $F(\mathbf{R})$.

(1.1.4) Examples

a) Since specifying a vector space means giving both a set of vectors and the two operations, one question that may occur to you is, given a set V, could there be more than one way to define a vector sum and a scalar multiplication to make V a vector space? For example, in $V = \mathbf{R}^2$, if we defined

$$(x_1, x_2) +' (y_1, y_2) = (x_1y_1, x_2y_2)$$

and

$$c(x_1, x_2) = (cx_1, cx_2)$$

would these operations give us another vector space?

If we look at the first three axioms, we see that they are satisfied for the new sum operation $+'$. (Do not be fooled, axiom 3 *is* satisfied if we take the identity element for $+'$ to be the vector $(1, 1)$. You should also note that this is the only vector that works here.) However, axiom 4 fails to hold here since there is no inverse for the vector $(0, 0)$ under the operation $+'$. (Why not?) Several of the other axioms fail to hold as well (which ones?). Hence we do *not* obtain a vector space in this example.

b) Similarly, we can ask if other "addition" and "scalar multiplication" operations might be defined in $F(\mathbf{R})$. Consider the addition operation $f +' g$ defined by $(f +' g)(x) = f(x) + 3g(x)$ and the usual scalar multiplication. We ask, is $F(\mathbf{R})$ a vector space with these operations? The answer is no, as the reader will see by checking the associative law for addition (axiom 1), for instance. (Other axioms fail as well. Which ones?)

c) As we know, $V = \mathbf{R}$ is a vector space. We can also make the subset $V' = \mathbf{R}^+ = \{r \in \mathbf{R} \mid r > 0\}$ into a vector space by defining our vector sum and scalar multiplication operations to be $x +' y = x \cdot y$, and $c \cdot' x = x^c$ for all x, y

$\in V'$ and all $c \in \mathbf{R}$. Since multiplication of real numbers (our *addition* operation in V') is associative and commutative, the first two axioms in Definition (1.1.1) are satisfied. The element $1 \in V'$ is an identity element for the operation $+'$, since $x +' 1 = x \cdot 1 = x$ for all $x \in V'$. Furthermore, each $x \in V'$ has an inverse $1/x \in V'$ under the operation $+'$, since $x +' (1/x) = x \cdot (1/x) = 1$.

The remaining axioms require some checking using the properties of exponents. To verify axiom 5, we compute

$$c \cdot' (x +' y) = (x \cdot y)^c = x^c \cdot y^c = x^c +' y^c = (c \cdot' x) +' (c \cdot' y)$$

Therefore axiom 5 is satisfied. The other distributive property of axiom 6 may be checked similarly. We have

$$(c+d) \cdot' x = x^{(c+d)} = x^c \cdot x^d = (c \cdot' x) +' (d \cdot' x)$$

Axiom 7 also holds since

$$(cd) \cdot' x = x^{cd} = (x^d)^c = c \cdot' (d \cdot' x)$$

Finally, $1 \cdot' x = x^1 = x$ for all $x \in V'$, so axiom 8 is satisfied as well.

We conclude this section by stating and proving some further properties of vector spaces to illustrate another benefit of introducing general definitions like Definition (1.1.1)—any proof that uses only the properties of *all* vector spaces expressed by the eight axioms is valid for all vector spaces. We do not have to reprove our results in each new example we encounter.

(1.1.5) Remark. In $\mathbf{R}^n$ there is clearly only one additive identity—the zero vector $(0, \ldots, 0) \in \mathbf{R}^n$. Moreover, each vector has only one additive inverse. We may ask whether these patterns are true in a general vector space. The fact that they are makes a general vector space somewhat simpler than it might appear at first.

(1.1.6) Proposition. Let V be a vector space. Then

a) The zero vector $\mathbf{0}$ is unique.

b) For all $\mathbf{x} \in V$, $0\mathbf{x} = \mathbf{0}$.

c) For each $\mathbf{x} \in \mathbf{V}$, the additive inverse $-x$ is unique.

d) For all $\mathbf{x} \in V$, and all $c \in \mathbf{R}$, $(-c)\mathbf{x} = -(c\mathbf{x})$.

Proof:

a) To prove that something is unique, a common technique is to assume we have two examples of the object in question, then show that those two examples must in fact be equal. So, suppose we had two vectors, $\mathbf{0}$ and $\mathbf{0}'$, both of which satisfy axiom 3 in the definition. Then, $\mathbf{0} + \mathbf{0}' = \mathbf{0}$, since $\mathbf{0}'$ is an additive identity. On the other hand, $\mathbf{0} + \mathbf{0}' = \mathbf{0}' + \mathbf{0} = \mathbf{0}'$, since addition is commutative and $\mathbf{0}$ is an additive identity. Hence $\mathbf{0} = \mathbf{0}'$, or, in other words, there is only one additive identity in V.

b) We have $0\mathbf{x} = (0 + 0)\mathbf{x} = 0\mathbf{x} + 0\mathbf{x}$, by axiom 6. Hence if we add the inverse of $0\mathbf{x}$ to both sides, we obtain $\mathbf{0} = 0\mathbf{x}$, as claimed.

c) We use the same idea as in the proof of part a. Given $\mathbf{x} \in V$, if $-\mathbf{x}$ and $(-\mathbf{x})'$ are two additive inverses of $\mathbf{x}$, then on one hand we have $\mathbf{x} + -\mathbf{x} + (-\mathbf{x})' = (\mathbf{x} + -\mathbf{x}) + (-\mathbf{x})' = \mathbf{0} + (-\mathbf{x})' = (-\mathbf{x})'$, by axioms 1, 4, and 3. On the other hand, if we use axiom 2 first before associating, we have $\mathbf{x} + -\mathbf{x} + (-\mathbf{x})' = \mathbf{x} + (-\mathbf{x})' + -\mathbf{x} = (\mathbf{x} + (-\mathbf{x})') + -\mathbf{x} = \mathbf{0} + -\mathbf{x} = -\mathbf{x}$. Hence $-\mathbf{x} = (-\mathbf{x})'$, and the additive inverse of $\mathbf{x}$ is unique.

d) We have $c\mathbf{x} + (-c)\mathbf{x} = (c + -c)\mathbf{x} = 0\mathbf{x} = \mathbf{0}$ by axiom 6 and part b. Hence $(-c)\mathbf{x}$ also serves as an additive inverse for the vector $c\mathbf{x}$. By part c, therefore, we must have $(-c)\mathbf{x} = -(c\mathbf{x})$. ■

EXERCISES

1. Let $\mathbf{x} = (1, 3, 2)$, $\mathbf{y} = (-2, 3, 4)$, $\mathbf{z} = (-3, 0, 3)$ in $\mathbf{R}^3$.
   a) Compute $3\mathbf{x}$.
   b) Compute $4\mathbf{x} - \mathbf{y}$.
   c) Compute $-\mathbf{x} + \mathbf{y} + 3\mathbf{z}$.

2. Let $f = 3e^{3x}$, $g = 4e^{3x} + e^x$, $h = 2e^x - e^{3x}$ in $F(\mathbf{R})$.
   a) Compute $5f$.
   b) Compute $2f + 3g$.
   c) Compute $-2f - g + 4h$.

3. Show that $\mathbf{R}^n$ with the vector sum and scalar multiplication operations given in Example (1.1.2) is a vector space.

4. Complete the proof that $P_n(\mathbf{R})$, with the operations given in Example (1.1.3), is a vector space.

5. Let $V = \{p: \mathbf{R} \to \mathbf{R} \mid p(x) = a_n x^n + \cdots + a_0$, where the $a_i \in \mathbf{R}$, and $a_n \neq 0\}$ (the set of polynomial functions of degree exactly n). Is V a vector space, using the operations given in Example (1.1.3)? Why or why not?

6. In each of the following parts, decide if the set $\mathbf{R}^2$, with the given operations, is a vector space. If this is not the case, say which of the axioms fail to hold.
   a) vector sum $(x_1, x_2) +' (y_1, y_2) = (x_1 + 2y_1, 3x_2 - y_2)$, and the usual scalar multiplication $c(x_1, x_2) = (cx_1, cx_2)$
   b) usual vector sum $(x_1, x_2) + (y_1, y_2) = (x_1 + y_1, x_2 + y_2)$, and scalar multiplication $c(x_1, x_2) = \begin{cases} (cx_1, (1/c)x_2) & \text{if } c \neq 0 \\ (0, 0) & \text{if } c = 0 \end{cases}$
   c) vector sum $(x_1, x_2) +' (y_1, y_2) = (0, x_1 + y_2)$, and the usual scalar multiplication

7. In each of the following parts, decide if the set $F(\mathbf{R})$, with the given operations, is a vector space. If this is not the case, say which of the axioms fail to hold.
   a) Sum operation defined by $f +' g = fg$, scalar multiplication given by $c \cdot f = c + f$, that is, the constant function c plus f

b) Sum defined by $f +' g = f - g$, scalar multiplication given by $(c \cdot f)(x) = f(cx)$
c) Sum defined by $f +' g = f \circ g$ (composition of functions), usual scalar multiplication

8. Show that in any vector space V
a) If $\mathbf{x}, \mathbf{y}, \mathbf{z} \in V$, then $\mathbf{x} + \mathbf{y} = \mathbf{x} + \mathbf{z}$ implies $\mathbf{y} = \mathbf{z}$.
b) If $\mathbf{x}, \mathbf{y} \in V$ and $a, b \in \mathbf{R}$, then $(a + b)(\mathbf{x} + \mathbf{y}) = a\mathbf{x} + b\mathbf{x} + a\mathbf{y} + b\mathbf{y}$.

9. a) What vector space might be used to describe (simultaneously) the position, velocity, and acceleration of an object moving along a path in the plane $\mathbf{R}^2$?
b) Same question for an object moving in three-dimensional space.

10. Let $[a, b]$ be the closed interval $\{x \in \mathbf{R} \mid a \leq x \leq b\} \subset \mathbf{R}$. Let $F([a, b])$ be the set of all functions $f: [a, b] \to \mathbf{R}$. Show that $F([a,b])$ is a vector space if we define the sum and scalar multiplication operations as in $F(\mathbf{R})$.

11. Let $V = \{a_1x^2 + a_2xy + a_3y^2 + a_4x + a_5y + a_6 \mid a_i \in \mathbf{R}\}$, the set of all polynomials in two variables x and y of total degree no larger than 2. Define sum and scalar multiplication operations in V as in the vector space $P_n(\mathbf{R})$; that is, the sum operation is ordinary addition of polynomials, and multiplication by a scalar multiplies each coefficient by that scalar. Show that V, with these operations, is a vector space.

12. Let $V = (\mathbf{R}^+)^n = \{(x_1, \ldots, x_n) \mid x_i \in \mathbf{R}^+ \text{ for each } i\}$. See Example (1.1.4c). In V define a vector sum operation $+'$ by $(x_1, \ldots, x_n) +' (y_1, \ldots, y_n) = (x_1y_1, \ldots, x_ny_n)$, and a scalar multiplication operation $\cdot'$ by $c \cdot' (x_1, \ldots, x_n) = (x_1^c, \ldots, x_n^c)$. Show that with these two operations V is a vector space.

§1.2. SUBSPACES

In Section 1.1 we saw several different types of examples of vector spaces, but those examples are far from a complete list of the vector spaces that arise in different areas of mathematics. In this section we begin by indicating some other important examples. Another type of example, vector spaces of matrices, is introduced in the exercises following this section.

Our first example deals with a special kind of subset of the vector space $\mathbf{R}^3$, which we introduced in Example (1.1.2).

(1.2.1) Example. Let $V = \{(x_1, x_2, x_3) \in \mathbf{R}^3 \mid 5x_1 - 2x_2 + x_3 = 0\}$. Geometrically, the set V is a plane passing through the origin in $\mathbf{R}^3$. See Figure 1.8. If we endow V with the operations of vector addition and scalar multiplication from the space $\mathbf{R}^3$, then V is also a vector space. Note first that if $\mathbf{x} = (x_1, x_2, x_3)$ and $\mathbf{y} = (y_1, y_2, y_3) \in V$, then the sum $\mathbf{x} + \mathbf{y} = (x_1 + y_1, x_2 + y_2, x_3 + y_3) \in V$

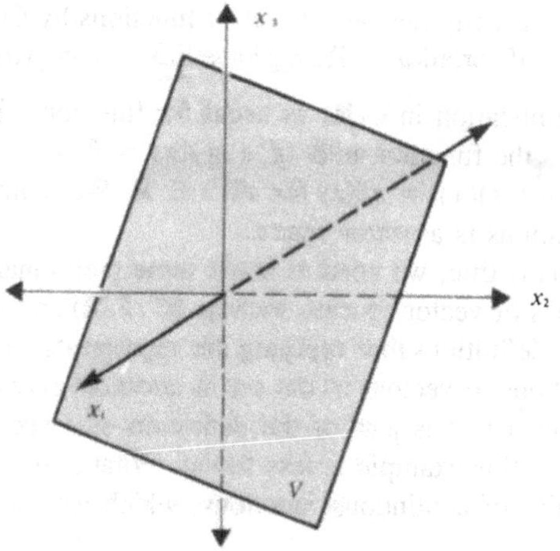

Figure 1.8

as well. This follows because the components of the sum also satisfy the defining equation of V:

$$\begin{aligned} 5(x_1 + y_1) - 2(x_2 + y_2) + (x_3 + y_3) &= (5x_1 - 2x_2 + x_3) + (5y_1 - 2y_2 + y_3) \\ &= 0 + 0 \\ &= 0 \end{aligned}$$

Similarly, for any $\mathbf{x} \in V$, and any $c \in \mathbf{R}$, $c\mathbf{x} \in V$, since again the components of the vector $c\mathbf{x}$ satisfy the defining equation of V:

$$5(cx_1) - 2(cx_2) + (cx_3) = c(5x_1 - 2x_2 + x_3) = 0$$

The zero vector (0, 0, 0) also satisfies the defining equation of V.

Now, since V is a subset of the vector space $\mathbf{R}^3$ and the vector sum and scalar multiplication are defined in the same way in both V and in $\mathbf{R}^3$, then it follows that the axioms for vector spaces are satisfied by V as well. Since they hold for all vectors in $\mathbf{R}^3$, they also hold for the vectors in the subset V.

(1.2.2) Example. The previous example may be generalized immediately. In $\mathbf{R}^n$, consider any set defined in the following way. Let $V = \{(x_1, \ldots, x_n) \in \mathbf{R}^n \mid a_1x_1 + \cdots + a_nx_n = 0$, where $a_i \in \mathbf{R}$ for all $i\}$. Then V is a vector space, if we define the vector sum and scalar multiplication to be the same as the operations in the whole space $\mathbf{R}^n$. V is sometimes called a hyperplane in $\mathbf{R}^n$.

(1.2.3) Example. Many of the sets of functions you have encountered in calculus give further examples of vector spaces. For example, consider the set $V = \{f\colon \mathbf{R} \to \mathbf{R}$

$\mid f$ is continuous}. We usually denote this set of functions by $C(\mathbf{R})$. Recall that f is continuous if and only if for all $a \in \mathbf{R}$, we have $\lim_{x \to a} f(x) = f(a)$. We define vector sum and scalar multiplication in $C(\mathbf{R})$ as usual for functions. If $f, g \in C(\mathbf{R})$ and $c \in \mathbf{R}$, then $f + g$ is the function with $(f + g)(x) = f(x) + g(x)$ for all $x \in \mathbf{R}$, and cf is the function $(cf)(x) = cf(x)$ for all $x \in \mathbf{R}$. We claim that the set $C(\mathbf{R})$ with these two operations is a vector space.

To see why this is true, we need to make some preliminary observations. In our previous examples of vector spaces, such as $\mathbf{R}^n$, $F(\mathbf{R})$ and $P_n(\mathbf{R})$, it was more or less clear from the definitions that applying the appropriate vector sum and scalar multiplication operations to vectors in the set in question gave us elements of the same set, a requirement that is part of the definition of a vector space. The fact that this is also true in this example is less trivial. What is involved here is a pair of important properties of continuous functions, which we summarize in the following result.

(1.2.4) Lemma. Let $f, g \in C(\mathbf{R})$, and let $c \in \mathbf{R}$. Then

a) $f + g \in C(\mathbf{R})$, and

b) $cf \in C(\mathbf{R})$.

Proof:

a) By the limit sum rule from calculus, for all $a \in \mathbf{R}$ we have

$$\lim_{x \to a} (f + g)(x) = \lim_{x \to a} (f(x) + g(x)) = \lim_{x \to a} f(x) + \lim_{x \to a} g(x)$$

Since f and g are continuous, this last expression is equal to $f(a) + g(a) = (f + g)(a)$. Hence $f + g$ is continuous.

b) By the limit product rule, we have

$$\lim_{x \to a} (cf)(x) = \lim_{x \to a} cf(x) = (\lim_{x \to a} c) \cdot (\lim_{x \to a} f(x)) = cf(a) = (cf)(a)$$

so cf is also continuous. ■

That the eight axioms in Definition (1.1.1) hold in $C(\mathbf{R})$ may now be verified in much the same way that we verified them for the vector space $F(\mathbf{R})$ in Section 1.1. Alternately, we may notice that these verifications are actually unnecessary—since we have already established those properties for all the functions in $F(\mathbf{R})$, they must also hold for the functions in the subset $C(\mathbf{R})$. Of course, we must convince ourselves that the zero function from $F(\mathbf{R})$ also serves as an additive identity element for $C(\mathbf{R})$, but this is clear.

(1.2.5) Example. Now, let $V = \{f: \mathbf{R} \to \mathbf{R} \mid f$ differentiable everywhere, and $f'(x) \in C(\mathbf{R})\}$. The elements of V are called continuously differentiable functions and the set V is usually denoted by $C^1(\mathbf{R})$. If we define sums of functions and scalar multiples as in $F(\mathbf{R})$, then $C^1(\mathbf{R})$ is also a vector space.

To see that this is true, once again we should start by checking that sums and scalar multiples of functions in $C^1(\mathbf{R})$ are indeed in $C^1(\mathbf{R})$. This follows from properties of differentiation that you learned in calculus. Note that if $f, g \in C^1(\mathbf{R})$, the sum rule for derivatives implies that $f + g$ is also differentiable and that $(f + g)' = f' + g'$. Then, since both f' and g' are in $C(\mathbf{R})$, by part (a) of Lemma (1.2.4) we have that $f' + g' \in C(\mathbf{R})$. Hence $f + g \in C^1(\mathbf{R})$. Similarly, if $f \in C^1(\mathbf{R})$ and $c \in \mathbf{R}$, then the product rule for derivatives implies that cf is differentiable and $(cf)' = cf'$. Then, by part (b) of Lemma (1.2.4), since $f' \in C(\mathbf{R})$, we see that $cf' \in C(\mathbf{R})$. Hence $cf \in C^1(\mathbf{R})$ as well.

It is clear that the zero function is continuously differentiable so $C^1(\mathbf{R})$ does have an additive identity element. The fact that the remaining axioms for vector spaces hold in $C^1(\mathbf{R})$ is now actually a direct consequence of our previous verifications in $C(\mathbf{R})$ or $F(\mathbf{R})$. Since those properties hold for all functions in $F(\mathbf{R})$, they also hold for the functions in the subset $C^1(\mathbf{R})$.

In all these examples, you should note that we were dealing with subsets of vector spaces that were vector spaces in their own right. In general, we will use the following definition.

(1.2.6) Definition. Let V be a vector space and let $W \subseteq V$ be a subset. Then W is a (vector) *subspace* of V if W is a vector space itself under the operations of vector sum and scalar multiplication from V.

(1.2.7) Examples

a) From Example (1.2.2), for each particular set of coefficients $a_i \in \mathbf{R}$, the vector space $W = \{(x_1, \ldots, x_n) \in \mathbf{R}^n \mid a_1x_1 + \cdots + a_nx_n = 0\}$ is a subspace of $\mathbf{R}^n$.

b) $C(\mathbf{R})$, $P_n(\mathbf{R})$, $C^1(\mathbf{R})$ are all subspaces of $F(\mathbf{R})$.

c) $C^1(\mathbf{R})$ is a subspace of $C(\mathbf{R})$, since differentiability implies continuity.

d) For each n, the vector space $P_n(\mathbf{R})$ is a subspace of $C^1(\mathbf{R})$.

e) The vector space structure we defined on the subset $\mathbf{R}^+ \subset \mathbf{R}$ in Example (1.1.4c) does *not* make $\mathbf{R}^+$ a subspace of $\mathbf{R}$. The reason is that the vector sum and scalar multiplication operations in $\mathbf{R}^+$ are different from those in $\mathbf{R}$, while in a subspace, we must use the same operations as in the larger vector space.

f) On a more general note, in every vector space, the subsets V and $\{\mathbf{0}\}$ are subspaces. This is clear for V itself. It is also easy to verify that the set containing the one vector $\mathbf{0}$ is a subspace of V.

Although it is always possible to determine if a subset W of a vector space is a subspace directly [by checking the axioms from Definition (1.1.1)], it would be desirable to have a more economical way of doing this. The observations we have made in the previous examples actually indicate a general criterion that can be used to tell when a subset of a vector space is a subspace. Note that if W is to be a vector space in its own right, then the following properties must be true:

1. For all $\mathbf{x}, \mathbf{y} \in W$, we must have $\mathbf{x} + \mathbf{y} \in W$. If W has this property, we say that the set W is *closed under addition*.

2. For all $\mathbf{x} \in W$ and all $c \in \mathbf{R}$, we must have $c\mathbf{x} \in W$. If W has this property, we say that W is *closed under scalar multiplication*.

3. The zero vector of V must be contained in W.

Note that since the vector sum operation is the same in W as in V, the additive identity element in W must be the same as it is in V, by Proposition (1.1.6a). In fact, it is possible to *condense* these three conditions into a single condition that is easily checked and that characterizes the subspaces of V completely.

(1.2.8) Theorem. Let V be a vector space, and let W be a nonempty subset of V. Then W is a subspace of V if and only if for all $\mathbf{x}, \mathbf{y} \in W$, and all $c \in \mathbf{R}$, we have $c\mathbf{x} + \mathbf{y} \in W$.

(1.2.9) Remarks

a) By Definition (1.1.1) a vector space must contain at least an additive identity element, hence the requirement that W be nonempty is certainly necessary.

b) Before we begin the proof itself, notice that any statement of the form "p if and only if q" is equivalent to the statement "if p then q and if q then p." (The "only if" part is the statement "if p then q.") Thus, to prove an if and only if statement, we must prove both the *direct implication* "if p then q," and the *reverse implication* or *converse* "if q then p." To signal these two sections in the proof of an if and only if statement, we use the symbol $\rightarrow$ to indicate proof of the direct implication, and the symbol $\leftarrow$ to indicate the proof of the reverse implication. (The reader may wish to consult Appendix 1, section b for further comments on the logic involved here.)

Proof: $\rightarrow$: If W is a subspace of V, then for all $\mathbf{x} \in W$, and all $c \in \mathbf{R}$, we have $c\mathbf{x} \in W$, and hence for all $\mathbf{y} \in W$, $c\mathbf{x} + \mathbf{y} \in W$ as well, because by the definition a subspace W of V must be closed under vector sums and scalar multiples.

$\leftarrow$: Let W be any subset of V satisfying the condition of the theorem. First, note that since $c\mathbf{x} + \mathbf{y} \in W$ for all choices of $\mathbf{x}, \mathbf{y} \in W$ and $c \in \mathbf{R}$, we may specialize to the case $c = 1$. Then we see that $1\mathbf{x} + \mathbf{y} = \mathbf{x} + \mathbf{y} \in W$, so that W is closed under sums. Next, let $\mathbf{x} = \mathbf{y}$ be any vector in W and $c = -1$. Then $-1\mathbf{x} + \mathbf{x} = (-1 + 1)\mathbf{x} = 0\mathbf{x} = \mathbf{0} \in W$. Now let $\mathbf{x}$ be any vector in W and let $\mathbf{y} = 0$. Then $c\mathbf{x} + \mathbf{0} = c\mathbf{x} \in W$, so W is closed under scalar multiplication. To see that these observations imply that W is a vector space, note that the axioms 1, 2, and 5 through 8 in Definition (1.1.1) are satisfied automatically for vectors in W, since they hold for all vectors in V. Axiom 3 is satisfied, since as we have seen $\mathbf{0} \in W$. Finally, for each $\mathbf{x} \in W$, by Proposition (1.1.6d) $(-1)\mathbf{x} = -\mathbf{x} \in W$ as well. Hence W is a vector space. ■

To see how the condition of the theorem may be applied, we consider several examples.

(1.2.10) Example. In $V = \mathbf{R}^3$ consider the subset

$$W = \{(x_1, x_2, x_3) \mid 4x_1 + 3x_2 - 2x_3 = 0 \text{ and } x_1 - x_3 = 0\}$$

By Theorem (1.2.8), W is a subspace of $\mathbf{R}^3$ since if $\mathbf{x}, \mathbf{y} \in W$ and $c \in \mathbf{R}$, then writing $\mathbf{x} = (x_1, x_2, x_3)$ and $\mathbf{y} = (y_1, y_2, y_3)$, we have that the components of the vector $c\mathbf{x} + \mathbf{y} = (cx_1 + y_1, cx_2 + y_2, cx_3 + y_3)$ satisfy the defining equations of the set W:

$$\begin{aligned} 4(cx_1 + y_1) + 3(cx_2 + y_2) - 2(cx_3 + y_3) &= c(4x_1 + 3x_2 - 2x_3) + \\ &\quad (4y_1 + 3y_2 - 2y_3) \\ &= c0 + 0 = 0 \end{aligned}$$

Similarly,

$$\begin{aligned} (cx_1 + y_1) - (cx_3 + y_3) &= c(x_1 - x_3) + (y_1 - y_3) \\ &= c0 + 0 = 0 \end{aligned}$$

Hence $c\mathbf{x} + \mathbf{y} \in W$. W is nonempty since the zero vector $(0, 0, 0)$ satisfies both equations.

(1.2.11) Examples

a) In $V = C(\mathbf{R})$, consider the set $W = \{f \in C(\mathbf{R}) \mid f(2) = 0\}$. First, W is nonempty, since it contains functions such as $f(x) = x - 2$. W is a subspace of V, since if $f, g \in W$, and $c \in \mathbf{R}$, then we have $(cf + g)(2) = cf(2) + g(2) = c0 + 0 = 0$. Hence $cf + g \in W$ as well.

b) Let $C^2(\mathbf{R})$ denote the set of functions $f \in F(\mathbf{R})$ such that f is twice differentiable and $f'' \in C(\mathbf{R})$. We will show that $C^2(\mathbf{R})$ is a subspace of $C(\mathbf{R})$. First, since every differentiable function on $\mathbf{R}$ is continuous everywhere, we have that $C^2(\mathbf{R}) \subset C(\mathbf{R})$. Furthermore, $C^2(\mathbf{R})$ is nonempty since it certainly contains all polynomial functions. We will show that the criterion given in Theorem (1.2.8) is satisfied. Let f, g be any functions in $C^2(\mathbf{R})$ and consider the function $cf + g$. What we must show is that $cf + g$ is also twice differentiable with continuous second derivative.

First, since f and g can be differentiated twice, $cf + g$ is also twice differentiable, and the sum and scalar product rules for derivatives show that $(cf + g)'' = (cf' + g')' = cf'' + g''$. Second, since f'' and g'' are continuous functions, by Lemma (1.2.4), $cf'' + g''$ is also continuous. Therefore $cf + g \in C^2(\mathbf{R})$, so $C^2(\mathbf{R})$ is a subspace of $C(\mathbf{R})$.

Since you may not have seen examples of this kind before, we include an example of a function $f \in C^1(\mathbf{R})$ which is *not* in $C^2(\mathbf{R})$. This shows that the vector space $C^2(\mathbf{R})$ is contained in, but is not equal to $C^1(\mathbf{R})$. Let

$$f(x) = \begin{cases} x^2 & \text{if } x > 0 \\ -x^2 & \text{if } x \leq 0 \end{cases}.$$

Then f is differentiable everywhere, and

$$f'(x) = \begin{cases} 2x \text{ if } x > 0 \\ 0 \text{ if } x = 0 \\ -2x \text{ if } x < 0 \end{cases}.$$

(You should check that in fact $f'(0) = \lim_{h \to 0} [f(0+h) - f(0)]/h = 0$, as we claim.) Furthermore, f' is a continuous function, which shows that $f \in C^1(\mathbf{R})$. However, f' is not differentiable at $x = 0$, since the graph of f' has a corner there—indeed, the formulas for f' given above show that f' is the same function as $g(x) = 2|x|$. Hence $f \notin C^2(\mathbf{R})$.

Theorem (1.2.8) may also be used to show that subsets of vector spaces are not subspaces.

(1.2.12) Example. In $V = \mathbf{R}^2$, consider $W = \{(x_1, x_2) \mid x_1^3 - x_2^2 = 0\}$. This W is not a subspace of $\mathbf{R}^2$, since, for instance, we have $(1, 1)$ and $(4, 8) \in W$, but the sum $(1, 1) + (4, 8) = (5, 9) \notin W$. The components of the sum do not satisfy the defining equation of W: $5^3 - 9^2 = 44 \neq 0$.

Let us return now to the subspace W of $\mathbf{R}^3$ given in Example (1.2.10). Note that if we define

$$W_1 = \{(x_1, x_2, x_3) \in \mathbf{R}^3 \mid 4x_1 + 3x_2 - 2x_3 = 0\}$$

and

$$W_2 = \{(x_1, x_2, x_3) \in \mathbf{R}^3 \mid x_1 - x_3 = 0\}$$

then W is the set of vectors in both W_1 and W_2. In other words, we have an equality of sets $W = W_1 \cap W_2$. In this example we see that the *intersection* of these two subspaces of $\mathbf{R}^3$ is also a subspace of $\mathbf{R}^3$. This is a property of intersections of subspaces, which is true in general.

(1.2.13) Theorem. Let V be a vector space. Then the intersection of any collection of subspaces of V is a subspace of V.

Proof: Consider any collection of subspaces of V. Note first that the intersection of the subspaces is nonempty, since it contains at least the zero vector from V. Now, let $\mathbf{x}$, $\mathbf{y}$ be any two vectors in the intersection of all the subspaces in the collection (i.e., $\mathbf{x}, \mathbf{y} \in W$ for all W in the collection). Since each W in the collection is a subspace of V, $c\mathbf{x} + \mathbf{y} \in W$. Since this is true for all the W in the collection, $c\mathbf{x} + \mathbf{y}$ is in the intersection of all the subspaces in the collection. Hence the intersection is a subspace of V by Theorem (1.2.8). ∎

One important application of this theorem deals with general subspaces of the space $\mathbf{R}^n$ of the form seen in Example (1.2.10). Namely, we will show that the set of all solutions of any *simultaneous system* of equations of the form

$$a_{11}x_1 + a_{12}x_2 + \cdots + a_{1n}x_n = 0$$
$$a_{21}x_1 + a_{22}x_2 + \cdots + a_{2n}x_n = 0$$
$$\vdots$$
$$a_{m1}x_1 + a_{m2}x_2 + \cdots + a_{mn}x_n = 0$$

is a subspace of $\mathbf{R}^n$. (Here the notation a_{ij} means the coefficient of x_j in the ith equation. For example, a_{23} is the coefficient of x_3 in the second equation in the system.)

(1.2.14) Corollary. Let a_{ij} $(1 \leq i \leq m,\ 1 \leq j \leq n)$ be any real numbers and let $W = \{(x_1, \ldots, x_n) \in \mathbf{R}^n \mid a_{i1}x_1 + \cdots + a_{in}x_n = 0 \text{ for all } i,\ 1 \leq i \leq m\}$. Then W is a subspace of $\mathbf{R}^n$.

Proof: For each i, $1 \leq i \leq m$, let $W_i = \{(x_1, \ldots, x_n) \mid a_{i1}x_1 + \cdots + a_{in}x_n = 0\}$. Then since W is precisely the set of solutions of the simultaneous system formed from the defining equations of all the W_i, we have $W = W_1 \cap W_2 \cap \cdots \cap W_m$. Each W_i is a subspace of $\mathbf{R}^n$ [see Example (1.2.2)], so by Theorem (1.2.13) W is also a subspace of $\mathbf{R}^n$. ■

Historically, finding methods for solving systems of equations of this type was the major impetus for the development of linear algebra. In addition, many of the applications of linear algebra come down to solving systems of linear equations (and describing the set of solutions in some way). We will return to these matters in Section 1.5 of this chapter.

EXERCISES

1. Show that $\{\mathbf{0}\} \subset V$ is a subspace of each vector space V.

2. a) Let $V_1 = \{f\colon \mathbf{R} \to \mathbf{R} \mid f(x) = f(-x) \text{ for all } x \in \mathbf{R}\}$. ($V_1$ is called the set of *even* functions.) Show that $\cos(x)$ and x^2 define functions in V_1.
   b) Show that V_1 is a subspace of $F(\mathbf{R})$ using the same operations given in Example (1.2.1).
   c) Let $V_2 = \{f\colon \mathbf{R} \to \mathbf{R} \mid f(-x) = -f(x) \text{ for all } x \in \mathbf{R}\}$. ($V_2$ is called the set of *odd* functions.) Give three examples of functions in V_2.
   d) Show that V_2 is also a subspace of $F(\mathbf{R})$.

3. For each of the following subsets W of a vector space V, determine if W is a subspace of V. Say why or why not in each case:
   a) $V = \mathbf{R}^3$, and $W = \{(a_1, a_2, a_3) \mid a_1 - 3a_2 + 4a_3 = 0, \text{ and } a_1 = a_2\}$
   b) $V = \mathbf{R}^2$, and $W = \{(a_1, a_2) \mid \sin(a_1) = a_2\}$
   c) $V = \mathbf{R}^3$, and $W = \{(a_1, a_2, a_3) \mid (a_1 + a_2 + a_3)^2 = 0\}$

d) $V = \mathbf{R}^3$, and $W = \{(a_1, a_2, a_3) \mid a_3 \geq 0\}$
e) $V = \mathbf{R}^3$, and $W = \{(a_1, a_2, a_3) \mid a_1, a_2, a_3$ all integers$\}$
f) $V = C^1(\mathbf{R})$, and $W = \{f \mid f'(x) + 4f(x) = 0$ for all $x \in \mathbf{R}\}$
g) $V = C^1(\mathbf{R})$, and $W = \{f \mid \sin(x) \cdot f'(x) + f(x) = 6$ for all $x \in \mathbf{R}\}$
h) $V = P_n(\mathbf{R})$, and $W = \{p \mid p(\sqrt{2}) = 0\}$
i) $V = P_n(\mathbf{R})$, and $W = \{p \mid p(1) = 1$ and $p(2) = 0\}$.
j) $V = P_3(\mathbf{R})$, and $W = \{p \mid p'(x) \in P_1(\mathbf{R})\}$
k) $V = F(\mathbf{R})$, and $W = \{f \mid f$ is periodic with period 2π: $f(x + 2\pi) = f(x)$ for all $x \in \mathbf{R}\}$

4. a) If W is a subspace of a vector space V, show that for all vectors $\mathbf{x}_1, \ldots, \mathbf{x}_n \in W$, and all scalars $a_1, \ldots, a_n \in \mathbf{R}$, the vector $a_1\mathbf{x}_1 + \cdots + a_n\mathbf{x}_n \in W$.
b) Is the converse of the statement in part a true?

5. Let W be a subspace of a vector space V, let $\mathbf{y} \in V$, and define the set $\mathbf{y} + W = \{\mathbf{x} \in V \mid \mathbf{x} = \mathbf{y} + \mathbf{w}$ for some $\mathbf{w} \in W\}$. Show that $\mathbf{y} + W$ is a subspace of V if and only if $\mathbf{y} \in W$.

6. If W_1 and W_2 are subspaces of a vector space V, is $W_1 \setminus W_2$ ever a subspace of V? Why or why not? (Here $W_1 \setminus W_2$ denotes the *set difference* of W_1 and W_2: $W_1 \setminus W_2 = \{\mathbf{w} \in W \mid \mathbf{w} \in W_1$ but $\mathbf{w} \notin W_2\}$.)

7. a) Show that in $V = \mathbf{R}^2$, each line containing the origin is a subspace.
b) Show that the only subspaces of $V = \mathbf{R}^2$ are the zero subspace, $\mathbf{R}^2$ itself, and the lines through the origin. (*Hint:* Show that if W is a subspace of $\mathbf{R}^2$ that contains two nonzero vectors lying along different lines through the origin, then W must be all of $\mathbf{R}^2$.)

8. Let $C([a, b])$ denote the set of continuous functions on the closed interval $[a, b] \subset \mathbf{R}$. Show that $C([a, b])$ is a subspace of the vector space $F([a, b])$ introduced in Exercise 10 of Section 1.1.

9. Let $C^\infty(\mathbf{R})$ denote the set of functions in $F(\mathbf{R})$ that have derivatives of all orders. Show that $C^\infty(\mathbf{R})$ is a subspace of $F(\mathbf{R})$.

10. Show that if V_1 is a subspace of V_2 and V_2 is a subspace of V_3, then V_1 is a subspace of V_3.

The following group of exercises introduces new examples of vector spaces that will be used extensively later in the text. Let $m, n \geq 1$ be integers. An *m by n matrix* is a rectangular array of real numbers with m (horizontal) rows and n (vertical) columns:

$$\begin{bmatrix} a_{11} \, a_{12} & \cdots & a_{1n} \\ a_{21} \, a_{22} & \cdots & a_{2n} \\ \vdots & & \vdots \\ a_{m1} a_{m2} & \cdots & a_{mn} \end{bmatrix}.$$

Here a_{ij} represents the entry in the ith row and the jth column of the matrix. The set of all m by n matrices with real entries will be denoted by $M_{m\times n}(\mathbf{R})$. We usually use the shorthand notation $A = (a_{ij})$ to indicate the matrix whose entries are the a_{ij}.

If $A = (a_{ij})$ and $B = (b_{ij})$ are both matrices in $M_{m\times n}(\mathbf{R})$, we can define their *sum*, denoted $A + B$, to be the m by n matrix whose entries are the sums of the corresponding entries from A and B. That is, $A + B$ is the matrix whose entries are $(a_{ij} + b_{ij})$. (This sum operation is not defined if the matrices A and B have different sizes.) In addition, given a matrix $A = (a_{ij}) \in M_{m\times n}(\mathbf{R})$ and a scalar $c \in \mathbf{R}$, we define the product of c and A to be the matrix cA whose entries are obtained by multiplying each entry of A by c: $cA = (ca_{ij})$.

11. Using the definitions of the matrix sum and scalar product operations given earlier, compute:

a) $\begin{bmatrix} 2 & 4 & -2 \\ 3 & 1 & 9 \end{bmatrix} + \begin{bmatrix} -1 & 1 & 0 \\ 7 & 1 & 4 \end{bmatrix}$ in $M_{2\times 3}(\mathbf{R})$

b) $4\cdot\begin{bmatrix} 2 & -2 \\ 6 & -1 \end{bmatrix} + 3\cdot\begin{bmatrix} -2 & -9 \\ 1 & 4 \end{bmatrix} - 2\cdot\begin{bmatrix} 1 & 1 \\ 5 & -3 \end{bmatrix}$ in $M_{2\times 2}(\mathbf{R})$

12. Show that $M_{m\times n}(\mathbf{R})$ is a vector space, using the sum and scalar product operations defined earlier.

13. Show that the subset $W = \{\begin{bmatrix} a_{11} & a_{12} \\ a_{21} & a_{22} \end{bmatrix} \in M_{2\times 2}(\mathbf{R}) \mid 3a_{11} - 2a_{22} = 0\}$ is a subspace of $M_{2\times 2}(\mathbf{R})$.

14. Show that the subset $W = \{\begin{bmatrix} a_{11} & a_{12} \\ a_{21} & a_{22} \end{bmatrix} \in M_{2\times 2}(\mathbf{R}) \mid a_{12} = a_{21}\}$ (called the set of *symmetric* 2 by 2 matrices) is a subspace of $M_{2\times 2}(\mathbf{R})$.

15. Show that the subset $W = \{\begin{bmatrix} a_{11} & a_{12} \\ a_{21} & a_{22} \end{bmatrix} \in M_{2\times 2}(\mathbf{R}) \mid a_{11} = a_{22} = 0$ and $a_{12} = -a_{21}\}$ (called the set of *skew-symmetric* 2 by 2 matrices) is a subspace of $M_{2\times 2}(\mathbf{R})$.

16. Ignoring the way the elements are written, do you see any similarities between the vector space $M_{m\times n}(\mathbf{R})$ and other vector spaces we have studied? Try to construct a one-to-one correspondence between the m by n matrices and the elements of another vector space.

§1.3. LINEAR COMBINATIONS

If we apply the operations of vector addition and multiplication by scalars repeatedly to vectors in a vector space V, the most general expressions we can produce have the form $a_1\mathbf{x}_1 + \cdots + a_n\mathbf{x}_n$, where the $a_i \in \mathbf{R}$, and the $\mathbf{x}_i \in V$. Frequently, the vectors involved will come from some specified subset of V. To discuss this situation, we introduce the following terminology.

(1.3.1) Definitions. Let S be a subset of a vector space V.

a) A *linear combination* of vectors in S is any sum $a_1\mathbf{x}_1 + \cdots + a_n\mathbf{x}_n$, where the $a_i \in \mathbf{R}$, and the $\mathbf{x}_i \in S$.

b) If $S \neq \phi$ (the empty subset of V), the set of all linear combinations of vectors in S is called the (linear) *span* of S, and denoted Span(S). If $S = \phi$, we define $\text{Span}(S) = \{\mathbf{0}\}$.

c) If $W = \text{Span}(S)$, we say S *spans* (or *generates*) W.

We think of the span of a set S as the set of all vectors that can be "built up" from the vectors in S by forming linear combinations.

(1.3.2) Example. In $V = \mathbf{R}^3$, let $S = \{(1, 0, 0), (0, 1, 0)\}$. Then a typical linear combination of the vectors in S is a vector

$$a_1(1, 0, 0) + a_2(0, 1, 0) = (a_1, a_2, 0)$$

The span of S is the set of all such vectors (i.e., the vectors produced for all choices of $a_1, a_2 \in \mathbf{R}$). We have $\text{Span}(S) = \{(a_1, a_2, 0) \in \mathbf{R}^3 \mid a_1, a_2 \in \mathbf{R}\}$. Geometrically, Span($S$) is just the $x_1 - x_2$-plane in $\mathbf{R}^3$. See Figure 1.9. Note that in this example, we can also describe Span(S) as the set of all vectors in $\mathbf{R}^3$ whose third components are 0, that is, $\text{Span}(S) = \{(a_1, a_2, a_3) \in \mathbf{R}^3 \mid a_3 = 0\}$. Hence, by Corollary (1.2.14), Span(S) is a subspace of $\mathbf{R}^3$.

(1.3.3) Example. In $V = C(\mathbf{R})$, let $S = \{1, x, x^2, \ldots, x^n\}$. Then we have $\text{Span}(S) = \{f \in C(\mathbf{R}) \mid f(x) = a_0 + a_1x + \cdots + a_nx^n \text{ for some } a_0, \ldots, a_n \in \mathbf{R}\}$. Thus Span($S$) is the subspace $P_n(\mathbf{R}) \subset C(\mathbf{R})$.

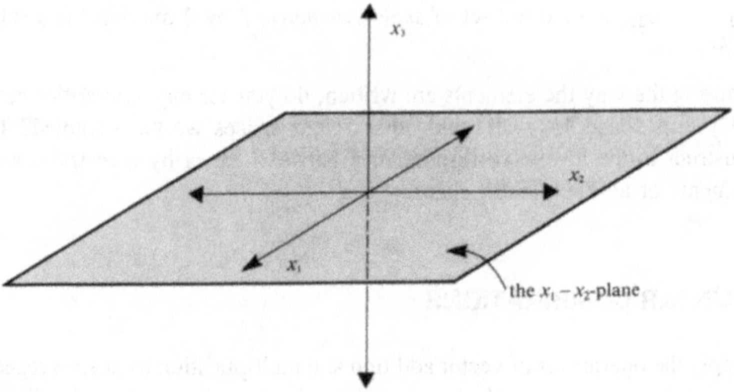

Figure 1.9.

The fact that the span of a set of vectors is a subspace of the vector space from which the vectors are chosen is true in general.

(1.3.4) Theorem. Let V be a vector space and let S be any subset of V. Then Span(S) is a subspace of V.

Proof: We prove this by applying Theorem (1.2.8) once again. Span(S) is nonempty by definition. Furthermore, let $\mathbf{x}, \mathbf{y} \in$ Span(S), and let $c \in \mathbf{R}$. Then we can write $\mathbf{x} = a_1\mathbf{x}_1 + \cdots + a_n\mathbf{x}_n$, with $a_i \in \mathbf{R}$ and $\mathbf{x}_i \in S$. Similarly, we can write $\mathbf{y} = b_1\mathbf{x}'_1 + \cdots + b_m\mathbf{x}'_m$, with $b_i \in \mathbf{R}$ and $\mathbf{x}'_i \in S$. Then for any scalar c we have

$$\begin{aligned} c\mathbf{x} + \mathbf{y} &= c(a_1\mathbf{x}_1 + \cdots + a_n\mathbf{x}_n) + b_1\mathbf{x}'_1 + \cdots + b_m\mathbf{x}'_m \\ &= ca_1\mathbf{x}_1 + \cdots + ca_n\mathbf{x}_n + b_1\mathbf{x}'_1 + \cdots + b_m\mathbf{x}'_m \end{aligned}$$

Since this is also a linear combination of the vectors in the set S, we have that $c\mathbf{x} + \mathbf{y} \in$ Span(S). Hence Span (S) is a subspace of V. (Note that there could be some duplications in the vectors appearing in the two expansions, but that does not affect the conclusion.) ■

Theorem (1.3.4) is very important because it gives another one of the frequently used methods for constructing subspaces of a given vector space. Indeed, later in this chapter we will see that every subspace may be obtained in this way.

Closely related to the idea of linear combinations in a vector space is an operation on subspaces of a given vector space.

(1.3.5) Definition. Let W_1 and W_2 be subspaces of a vector space V. The *sum* of W_1 and W_2 is the set

$$W_1 + W_2 = \{\mathbf{x} \in V \mid \mathbf{x} = \mathbf{x}_1 + \mathbf{x}_2, \text{ for some } \mathbf{x}_1 \in W_1 \text{ and } \mathbf{x}_2 \in W_2\}.$$

We think of $W_1 + W_2$ as the set of vectors that can be "built up" from the vectors in W_1 and W_2 by linear combinations. Conversely, the vectors in $W_1 + W_2$ are precisely those vectors in V that can be "broken down" into the sum of a vector in W_1 and a vector in W_2.

(1.3.6) Example. If $W_1 = \{(a_1, a_2) \in \mathbf{R}^2 \mid a_2 = 0\}$ and $W_2 = \{(a_1, a_2) \in \mathbf{R}^2 \mid a_1 = 0\}$, then $W_1 + W_2 = \mathbf{R}^2$, since every vector in $\mathbf{R}^2$ can be written as the sum of a vector in W_1 and a vector in W_2. For instance, we have $(5, -6) = (5, 0) + (0, -6)$, and $(5, 0) \in W_1$, while $(0, -6) \in W_2$.

(1.3.7) Example. In $V = C(\mathbf{R})$, consider $W_1 = \text{Span}(\{1, e^x\})$, and $W_2 = \text{Span}(\{\sin(x), \cos(x)\})$. Then $W_1 + W_2 = \text{Span}(\{1, e^x, \sin(x), \cos(x)\})$. To see why this is so, note that on one hand every function in $W_1 + W_2$ is in the span of the set $\{1, e^x, \sin(x), \cos(x)\}$ by definition. On the other hand, every linear combination in Span ($\{1, e^x, \sin(x), \cos(x)\}$) can be broken up as follows:

$$a_1 + a_2e^x + a_3\sin(x) + a_4\cos(x) = (a_1 + a_2e^x) + (a_3\sin(x) + a_4\cos(x))$$

so we have that every function in Span($\{1, e^x, \sin(x), \cos(x)\}$) is the sum of a function in W_1 and one in W_2.

In general, it is true that if $W_1 = \text{Span}(S_1)$ and $W_2 = \text{Span}(S_2)$, then $W_1 + W_2$ is the span of the union of the spanning sets S_1 and S_2.

(1.3.8) Proposition. Let $W_1 = \text{Span}(S_1)$ and $W_2 = \text{Span}(S_2)$ be subspaces of a vector space V. Then $W_1 + W_2 = \text{Span}(S_1 \cup S_2)$.

Proof: To show that two sets A and B are equal, we can show that each set is contained in the other.

To see that $W_1 + W_2 \subseteq \text{Span}(S_1 \cup S_2)$, let $\mathbf{v} \in W_1 + W_2$. Then $\mathbf{v} = \mathbf{v}_1 + \mathbf{v}_2$, where $\mathbf{v}_1 \in W_1$ and $\mathbf{v}_2 \in W_2$. Since $W_1 = \text{Span}(S_1)$, we can write $\mathbf{v}_1 = a_1\mathbf{x}_1 + \cdots + a_m\mathbf{x}_m$, where each $\mathbf{x}_i \in S_1$ and each $a_i \in \mathbf{R}$. Similarly, we can write $\mathbf{v}_2 = b_1\mathbf{y}_1 + \cdots + b_n\mathbf{y}_n$, where each $\mathbf{y}_j \in S_2$ and each $b_j \in \mathbf{R}$. Hence we have $\mathbf{v} = a_1\mathbf{x}_1 + \cdots + a_m\mathbf{x}_m + b_1\mathbf{y}_1 + \cdots + b_n\mathbf{y}_n$. This is a linear combination of vectors that are either in S_1 or in S_2. Hence $\mathbf{v} \in \text{Span}(S_1 \cup S_2)$, and since this is true for all such $\mathbf{v}$, $W_1 + W_2 \subseteq \text{Span}(S_1 \cup S_2)$.

Conversely, to see that $\text{Span}(S_1 \cup S_2) \subseteq W_1 + W_2$, we note that if $\mathbf{v} \in \text{Span}(S_1 \cup S_2)$, then $\mathbf{v} = c_1\mathbf{z}_1 + \cdots + c_l\mathbf{z}_l$, where each $\mathbf{z}_k \in S_1 \cup S_2$ and each $c_k \in \mathbf{R}$. Each $\mathbf{z}_k$ is in S_1 or in S_2, so by renaming the vectors and regrouping the terms, we have $\mathbf{v} = a_1\mathbf{x}_1 + \cdots + a_m\mathbf{x}_m + b_1\mathbf{y}_1 + \cdots + b_n\mathbf{y}_n$, where each $\mathbf{x}_i \in S_1$ and each $\mathbf{y}_j \in S_2$. Hence, by definition, we have written $\mathbf{v}$ as the sum of a vector in W_1 and a vector in W_2, so $\mathbf{v} \in W_1 + W_2$. Since this is true for all $\mathbf{v} \in \text{Span}(S_1 \cup S_2)$, we have $\text{Span}(S_1 \cup S_2) \subseteq W_1 + W_2$. Combining the two parts of the proof, we have the claim. ■

We now show that, in general, the sum of two subspaces is also a subspace.

(1.3.9) Theorem. Let W_1 and W_2 be subspaces of a vector space V. Then $W_1 + W_2$ is also a subspace of V.

Proof: It is clear that $W_1 + W_2$ is nonempty, since W_1 and W_2 are nonempty. Let $\mathbf{x}$, $\mathbf{y}$ be any two vectors in $W_1 + W_2$ and let $c \in \mathbf{R}$. Since $\mathbf{x}$ and $\mathbf{y} \in W_1 + W_2$, we can write $\mathbf{x} = \mathbf{x}_1 + \mathbf{x}_2$, $\mathbf{y} = \mathbf{y}_1 + \mathbf{y}_2$, where $\mathbf{x}_1, \mathbf{y}_1 \in W_1$ and $\mathbf{x}_2, \mathbf{y}_2 \in W_2$. Then we have

$$\begin{aligned} c\mathbf{x} + \mathbf{y} &= c(\mathbf{x}_1 + \mathbf{x}_2) + (\mathbf{y}_1 + \mathbf{y}_2) \\ &= (c\mathbf{x}_1 + \mathbf{y}_1) + (c\mathbf{x}_2 + \mathbf{y}_2) \end{aligned}$$

Since W_1 and W_2 are subspaces of V, we have $c\mathbf{x}_1 + \mathbf{y}_1 \in W_1$ and $c\mathbf{x}_2 + \mathbf{y}_2 \in W_2$. Hence by the definition, $c\mathbf{x} + \mathbf{y} \in W_1 + W_2$. By Theorem (1.2.8), $W_1 + W_2$ is a subspace of V. ■

(1.3.10) Remark. In general, if W_1 and W_2 are subspaces of V, then $W_1 \cup W_2$ will not be a subspace of V. For example, consider the two subspaces of $\mathbf{R}^2$ given in Example (1.3.6). In that case $W_1 \cup W_2$ is the union of two lines through the

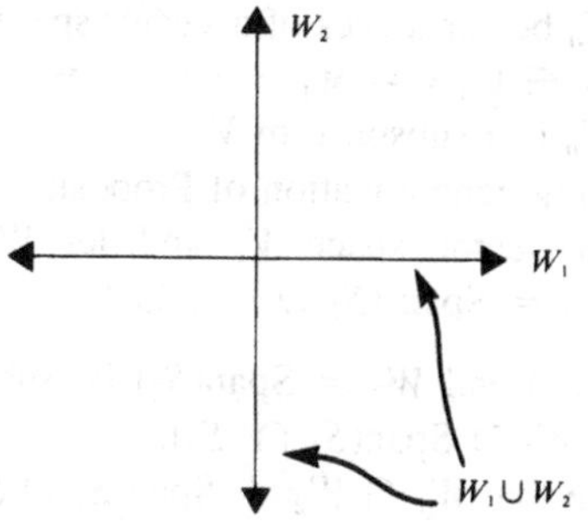

Figure 1.10.

origin in $\mathbf{R}^2$. See Figure 1.10. In this case $W_1 \cup W_2$ is *not* a subspace of $\mathbf{R}^2$, since, for example, we have $(1, 0) + (0, 1) = (1, 1) \notin W_1 \cup W_2$, even though $(1, 0) \in W_1$, and $(0, 1) \in W_2$.

It follows from Proposition (1.3.8) that $W_1 + W_2$ is a subspace of V containing $W_1 \cup W_2$, however. In fact, $W_1 + W_2$ is the smallest subspace of V containing $W_1 \cup W_2$ in the sense that any other subspace of V that contains $W_1 \cup W_2$ contains $W_1 + W_2$.

(1.3.11) Proposition. Let W_1 and W_2 be subspaces of a vector space V and let W be a subspace of V such that $W \supseteq W_1 \cup W_2$. Then $W \supseteq W_1 + W_2$.

Proof: Let $\mathbf{v}_1 \in W_1$ and $\mathbf{v}_2 \in W_2$ be any vectors. Since $\mathbf{v}_1 \in W_1 \subseteq W_1 \cup W_2$, $\mathbf{v}_1 \in W$ as well. Similarly, $\mathbf{v}_2 \in W$. Hence, since W is a subspace of V, $\mathbf{v}_1 + \mathbf{v}_2 \in W$. But this shows that every vector in $W_1 + W_2$ is contained in W, so $W \supseteq W_1 + W_2$ as claimed. ■

(1.3.12) Question. Are there any cases where $W_1 \cup W_2$ *is* a subspace of V? Can you characterize those cases completely?

EXERCISES

1. a) Let $S = \{(1, 0, 0), (0, 0, 2)\}$ in $\mathbf{R}^3$. Which vectors are in Span(S)? Describe this set geometrically.
   b) Same question for $S = \{(1, 4, 0, 0), (2, 3, 0, 0)\}$ in $\mathbf{R}^4$
   c) Same question for $S = \{(1, 1, 1)\}$ in $\mathbf{R}^3$
   d) Same question for $S = \{1, x, x^2\}$ in $P_4(\mathbf{R})$
2. In $V = C(\mathbf{R})$, let $S_1 = \{\sin(x), \cos(x), \sin^2(x), \cos^2(x)\}$ and $S_2 = \{1, \sin(2x), \cos(2x)\}$. Is Span($S_1$) = Span($S_2$)? Why or why not?
3. In $V = P_2(\mathbf{R})$. Let $S = \{1, 1 + x, 1 + x + x^2\}$. Show that Span ($S$) = $P_2(\mathbf{R})$.
4. Show that a subset W of a vector space is a subspace if and only if Span (W) = W.

5. a) Let $W_1, \ldots, W_n$ be subspaces of a vector space V. Define the sum $W_1 + \cdots + W_n = \{\mathbf{v} \in V \mid \mathbf{v} = \mathbf{w}_1 + \cdots + \mathbf{w}_n$, where $\mathbf{w}_i \in W_i\}$. Show that $W_1 + \cdots + W_n$ is a subspace of V.
   b) Prove the following generalization of Proposition (1.3.8). Let $S_1, \ldots, S_n$ be subsets of a vector space V, and let $W_i = \text{Span}(S_i)$. Show that $W_1 + \cdots + W_n = \text{Span}(S_1 \cup \cdots \cup S_n)$.

6. a) Let $W_1 = \text{Span}(S_1)$ and $W_2 = \text{Span}(S_2)$ be subspaces of a vector space. Show that $W_1 \cap W_2 \supset \text{Span}(S_1 \cap S_2)$.
   b) Show by example that $W_1 \cap W_2 \neq \text{Span}(S_1 \cap S_2)$ in general.

7. Show that if S is a subset of a vector space V and W is a subspace of V with $W \supset S$, then $W \supset \text{Span}(S)$.

8. Show that if W_1 and W_2 are subspaces of a vector space with $W_1 \cap W_2 = \{\mathbf{0}\}$, then for each vector $\mathbf{x} \in W_1 + W_2$ there are *unique* vectors $\mathbf{x}_1 \in W_1$ and $\mathbf{x}_2 \in W_2$ such that $\mathbf{x} = \mathbf{x}_1 + \mathbf{x}_2$.

9. If $V = W_1 + W_2$ and $W_1 \cap W_2 = \{\mathbf{0}\}$, we say that V is the *direct sum* of W_1 and W_2, written $V = W_1 \oplus W_2$.
   a) Show that $\mathbf{R}^2$ is the direct sum of $W_1 = \{(x_1, x_2) \mid x_1 = 0\}$ and $W_2 = \{(x_1, x_2) \mid x_2 = 0\}$.
   b) Show that $\mathbf{R}^3$ is the direct sum of $W_1 = \text{Span}(\{(1, 1, 1)\})$ and $W_2 = \text{Span}(\{(1, 0, 0), (1, 1, 0)\})$.
   c) Show that $F(\mathbf{R})$ is the direct sum of $W_1 = \{f \mid f(-x) = f(x) \text{ for all } x \in \mathbf{R}\}$ (the even functions) and $W_2 = \{f \mid f(-x) = -f(x) \text{ for all } x \in \mathbf{R}\}$ (the odd functions). [*Hint:* if $f(x) \in F(\mathbf{R})$ begin by showing that $f(x) + f(-x)$ is always an even function and that $f(x) - f(-x)$ is always odd.]
   d) Show that, in general, if $V = W_1 \oplus W_2$ then for each $\mathbf{v} \in V$, there exist unique vectors $\mathbf{w}_1 \in W_1$ and $\mathbf{w}_2 \in W_2$ such that $\mathbf{v} = \mathbf{w}_1 + \mathbf{w}_2$.

10. Let S_1 and S_2 be subsets of a vector space V. Assume that $\text{Span}(S_1) = V$ and that every vector in S_1 is in $\text{Span}(S_2)$. Show that $V = \text{Span}(S_2)$ as well.

The following exercises refer to the vector space $M_{m \times n}(\mathbf{R})$ defined in the exercises accompanying Section 1.2.

11. a) Let $S = \left\{\begin{bmatrix} 1 & 0 \\ 0 & 0 \end{bmatrix}, \begin{bmatrix} 0 & 1 \\ 0 & 0 \end{bmatrix}\right\}$ in $M_{2 \times 2}(\mathbf{R})$. Describe the subspace $\text{Span}(S)$.
    b) Same question for $S = \left\{\begin{bmatrix} 2 & -1 \\ 0 & 0 \end{bmatrix}, \begin{bmatrix} 0 & 0 \\ -1 & 1 \end{bmatrix}\right\}$.

12. Let W_1 be the subspace of symmetric 2 by 2 matrices and let W_2 be the subspace of skew-symmetric 2 by 2 matrices defined in the exercises in Section 1.2. Show that $M_{2 \times 2}(\mathbf{R}) = W_1 \oplus W_2$.

§1.4. LINEAR DEPENDENCE AND LINEAR INDEPENDENCE

Because of the importance of constructing subspaces by giving spanning sets, we now examine this process in somewhat greater detail. In our study of linear algebra, we frequently need to consider different spanning sets for the same subspace of some vector space.

(1.4.1) Example. In $V = \mathbf{R}^3$, consider the subspace W spanned by the set

$$S = \{(1, 2, 1), (0, -1, 3), (1, 0, 7)\}$$

By Definition (1.3.1) we have

$$W = \{(x_1, x_2, x_3) \in \mathbf{R}^3 \mid (x_1, x_2, x_3) = a_1(1, 2, 1) + a_2(0, -1, 3) + a_3(1, 0, 7) \quad \text{for some } a_i \in \mathbf{R}\}$$

We already have one set (S itself) that spans W. Nevertheless, we can ask, are there other sets that also span W? In particular, is using the vectors in S the most *economical* way to generate all the vectors in W by forming linear combinations? Might it be true that some *subset* of S will also span W?

In this case there is such a subset of the set S, since the vectors in S are related to the following way:

$$(1, 2, 1) + 2(0, -1, 3) + (-1)(1, 0, 7) = (0, 0, 0) \tag{1.1}$$

In other words, there is a linear combination of the vectors in S with nonzero coefficients that adds up to the zero vector in $\mathbf{R}^3$. This allows us to "solve for" one of the vectors in terms of the other two. For example, from Eq. (1.1) we have

$$(1, 0, 7) = 1(1, 2, 1) + 2(0, -1, 3)$$

This apparently innocuous fact has an interesting consequence: In any linear combination of the vectors in S we can substitute for the vector $(1, 0, 7)$ as follows:

$$\begin{aligned} a_1(1, 2, 1) + a_2(0, 1, -3) + a_3(1, 0, 7) &= a_1(1, 2, 1) + a_2(0, -1, 3) \\ &\quad + a_3((1, 2, 1) \\ &\quad + 2(0, -1, 3)) \\ &= (a_1 + a_3) \cdot (1, 2, 1) \\ &\quad + (a_2 + 2a_3) \cdot (0, -1, 3) \end{aligned}$$

As a result, we see that every vector in Span(S) is also in the span of the subset $S' = \{(1, 2, 1), (0, -1, 3)\}$. Since the opposite inclusion is clear, we have Span(S) = Span(S'). The existence of the relation (1.1) in effect makes one of the vectors in the original set S *redundant*.

In general, if we have any vector space V and a set of vectors $S \subset V$, and there is a relation

$$a_1\mathbf{x}_1 + \cdots + a_n\mathbf{x}_n = \mathbf{0} \tag{1.2}$$

where $a_i \in \mathbf{R}$ and the $\mathbf{x}_i \in S$, and at least one of the a_i, say, $a_n \neq 0$, then the situation is the same as in the example. We can solve equation (1.2) for $\mathbf{x}_n$, and we see that

$$\mathbf{x}_n = (-a_1/a_n)\mathbf{x}_1 + \cdots + (-a_{n-1}/a_n)\mathbf{x}_{n-1}$$

Hence $\mathbf{x}_n$ is a linear combination of the vectors $\mathbf{x}_1, \ldots, \mathbf{x}_{n-1}$, and in any linear combination involving $\mathbf{x}_n$, we could replace $\mathbf{x}_n$ by that linear combination of $\mathbf{x}_1, \ldots, \mathbf{x}_{n-1}$. This shows that a smaller set spans the same space. If there were any further relations such as Eq. (1.2) between the vectors in $\{\mathbf{x}_1, \ldots, \mathbf{x}_{n-1}\}$, we could find an even smaller spanning set.

The foregoing discussion leads us to the following definitions.

(1.4.2) Definitions. Let V be a vector space, and let S be a subset of V.

a) A *linear dependence* among the vectors of S is an equation

$$a_1\mathbf{x}_1 + \cdots + a_n\mathbf{x}_n = \mathbf{0}$$

where the $\mathbf{x}_i \in S$, and the $a_i \in \mathbf{R}$ are not all zero (i.e., at least one of the $a_i \neq 0$).

b) the set S is said to be *linearly dependent* if there exists a linear dependence among the vectors in S.

(1.4.3) Examples

a) $S = \{\mathbf{0}\}$ is linearly dependent in any vector space, since the equation $a\mathbf{0} = \mathbf{0}$ (any $a \neq 0$ in $\mathbf{R}$) is a linear dependence.

b) More generally, if S is any set containing the vector $\mathbf{0}$ then S is linearly dependent for the same reason.

c) To decide whether a given set of vectors is linearly dependent or not, we must determine if there are any linear dependences among the vectors in the set. For example, if

$$S = \{(1, 5, 0, 0), (2, 1, 0, 0), (-1, 1, 0, 0)\}$$

in $\mathbf{R}^4$, then we ask, are there any scalars a_1, a_2, a_3 that are not all zero such that

$$a_1(1, 5, 0, 0) + a_2(2, 1, 0, 0) + a_3(-1, 1, 0, 0) = (0, 0, 0, 0)$$

After adding the vectors on the left, this yields the vector equation

$$(a_1 + 2a_2 - a_3, 5a_1 + a_2 + a_3, 0, 0) = (0, 0, 0, 0)$$

Equating components of the vectors, we obtain a system of two simultaneous equations in the a_i:

$$\begin{aligned} a_1 + 2a_2 - a_3 &= 0 \\ 5a_1 + a_2 + a_3 &= 0 \end{aligned} \tag{1.3}$$

(and two trivial equations $0 = 0$, which we will ignore).

By inspection, it is clear that setting a_3 equal to any (nonzero) real number, say, $a_3 = 1$, we obtain a system of equations that can be solved for a_1 and a_2 by elementary means. If we set $a_3 = 1$, we obtain

$$\begin{aligned} a_1 + 2a_2 &= 1 \\ 5a_1 + a_2 &= -1 \end{aligned}$$

Hence one solution of the original Eq. (1.3) is $a_1 = -1/3$, $a_2 = 2/3$, and $a_3 = 1$. (There are many other solutions as well.) Hence the set S is indeed linearly dependent.

d) In $V = C^1(\mathbf{R})$, consider the set $S = \{\sin(x), \cos(x)\}$. We ask, is S linearly dependent? Once again, this means, do there exist scalars a_1 and a_2, at least one of which is nonzero, such that

$$a_1 \sin(x) + a_2 \cos(x) = 0 \tag{1.4}$$

in $C^1(\mathbf{R})$? If Eq. (1.4) is valid for all $x \in \mathbf{R}$, then it is valid for any specific value of x. For example, if $x = 0$, we obtain $a_1 \sin(0) + a_2 \cos(0) = 0$. Since $\sin(0) = 0$ and $\cos(0) = 1$, this implies $a_2 = 0$. Similarly, if we substitute $x = \pi/2$, then since $\sin(\pi/2) = 1$ and $\cos(\pi/2) = 0$, we obtain $a_1 = 0$. As a result, there are no linear dependences among the functions in S, and S is not linearly dependent.

In Example (1.4.3d) we saw a set of vectors that was not linearly dependent. Such sets are called linearly independent sets. Now, it is possible to define the property of linear independence in precisely this manner—S is linearly independent if S is not linearly dependent. However, another (logically equivalent) definition is used more often, because this alternate definition makes it clearer as to exactly what must be shown to see that a given S is linearly independent.

(1.4.4) Definition. A subset S of a vector space V is *linearly independent* if whenever we have $a_i \in \mathbf{R}$ and $\mathbf{x}_i \in S$ such that $a_1\mathbf{x}_1 + \cdots + a_n\mathbf{x}_n = \mathbf{0}$, then $a_i = 0$ for all i.

(1.4.5) Remark. This is easily the subtlest concept we have encountered so far in our study of linear algebra. The student should give special attention to this definition and its ramifications. For anyone who wishes to see where the alternate form of the definition comes from, let us analyze the statement "S is linearly dependent" from the point of view of logic. (The interested reader may wish to consult Appendix 1 for some of the facts from propositional logic that we use here.) "S is linearly dependent" is equivalent to the assertion "there exist vectors $\mathbf{x}_i \in S$ and scalars $a_i \in \mathbf{R}$ such that $a_1\mathbf{x}_1 + \cdots + a_n\mathbf{x}_n = \mathbf{0}$ and there exists i such that $a_i \neq 0$." In symbolic form:

$$(\exists \mathbf{x}_i \in S)(\exists a_i \in \mathbf{R})\ (a_1\mathbf{x}_1 + \cdots + a_n\mathbf{x}_n = \mathbf{0}) \wedge (\exists i)(a_i \neq 0))$$

Using the rules for negation of statements with quantifiers and DeMorgan's Law, the negation of this statement is

$$(\forall \mathbf{x}_i \in S)(\forall a_i \in \mathbf{R})(\sim(a_1\mathbf{x}_1 + \cdots + a_n\mathbf{x}_n = \mathbf{0}) \vee \sim (\exists i)(a_i \neq 0))$$

This is in turn equivalent to

$$(\forall \mathbf{x}_i \in S)(\forall a_i \in \mathbf{R})((a_1\mathbf{x}_1 + \cdots + a_n\mathbf{x}_n = \mathbf{0}) \rightarrow (\forall i)(a_i = 0))$$

by the definition of the logical implication operator, and the rules for negating statements with quantifiers once again. In words, "If $a_1\mathbf{x}_1 + \cdots + a_n\mathbf{x}_n = \mathbf{0}$, then all the $a_i = 0$." This is the form in the definition.

(1.4.6) Examples

a) The set $\{\sin(x), \cos(x)\}$ in $C^1(\mathbf{R})$ is linearly independent. We saw in Example (1.4.3d) that the only linear combination of the two functions that adds up to zero is the one where both scalars $a_i = 0$.

b) In $V = \mathbf{R}^3$, consider $S = \{(1, 0, 0), (0, 1, 0), (0, 0, 1)\}$. S is linearly independent, since if

$$a_1(1, 0, 0) + a_2(0, 1, 0) + a_3(0, 0, 1) = (0, 0, 0)$$

then equating components on the two sides of the equation, we have $a_1 = a_2 = a_3 = 0$.

c) Generalizing part b, in $V = \mathbf{R}^n$, take $S = \{\mathbf{e}_1, \ldots, \mathbf{e}_n\}$, where by definition, $\mathbf{e}_i$ is the vector with a 1 in the ith component and 0 in every other component (e.g., $\mathbf{e}_1 = (1, 0, \ldots, 0)$, $\mathbf{e}_2 = (0, 1, 0, \ldots, 0)$, and so forth). Then S is linearly independent in $\mathbf{R}^n$, by an argument along the lines of the one given in part b. In addition to being linearly independent, S also spans $\mathbf{R}^n$. (Why?) This special set of vectors is known as the *standard basis* of $\mathbf{R}^n$.

d) In any vector space the empty subset ϕ is linearly independent. This fact may seem strange at first, but it can be explained as follows. The definition of linear independence in effect says that if there are any linear combinations among the vectors in the set that add up to zero, then all the coefficients must be zero. That condition is satisfied ("vacuously") here—there are no linear combinations to check!

We conclude this section by giving some general properties of linearly dependent sets and linearly independent sets that follow from the definitions.

(1.4.7) Proposition

a) Let S be a linearly dependent subset of a vector space V, and let S' be another subset of V that contains S. Then S' is also linearly dependent.

b) Let S be a linearly independent subset of a vector space V and let S' be another subset of V that is contained in S. Then S' is also linearly independent.

Proof:

a) Since S is linearly dependent, there exists a linear dependence among the vectors in S, say, $a_1\mathbf{x}_1 + \cdots + a_n\mathbf{x}_n = \mathbf{0}$. Since S is contained in S', this is also a linear dependence among the vectors in S'. Hence S' is linearly dependent.

b) Consider any equation $a_1\mathbf{x}_1 + \cdots + a_n\mathbf{x}_n = \mathbf{0}$, where the $a_i \in \mathbf{R}$ and the $\mathbf{x}_i \in S'$. Since S' is contained in S, we can also view this as a potential linear dependence among vectors in S. However, S is linearly independent, so it follows that all the $a_i = 0$. Hence S' is also linearly independent. ■

EXERCISES

1. Determine whether each of the following sets of vectors is linearly dependent or linearly independent:
   a) $S = \{(1, 1), (1, 3), (0, 2)\}$ in $\mathbf{R}^2$
   b) $S = \{(1, 2, 1), (1, 3, 0)\}$ in $\mathbf{R}^3$
   c) $S = \{(0, 0, 0), (1, 1, 1)\}$ in $\mathbf{R}^3$
   d) $S = \{(1, 1, 1, 1), (1, 0, 0, 1), (3, 0, 0, 0)\}$ in $\mathbf{R}^4$
   e) $S = \{x^2 + 1, x - 7\}$ in $P_2(\mathbf{R})$
   f) $S = \{x^4, x^4 + x^3, x^4 + x^3 + x^2\}$ in $P_4(\mathbf{R})$
   g) $S = \{e^x, e^{2x}\}$ in $C(\mathbf{R})$
   h) $S = \{e^x, \cos(x)\}$ in $C(\mathbf{R})$
   i) $S = \{\sqrt{3}, e, \pi\}$ in $\mathbf{R}$ ($= \mathbf{R}^1$)
   j) $S = \{\sin^2(x), \cos^2(x), -3\}$ in $C(\mathbf{R})$
   k) $S = (e^x, \sinh(x), \cosh(x)\}$ in $C(\mathbf{R})$

2. Let S_1 and S_2 be linearly independent subsets of a vector space V.
   a) Is $S_1 \cup S_2$ always linearly independent? Why or why not?
   b) Is $S_1 \cap S_2$ always linearly independent? Why or why not?
   c) Is $S_1 \setminus S_2$ always linearly independent? Why or why not?

3. Repeat the three parts of Question 2, replacing linearly independent by linearly dependent.

4. a) Show that if $\mathbf{v}, \mathbf{w} \in V$, then $\{\mathbf{v},\mathbf{w}\}$ is linearly dependent if an only if $\mathbf{v}$ is a scalar multiple of $\mathbf{w}$, or $\mathbf{w}$ is a scalar multiple of $\mathbf{v}$.
   b) Show by example, however, that there are linearly dependent sets of three vectors such that no pair are scalar multiples of eaçh other.

5. Let $\mathbf{v}, \mathbf{w} \in V$. Show that $\{\mathbf{v},\mathbf{w}\}$ is linearly independent if and only if $\{\mathbf{v} + \mathbf{w}, \mathbf{v} - \mathbf{w}\}$ is linearly independent.

6. Let $\mathbf{v}_1, \ldots, \mathbf{v}_n \in V$. Show that if $\{\mathbf{v}_1, \ldots, \mathbf{v}_n\}$ is linearly independent, then $\{\mathbf{v}_1, \mathbf{v}_1 + \mathbf{v}_2, \mathbf{v}_1 + \mathbf{v}_2 + \mathbf{v}_3, \ldots, \mathbf{v}_1 + \mathbf{v}_2 + \cdots + \mathbf{v}_n\}$ is linearly independent.

7. If S is linearly independent, show that there is no proper subset $S' \subset S$ with $\text{Span}(S') = \text{Span}(S)$. (A *proper* subset is a subset $S' \neq S$.)

8. Let W_1 and W_2 be subspaces of a vector space satisfying $W_1 \cap W_2 = \{\mathbf{0}\}$. Show that if $S_1 \subset W_1$ and $S_2 \subset W_2$ are linearly independent, then $S_1 \cup S_2$ is linearly independent.

9. a) Let S be a subset of a vector space V. Show that if $\text{Span}(S) = V$, then for all $\mathbf{v} \in V$, $\{\mathbf{v}\} \cup S$ is linearly dependent.
   b) Is the converse of this statement true? Why or why not?

10. a) Show that in $V = \mathbf{R}^2$, any set of three or more vectors is linearly dependent.
    b) What is the analogous statement for $V = \mathbf{R}^3$? Prove your assertion.

11. Let S be a nonempty subset of a vector space V, and assume that each vector in Span(S) can be written in one and only one way as a linear combination of vectors in S. Show that S is linearly independent.

12. a) Show that $S = \{\begin{bmatrix} 1 & 0 \\ 1 & 1 \end{bmatrix}, \begin{bmatrix} -1 & 3 \\ 2 & 1 \end{bmatrix}\}$ is a linearly independent subset of $V = M_{2\times 2}(\mathbf{R})$.

b) Show that $S = \{\begin{bmatrix} 1 & 0 \\ 0 & 0 \end{bmatrix}, \begin{bmatrix} 0 & 1 \\ 0 & 0 \end{bmatrix}, \begin{bmatrix} 0 & 0 \\ 1 & 0 \end{bmatrix}, \begin{bmatrix} 0 & 0 \\ 0 & 1 \end{bmatrix}\}$ is a linearly independent subset of $M_{2\times 2}(\mathbf{R})$ and that Span(S) $= M_{2\times 2}(\mathbf{R})$.

§1.5. INTERLUDE ON SOLVING SYSTEMS OF LINEAR EQUATIONS

In working with vector spaces and their subspaces, we will find that most of the computational problems to be solved come down to some combination of the following three basic problems.

Basic Problems:

1. Given a subspace W of a vector space V, defined by giving some condition on the vectors [e.g., a subspace of $\mathbf{R}^n$ defined by a system of linear equations as in Corollary (1.2.14)], find a set of vectors S such that $W =$ Span(S).

2. Given a set of vectors S and a vector $\mathbf{x} \in V$, determine if $\mathbf{x} \in$ Span(S). [Note that this is, in effect, the opposite problem of (1), since if we can do this for a general vector $\mathbf{x} \in V$, then we obtain conditions on $\mathbf{x}$ that are necessary and sufficient for $\mathbf{x}$ to lie in the subspace $W =$ Span(S)].

3. Given a set of vectors, S, determine if S is linearly dependent or linearly independent.

In practice, the subspaces we deal with in problems of the first type are usually subspaces of $\mathbf{R}^n$ defined by systems of equations as in Corollary (1.2.14). Given scalars a_{ij} with $1 \leq i \leq m$ and $1 \leq j \leq n$, if

$$W = \{(x_1, \ldots, x_n) \in \mathbf{R}^n \mid a_{i1}x_1 + \cdots + a_{in}x_n = 0 \text{ for all } 1 \leq i \leq m\},$$

then solving problems of the first type means finding a set of solutions of the system of equations with the property that any other solution may be written as a linear combination of these solutions. The key issue, then, is to find a way to describe or parametrize all solutions of the equations.

Now, let us consider problems of the second type. If $V = \mathbf{R}^m$, S is a set of n vectors in V, say, $S = \{(a_{11}, \ldots, a_{m1}), \ldots, (a_{1n}, \ldots, a_{mn})\}$, and $\mathbf{b} = (b_1, \ldots, b_m) \in \mathbf{R}^m$, then $\mathbf{b} \in$ Span(S) if and only if there are scalars $x_1, \ldots, x_n$ such that

$$(b_1, \ldots, b_m) = x_1(a_{11}, \ldots, a_{m1}) + \cdots + x_n(a_{1n}, \ldots, a_{mn})$$
$$= (a_{11}x_1 + \cdots + a_{1n}x_n, \ldots, a_{m1}x_1 + \cdots + a_{mn}x_n)$$

If we equate the corresponding components in these two vectors, we obtain a system of equations:

$$\begin{aligned} a_{11}x_1 + \cdots + a_{1n}x_n &= b_1 \\ a_{21}x_1 + \cdots + a_{2n}x_n &= b_2 \\ &\ \vdots \\ a_{m1}x_1 + \cdots + a_{mn}x_n &= b_m \end{aligned}$$

We ask, are there any solutions to the system? In addition, if there are *no* solutions for some vectors $(b_1, \ldots, b_m)$, what conditions must the b_i satisfy for solutions to exist? Note that except for the possible nonzero right-hand sides b_i, the form of these equations is the *same* as that of the ones previously considered.

Finally, let us consider problems of the third type. [The reader should compare the following general discussion with the calculation in Example (1.4.3c).] Again, let $V = \mathbf{R}^m$ and let $S = \{(a_{11}, \ldots, a_{m1}), \ldots, (a_{1n}, \ldots, a_{mn})\}$. To determine if S is linearly independent, we must consider linear combinations:

$$x_1(a_{11}, \ldots, a_{m1}) + \cdots + x_n(a_{1n}, \ldots, a_{mn}) = (0, \ldots, 0)$$

or

$$(a_{11}x_1 + \cdots + a_{1n}x_n, \ldots, a_{m1}x_1 + \cdots + a_{mn}x_n) = (0, \ldots, 0)$$

Once again, we obtain a system of equations in the x_i by equating components:

$$\begin{aligned} a_{11}x_1 + \cdots + a_{1n}x_n &= 0 \\ a_{21}x_1 + \cdots + a_{2n}x_n &= 0 \\ &\ \vdots \\ a_{m1}x_1 + \cdots + a_{mn}x_n &= 0 \end{aligned}$$

Here the question is, do there exist any solutions other than the obvious solution $(x_1, \ldots, x_n) = (0, \ldots, 0)$? If there are other solutions, then those n-tuples of scalars yield linear dependences among the vectors in S, and consequently S is linearly dependent. On the other hand, if $(0, \ldots, 0)$ is the only solution of the system, then S is linearly independent.

Although our discussion has centered on subspaces of the vector spaces $\mathbf{R}^n$, these same basic problems arise in other spaces as well, and they can be treated by the same methods if we have finite spanning sets for the subspaces in question. The linear equations in these other cases are obtained by looking at conditions on the scalar coefficients in linear combinations of vectors in the spanning set.

In short, all three basic problems lead us to the even more basic problem of finding all solutions of a simultaneous system of linear equations. In the remainder of this section, we present a general, systematic method for doing this. We begin with some terminology.

(1.5.1) Definitions. A system of m equations in n unknowns $x_1, \ldots, x_n$ of the form

$$\begin{aligned} a_{11}x_1 + \cdots + a_{1n}x_n &= b_1 \\ a_{21}x_1 + \cdots + a_{2n}x_n &= b_2 \\ &\vdots \\ a_{m1}x_1 + \cdots + a_{mn}x_n &= b_m \end{aligned} \tag{1.5}$$

where the a_{ij} and the $b_i \in \mathbf{R}$, is called a *system of linear equations*. The a_{ij} are called the *coefficients* of the system. The system is said to be *homogeneous* if all the $b_i = 0$, and *inhomogeneous* otherwise. A *solution* (vector) of the system (1.5) is a vector $(x_1, \ldots, x_n) \in \mathbf{R}^n$ whose components solve all the equations in the system. A homogeneous system always has the *trivial solution* $(0, \ldots, 0)$.

In high school algebra you undoubtedly dealt with some simple cases of the problem of finding all solutions of a system of linear equations. The method you learned most likely involved adding multiples of one equation to other equations to eliminate variables and yield simpler equations that could be solved by inspection. The same idea may be used in general. We will also try to produce simpler systems of equations by eliminating variables.

Of course, in the process, we must be sure that we have neither introduced new solutions nor lost any solutions we had before.

(1.5.2) Definition. Two systems of linear equations are said to be *equivalent* if their sets of solutions are the same (i.e., every solution of one is a solution of the other, and vice versa).

The following proposition gives three basic classes of operations on systems of equations that yield new, equivalent systems.

(1.5.3) Proposition. Consider a system of linear equations as in (1.5).

a) The system obtained by adding any multiple of any one equation to any second equation, while leaving the other equations unchanged, is an equivalent system.

b) The system obtained by multiplying any one equation by a nonzero scalar and leaving the other equations unchanged is an equivalent system.

c) The system obtained by interchanging any two equations is an equivalent system.

Before we prove the proposition, here are concrete examples of each of these operations. In the future we will refer to them as *elementary operations* of types a, b, and c, respectively.

(1.5.4) Example. Consider the inhomogeneous system of two equations in three unknowns:

$$\begin{aligned} 5x_1 + 3x_2 - x_3 &= 1 \\ x_1 + 2x_2 + 2x_3 &= 2 \end{aligned} \tag{1.6}$$

a) The system obtained from Eq. (1.6) by subtracting 5 times the second equation from the first, leaving the second unchanged

$$\begin{aligned} -7x_2 - 11x_3 &= -9 \\ x_1 + 2x_2 + x_3 &= 2 \end{aligned} \tag{1.7}$$

is a system equivalent to (1.6).

b) The system obtained from Eq. (1.7) by multiplying the first equation by -1

$$\begin{aligned} 7x_2 + 11x_3 &= 9 \\ x_1 + 2x_2 + x_3 &= 2 \end{aligned} \tag{1.8}$$

is another equivalent system.

c) The system obtained from (1.8) by interchanging the two equations

$$\begin{aligned} x_1 + 2x_2 + x_3 &= 2 \\ 7x_2 + 11x_3 &= 9 \end{aligned} \tag{1.9}$$

is another equivalent system.

Proof [of Proposition (1.5.3)]:

Let X be the set of solutions of a system of linear equations as in Eq. (1.5). First, note that if we apply any of the elementary operations to the system, every vector in X is also a solution of the new system. That is, if X' is the set of solutions of the new system, we have $X \subseteq X'$. This is clear for the elementary operations of types b and c. To see that it is the case for the operations of type a, note that if $(x_1, \ldots, x_n)$ is a solution of the original system of equations, then if we add, say, c times the jth equation to the ith equation, the new ith equation is

$$c(a_{j1}x_1 + \cdots + a_{jn}x_n) + (a_{i1}x_1 + \cdots + a_{in}x_n) = cb_j + b_i$$

This equation is also satisfied by the x_i since $a_{j1}x_1 + \cdots + a_{jn}x_n = b_j$, and $a_{i1}x_1 + \cdots + a_{in}x_n = b_i$.

Next, note that each elementary operation is *reversible* in the sense that we can apply another elementary operation and recover the original system of equations. In case a, if we added c times equation j to equation i, then subtracting c times equation j from equation i in the new system will reproduce the original system. In case b, if we multiplied equation i by a scalar c, then multiplying equation i in the new system by c^{-1} (recall, $c \neq 0$) will reproduce the original system. Finally, in case c, if equations i and j were interchanged, then interchanging equations i and j in the new system will reproduce the original system. Hence, by the same reasoning as before, we now have $X' \subseteq X$. Therefore $X = X'$ and the proposition is proved. ■

It should be clear that by applying elementary operations of type a, we can eliminate occurrences of variables in equations, and that by applying elementary operations of type b, we can simplify the coefficients appearing in equations. The usefulness of the operations of type c is not so obvious. However, it should become clearer if we consider the following example.

(1.5.5) Example. Let us "follow our noses" and try to transform the system

$$\begin{aligned} x_1 - x_2 + x_3 + x_4 &= 1 \\ -2x_1 + x_2 + x_4 &= 0 \\ x_2 + 3x_3 - x_4 &= 3 \end{aligned} \qquad (1.10)$$

into as simple a form as possible. First, we can add 2 times the first equation in Eq. (1.10) to the second equation, yielding

$$\begin{aligned} x_1 - x_2 + x_3 + x_4 &= 1 \\ -x_2 + 2x_3 + 3x_4 &= 2 \\ x_2 + 3x_3 - x_4 &= 3 \end{aligned} \qquad (1.11)$$

At this point, notice that the variable x_1 has been eliminated from all the equations other than the first. For this reason, we cannot expect any more simplification in the x_1 terms, and we *should not use the first equation again,* since adding a multiple of the first equation to either of the other equations would reintroduce the x_1 terms that we have worked so hard to eliminate.

Notice that we can go on and try to simplify the x_2 and x_3 terms appearing in Eq. (1.11), though. If we add the third equation to the first equation, then to the second equation, we eliminate the occurrences of x_2 in all equations except the third:

$$\begin{aligned} x_1 + 4x_3 &= 4 \\ 5x_3 + 2x_4 &= 5 \\ x_2 + 3x_3 - x_4 &= 3 \end{aligned}$$

If we now interchange the second and third equations here, the resulting system has a somewhat neater appearance (this interchange could also have been done before eliminating the x_2 terms—the result would be the same):

$$\begin{aligned} x_1 \quad + 4x_3 \quad &= 4 \\ x_2 + 3x_3 - x_4 &= 3 \\ 5x_3 + 2x_4 &= 5 \end{aligned} \tag{1.12}$$

At this point no further simplifications are possible in the x_2 terms, so we go on to the x_3 terms. Here we can use the x_3 term in the last equation to eliminate x_3 from the other two equations. To do this, we can first multiply the third equation in Eq. (1.12) by 1/5, then subtract 4 times the third equation from the first equation, and subtract 3 times the third equation from the second equation, yielding:

$$\begin{aligned} x_1 \quad - (8/5)x_4 &= 0 \\ x_2 \quad - (11/5)x_4 &= 0 \\ x_3 + (2/5)x_4 &= 1 \end{aligned} \tag{1.13}$$

We cannot eliminate any occurrences of x_4 without reintroducing x_1, x_2, or x_3 terms in the equations, so we cannot really make the system any simpler. We will stop here.

Our final system of equations (1.13) is special in three ways:

1. In each equation, there is a "leading term" with coefficient 1. [e.g., in the second equation in (1.13) the leading term is the x_2 term].
2. The variable occurring in the leading term of an equation occurs in no other equation.
3. The indices of the variables occurring in leading terms increase as we "go down" the system. For example, the leading term in the first equation in (1.13) involves x_1, the leading term in the second equation involves x_2, and so on.]

In general, we call any system satisfying these three conditions an echelon form system. Slightly more formally we have Definition (1.5.6).

(1.5.6) Definition. A system of linear equations as in Eq. (1.5) is in *echelon form* if it has all three of the following properties:

1) In each equation, all the coefficients are 0, or the first nonzero coefficient counting from the left of the equation is a 1. The corresponding term in the equation is called the *leading term*. In the ith equation, we will call the subscript of the variable in the leading term $j(i)$.

2) For each i the coefficient of $x_{j(i)}$ is zero in every equation other than the ith.

3) For each i (for which equations i and $i + 1$ have some nonzero coefficients) $j(i + 1) > j(i)$.

(1.5.7) Remark. The name "echelon form" comes from the *step-* or *ladder-like* pattern of an echelon form system when space is left to indicate zero coefficients and line up the terms containing each variable. For example, consider this echelon form system:

$$\begin{array}{llll} x_1 + 2x_2 & & + x_5 & = 1 \\ & x_3 & - 2x_5 & = 1 \\ & & x_4 - x_5 & = 0 \end{array} \qquad (1.14)$$

From Example (1.5.5) the following statement should be extremely plausible.

(1.5.8) Theorem. Every system of linear equations is equivalent to a system in echelon form. Moreover this echelon form system may be found by applying a sequence of elementary operations to the original system.

Proof: We will prove the theorem by an important technique we have not used before called mathematical induction. The reader who has not seen mathematical induction before or who wishes to review the ideas involved should consult Appendix 2. The proof will proceed by induction on m, the number of equations in the system. If $m = 1$, and the equation is not of the form $0 = c$, then we have some nonzero leading coefficient in the equation. Applying the elementary operation of type b, we can make the leading coefficient 1, and when this is done, the system is in echelon form. If the equation is of the form $0 = c$, it is in echelon form already.

Now assume the theorem is true for systems of k equations, and consider a system of $k + 1$ equations:

$$\begin{array}{l} a_{11}x_1 + \cdots + a_{1n} \quad x_n = b_1 \\ a_{21}x_1 + \cdots + a_{2n} \quad x_n = b_2 \\ \vdots \\ a_{k+1,1}x_1 + \cdots + a_{k+1,n}x_n = b_{k+1} \end{array}$$

If this system does not consist of $k + 1$ equations of the form $0 = c$, then there is some nonzero coefficient, hence at least one nonzero coefficient "farthest to the left" in the system (i.e., containing a variable with the smallest index among the indices of the variables appearing explicitly in the equations). Call this coefficient a_{ij}. Note that by the choice of j, none of the variables $x_1, \ldots, x_{j-1}$ can appear explicitly.

First, we apply an elementary operation of type c and interchange the ith equation and the first equation. Next, we apply an elementary operation of type b and multiply (the new) first equation by $1/a_{ij}$ to make the leading coefficient equal to 1. Then, applying elementary operations of type a, we can eliminate any other occurrences of the variable x_j, leaving a system

$$\begin{aligned} x_j + a'_{1,j+1}x_{j+1} + \cdots + a'_{1n}x_n &= b'_1 \\ a'_{2,j+1}x_{j+1} + \cdots + a'_{2n}x_n &= b'_2 \\ &\ \vdots \\ a'_{k+1,j+1}x_{j+1} + \cdots + a'_{k+1,n}x_n &= b_{k+1}' \end{aligned} \tag{1.15}$$

(The primes indicate that the coefficients here are not necessarily the same as those in the original equations.)

Now, we apply the induction hypothesis to the system of k equations formed by equations 2 through $k + 1$ in (1.1.5). We obtain an echelon form system of k equations in this way. Finally, we can eliminate any terms in the first equation containing variables appearing in leading terms in the other k equations, using operations of type a again. We thus have an echelon form system, which [by Proposition (1.5.3)] is equivalent to our original system of equations. ■

(1.5.9) Remarks

a) It may also be shown that the echelon form system whose existence is guaranteed by the theorem is *unique*. We will not need this fact, however, so we omit the proof.

b) The process outlined in the proof of Theorem (1.5.8) (or rather a slight modification thereof) may be elaborated into an *algorithm* or step-by-step procedure for producing the echelon form system. What must be done to produce the echelon form system may be summarized as follows. [The reader may wish to refer to Example (1.5.5) again to see this procedure in action.]

1. Begin with the entire system of m equations.

2. Do the following steps for each i between 1 and m:

   a) Among the equations numbered i through m in the "updated," modified system of equations, pick one equation containing the variable with the smallest index among the variables appearing with nonzero coefficients in those equations. [As a practical matter, if there is more than one such equation, when doing the procedure by hand it will pay to choose the equation with the simplest leading coefficient (e.g., 1 or -1).]

   b) If necessary, apply an elementary operation of type c to interchange the equation chosen in step 2a and the ith equation.

c) If necessary, use an elementary operation of type b to make the leading coefficient in the (new) ith equation equal to 1.

d) If necessary, using elementary operations of type a, eliminate all other occurrences of the leading variable in the ith equation.

3. The process terminates when we complete step 2d with $i = m$, or possibly sooner if all the terms on the left-hand sides of the remaining equations have already been eliminated.

We refer to this procedure as the *elimination algorithm*. (A second algorithm, that accomplishes the same reduction to echelon form in a slightly more efficient way, is discussed in the exercises.)

Due to its essentially mechanical nature, the elimination algorithm can serve as the basis for a rudimentary computer program for solving systems of linear equations. Readers with some programming experience in a higher-level language such as BASIC, FORTRAN, or Pascal are warmly encouraged to try to write such a program to test their understanding of the elimination algorithm. Typically, other methods are used to solve large or complicated systems of linear equations arising from applied problems by computer, however. One reason for this is that our method, while perfectly reasonable if the arithmetic involved can be carried out exactly (as when we do the computations by hand and the coefficients are rational numbers), can yield inaccurate results if there are small errors introduced by representing the values of the coefficients by real numbers with a fixed number of decimal places, as is done in a computer. In solving systems of linear equations by computer, care must be taken to ensure that these round-off errors do not compromise the accuracy of the computed solution, and this requires more sophisticated methods.

After a system of linear equations has been reduced to echelon form, we may exploit the most important feature of echelon form systems—their form allows us to write down the set of solutions almost immediately.

(1.5.10) Example. Consider the echelon form system in Eq. (1.13). It is clear that the equations may be rewritten in the form

$$\begin{aligned} x_1 &= (8/5)x_4 \\ x_2 &= (11/5)x_4 \\ x_3 &= (-2/5)x_4 + 1 \end{aligned}$$

and from this we see that for any fixed value of x_4 we obtain a solution vector by substituting that value and solving for the other variables. For example, for $x_4 = 1$ we obtain $x_1 = 8/5$, $x_2 = 11/5$, $x_3 = 3/5$, so that $(8/5, 11/5, 3/5, 1)$ is one solution. However, x_4 is completely arbitrary, so if we set $x_4 = t$ for any $t \in \mathbf{R}$, we obtain a corresponding solution vector $((8/5)t, (11/5)t, (-2/5)t + 1, t)$. The set of all solutions of the system is the set of all such vectors:

$$X = \{((8/5)t,\ (11/5)t,\ (-2/5)t + 1,\ t) \in \mathbf{R}^4 |\ t \in \mathbf{R}\}.$$

Geometrically, the set of solutions is a line in $\mathbf{R}^4$ that does not pass through the origin. The set of solutions of this inhomogeneous system is *not* a vector subspace of $\mathbf{R}^4$.

In general, for any echelon form system, we will be able to follow a similar procedure to determine the set of solutions of the system. First, if the system contains an equation of the form $0 = c$ for some $c \neq 0$, then there are clearly no solutions, and the system was *inconsistent* to start with. To give a simple example of such a system, consider

$$\begin{aligned} x_1 + x_2 &= 1 \\ 2x_1 + 2x_2 &= 3 \end{aligned}$$

If we subtract 2 times the first equation from the second, we obtain the echelon form system

$$\begin{aligned} x_1 + x_2 &= 1 \\ 0 &= 1 \end{aligned}$$

There are no solutions of this (or the original) system. If there are no equations of this form in the echelon form system, then the system is said to be *consistent,* and solutions do exist. To describe them, we introduce some terminology.

(1.5.11) Definitions

a) In an echelon form system the variables appearing in leading terms of equations are called the *basic variables* of the system.

b) All the other variables are called *free variables.*
[*Caution:* In some systems of equations it will be true that there are some variables that do not appear explicitly in the equations (i.e., only with zero coefficients). They are also free variables.]

As in Example (1.5.10), we can express each of the basic variables in terms of the free variables by rewriting the equations to solve for those basic variables. We obtain one solution of the equations for each set of specific values that are substituted for the free variables, and this gives us a way to write down all solutions of an echelon form system.

(1.5.12) Example. Consider the echelon form system

$$\begin{aligned} x_1 \quad + x_3 \qquad\quad &= 0 \\ x_2 - x_3 \quad + x_5 &= 0 \\ x_4 + x_5 &= 0 \end{aligned}$$

The basic variables are x_1, x_2, and x_4. The free variables are x_3 and x_5. We have

$$
\begin{aligned}
x_1 &= -x_3 \\
x_2 &= \quad x_3 - x_5 \\
x_4 &= \qquad\; - x_5
\end{aligned}
\tag{1.16}
$$

Hence the set of solutions is the set of all vectors obtained by substituting values for x_3 and x_5. If we set $x_3 = 1$ and $x_5 = 0$, we obtain one solution $(-1, 1, 1, 0, 0)$. If we set $x_3 = 0$ and $x_5 = 1$, we obtain a second—$(0, -1, 0, -1, 1)$. The general solution of the system, obtained by setting $x_3 = t_1$ and $x_5 = t_2$ in $\mathbf{R}$, is a general linear combination of these two vectors, as may be seen easily by looking at Eq. (1.16). Indeed, the set of all solutions is

$$W = \{t_1(-1, 1, 1, 0, 0) + t_2(0, -1, 0, -1, 1) | t_1, t_2, \in \mathbf{R}\}$$

This kind of description is called a *parametrization* of the set of solutions. [What we have actually done is to define a mapping $F: \mathbf{R}^2 \to \mathbf{R}^5$ whose image is the set of solutions. Here the mapping F is given by $F(t_1, t_2) = t_1(-1, 1, 1, 0, 0) + t_2(0, -1, 0, -1, 1) = (-t_1, t_1 - t_2, t_1, -t_2, t_2)$. Geometrically, the set W is a plane containing the origin in the space $V = \mathbf{R}^5$.]

In general, for any echelon form system, we can obtain a parametrization of the set of solutions in the same way. Since each free variable can take on any real values, and the free variables are independent of each other, we can obtain all solutions by substituting t_i for the ith free variable in the system. If k is the number of free variables, this gives a description of the set of solutions as the image of a mapping from $\mathbf{R}^k$ into $\mathbf{R}^n$ (where n was the total number of variables in the original system).

This description of the set of solutions of systems of linear equations in terms of the free variables and the basic variables of the associated echelon form system allows us to make some general observations about the behavior of such systems. Note that in Example (1.5.12), if $t_1 \neq 0$ or $t_2 \neq 0$, then we get a nontrivial solution of the original homogeneous system of equations. The same will be true for any homogeneous system in which there is at least one free variable. One case in which this *always* happens is given in the following corollary of Theorem (1.5.8).

(1.5.13) Corollary. If $m < n$, every homogeneous system of m linear equations in n unknowns has a nontrivial solution.

Proof: By Theorem (1.5.8) any such system is equivalent to one in echelon form. In the echelon form system there will again be more variables than equations (the number of nontrivial equations can only decrease), so the number of basic variables is no bigger than m. Hence there will be at least one free variable. If we set that free variable equal to some nonzero real number, the resulting solution will be nontrivial. ■

Finally, to summarize, we indicate how the three basic problems discussed at the beginning of this section may be solved by using the techniques we have developed.

1) Given a subspace W of $\mathbf{R}^n$ defined as the set of solutions of a system of homogeneous linear equations in n variables (the components of the solution vectors), to find a set S such that $W = \text{Span}(S)$, we can proceed as follows: First, reduce the system of equations to echelon form, using elementary operations as in Theorem (1.5.8). Then the set of solutions may be found exactly as in Example (1.5.12). A set S (with as many vectors as there are free variables in the echelon form system) may be obtained by, for each free variable in turn, setting that free variable equal to 1 and all the other free variables equal to 0.

2) Given a finite set S of vectors in $\mathbf{R}^m$, and a vector $\mathbf{x}$ in $\mathbf{R}^m$, to determine if $\mathbf{x} \in \text{Span}(S)$, set up the appropriate system of linear equations and reduce them to echelon form, using Theorem (1.5.8). If the system is inconsistent, then $\mathbf{x} \notin \text{Span}(S)$; if it is consistent, then $\mathbf{x} \in \text{Span}(S)$, and the solution vectors of the system give the scalars in the linear combinations of the vectors in S, which yield the vector $\mathbf{x}$.

This method may also be applied to produce a new set of linear equations whose solutions are the vectors in the subspace Span(S). As before, we write

$$S = \{(a_{11}, \ldots, a_{m1}), \ldots, (a_{1n}, \ldots, a_{mn})\}$$

If we think of the components of the vector $\mathbf{x} = (b_1, \ldots, b_m)$ as *another set of variables*, then when the system

$$\begin{aligned} a_{11}x_1 + \cdots + a_{1n}x_n &= b_1 \\ a_{21}x_1 + \cdots + a_{2n}x_n &= b_2 \\ &\vdots \\ a_{m1}x_1 + \cdots + a_{mn}x_n &= b_m \end{aligned}$$

is reduced to echelon form, any resulting equation

$$0 = c_1b_1 + \cdots + c_mb_m$$

must be satisfied if the vector $\mathbf{x} \in \text{Span}(S)$. (Otherwise, the system would be inconsistent.) Conversely if all these equations are satisfied, then there will be at least one solution of the system, so $\mathbf{x}$ will be in Span(S). The set of all equations of this form obtained from the echelon form system will be a set of defining equations for the subspace Span(S).

(1.5.14) Example. In $\mathbf{R}^4$ consider $S = \{(1, 3, 1, 1), (1, -1, 6, 2)\}$. Then we have $\mathbf{x} = (b_1, b_2, b_3, b_4) \in \text{Span}(S)$ if and only if the system

$$\begin{aligned} x_1 + x_2 &= b_1 \\ 3x_1 - x_2 &= b_2 \end{aligned}$$

$$x_1 + 6x_2 = b_3$$

$$x_1 + 2x_2 = b_4$$

has solutions. This system is equivalent to the echelon form system

$$\begin{aligned} x_1 &= (1/4)b_1 + (1/4)b_2 \\ x_2 &= (3/4)b_1 - (1/4)b_2 \\ 0 &= (-19/4)b_1 + (5/4)b_2 + b_3 \\ 0 &= (-7/4)b_1 + (1/4)b_2 + b_4 \end{aligned} \tag{1.17}$$

(check this!). Hence every vector (b_1, b_2, b_3, b_4) in Span(S) must satisfy the system of equations obtained from the two equations in (1.17) with zero on the left-hand side:

$$(-19/4)b_1 + (5/4)b_2 + b_3 = 0$$

$$(-7/4)b_1 + (1/4)b_2 + b_4 = 0$$

[It is easy to see that Span(S) is the set of all solutions of this system.]

3) To determine if a set of vectors is linearly independent, we set up the appropriate system of homogeneous linear equations, reduce to echelon form, and count the number of free variables. If this is ≥ 1, then there are nontrivial solutions of the system, so the original set of vectors was linearly dependent. Otherwise, the set is independent.

(1.5.15) Example. We work with another vector space $V = P_n(\mathbf{R})$ and show that determining if a set of polyomials is linearly independent or not may be accomplished by the same techniques we have discussed in $\mathbf{R}^n$. Consider the set $S = \{x^2 + 2x + 1, x^2 + 4x + 3, x^2 + 6x + 5\} \subset P_2(\mathbf{R})$. To determine if S is linearly independent, we consider a potential linear dependence

$$a_1(x^2 + 2x + 1) + a_2(x^2 + 4x + 3) + a_3(x^2 + 6x + 5) = 0x^2 + 0x + 0,$$

or, after collecting like terms in x,

$$(a_1 + a_2 + a_3)x^2 + (2a_1 + 4a_2 + 6a_3)x + (a_1 + 3a_2 + 5a_3) = 0x^2 + 0x + 0.$$

Since polynomials are equal if and only if they have the same coefficients, we must have

$$a_1 + a_2 + a_3 = 0$$

$$2a_1 + 4a_2 + 6a_3 = 0$$

$$a_1 + 3a_2 + 5a_3 = 0$$

This is a homogeneous system of three equations in the unknowns a_1, a_2, a_3. From this point we may proceed exactly as we would in $\mathbf{R}^3$. After reducing the system to echelon form we have

$$\begin{aligned} a_1 \quad - \ a_3 &= 0 \\ a_2 + 2a_3 &= 0 \\ 0 &= 0 \end{aligned}$$

Since a_3 is a free variable, there are nontrivial solutions, such as $(a_1,a_2,a_3) = (1,-2,1)$, and this implies that S is linearly dependent. The reader should check that these scalars actually do give us a linear dependence among the three polynomials in the set S.

EXERCISES

1. Find a parametrization of the set of solutions (if there are any) of each of the following systems of linear equations by reducing to echelon form and applying the method discussed in the section:

a)
$$\begin{aligned} x_1 - 3x_2 &= 1 \\ 2x_1 + 4x_2 &= 2 \end{aligned}$$

b)
$$\begin{aligned} x_1 + 4x_2 &= 0 \\ 2x_1 + 3x_2 &= 1 \\ -x_1 + 6x_2 &= -2 \end{aligned}$$

c)
$$\begin{aligned} 2x_1 + 3x_2 - 3x_3 &= 0 \\ 3x_1 \qquad + x_3 &= 0 \end{aligned}$$

d)
$$\begin{aligned} x_1 + 2x_2 + x_3 &= 1 \\ 2x_1 - x_2 \qquad &= 4 \\ x_2 + x_3 &= 1 \\ x_1 + 7x_2 - x_3 &= 5 \end{aligned}$$

e)
$$\begin{aligned} 4x_1 + 2x_2 - x_3 + x_4 &= -1 \\ x_1 + x_2 \qquad + 2x_4 &= 2 \\ 6x_1 + 4x_2 - x_3 + 5x_4 &= 3 \end{aligned}$$

f)
$$\begin{aligned} x_1 + \qquad + 2x_3 + \qquad x_5 &= 0 \\ -3x_1 + \qquad + x_4 \qquad &= 1 \end{aligned}$$

2. Find a set of vectors that spans the subspace W of V defined by the each of the following sets of conditions:

a) $V = \mathbf{R}^4$
$$\begin{aligned} x_1 + 2x_2 - x_3 + x_4 &= 0 \\ -3x_1 \qquad + x_3 + 2x_4 &= 0 \end{aligned}$$

b) $V = \mathbf{R}^4$
$$x_1 \qquad + 2x_3 = 0$$

c) $V = \mathbf{R}^5$
$$\begin{aligned} x_1 + 2x_2 \qquad + \qquad x_5 &= 0 \\ -x_1 \qquad + x_3 \qquad + x_5 &= 0 \\ x_1 \qquad - 3x_3 + x_4 \qquad &= 0 \end{aligned}$$

d) $V = P_3(\mathbf{R})$ $\quad W = \{p \mid p(1) = p(3) = 0\}$

[*Hint:* First, use the conditions $p(1) = p(3) = 0$ to get a system of linear equations on the coefficients of p.]

e) $V = P_4(\mathbf{R})$, $W = \{p | p(1) = p'(0) = p''(0) = 0\}$

f) $V = \text{Span}(\{\sin(x), \cos(x), \sin(2x), \cos(2x)\}) \subset C^1 (\mathbf{R})$,
$W = \{f \in V | f(\pi/4) = 0 \text{ and } f'(\pi/4) = 0\}$

3. For each set of vectors S and given vector $\mathbf{v}$, determine if $\mathbf{v} \in \text{Span}(S)$.
   a) $\mathbf{v} = (1, 1, 2)$ $S = \{(1, 1, 4), (-1, 1, 3), (0, 1, 0)\}$ in $\mathbf{R}^3$
   b) $\mathbf{v} = (0, -1, 0)$ $S = \{(1, -1, 2), (4, 0, 1)\}$ in $\mathbf{R}^3$
   c) $\mathbf{v} = (1, 1, 0, 0)$ $S = \{(1, 1, 1, 1), (1, -1, 1, 0), (0, 0, 0, 1)\}$ in $\mathbf{R}^4$
   d) $\mathbf{v} = 2x^3 + x + 1$ $S = \{x^3 + 1, x^2 + 1, x + 1\}$ in $P_3 (\mathbf{R})$
   e) $\mathbf{v} = 3x^2 + 2x + 1$ $S = \{x^3, x^3 + x, x^3 + 3x^2 + 2, 1\}$ in $P_3 (\mathbf{R})$

4. a) Find a system of homogeneous linear equations whose set of solutions is the subspace $W = \text{Span}\{(1, 1, 2, 3), (1, 4, 2, 1)\}$ in $\mathbf{R}^4$.
   b) Same question for $W = \text{Span}\{1, 0, -1, 0, 1), (2, 1, 2, 1, 2), (1, 0, 0, 0, 0)\}$ in $\mathbf{R}^5$.
   c) Same question for $W = \text{Span}\{(1, 1, 1, -1), (1, -1, 1, 1), (-1, 1, 1, 1)\}$ in $\mathbf{R}^4$.

5. Determine if the given set of vectors is linearly independent.
   a) $S = \{(1, 1, 2, 0), (1, -1, 1, 1), (3, 1, 0, 0)\}$ in $\mathbf{R}^4$
   b) $S = \{(1, 1, 1), (-1, 2, 0), (1, 0, -1), (1, 1, 0)\}$ in $\mathbf{R}^3$
   c) $S = \{x^2 + x + 1, -x^2 + 2x, x^2 + 2, x^2 - x\}$ in $P_2 (\mathbf{R})$.
   d) $S = \{\sin(x), \sin(2x), \sin(3x)\}$ in $C(\mathbf{R})$.

6. Suppose we have a system of m inhomogeneous linear equations in n unknowns with $m > n$. Is the system necessarily inconsistent? Why or why not? Is it always the case that two equivalent systems of equations have the same number of equations? Why or why not?

To deal with large or complicated systems of equations, there are variants of the elimination algorithm presented in the text that can reduce the number of numerical computations required to put the system into echelon form. Here is one such algorithm, which uses the same elementary operations, but which eliminates occurrences of a leading variable in the equations "below" the equation containing that variable first, then eliminates the other occurrences of leading variables in a separate step, starting from the right of the system.

Elimination Algorithm 2

1. Begin with the entire system of m equations.

2. For each i from 1 to m:

   a) Among the equations numbered i to m pick one equation containing the variable with the smallest index among the variables occurring in those equations.

   b) If necessary, interchange that equation with the ith equation, using an elementary operation of type c.

c) Make the leading coefficient of the (new) ith equation equal to 1, using an elementary operation of type b.

d) Eliminate the occurrences of the leading variable in equations numbered $i + 1$ through m, using elementary operations of type a.

3. For each i from m to 2 in reverse order:
Eliminate the occurrences of the leading variable in equation i in equations 1 through $i - 1$, using elementary operations of type a.

7. Explain why the system of equations produced by Elimination Algorithm 2 will always be a system in echelon form.

8. Repeat Exercise 1, using Elimination Algorithm 2. Do you see any advantages?

§1.6. BASES AND DIMENSION

Armed with our new technique for solving systems of linear equations, we now return to the questions about economical spanning sets for subspaces that we raised in Section 1.4. We saw there that any linear dependences among the vectors in a spanning set S for a vector space V led to redundancies in the linear combinations formed from the vectors in S. In fact, in this case a subset of S could be used to span the same space. This suggests that the most economical spanning sets should be the ones that are also linearly independent. We give such sets a special name.

(1.6.1) Definition. A subset S of a vector space V is called a *basis* of V if $V = \mathrm{Span}(S)$ and S is linearly independent.

(1.6.2) Examples. **a)** Consider the standard basis $S = \{\mathbf{e}_1, \ldots, \mathbf{e}_n\}$ in $\mathbf{R}^n$ introduced in Example (1.4.5c). S is indeed a basis of $\mathbf{R}^n$ by our definition, since it is linearly independent, and every vector $(a_1, \ldots, a_n) \in \mathbf{R}^n$ may be written as the linear combination

$$(a_1, \ldots, a_n) = a_1\mathbf{e}_1 + \cdots + a_n\mathbf{e}_n$$

[Note that this is also the only way to produce the vector $(a_1, \ldots, a_n)$ as a linear combination of the standard basis vectors.]

b) The vector space $\mathbf{R}^n$ has many other bases as well. For example, in $\mathbf{R}^2$, consider the set $S = \{(1, 2), (1, -1)\}$. To see that S is linearly independent, we proceed as in Section 1.5. Consider the system

$$\begin{aligned} x_1 + x_2 &= 0 \\ 2x_1 - x_2 &= 0 \end{aligned}$$

(Where does this come from?) Reducing to echelon form, we have

$$\begin{aligned} x_1 &= 0 \\ x_2 &= 0 \end{aligned}$$

so, since the trivial solution is the only solution, S is linearly independent.

To see that S spans $\mathbf{R}^2$ we proceed as in Example (1.5.14). We have $(b_1, b_2) \in \text{Span}(S)$ if and only if the system

$$\begin{aligned} x_1 + x_2 &= b_1 \\ 2x_1 - x_2 &= b_2 \end{aligned} \tag{1.18}$$

has a solution. Reducing this system to echelon form, we have

$$\begin{aligned} x_1 \quad &= (1/3)b_1 + (1/3)b_2 \\ x_2 &= (2/3)b_1 - (1/3)b_2 \end{aligned} \tag{1.19}$$

Since there are no equations of the form $0 = c_1b_1 + c_2b_2$, we see that there are solutions (x_1, x_2) for all $(b_1, b_2) \in \mathbf{R}^2$. Hence $\text{Span}(S) = \mathbf{R}^2$.

Although we have presented the two verifications as separate calculations, as a practical matter, some work could be saved by combining them. Namely, if we start from the general system in Eq. (1.18) and reduce to the echelon form in Eq. (1.19) then the fact that S is linearly independent may be "read off" by noting that if $b_1 = b_2 = 0$, then the only solution of Eq. (1.18) is the trivial solution $x_1 = x_2 = 0$. Then, from the same reduced system, it may be seen that $\text{Span}(S) = \mathbf{R}^2$, as before. Note that here, too, given $(b_1, b_2) \in \mathbf{R}^2$, there is only one solution of the system of Eq. (1.19).

c) Let $V = P_n(\mathbf{R})$ and consider $S = \{1, x, x^2, \ldots, x^n\}$. It is clear that S spans V. Furthermore, S is a linearly independent set of functions. To see this, consider a potential linear dependence

$$a_0 + a_1x + a_2x^2 + \cdots + a_nx^n = 0 \tag{1.20}$$

in $P_n(\mathbf{R})$. If this equality of functions holds, then we can substitute any real value for x and obtain another true equation. However, from high school algebra, recall that an actual polynomial equation of degree k (i.e., one in which the coefficient of the highest degree term, $a_k \neq 0$) has at most k roots in $\mathbf{R}$. Thus Eq. (1.20) can be satisfied for all real x if and only if $a_i = 0$ for all i. Hence S is a basis of $P_n(\mathbf{R})$.

d) The empty subset, ϕ, is a basis of the vector space consisting only of a zero vector, $\{\mathbf{0}\}$. See Example (1.4.6d) and Definition (1.3.1b).

In each of the examples, it is true that given a basis S of V and a vector $\mathbf{x} \in V$, there is only one way to produce $\mathbf{x}$ as a linear combination of the vectors in S. This is true in general and is the most important and characteristic property of bases.

(1.6.3) Theorem. Let V be a vector space, and let S be a nonempty subset of V. Then S is a basis of V if and only if every vector $\mathbf{x} \in V$ may be written uniquely as a linear combination of the vectors in S.

Proof: $\rightarrow$: If S is a basis, then by definition, given $\mathbf{x} \in V$, there are scalars $a_i \in \mathbf{R}$, and vectors $\mathbf{x}_i \in S$ such that $\mathbf{x} = a_1\mathbf{x}_1 + \cdots + a_n\mathbf{x}_n$. To show this linear combination is unique, consider a possible second linear combination of vectors in S, which also adds up to $\mathbf{x}$. By introducing terms with zero coefficients if necessary, we may assume that both linear combinations involve vectors in the same subset $\{\mathbf{x}_1, \ldots, \mathbf{x}_n\}$ of S. Hence the second linear combination will have the form $\mathbf{x} = a_1'\mathbf{x}_1 + \cdots + a_n'\mathbf{x}_n$ for some $a_i' \in \mathbf{R}$. Subtracting these two expressions for $\mathbf{x}$, we find that

$$\begin{aligned}\mathbf{0} &= (a_1\mathbf{x}_1 + \cdots + a_n\mathbf{x}_n) - (a_1'\mathbf{x}_1 + \cdots + a_n'\mathbf{x}_n) \\ &= (a_1 - a_1')\mathbf{x}_1 + \cdots + (a_n - a_n')\mathbf{x}_n\end{aligned}$$

Since S is linearly independent, this last equation implies that $a_i - a_i' = 0$ for all i. In other words, there is only one set of scalars that can be used to write $\mathbf{x}$ as a linear combination of the vectors in S.

$\leftarrow$: Now assume that every vector in V may be written in one and only one way as a linear combination of the vectors in S. This implies immediately that $\mathrm{Span}(S) = V$. So in order to prove that S is a basis of V, we must show that S is also linearly independent. To see this, consider an equation

$$a_1\mathbf{x}_1 + \cdots + a_n\mathbf{x}_n = \mathbf{0}$$

where the $\mathbf{x}_i \in S$, and the $a_i \in \mathbf{R}$. Note that it is also clearly the case that

$$0\mathbf{x}_1 + \cdots + 0\mathbf{x}_n = \mathbf{0}$$

using part b of Proposition (1.1.6). Hence by our hypothesis, it must be true that $a_i = 0$ for all i. Consequently, the set S is linearly independent. ■

Hence if we have any basis for a vector space, it is possible to express each vector in the space in one and only one way as a linear combination of the basis vectors. If the basis is a finite set, say, $S = \{\mathbf{x}_1, \ldots, \mathbf{x}_n\}$, then each $\mathbf{x} \in V$ can be written uniquely as $\mathbf{x} = a_1\mathbf{x}_1 + \cdots + a_n\mathbf{x}_n$. The scalars appearing in the linear combination may be thought of as the *coordinates* of the vector $\mathbf{x}$ with respect to the basis S. They are the natural generalization of the components of a vector in $\mathbf{R}^n$ [recall the components are just the scalars that are required to expand a vector in $\mathbf{R}^n$ in terms of the standard basis—see Example (1.6.2a).]

(1.6.4) Example. Consider the basis $S' = \{(1, 2), (1, -1)\}$ for $\mathbf{R}^2$ from Example (1.6.2b). If we have any vector $(b_1, b_2) \in \mathbf{R}^2$, then we saw in that example that $(b_1, b_2) = ((1/3)b_1 + (1/3)b_2)(1, 2) + ((2/3)b_1 - (1/3)b_2)(1, -1)$. Thus, with respect to the basis S', our vector is described by the vector of coordinates $((1/3)b_1 + (1/3)b_2, (2/3)b_1 - (1/3)b_2)$.

In the future we will frequently want to use this observation to be able to use the basis that is best adapted to the situation at hand.

The examples in (1.6.2) and Theorem (1.6.3) should convince you of the usefulness and importance of bases, so the next question that presents itself is, does every vector space have a basis?

We obtain results concerning this question only in the special case in which the vector space V has some finite spanning set S. We do this to avoid some technical complications that would lead us too far afield. Most of the results we prove are also valid for more general vector spaces, but they must be proved by different methods. Hence, except when we explicitly say otherwise, from this point on in this section we will consider only vector spaces that have some finite spanning set, and all the sets of vectors we will consider in the course of our proofs will be finite sets.

(1.6.5) Remark. Some examples of vector spaces that do have finite spanning sets are the spaces $\mathbf{R}^n$, $P_n(\mathbf{R})$, and all their subspaces. Some of the other vector spaces that we have considered, such as $C^1(\mathbf{R})$ and $C(\mathbf{R})$, do not have this property, however. To see why this is true, consider first the vector space of all polynomial functions (with no restriction on the degree). We denote this vector space by $P(\mathbf{R})$. $P(\mathbf{R})$ has as one basis the set $\{1, x, x^2, \ldots, x^n, \ldots\}$ (an infinite set). However, no finite set of vectors will be a basis, for the following reason. It is clear that in any finite set of polynomials there is (at least) one polynomial of maximal degree, say, n_0. Furthermore, by forming linear combinations of the polynomials in the set, it is not possible to produce polynomials of degree any higher than n_0. Hence no finite set can span $P(\mathbf{R})$ since $P(\mathbf{R})$ contains polynomials of arbitrarily high degree. Now, clearly $P(\mathbf{R}) \subset C^1(\mathbf{R}) \subset C(\mathbf{R})$ (every polynomial function is differentiable and hence continuous). Since no finite set spans the subspace $P(\mathbf{R})$, no finite set can span $C^1(\mathbf{R})$ or $C(\mathbf{R})$ either.

Indeed, it is difficult even to describe spanning sets (or bases) for $C^1(\mathbf{R})$ or $C(\mathbf{R})$, although such spanning sets (and bases) do exist. It can be shown, for example, that any basis for $C^1(\mathbf{R})$ must be an uncountable set—that is, there is no one-to-one correspondence between the elements of the basis and the set of natural numbers. For the reader who is interested in learning more about the (more) advanced topics in set theory that are necessary to treat these more complicated spaces, we recommend the book *Naive Set Theory* by Paul Halmos (New York: D. Van Nostrand, 1960).

Our first result answers one question about the existence of bases.

(1.6.6) Theorem. Let V be a vector space that has a finite spanning set, and let S be a linearly independent subset of V. Then there exists a basis S' of V, with $S \subseteq S'$.

(1.6.7) Remark. The content of the theorem is usually summarized by saying that every linearly independent set may be *extended to a basis* (by adjoining further vectors).

To prove Theorem (1.6.6), we need the following result, which is important in its own right.

(1.6.8) Lemma. Let S be a linearly independent subset of V and let $\mathbf{x} \in V$, but $\mathbf{x} \notin S$. Then $S \cup \{\mathbf{x}\}$ is linearly independent if and only if $\mathbf{x} \notin \text{Span}(S)$.

Proof (of the lemma): $\rightarrow$: (by contradiction) Suppose $S \cup \{\mathbf{x}\}$ is linearly independent, but $\mathbf{x} \in \text{Span}(S)$. Then we have an equation $\mathbf{x} = a_1\mathbf{x}_1 + \cdots + a_n\mathbf{x}_n$, where $\mathbf{x}_i \in S$ and $a_i \in \mathbf{R}$. Rewriting this, we have $\mathbf{0} = (-1)\mathbf{x} + a_1\mathbf{x}_1 + \cdots + a_n\mathbf{x}_n$. Since the coefficient of $\mathbf{x}$ is $-1 \neq 0$, this is a linear dependence among the vectors in $S \cup \{\mathbf{x}\}$. The contradiction shows that $\mathbf{x} \notin \text{Span}(S)$, as claimed.

$\leftarrow$: Now suppose that $\mathbf{x} \notin \text{Span}(S)$. Recall that we are also assuming S is linearly independent, and we must show that $S \cup \{\mathbf{x}\}$ is linearly independent. Consider any potential linear dependence among the vectors in $S \cup \{\mathbf{x}\}$: $a\mathbf{x} + a_1\mathbf{x}_1 + \cdots + a_n\mathbf{x}_n = \mathbf{0}$, with $a, a_i \in \mathbf{R}$ and $\mathbf{x}_i \in S$. If $a \neq 0$, then we could write

$$\mathbf{x} = (-a_1/a)\mathbf{x}_1 + \cdots + (-a_n/a)\mathbf{x}_n$$

But this would contradict our hypothesis that $\mathbf{x} \notin \text{Span}(S)$. Hence, $a = 0$. Now, since S is linearly independent, it follows that $a_i = 0$ for all i, as well. Hence $S \cup \{\mathbf{x}\}$ is linearly independent. ■

We are now ready to prove Theorem (1.6.6).

Proof: The idea of this proof is to use the vectors in a finite spanning set for V to augment the vectors in the given linearly independent set and produce a basis S' one step at a time.

Let $T = \{\mathbf{y}_1, \ldots, \mathbf{y}_n\}$ be a finite set that spans V, and let $S = \{\mathbf{x}_1, \ldots, \mathbf{x}_k\}$ be a linearly independent set in V. We claim the following process will produce a basis of V. First, start by setting $S' = S$. Then, for each $\mathbf{y}_i \in T$ in turn do the following: If $S' \cup \{\mathbf{y}_i\}$ is linearly independent, replace the current S' by $S' \cup \{\mathbf{y}_i\}$. Otherwise, leave S' unchanged. Then go on to the next $\mathbf{y}_i$. When the "loop" is completed, S' will be a basis of V.

To see why this works, note first that we are only including the $\mathbf{y}_i$ such that $S' \cup \{\mathbf{y}_i\}$ *is* linearly independent at each stage. Hence the final set S' will also be a linearly independent set. Second, note that every $\mathbf{y}_i \in T$ is in the span of the final set S', since that set contains all the $\mathbf{y}_i$ that are adjoined to the original S. On the other hand, by Lemma (1.6.8), each time the current $S' \cup \{\mathbf{y}_i\}$ is not linearly independent, that $\mathbf{y}_i \in \text{Span}(S')$ already. Since T spans V, and every vector in T is in $\text{Span}(S')$, it follows that S' spans V as well. Hence S' is a basis of V. ■

It is an immediate consequence that every vector space (which has some finite spanning set) has a basis that is also a finite set of vectors. With this fundamental existence question settled, we now turn our attention to some of the information about vector spaces that may be deduced from the form of a basis for the space. If we reconsider some of the examples we have seen, it becomes rather plausible that *the number of vectors in a basis is, in a rough sense, a measure of "how big" the space is.*

(1.6.9) Example. Consider the subspace $W \subset \mathbf{R}^5$ we looked at in Example (1.5.12). W was defined by a system of equations

$$\begin{aligned} x_1 \quad + x_3 \quad\quad &= 0 \\ x_2 - x_3 \quad + x_5 &= 0 \\ x_4 + x_5 &= 0 \end{aligned}$$

We found that the set $S = \{(-1, 1, 1, 0, 0), (0, -1, 0, -1, 1)\}$ spans W. It is easily checked (do it!) that S is also linearly independent, so S is a basis of W. Now we know that $\mathbf{R}^5$ has a basis with five vectors (the standard basis). W is not the whole space $\mathbf{R}^5$, and W has a basis with only two vectors. The fact that the numbers of vectors in the two bases satisfy $2 < 5$ is a reflection of the fact that $W \subset \mathbf{R}^5$, but $W \neq \mathbf{R}^5$.

To make this precise, we must address the following question. Is it true that every basis of a vector space has the same number of elements? (If this were not necessarily true, then we could not really draw any conclusions from the number of elements in any given basis.)

In fact, it is true that every basis of a given vector space has the same number of elements, and we deduce this result from the following theorem.

(1.6.10) Theorem. Let V be a vector space and let S be a spanning set for V, which has m elements. Then no linearly independent set in V can have more than m elements.

Proof: It suffices to prove that every set in V with more than m elements is linearly dependent. Write $S = \{\mathbf{y}_1, \ldots, \mathbf{y}_m\}$ and suppose $S' = \{\mathbf{x}_1, \ldots, \mathbf{x}_n\}$ is a subset of V containing $n > m$ vectors. Consider an equation

$$a_1\mathbf{x}_1 + \cdots + a_n\mathbf{x}_n = \mathbf{0} \tag{1.21}$$

where the $\mathbf{x}_i \in S'$ and the $a_i \in \mathbf{R}$. Since S spans V, there are scalars b_{ij} such that for each i $(1 \leqslant i \leqslant m)$

$$\mathbf{x}_i = b_{i1}\mathbf{y}_1 + \cdots + b_{im}\mathbf{y}_m$$

Substituting these expressions into Eq. (1.21) we have

$$a_1(b_{11}\mathbf{y}_1 + \cdots + b_{m1}\mathbf{y}_m) + \cdots + a_n(b_{1n}\mathbf{y}_1 + \cdots + b_{mn}\mathbf{y}_m) = \mathbf{0}$$

Collecting terms and rearranging, this becomes

$$(b_{11}a_1 + \cdots + b_{1n}a_n)\mathbf{y}_1 + \cdots + (b_{m1}a_1 + \cdots + b_{mn}a_n)\mathbf{y}_m = \mathbf{0}$$

If we can find $a_1, \ldots, a_n$ not all 0, solving the system of equations

$$\begin{aligned} b_{11}a_1 + \cdots + b_{1n}a_n &= 0 \\ &\vdots \\ b_{m1}a_1 + \cdots + b_{mn}a_n &= 0 \end{aligned} \tag{1.22}$$

then those scalars will give us a linear dependence among the vectors in S' in Eq. (1.21). But Eq. (1.22) is a system of m homogeneous linear equations in the n unknowns a_i. Hence since $m < n$, by Corollary (1.5.13), there is a nontrivial solution. Hence S' is linearly dependent. ■

Using a clever argument, we can now show that Theorem (1.6.10) implies that the number of vectors in a basis is an invariant of the space V.

(1.6.11) Corollary. Let V be a vector space and let S and S' be two bases of V, with m and m' elements, respectively. Then $m = m'$.

Proof: Since S spans V and S' is linearly independent, by Theorem (1.6.10) we have that $m \geq m'$. On the other hand, since S' spans V and S is linearly independent, by Theorem (1.6.10) again, $m' \geq m$. It follows that $m = m'$. ■

Corollary (1.6.11) justifies the following definitions.

(1.6.12) Definitions

a) If V is a vector space with some finite basis (possibly empty), we say V is *finite-dimensional*.

b) Let V be a finite-dimensional vector space. The *dimension* of V, denoted $\dim(V)$, is the number of vectors in a (hence any) basis of V.

c) If $V = \{\mathbf{0}\}$, we define $\dim(V) = 0$. See Example (1.6.2d).

(1.6.13) Examples. If we have a basis of V, then computing $\dim(V)$ is simply a matter of counting the number of vectors in the basis.

a) For each n, $\dim(\mathbf{R}^n)$, $= n$, since the standard basis $\{\mathbf{e}_1, \ldots, \mathbf{e}_n\}$ contains n vectors.

b) $\dim(P_n(\mathbf{R})) = n + 1$, since by Example (1.6.2c) a basis for $P_n(\mathbf{R})$ is the set $\{1, x, x^2, \ldots, x^n\}$, which contains $n + 1$ functions in all.

c) the subspace W in $\mathbf{R}^5$ considered in Example (1.6.9) has dimension 2.

d) The vector spaces $P(\mathbf{R})$, $C^1(\mathbf{R})$, and $C(\mathbf{R})$ are not finite-dimensional. We say that such spaces are *infinite-dimensional*.

We conclude this section with several general observations about dimensions.

(1.6.14) Corollary. Let W be a subspace of a finite-dimensional vector space V. Then $\dim(W) \leq \dim(V)$. Furthermore, $\dim(W) = \dim(V)$ if and only if $W = V$.

Proof: We leave the proof of this corollary as an exercise for the reader. ■

Given a subspace W of $\mathbf{R}^n$ defined by a system of homogeneous linear equations, the technique for solving systems of equations presented in Section 1.5 also gives a method for computing the dimension of W. This is so because the spanning set for W, obtained by reducing the system to echelon form, then setting each free variable equal to 1 and the others to 0 in turn, always yields a basis for W. In Exercise 1 you will show that any set of vectors constructed in this fashion is linearly independent. [See also Example (1.6.16) that follows.] As a result, we have Corollary (1.6.15).

(1.6.15) Corollary. Let W be a subspace of $\mathbf{R}^n$ defined by a system of homogeneous linear equations. Then $\dim(W)$ is equal to the number of free variables in the corresponding echelon form system.

(1.6.16) Example. In $\mathbf{R}^5$, consider the subspace defined by the system

$$\begin{aligned} x_1 + 2x_2 \qquad + x_4 \qquad &= 0 \\ x_1 \qquad + x_3 \qquad + x_5 &= 0 \\ x_2 + x_3 + x_4 - x_5 &= 0 \end{aligned}$$

After reducing to echelon form, you should check that we obtain

$$\begin{aligned} x_1 \qquad - (1/3)x_4 + (4/3)x_5 &= 0 \\ x_2 \qquad + (2/3)x_4 - (2/3)x_5 &= 0 \\ x_3 + (1/3)x_4 - (1/3)x_5 &= 0 \end{aligned}$$

The free variables are x_4 and x_5. Hence $\dim(W) = 2$, and a basis of W may be obtained by setting $x_4 = 1$ and $x_5 = 0$ to obtain $(1/3, -2/3, -1/3, 1, 0)$, then $x_4 = 0$ and $x_5 = 1$ to obtain $(-4/3, 2/3, 1/3, 0, 1)$. Note that we *do* obtain a linearly independent set of vectors here, since if

$$a_1(1/3, -2/3, -1/3, 1, 0) + a_2(-4/3, 2/3, 1/3, 0, 1) = (0, 0, 0, 0, 0)$$

then looking at the 4th components gives $a_1 = 0$, and similarly looking at the 5th components gives $a_2 = 0$. [This observation may be generalized to give a proof of Corollary (1.6.15).]

Finally, we prove another general result that gives a formula for the dimension of the sum of two subspaces of a given vector space.

(1.6.17) Example. In $\mathbf{R}^4$ consider the subspaces $V_1 = \text{Span}\{(1, 1, 1, 1), (1, 0, 0, 0)\}$ and $V_2 = \text{Span}\{(1, 0, 1, 0), (0, 1, 0, 1)\}$. Since these sets of two vectors are linearly independent, we have $\dim(V_1) = \dim(V_2) = 2$. From this, we might expect naively that $\dim(V_1 + V_2) = \dim(V_1) + \dim(V_2) = 4$. However, this is *not* true! From Exercise 5 of Section 1.3, we do have that $V_1 + V_2 = \text{Span}\{(1, 1, 1, 1), (1, 0, 0, 0), (1, 0, 1, 0), (0, 1, 0, 1)\}$, but we recognize that this set is linearly dependent. Indeed, we have a linear dependence:

$$(1, 1, 1, 1) - (1, 0, 1, 0) - (0, 1, 0, 1) = (0, 0, 0, 0)$$

If we remove one of the vectors appearing in this equation from our spanning set, we do obtain a linearly independent set, however. Therefore $V_1 + V_2$ has a basis with three elements, and hence $\dim(V_1 + V_2) = 3$.

We may ask why $\dim(V_1 + V_2) < \dim(V_1) + \dim(V_2)$ in this case, answer is clearly to be seen in the linear dependence found earlier. From equation it follows that $(1, 1, 1, 1) \in V_1 \cap V_2$. (In fact, you should verify th every vector in $V_1 \cap V_2$ is a scalar multiple of this one.) Since our two two-dimensional subspaces have more than the zero vector in common, they cannot together span a four-dimensional space. The overlap *decreases* the dimension of the sum. Our final result in this chapter gives a quantitative statement about how much the dimension of the sum of two subspaces is decreased in general, in the case that the subspaces do overlap in this way.

(1.6.18) Theorem. Let W_1 and W_2 be finite-dimensional subspaces of a vector space V. Then

$$\dim(W_1 + W_2) = \dim(W_1) + \dim(W_2) - \dim(W_1 \cap W_2)$$

Proof: The result is clear if $W_1 = \{\mathbf{0}\}$ or $W_2 = \{\mathbf{0}\}$ so we may assume that both our subspaces contain some nonzero vectors. The idea of the proof is to build up to a basis of $W_1 + W_2$ starting from a basis of $W_1 \cap W_2$. When we have done this, we will be able to count the vectors in our basis to derive the desired result.

Let S be any basis of $W_1 \cap W_2$. In Exercise 6a of this section you will show that we can always find sets T_1 and T_2 (disjoint from S) such that $S \cup T_1$ is a basis for W_1 and $S \cup T_2$ is a basis for W_2. We claim that $U = S \cup T_1 \cup T_2$ is a basis for $W_1 + W_2$.

You will show in Exercise 6b that U spans $W_1 + W_2$. Hence it remains to prove that U is linearly independent. Any potential linear dependence among the vectors in U must have the form

$$\mathbf{v} + \mathbf{w}_1 + \mathbf{w}_2 = \mathbf{0}$$

where $\mathbf{v} \in \text{Span}(S) = W_1 \cap W_2$, $\mathbf{w}_1 \in \text{Span}(T_1) \subset W_1$, and $\mathbf{w}_2 \in \text{Span}(T_2) \subset W_2$. To show that $S \cup T_1 \cup T_2$ is linearly independent, it suffices to prove that in any such potential linear dependence, we must have $\mathbf{v} = \mathbf{w}_1 = \mathbf{w}_2 = \mathbf{0}$. (Why?) We claim first that from the preceding equation, $\mathbf{w}_2$ must be zero. If not, then we would have that $\mathbf{w}_2 = -\mathbf{v} - \mathbf{w}_1$, so $\mathbf{w}_2 \in W_1 \cap W_2$, and $\mathbf{w}_2 \neq \mathbf{0}$. This leads immediately to a contradiction. By definition, $\mathbf{w}_2 \in \text{Span}(T_2)$. But $S \cup T_2$ is a basis for W_2, and $S \cap T_2 = \phi$. Hence $\text{Span}(S) \cap \text{Span}(T_2) = \{\mathbf{0}\}$. Therefore in our potential linear dependence, we must have $\mathbf{w}_2 = \mathbf{0}$. But now from this observation, it follows that $\mathbf{v} = \mathbf{w}_1 = \mathbf{0}$ as well, since our equation reduces to $\mathbf{v} + \mathbf{w}_1 = \mathbf{0}$, while $S \cup T_1$ is linearly independent. The remainder of the proof is left to the reader as Exercise 6c. ■

EXERCISES

1. Using the observation made in Example (1.6.16), show that the set of vectors obtained from an echelon form system of homogeneous linear equations by setting each free variable equal to 1 and the others equal to 0 in turn always yields a basis of the subspace of solutions of the system.

2. Find a basis for and the dimension of the subspaces defined by each of the following sets of conditions:

a)
$$\begin{aligned} x_1 \phantom{{}+x_2} + x_4 &= 0 \\ 3x_1 + x_2 + x_4 &= 0 \end{aligned} \quad \text{in } \mathbf{R}^4$$

b)
$$\begin{aligned} 2x_1 - x_2 + 3x_3 &= 0 \\ x_1 + 4x_2 - x_5 &= 0 \\ x_1 + x_2 - x_3 + x_4 + x_5 &= 0 \\ 2x_1 + 2x_2 + x_3 + (1/2)x_4 &= 0 \end{aligned} \quad \text{in } \mathbf{R}^5$$

c)
$$\begin{aligned} x_1 + 2x_2 + x_3 &= 0 \\ 2x_1 - 4x_2 + 2x_3 &= 0 \\ -x_1 - 2x_2 - x_3 &= 0 \end{aligned} \quad \text{in } \mathbf{R}^3$$

d) $\{p \in P_3(\mathbf{R}) \mid p(2) = p(-1) = 0\}$

e) $\{f \in \text{Span}\{e^x, e^{2x}, e^{3x}\} \mid f(0) = f'(0) = 0\}$

f) $\{f \in \text{Span}\{e^x, e^{-x}, \cos(x), \sin(x), 1, x\} \mid f(0) = f'(0) = 0\}$

3. Prove Corollary (1.6.14).

4. Let $W_1 = \text{Span}\{(1, 1, 2, 1), (3, 1, 0, 0)\}$ and $W_2 = \text{Span}\{(-1, -2, 0, 1), (-4, -2, -2, -1)\}$ in $\mathbf{R}^4$. Compute $\dim(W_1)$, $\dim(W_2)$, $\dim(W_1 + W_2)$, and $\dim(W_1 \cap W_2)$, and verify that the formula of Theorem (1.6.18) holds in this example.

5. Let W_1 and W_2 be subspaces of a finite-dimensional vector space V, and let $\dim(W_1) = n_1$, and $\dim(W_2) = n_2$.
   a) Show that $\dim(W_1 \cap W_2) \leq$ the smaller of n_1 and n_2.
   b) Show by examples that if $\dim(W_1) = 2$ and $\dim(W_2) = 2$, all the values $\dim(W_1 \cap W_2) = 0, 1, 2$ are possible.
   c) Say $n_1 \leq n_2$ so that $\dim(W_1 \cap W_2) \leq n_1$. When will it be true that $\dim(W_1 \cap W_2) = n_1$? Prove your assertion.

6. a) Let W_1 and W_2 be subspaces of a finite-dimensional vector space V. Let S be a basis for the subspace $W_1 \cap W_2$. Show that there are sets of vectors T_1 and T_2 such that $S \cup T_1$ is a basis for W_1 and $S \cup T_2$ is a basis for W_2.
   b) Show that if S, T_1, and T_2 are as in part a, then $S \cup T_1 \cup T_2$ spans $W_1 + W_2$.
   c) In the text, we proved that $S \cup T_1 \cup T_2$ is also linearly independent, hence a basis for $W_1 + W_2$. Deduce that $\dim(W_1 + W_2) = \dim(W_1) + \dim(W_2) - \dim(W_1 \cap W_2)$.

7. a) Find a basis of $\mathbf{R}^4$ containing the linearly independent set $S = \{(1, 2, 3, 4), (-1, 0, 0, 0)\}$ using the method of the proof of Theorem (1.6.6).
   b) Find a basis of $\mathbf{R}^5$ containing the linearly independent set $S = \{(0, 0, 2, 0, 1), (-1, 0, 0, 1, 0)\}$

c) Find a basis of $P_5(\mathbf{R})$ containing the linearly independent set $S = \{x^5 - 2x, x^4 + 3x^2, 6x^5 + 2x^3\}$.

8. In the text we showed that every linearly independent set of vectors in a finite-dimensional vector space can be extended to a basis of V. Prove the following parallel result: If S is a subset of V with $\text{Span}(S) = V$, then there is a subset $S' \subseteq S$ that is a basis of V. [*Hint:* Use Lemma (1.6.8) to "build up" a linearly independent subset of S, which also spans V, one vector at a time.]

9. a) Using the result of Exercise 8, find a basis of $\mathbf{R}^3$, which is contained in the spanning set $S = \{(1, 2, 1), (-1, 3, 1), (0, 5, 2), (1, 1, 1), (0, 4, 2)\}$.

   b) Find a basis of $P_3(\mathbf{R})$ that is contained in the spanning set

$$S = \{x^3 + x, 2x^3 + 3x, 3x^3 - x - 1, x + 2, x^3 + x^2, x^2 - 8\}$$

10. a) Let W be a subspace of a finite-dimensional vector space V. Show that there always exists a subspace W' with the property that $V = W \oplus W'$. (Such a subspace is called a *complementary subspace* for W. *Caution:* complementary subspaces are not unique.)

    b) Find two different complementary subspaces for $W = \text{Span}(\{(1, 1, 1)\})$ in $V = \mathbf{R}^3$.

11. Show that if S is a linearly independent subset of a vector space V and $S = S_1 \cup S_2$ with $S_1 \cap S_2 = \phi$ then $\text{Span}(S_1) \cap \text{Span}(S_2) = \{\mathbf{0}\}$.

12. Let V be a finite-dimensional vector space and let W_1, W_2 be subspaces of V such that $V = W_1 \oplus W_2$.

    a) Show that if S_1 is a basis of W_1 and S_2 is a basis of W_2, then $S_1 \cup S_2$ is a basis for V.

    b) Conclude that in this case $\dim(V) = \dim(W_1) + \dim(W_2)$. [This also follows from Theorem (1.6.18).]

13. Give a list of all the different (kinds of) subspaces of $\mathbf{R}^3$, of $\mathbf{R}^4$, of $\mathbf{R}^n$ in general. (*Hint:* What are their possible dimensions?)

14. a) Show that $\left\{\begin{bmatrix} 1 & 0 \\ 0 & 0 \end{bmatrix}, \begin{bmatrix} 0 & 1 \\ 0 & 0 \end{bmatrix}, \begin{bmatrix} 0 & 0 \\ 1 & 0 \end{bmatrix}, \begin{bmatrix} 0 & 0 \\ 0 & 1 \end{bmatrix}\right\}$ is a basis for the vector space $M_{2\times 2}(\mathbf{R})$. Hence $\dim(M_{2\times 2}(\mathbf{R})) = 4$.

    b) Find a basis for and the dimension of the vector space $M_{2\times 3}(\mathbf{R})$. (*Hint:* Try to follow the pattern from part a of this exercise.)

    c) Find a basis for and the dimension of the vector space $M_{m\times n}(\mathbf{R})$.

15. Find a basis for and the dimension of each of the following subspaces of $M_{2\times 2}(\mathbf{R})$.

    a) $W = \left\{\begin{bmatrix} a & b \\ c & d \end{bmatrix} \in M_{2\times 2}(\mathbf{R}) \mid a + 2b = c - 3d = 0\right\}$.

    b) The subspace of symmetric 2 by 2 matrices. (See Exercise 14 of Section 1.2.)

c) The subspace of skew-symmetric 2 by 2 matrices. (See Exercise 15 of Section 1.2.)

16. a) Show that there is a unique basis $\{p_1, p_2, p_3\}$ of $P_2(\mathbf{R})$ with the property that $p_1(0) = 1$, $p_1(1) = p_1(2) = 0$, $p_2(1) = 1$, $p_2(0) = p_2(2) = 0$, and $p_3(2) = 1$, $p_3(0) = p_3(1) = 0$.

b) Show more generally that, given any three distinct real numbers x_1, x_2, x_3, there is a unique basis $\{p_1, p_2, p_3\}$ of $P_2(\mathbf{R})$ such that $p_i(x_j) = \begin{cases} 1 & \text{if } j=i \\ 0 & \text{if } j\neq i \end{cases}$ Find explicit formulas for the polynomials p_i.

c) Show that given any n distinct real numbers $x_1, \ldots, x_n$ there exists a unique basis $\{p_1, \ldots, p_n\}$ of $P_{n-1}(\mathbf{R})$ such that $p_i(x_j) = \begin{cases} 1 & \text{if } j=i \\ 0 & \text{if } j\neq i \end{cases}$

d) Using the basis of $P_{n-1}(\mathbf{R})$ constructed in part c, show that given any n distinct real numbers $x_1, \ldots, x_n$ and arbitrary real numbers $a_1, \ldots, a_n$, there exists a unique polynomial $p \in P_{n-1}(\mathbf{R})$ such that $p(x_i) = a_i$ for $1 \leq i \leq n$. (The formula you obtain for p is called the *Lagrange interpolation formula.*)

CHAPTER SUMMARY

In this chapter we have considered, in some detail, the structure of *vector spaces* [Definition (1.1.1)]. We began by introducing basic examples such as the spaces $\mathbf{R}^n$, the vector spaces of functions $F(\mathbf{R})$, $C(\mathbf{R})$, and $C^1(\mathbf{R})$, and, in exercises, the vector spaces of matrices $M_{m\times n}(\mathbf{R})$. Next, we saw how other examples of vector spaces arise as *subspaces* [Definition (1.2.6)] of known spaces. The two basic methods to define subspaces we have seen are as follows:

1. Taking the set of solutions of a system of homogeneous linear equations in $\mathbf{R}^n$ [Corollary (1.2.14)], or a subset of another vector space defined by analogous equations—the reader should note the similarities between a system of linear equations and the equations defining subspaces of function spaces as in Example (1.2.11a), and in numerous exercises. In cases like these we are *cutting out* subspaces of a given vector space by equations.

2. Taking the *span* of a set of vectors in any vector space [Definition (1.3.1)]. Closely related is the notion of a *sum* of subspaces [Definition (1.3.5)]. In both cases we are *building up* subspaces from smaller sets by linear combinations.

Next, we considered the question of whether, given a spanning set for a vector space, some subset of the given set might also span the same space, in effect making some of the vectors redundant. This led us to introduce the notions of *linear dependence* [Definition (1.4.2)] and *linear independence* [Definition (1.4.4)]. The linearly independent spanning sets are the economical ones, in the sense that no subset of a linearly independent set spans the same space that the whole set does.

To decide whether a given set of vectors is linearly independent, or whether a given vector is in the span of a given set of vectors, we saw that we usually must solve *systems of linear equations* [Definition (1.5.1)]. This observation, coupled with method 1 for defining subspaces, led us to consider a general method for solving systems, called the *elimination algorithm*. After reducing a system of linear equations to the equivalent *echelon form* system [Definition (1.5.6)], we saw that we may construct all solutions of the equations by substituting values for the *free variables* and solving for the *basic variables* [Definitions (1.5.11)]. Using the elimination algorithm, we also saw how to relate the two methods given earlier for defining subspaces in $\mathbf{R}^n$. Any subspace defined by either method may also be defined by the other, and we saw how to move from one description to the other.

Finally, we defined a *basis* of a vector space [Definition (1.6.1)] and showed that if a vector space V has a finite spanning set, then

1. V has a finite basis [Theorem (1.6.6)], and furthermore

2. all bases of V have the same number of elements.

These two results taken together are the most important and profound statements we have seen. They show that there is a well-defined integer, the *dimension* of V [Definition (1.6.12)], which measures how big the space is in a certain sense. For example, if we have a subspace of $\mathbf{R}^n$ defined as the set of solutions of a system of homogeneous linear equations, then the dimension is the number of free variables in the equivalent echelon form system, and this tells us how many vectors we need to span the subspace.

We will see in Chapter 2 just how similar vector spaces of the same dimension are (and how different spaces of different dimensions are).

SUPPLEMENTARY EXERCISES

1. **True-False.** For each true statement, give a short proof or reason. For each false statement give a counterexample.
   a) The set of solutions of each system of linear equations is a subspace of $\mathbf{R}^n$ (n = number of variables).
   b) If V is a finite-dimensional vector space and W is a subspace of V, then $\dim(W) \leq \dim(V)$.
   c) If V is a vector space of dimension n and S is a subset of V containing $k < n$ vectors, then S is linearly independent.
   d) If $W_1 = \text{Span}\{(1, 1)\}$ and $W_2 = \text{Span}\{(-1, 3)\}$ in $\mathbf{R}^2$, then $\mathbf{R}^2 = W_1 + W_2$.
   e) Every system of linear equations has at least one free variable.
   f) Every system of homogeneous linear equations has at least one solution.
   g) Every system of m linear equations in n variables has a solution when $m < n$.
   h) $\{f \mid 4f' + 3f = 0\}$ is a subspace of $C^1(\mathbf{R})$.
   i) $S = \{\cos(2x), \sin^2(x), 1, e^x\}$ is a linearly independent set in $C(\mathbf{R})$.

j) If S is a linearly independent set, then each vector in Span(S) can be written in only one way as a linear combination of vectors in S.

k) The set $V = \mathbf{R}^2$, with the operations $(x_1, x_2) +' (y_1, y_2) = (x_1 + y_1, 0)$ and $c \cdot (x_1, x_2) = (cx_1, 0)$ is a vector space.

2. a) Find a basis for and the dimension of the subspace of $\mathbf{R}^4$ defined by:

$$\begin{aligned} 2x_1 + x_3 + x_4 &= 0 \\ x_1 + 2x_2 - x_4 &= 0 \\ x_1 - 6x_2 + 2x_3 + 5x_4 &= 0 \end{aligned}$$

b) Find a parametrization of the set of solutions of the system of equations obtained by replacing the right-hand sides of the equations in part a by 1, -1, 5, and leaving the left-hand sides the same. Do you see a connection between your answers to parts a and b? We will examine this relation in more detail in Chapter 2.

3. Find a basis for and the dimension of the subspace W of $P_4(\mathbf{R})$ defined by $W = \{p \in P_4(\mathbf{R}) \mid p(1) = p(-1) = 0\}$.

4. Same question for the subspace $W = \left\{\begin{bmatrix} a & b \\ c & d \end{bmatrix} \in M_{2\times 2}(\mathbf{R}) \mid a + d = 0\right\}$

5. Let $\{\mathbf{u}, \mathbf{v}, \mathbf{w}\}$ be a linearly independent subset of a vector space. Show that $\{\mathbf{v} + \mathbf{w}, \mathbf{u} + \mathbf{w}, \mathbf{v} + \mathbf{w}\}$ is also linearly independent.

6. a) Is $W = \{(x_1, x_2, x_3) \in \mathbf{R}^3 \mid x_1^3 + x_2^3 + x_3^3 = 0\}$ a subspace of $\mathbf{R}^3$? If not, does it contain any subspaces of $\mathbf{R}^3$?

b) Is $W = \{f \in F(\mathbf{R}) \mid 2f(x) = 3f(x + 2) \text{ for all } x \in \mathbf{R}\}$ *a* subspace of $F(\mathbf{R})$? Try to find an element of W other than $f = 0$.

7. Is $\{1 + x, 2 - x^2, 3 + 5x^2, 7 - 2x\}$ a linearly independent subset of $P_2(\mathbf{R})$? (You can answer this without calculations!)

8. a) Is $(3, 2, 1, -1)$ an element of $W = \text{Span}\{(1, -1, 1, 1), (0, 0, 1, 1), (-3, 1, 1, 2)\}$? (Before calculating, read part b of this exercise, and see if there is a way to answer both questions with only one calculation!)

b) Find a system of linear equations whose set of solutions is the subspace W from part a.

9. a) Show that $S = \{(1, 1, 1), (1, 1, 0), (1, 0, 0)\}$ and $S' = \{(1, 1, 1), (0, 1, 1), (0, 0, 1)\}$ are both bases of $\mathbf{R}^3$.

b) Express each vector in S as a linear combination of the vectors in S'.

10. Explain why in any system of m linear equations in n variables, when we reduce to echelon form, the number of free variables plus the number of basic variables must equal n.

11. Is there *any* other way to define a vector space structure on the set $V = \mathbf{R}^2$ (or $\mathbf{R}^n$) different from the one given in Example (1.1.2) [i.e., are there any

other sum and product operations on vectors in $\mathbf{R}^n$ that satisfy the eight axioms of Definition (1.1.1) other than those given in Example (1.1.2)]?

12. Let $S = \{\mathbf{x}_1, \ldots, \mathbf{x}_n\}$ be any set of n vectors in an n-dimensional vector space V. Show that S is linearly independent if and only if Span$(S) = V$. [In particular, this says that if there exists a vector $\mathbf{v} \in V$ such that $\mathbf{v} \notin$ Span(S), then S must be linearly dependent.]

CHAPTER 2

Linear Transformations

Introduction

In this chapter we begin our study of functions between vector spaces. In one form or another this will also be the focus of the remainder of the text. This is a good point to step back and think of the study of linear algebra as a whole, for we are on the verge of making the transition from the study of the objects of linear algebra—vector spaces (not vectors!)—to the study of functions between the objects—linear transformations. Every branch of mathematics can be broken down in this way and comparing various branches of mathematics based on this breakdown can be fruitful. For example, after completing the study of linear algebra, it would be a revealing exercise to compare linear algebra and calculus in this fashion.

The functions of interest for us are those functions that preserve the algebraic structure of vector spaces, that is, preserve the operations of addition and scalar multiplication of vectors. These functions are called linear transformations. Our study of linear transformations will begin with several examples that will often provide test cases for the work that follows. Then we address the obvious but important question: How do we write down or specify a linear transformation? If the vector spaces in question are finite-dimensional, then we are led very naturally to study the effect of a linear transformation on a basis, and consequently, to the concept of the matrix of a transformation, a concise way of specifying a transformation.

We begin to see the interplay between functions and objects when we introduce the kernel and image of a linear transformation. For any $T: V \to W$, the kernel of T is a subspace of V and the image of T is a subspace of W. Using both the structure of vector spaces and the behavior of linear transformations, we will be able to compute the dimensions of these subspaces and with the Dimension theorem to see the relationship between their dimensions. This simple but profound relationship will provide us with an algebraic means of determining if a linear transformation is injective or surjective.

There are other properties of functions, like injectivity and surjectivity, which occur in every branch of mathematics. We will be interested in the composition of functions and the invertibility of functions, and how the representation of a function, that is, the matrix of a transformation, depends on the particular choices that we make in the representing process. The particular branch of mathematics determines how these properties are approached. We will use the mechanics of matrix algebra where possible—matrices only make sense in finite dimensions—to explore these ideas. But again, we should keep in mind that the algebra of matrices is a tool that facilitates the study of linear transformations; it is not the end in itself.

---

§2.1. LINEAR TRANSFORMATIONS

This section introduces the basic concepts concerning a special class of functions between vector spaces. First, let us review the basic notation for functions. Let V and W be vector spaces. A function T from V to W is denoted by $T: V \to W$. T assigns to each vector $\mathbf{v} \in V$ a unique vector $\mathbf{w} \in W$, which is denoted by $\mathbf{w} = T(\mathbf{v})$. The vector $\mathbf{w} = T(\mathbf{v})$ in W is called the *image* of $\mathbf{v}$ under the function T. To specify a function T, we give an expression for $\mathbf{w}$ in terms of $\mathbf{v}$ just as in calculus we express y in terms of x if $y = f(x)$ for a function $f: \mathbf{R} \to \mathbf{R}$.

We begin by defining the collection of functions $T: V \to W$, which will be of interest to us. From the development in Chapter 1, the reader might conjecture that we will not be concerned with all functions $T: V \to W$, but rather, with those functions that preserve the algebraic structure of vector spaces. Loosely speaking, we want our functions to turn the algebraic operations of addition and scalar multiplication in V into addition and scalar multiplication in W. More precisely, we have Definition (2.1.1).

(2.1.1) Definition. A function $T: V \to W$ is called a *linear mapping* or a *linear transformation* if it satisfies

(i) $T(\mathbf{u} + \mathbf{v}) = T(\mathbf{u}) + T(\mathbf{v})$ for all $\mathbf{u}$ and $\mathbf{v} \in V$

(ii) $T(a\mathbf{v}) = aT(\mathbf{v})$ for all $a \in \mathbf{R}$ and $\mathbf{v} \in V$.

V is called the *domain* of T and W is called the *target* of T.

Notice that two different operations of addition are used in condition (i) of the definition. The first addition, $\mathbf{u} + \mathbf{v}$, takes place in the vector space V and the second addition, $T(\mathbf{u}) + T(\mathbf{v})$, takes place in the vector space W. Similarly, two different operations of scalar multiplication are used in condition (ii). The first, $a\mathbf{v}$, takes place in the vector space V and the second, $aT(\mathbf{v})$, takes place in the vector space W. Thus, we say that a linear transformation preserves the operations of addition and scalar multiplication.

If a function $T: V \to W$ is a linear transformation, we can draw some conclusions about its behavior directly from Definition (2.1.1). For example, we can ask where the zero vector in V is mapped by a linear mapping T. To see the answer, note that if we write $\mathbf{0}_V$ for the zero vector in V and $\mathbf{0}_W$ for the zero vector in W, then

$$\begin{aligned} T(\mathbf{0}_V) &= T(\mathbf{0}_V + \mathbf{0}_V) \\ &= T(\mathbf{0}_V) + T(\mathbf{0}_V) \qquad \text{[by property (i)]} \end{aligned}$$

Hence, if we add the additive inverse of $T(\mathbf{0}_V)$ in W to both sides of the equation, we obtain

$$\mathbf{0}_W = T(\mathbf{0}_V)$$

A linear mapping always takes the zero vector in the domain vector space to the zero vector in the target vector space.

The following proposition and its corollary give useful variations on this definition. Their proofs are simple consequences of Definition (2.1.1) and will be left as exercises. These two results will be used repeatedly throughout the text and their use should become second nature to the reader.

(2.1.2) Proposition. A function $T: V \to W$ is a linear transformation if and only if for all a and $b \in \mathbf{R}$ and all $\mathbf{u}$ and $\mathbf{v} \in V$

$$T(a\mathbf{u} + b\mathbf{v}) = aT(\mathbf{u}) + bT(\mathbf{v})$$

Proof: Exercise 1. ■

(2.1.3) Corollary. A function $T: V \to W$ is a linear transformation if and only if for all $a_1, \ldots, a_k \in \mathbf{R}$ and for all $\mathbf{v}_1, \ldots, \mathbf{v}_k \in V$:

$$T\left(\sum_{i=1}^{k} a_i\mathbf{v}_i\right) = \sum_{i=1}^{k} a_iT(\mathbf{v}_i)$$

Proof: Exercise 2. ■

In order to become familiar with these notions, we begin our discussion of linear transformations with several examples. The first examples [(2.1.4) to (2.1.8)] are algebraic in nature. The remaining two examples are presented in a geometric manner, which will require a digression in plane geometry.

(2.1.4) Examples

a) Let V be any vector space, and let $W = V$. The *identity transformation* $I: V \to V$ is defined by $I(\mathbf{v}) = \mathbf{v}$ for all $\mathbf{v} \in V$. Let us verify (2.1.2). For a and $b \in \mathbf{R}$ and $\mathbf{u}$ and $\mathbf{v} \in V$, $I(a\mathbf{u} + b\mathbf{v}) = a\mathbf{u} + b\mathbf{v} = aI(\mathbf{u}) + bI(\mathbf{v})$. Hence, I is a linear transformation. If we want to specify that I is the identity transformation of V, we write $I_V: V \to V$.

b) Let V and W be any vector spaces, and let $T: V \to W$ be the mapping that takes every vector in V to the zero vector in W:

$$T(\mathbf{v}) = \mathbf{0}_W$$

for all $\mathbf{v} \in V$. Let us verify (2.1.2). For a and $b \in \mathbf{R}$ and $\mathbf{u}$ and $\mathbf{v} \in V$, $T(a\mathbf{u} + b\mathbf{v}) = \mathbf{0}_W$ and $aT(\mathbf{u}) + bT(\mathbf{v}) = a\mathbf{0}_W + b\mathbf{0}_W = \mathbf{0}_W$. Thus $T(a\mathbf{u} + b\mathbf{v}) = aT(\mathbf{u}) + bT(\mathbf{v})$. T is called the *zero transformation.*

(2.1.5) Example. Let $V = \mathbf{R}^2$ and $W = \mathbf{R}^3$. Define a function $T: V \to W$ by

$$T((x_1, x_2)) = (x_1 + x_2, 2x_2, 3x_1 - x_2)$$

T is linear transformation. To verify this it must be shown that the condition of Proposition (2.1.2) is satisfied; that is

$$T(a\mathbf{x} + b\mathbf{y}) = aT(\mathbf{x}) + bT(\mathbf{y})$$

for all a and $b \in \mathbf{R}$ and all $\mathbf{x}$ and $\mathbf{y} \in \mathbf{R}^2$. By the definition of T

$$\begin{aligned}
T(a\mathbf{x} + b\mathbf{y}) &= T((ax_1 + by_1, ax_2 + by_2)) \\
&= ((ax_1 + by_1) + (ax_2 + by_2), 2(ax_2 + by_2), 3(ax_1 + by_1) \\
&\quad - (ax_2 + by_2)) \\
&= (a(x_1 + x_2) + b(y_1 + y_2), a2x_2 + b2y_2, a(3x_1 - x_2) \\
&\quad + b(3y_1 - y_2)) \\
&= (a(x_1 + x_2), a2x_2, a(3x_1 - x_2)) + (b(y_1 + y_2), b2y_2, \\
&\quad b(3y_1 - y_2)) \\
&= a(x_1 + x_2, 2x_2, 3x_1 - x_2) + b(y_1 + y_2, 2y_2, 3y_1 - y_2) \\
&= aT(\mathbf{x}) + bT(\mathbf{y})
\end{aligned}$$

Thus, we have shown that T is a linear transformation.

(2.1.6) Example. Let $V = \mathbf{R}^2$ and $W = \mathbf{R}^2$. Define $T: V \to W$ by

$$T((x_1, x_2)) = (x_1^2, x_2^2)$$

T is not a linear transformation since $T(a\mathbf{x}) = a^2T(\mathbf{x})$, which is different from $aT(\mathbf{x})$ in general [i.e., if $a \neq 0$, 1 and $T(\mathbf{x}) \neq \mathbf{0}$].

(2.1.7) Example. Let $V = \mathbf{R}^3$ and $W = \mathbf{R}$. Define $T: V \to W$ by

$$T((x_1, x_2, x_3)) = 2x_1 - x_2 + 3x_3$$

As in Example (2.1.5), it is straightforward to show that T is linear. More generally, let $V = \mathbf{R}^n$ and $W = \mathbf{R}$. Let $\mathbf{a} = (a_1, \ldots, a_n)$ be a fixed vector in V. Define a function $T: V \to W$ by

$$T(\mathbf{x}) = a_1x_1 + \cdots + a_nx_n$$

T is a linear transformation. The proof of this fact is left to the reader as Exercise 4. Note that if a different vector $\mathbf{a}$ is chosen, a different transformation is obtained.

(2.1.8) Examples. Several of the basic processes of calculus may be viewed as defining linear mappings on vector spaces of functions.

a) For instance, consider the operation of differentiation. Let V be the vector space $C^\infty(\mathbf{R})$ (of functions $f: \mathbf{R} \to \mathbf{R}$ with derivatives of all orders; see Exercise 9 of Chapter 1, Section 1.2). Let $D: C^\infty(\mathbf{R}) \to C^\infty(\mathbf{R})$ be the mapping that takes each function $f \in C^\infty(\mathbf{R})$ to its derivative function: $D(f) = f' \in C^\infty(\mathbf{R})$. Using the sum and scalar multiple rules for derivatives, we can show that D is a linear transformation. See Exercise 5a.

b) Definite integration may also be viewed as a linear mapping. For example, let V denote the vector space $C[a, b]$ of continuous functions on the closed interval $[a, b] \subset \mathbf{R}$, and let $W = \mathbf{R}$. Then we can define the integration mapping Int: $V \to W$ by the rule $\text{Int}(f) = \int_a^b f(x)\,dx \in \mathbf{R}$. In Exercise 5b you will show that the sum and scalar multiple rules for integration imply that Int is a linear mapping from V to W.

Before proceeding with the next two examples, we will review certain important facts from plane geometry and trigonometry in the context of vectors in the plane. If $\mathbf{a} = (a_1, a_2)$ and $\mathbf{b} = (b_1, b_2)$ are vectors in the plane, then $\mathbf{a}$ and $\mathbf{b}$ form two sides of a triangle whose third side is the line segment between (a_1, a_2) and

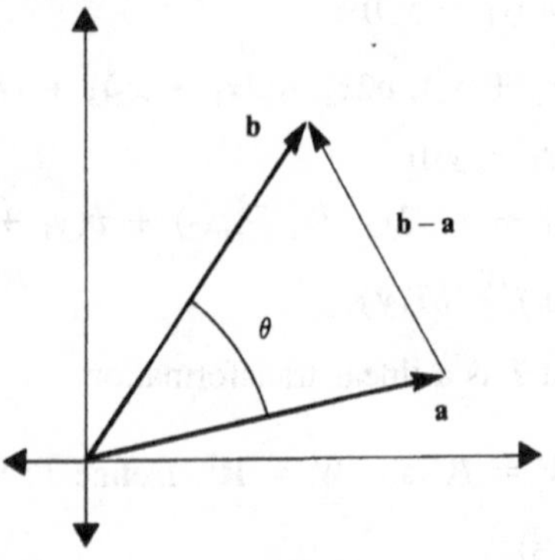

Figure 2.1

(b_1, b_2). In terms of vector addition, the third side is expressed as $\mathbf{b} - \mathbf{a}$. See Figure 2.1.

From the Pythagorean theorem we know that the length of a line segment is the square root of the sum of the squares of the differences of the coordinates of the endpoints. If the line segment is a vector $\mathbf{v}$, its length is denoted $\|\mathbf{v}\|$. Therefore, the lengths of the sides of the preceding triangle are

$$\|\mathbf{a}\| = \sqrt{a_1^2 + a_2^2}$$

$$\|\mathbf{b}\| = \sqrt{b_1^2 + b_2^2}$$

$$\|\mathbf{b} - \mathbf{a}\| = \sqrt{(b_1 - a_1)^2 + (b_2 - a_2)^2}$$

The law of cosines applied to this triangle and the angle θ between the sides $\mathbf{a}$ and $\mathbf{b}$ yields

$$\|\mathbf{b} - \mathbf{a}\|^2 = \|\mathbf{a}\|^2 + \|\mathbf{b}\|^2 - 2\|\mathbf{a}\| \cdot \|\mathbf{b}\| \cdot \cos(\theta)$$

Using the previous expressions to simplify this formula, we see that

$$(b_1 - a_1)^2 + (b_2 - a_2)^2 = (a_1^2 + a_2^2) + (b_1^2 + b_2^2) - 2\|\mathbf{a}\| \cdot \|\mathbf{b}\| \cdot \cos(\theta)$$

After expanding and canceling terms on each side of the equality, we obtain

$$-2a_1b_1 - 2a_2b_2 = -2\|\mathbf{a}\| \cdot \|\mathbf{b}\| \cdot \cos(\theta)$$

or

$$a_1b_1 + a_2b_2 = \|\mathbf{a}\| \cdot \|\mathbf{b}\| \cdot \cos(\theta)$$

The quantity $a_1b_1 + a_2b_2$ is called the *inner* (or *dot*) *product* of the vectors $\mathbf{a}$ and $\mathbf{b}$ and is denoted $\langle \mathbf{a}, \mathbf{b} \rangle$.

Thus, we have proved the following proposition.

(2.1.9) Proposition. If $\mathbf{a}$ and $\mathbf{b}$ are nonzero vectors in $\mathbf{R}^2$, the angle θ between $\mathbf{a}$ and $\mathbf{b}$ satisfies

$$\cos(\theta) = \frac{\langle \mathbf{a}, \mathbf{b} \rangle}{\|\mathbf{a}\| \cdot \|\mathbf{b}\|}$$

(2.1.10) Corollary. If $\mathbf{a} \neq \mathbf{0}$ and $\mathbf{b} \neq \mathbf{0}$ are vectors in $\mathbf{R}^2$, the angle θ between them is a right angle if and only if $\langle \mathbf{a}, \mathbf{b} \rangle = 0$.

Proof: Since $\pm\pi/2$ are the only angles (up to addition of integer multiples of π) for which the cosine is zero, the result follows immediately. ■

The length of a vector can be expressed in terms of the inner product by $\|\mathbf{v}\| = \sqrt{\langle \mathbf{v}, \mathbf{v} \rangle}$. Every vector $\mathbf{v}$ in the plane may be expressed in terms of the angle φ it makes with the first coordinate axis. Since $\cos(\varphi) = v_1/\|\mathbf{v}\|$ and $\sin(\varphi) = v_2/\|\mathbf{v}\|$, we have $\mathbf{v} = \|\mathbf{v}\| (\cos(\varphi), \sin(\varphi))$.

(2.1.11) Remark. For further properties of the inner product see Exercise 6. We will return to the inner product and the accompanying geometry in Chapters 4 and 5. The inner product plays a crucial role in linear algebra in that it provides a bridge between algebra and geometry, which is the heart of the more advanced material that appears later in the text. Here, we apply this geometry to two examples in the plane. These are important examples and we refer to them virtually every time we develop a new concept relating to linear transformations.

(2.1.12) Example. Rotation through an angle θ. Let $V = W = \mathbf{R}^2$, and let θ be a fixed real number that represents an angle in radians. Define a function R_θ: $V \rightarrow V$ by

$$R_\theta(\mathbf{v}) = \text{the vector obtained by rotating the vector } \mathbf{v} \text{ through an angle } \theta \text{ while preserving its length}$$

(Note that a positive angle is measured in a counterclockwise manner, whereas a negative angle is measured in a clockwise manner.)

If $\mathbf{w} = R_\theta(\mathbf{v})$, then the expression for $\mathbf{w}$ in terms of its length and the angle it makes with the first coordinate axis is

$$\mathbf{w} = \|\mathbf{v}\|(\cos(\varphi + \theta), \sin(\varphi + \theta))$$

where φ is the angle $\mathbf{v}$ makes with the first coordinate axis. See Figure 2.2.

Using the formulas for cosine and sine of a sum of angles, we obtain:

$$\begin{aligned}\mathbf{w} &= \|\mathbf{v}\|(\cos(\varphi) \cdot \cos(\theta) - \sin(\varphi) \cdot \sin(\theta), \cos(\varphi) \cdot \sin(\theta) \\ &\quad + \sin(\varphi) \cdot \cos(\theta)) \\ &= (v_1 \cos(\theta) - v_2 \sin(\theta), v_1 \sin(\theta) + v_2 \cos(\theta))\end{aligned}$$

Using this algebraic expression for $R_\theta(\mathbf{v})$ we can easily check that R_θ is a linear transformation. Again, we must verify (2.1.2), and the verification is similar to the

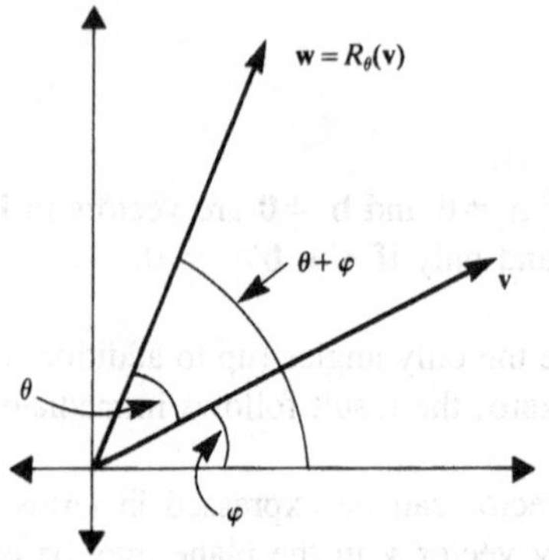

Figure 2.2

calculation of Example (2.1.5). Let a and $b \in \mathbf{R}$ and $\mathbf{u} = (u_1, u_2)$ and $\mathbf{v} = (v_1, v_2) \in \mathbf{R}^2$, then

$$\begin{aligned}
R_\theta(a\mathbf{u} + b\mathbf{v}) &= R_\theta((au_1 + bv_1, au_2 + bv_2)) \\
&= ((au_1 + bv_1)\cos(\theta) - (au_2 + bv_2)\sin(\theta), \\
&\quad (au_1 + bv_1)\sin(\theta) + (au_2 + bv_2)\cos(\theta)) \\
&= (a(u_1\cos(\theta) - u_2\sin(\theta)) + b(v_1\cos(\theta) - v_2\sin(\theta)), \\
&\quad a(u_1\sin(\theta) + u_2\cos(\theta)) + b(v_1\sin(\theta) + v_2\cos(\theta))) \\
&= a(u_1\cos(\theta) - u_2\sin(\theta), u_1\sin(\theta) + u_2\cos(\theta)) + \\
&\quad b(v_1\cos(\theta) - v_2\sin(\theta), v_1\sin(\theta) + v_2\cos(\theta)) \\
&= aR_\theta(\mathbf{u}) + bR_\theta(\mathbf{v}).
\end{aligned}$$

Therefore R_θ is a linear transformation.

(2.1.13) Example. Projection to a line in the plane. Let $V = W = \mathbf{R}^2$, let $\mathbf{a} = (a_1, a_2)$ be a nonzero vector in V, and let L denote the line spanned by the vector $\mathbf{a}$. The projection to L, denoted $P_\mathbf{a}$, is defined as follows:

Given $\mathbf{v} \in \mathbf{R}^2$, construct the line perpendicular to L passing through $\mathbf{v}$. This line intersects L in a point $\mathbf{w}$. $P_\mathbf{a}(\mathbf{v})$ is defined to be this vector $\mathbf{w}$.

$P_\mathbf{a}(\mathbf{v}) = \mathbf{w}$ is a multiple of the vector $\mathbf{a}$ since $\mathbf{w} \in L$. We can, therefore, write $\mathbf{w} = c\mathbf{a}$ for some scalar c. From plane geometry it follows that $\mathbf{v} - P_\mathbf{a}(\mathbf{v}) = \mathbf{v} - c\mathbf{a}$ must be perpendicular to the vector $\mathbf{a}$. See Figure 2.3. From the discussion of the law of cosines $P_\mathbf{a}(\mathbf{v})$ must satisfy

$$\langle \mathbf{a}, (\mathbf{v} - c\mathbf{a}) \rangle = 0$$

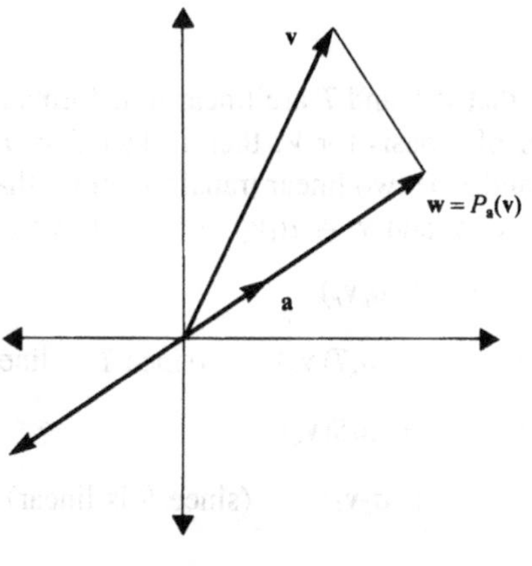

Figure 2.3

Therefore, $<\mathbf{a}, \mathbf{v}> = c \cdot <\mathbf{a}, \mathbf{a}>$ and $c = <\mathbf{a}, \mathbf{v}>/<\mathbf{a}, \mathbf{a}>$. Notice that since $\mathbf{a} \neq \mathbf{0}$, the denominator $<\mathbf{a}, \mathbf{a}> \neq 0$. Then

$$\begin{aligned} P_a(\mathbf{v}) &= c\mathbf{a} = \frac{<\mathbf{a}, \mathbf{v}>}{<\mathbf{a}, \mathbf{a}>}\,\mathbf{a} \\ &= \frac{(a_1v_1 + a_2v_2)}{(a_1^2 + a_2^2)}\,(a_1, a_2) \\ &= \frac{1}{(a_1^2 + a_2^2)}\,(a_1^2v_1 + a_1a_2v_2,\ a_1a_2v_1 + a_2^2v_2) \end{aligned}$$

To see that $P_\mathbf{a}$ is a linear transformation, we must verify (2.1.2). Since the calculation is similar to those in (2.1.5) and (2.1.12), we will leave it to the reader as an exercise.

Now let us return to the general theory of linear transformations. Let $T: V \rightarrow W$ be a linear transformation. If V is finite-dimensional, then we may choose a basis $\{\mathbf{v}_1, \ldots, \mathbf{v}_k\}$ for V. For each $\mathbf{v} \in V$ there is a unique choice of scalars $a_1, \ldots a_k$ so that $\mathbf{v} = a_1\mathbf{v}_1 + \cdots + a_k\mathbf{v}_k$. Then

$$\begin{aligned} T(\mathbf{v}) &= T(a_1\mathbf{v}_1 + \cdots + a_k\mathbf{v}_k) \\ &= T(a_1\mathbf{v}_1) + \cdots + T(a_k\mathbf{v}_k) \\ &= a_1T(\mathbf{v}_1) + \cdots + a_kT(\mathbf{v}_k) \end{aligned}$$

Therefore, if the values of T on the members of a basis for V are known, then all the values of T are known. In fact, we have Proposition (2.1.14).

(2.1.14) Proposition. If $T: V \rightarrow W$ is a linear transformation and V is finite-dimensional, then T is uniquely determined by its values on the members of a basis of V.

Proof: We will show that if S and T are linear transformations that take the same values on each member of a basis for V, then in fact $S = T$. Let $\{\mathbf{v}_1, \ldots, \mathbf{v}_k\}$ be a basis for V, and let S and T be two linear transformations that satisfy $T(\mathbf{v}_i) = S(\mathbf{v}_i)$ for $i = 1, \ldots, k$. If $\mathbf{v} \in V$ and $\mathbf{v} = a_1\mathbf{v}_1 + \cdots + a_k\mathbf{v}_k$, then

$$\begin{aligned} T(\mathbf{v}) &= T(a_1\mathbf{v}_1 + \cdots + a_k\mathbf{v}_k) \\ &= a_1T(\mathbf{v}_1) + \cdots + a_kT(\mathbf{v}_k) \qquad \text{(since } T \text{ is linear)} \\ &= a_1S(\mathbf{v}_1) + \cdots + a_kS(\mathbf{v}_k) \\ &= S(a_1\mathbf{v}_1 + \cdots + a_k\mathbf{v}_k) \qquad \text{(since } S \text{ is linear)} \\ &= S(\mathbf{v}) \end{aligned}$$

Therefore, S and T are equal as mappings from V to W. ■

We can turn this proposition around by saying that if $\{\mathbf{v}_1, \ldots, \mathbf{v}_k\}$ is a basis for V and $\{\mathbf{w}_1, \ldots, \mathbf{w}_k\}$ are any k vectors in W, then we can define a linear transformation $T: V \to W$ by insisting that $T(\mathbf{v}_i) = \mathbf{w}_i$, $i = 1, \ldots, k$ and that T satisfies the linearity condition

$$T(a_1\mathbf{u}_1 + a_2\mathbf{u}_2) = a_1T(\mathbf{u}_1) + a_2T(\mathbf{u}_2)$$

for all $\mathbf{u}_1$ and $\mathbf{u}_2 \in V$ and a_1 and $a_2 \in \mathbf{R}$. For an arbitrary $\mathbf{v} \in V$ expressed as $\mathbf{v} = a_1\mathbf{v}_1 + \cdots + a_k\mathbf{v}_k$, $T(\mathbf{v}) = a_1T(\mathbf{v}_1) + \cdots + a_kT(\mathbf{v}_k) = a_1\mathbf{w}_1 + \cdots + a_k\mathbf{w}_k$ defines T. Since the expression $\mathbf{v} = a_1\mathbf{v}_1 + \cdots + a_k\mathbf{v}_k$ in terms of the basis $\{\mathbf{v}_1, \ldots, \mathbf{v}_k\}$ is unique, $T(\mathbf{v})$ is specified uniquely.

We end this section with three examples related to Proposition (2.1.14).

(2.1.15) Example. Define $T: \mathbf{R}^2 \to \mathbf{R}^3$ by specifying $T((1, 0)) = (a_1, a_2, a_3)$ and $T((0, 1)) = (b_1, b_2, b_3)$ and insisting that T be linear. Then

$$\begin{aligned} T((v_1, v_2)) &= T(v_1(1, 0) + v_2(0, 1)) \\ &= v_1T((1, 0)) + v_2T((0, 1)) \\ &= v_1(a_1, a_2, a_3) + v_2(b_1, b_2, b_3) \\ &= (v_1a_1 + v_2b_1, v_1a_2 + v_2b_2, v_1a_3 + v_2b_3) \end{aligned}$$

(2.1.16) Example. Let $V = W = \mathbf{R}^2$, and let $\mathbf{u}_1 = (1, 1)$ and $\mathbf{u}_2 = (1, 2)$. The vectors $\mathbf{u}_1$ and $\mathbf{u}_2$ form a basis for V. Then specifying that $T(\mathbf{u}_1) = (1, 0)$ and $T(\mathbf{u}_2) = (3, 0)$ and requiring that T be linear defines a linear transformation $T: V \to W$. In order to evaluate $T(\mathbf{v})$ for an arbitrary vector $\mathbf{v}$, $\mathbf{v}$ must be expressed in terms of the basis vectors. A calculation shows that if $\mathbf{v} = (v_1, v_2)$, then

$$\mathbf{v} = (2v_1 - v_2)\mathbf{u}_1 + (-v_1 + v_2)\mathbf{u}_2$$

A further calculation along the lines of the previous example shows that

$$T(\mathbf{v}) = (-v_1 + 2v_2, 0)$$

(2.1.17) Example. Let $V = W = P_2(\mathbf{R})$. A basis for V is given by the polynomials 1, $1 + x$, and $1 + x + x^2$. Define T on this basis by $T(1) = x$, $T(1 + x) = x^2$, and $T(1 + x + x^2) = 1$. If we insist that T be linear, this defines a linear transformation. If $p(x) = a_2x^2 + a_1x + a_0$, then the equation

$$p(x) = (a_0 - a_1)1 + (a_1 - a_2)(1 + x) + a_2(1 + x + x^2)$$

expresses $p(x)$ in terms of the basis. Calculating as in the previous examples, we see that $T(p(x)) = (a_1 - a_2)x^2 + (a_0 - a_1)x + a_2$.

EXERCISES

1. Prove Proposition (2.1.2).
2. Prove Corollary (2.1.3).

3. Determine whether the following functions $T: V \rightarrow W$ define linear transformations.
   a) $V = \mathbf{R}^2$ and $W = \mathbf{R}^4$. $T((a_1, a_2)) = (a_2, a_1, a_1, a_2)$
   b) $V = \mathbf{R}^3$ and $W = \mathbf{R}^3$. $T((a_1, a_2, a_3)) = (1 + a_1, 2a_2, a_3 - a_2)$
   c) $V = \mathbf{R}^3$ and $W = \mathbf{R}$. $T((a_1, a_2, a_3)) = 3a_1 + 2a_2 + a_3$
   d) $V = \mathbf{R}^2$ and $W = \mathbf{R}^2$. $T((a_1, a_2)) = (\sin(a_1 + a_2), \cos(a_1 + a_2))$
   e) $V = \mathbf{R}^2$ and $W = \mathbf{R}^2$. Let (b_1, b_2) and (c_1, c_2) be fixed vectors in $\mathbf{R}^2$. Define $T(a_1, a_2) = (a_1b_1 + a_2c_1, a_1b_2 + a_2c_2)$.
   f) $V = \mathbf{R}^3$ and $W = \mathbf{R}$. $T((a_1, a_2, a_3)) = e^{a_1+a_2+a_3}$
   g) $V = \mathbf{R}$ with the usual vector space structure and $W = \mathbf{R}^+$ with the vector space structure given in Example (1.1.4c). $T(x) = e^x$.
   h) $V = \mathbf{R}^+$ with the vector space structure given in Example (1.1.4c) and $W = \mathbf{R}$ with the usual vector space structure. $T(x) = x$.

4. Prove that if $\mathbf{a} = (a_1, \ldots, a_n)$ is a fixed vector in $\mathbf{R}^n$, then $T(\mathbf{x}) = a_1x_1 + \cdots + a_nx_n$ defines a linear transformation $T: \mathbf{R}^n \rightarrow \mathbf{R}$.

5. a) Let $V = C^\infty(\mathbf{R})$, and let $D: V \rightarrow V$ be the mapping $D(f) = f'$. Show that D is a linear mapping. [See Example (2.1.8a).]
   b) Let $V = C[a,b]$, $W = \mathbf{R}$, and let Int: $V \rightarrow W$ be the mapping $\text{Int}(f) = \int_a^b f(x)\,dx$. Show that Int is a linear mapping. [See Example (2.1.8b).
   c) Let $V = C^\infty(\mathbf{R})$, and let Int: $V \rightarrow V$ be the mapping $\text{Int}(f) = \int_0^x f(t)\,dt$ [i.e., Int(f) is the antiderivative of f that takes the value 0 at $x = 0$]. Show that Int is a linear mapping. (Note that this is similar to but not the same as 5b.)

6. Let $\langle \mathbf{v}, \mathbf{w} \rangle = v_1w_1 + v_2w_2$ be the inner product on $\mathbf{R}^2$ defined in the text. Prove the following for all vectors $\mathbf{u}, \mathbf{v}, \mathbf{w} \in \mathbf{R}^2$ and scalars $a, b \in \mathbf{R}$.
   a) $\langle (a\mathbf{u} + b\mathbf{v}), \mathbf{w} \rangle = a\langle \mathbf{u}, \mathbf{w} \rangle + b\langle \mathbf{v}, \mathbf{w} \rangle$
   b) $\langle \mathbf{v}, \mathbf{w} \rangle = \langle \mathbf{w}, \mathbf{v} \rangle$
   c) $\langle \mathbf{v}, \mathbf{v} \rangle \geq 0$ and $\langle \mathbf{v}, \mathbf{v} \rangle = 0$ if and only if $\mathbf{v} = \mathbf{0}$.

7. Which of the following pairs of vectors in $\mathbf{R}^2$ are perpendicular?
   a) $(3, 0)$ and $(0, 4)$
   b) $(4, 2)$ and $(-1, 3)$
   c) $(-5, 8)$ and $(16, 10)$.

8. a) Write out the equations that define $R_{\pi/4}$ and $R_{-\pi/4}$.
   b) If $\mathbf{v}$ is a vector in $\mathbf{R}^2$ and $\mathbf{w} = R_{\pi/4}(\mathbf{v})$, what is $R_{-\pi/4}(\mathbf{w})$?
   c) If θ is any angle and $\mathbf{w} = R_\theta(\mathbf{v})$, what is $R_{-\theta}(\mathbf{w})$?

9. Let $\mathbf{a} = (1, 1)$ and $\mathbf{b} = (1, -1)$.
   a) What are the equations for $P_\mathbf{a}(\mathbf{v})$ and $P_\mathbf{b}(\mathbf{v})$?
   b) If $\mathbf{v} \in \mathbf{R}^2$ and $\mathbf{w} = P_\mathbf{a}(\mathbf{v})$, what is $P_\mathbf{b}(\mathbf{w})$?
   c) If $\mathbf{v} \in \mathbf{R}^2$ and $\mathbf{w} = P_\mathbf{a}(\mathbf{v})$, what is $P_\mathbf{a}(\mathbf{w})$?

10. Let $V = \mathbf{R}^3$ and $W = \mathbf{R}^2$. Define T on the basis $\mathbf{v}_1 = (1, 0, 1)$, $\mathbf{v}_2 = (0, 1, 1)$ and $\mathbf{v}_3 = (1, 1, 0)$ by $T(\mathbf{v}_1) = (1, 0)$, $T(\mathbf{v}_2) = (0, 1)$ and $T(\mathbf{v}_3) = (1, 1)$.
    a) Express a vector $\mathbf{x} \in V$ as a linear combination of the vectors $\mathbf{v}_1$, $\mathbf{v}_2$, and $\mathbf{v}_3$.
    b) What is $T(\mathbf{x})$?

11. Let $V = \mathbf{R}^2$ and $W = P_3(\mathbf{R})$. If T is a linear transformation that satisfies

    $T((1, 1)) = x + x^2$ and $T((3, 0)) = x - x^3$, what is $T((2, 2))$?

12. Let V and W be vector spaces, and let $S, T: V \rightarrow W$ be two linear transformations.
    a) Define a function $R: V \rightarrow W$ by $R(\mathbf{v}) = S(\mathbf{v}) + T(\mathbf{v})$. Prove that R is a linear transformation. (R is called the *sum* of S and T and is denoted $S + T$.)
    b) Define a function $R: V \rightarrow W$ by $R(\mathbf{v}) = aS(\mathbf{v})$ for a fixed real number a. Prove that R is a linear transformation. (R is called a *scalar multiple* of S and is denoted aS.)
    c) Show that the set of all linear transformations from V to W, which we denote by $L(V, W)$, is a real vector space (with the two operations defined in parts a and b of this problem). [*Hint:* The additive identity ("zero vector") in $L(V, W)$ is the zero transformation defined in Example (2.1.4b).]

13. Show that if $\mathbf{a}$ and $\mathbf{b}$ are the vectors in Exercise 9, then $P_{\mathbf{a}} + P_{\mathbf{b}} = I$, the identity mapping on $\mathbf{R}^2$.

14. Let $T: V \rightarrow W$ be a linear transformation and let $U \subset V$ be a vector subspace. Define a function $S: U \rightarrow W$ by $S(\mathbf{u}) = T(\mathbf{u})$. Prove that S is a linear transformation. (S is called the restriction of T to the subspace U and is usually denoted $T|_U$.)

15. Let U be the subspace of $\mathbf{R}^3$ spanned by $\mathbf{u}_1 = (1, 1, 1)$ and $\mathbf{u}_2 = (2, 1, 0)$. If $\mathbf{u} \in U$ is the vector $x_1\mathbf{u}_1 + x_2\mathbf{u}_2$, find a formula for $T|_U(\mathbf{u})$, where $T: \mathbf{R}^3 \rightarrow \mathbf{R}$ is the linear transformation of Exercise 3c.

§2.2. LINEAR TRANSFORMATIONS BETWEEN FINITE-DIMENSIONAL VECTOR SPACES

This section builds on the work of the previous section. We give a complete description of a linear transformation $T: V \rightarrow W$ if both V and W are finite-dimensional. This description will rely on the fact that in a finite-dimensional space, every vector can be expressed uniquely as a linear combination of the vectors in a finite set. We use this fact in both V and W.

Let $\dim(V) = k$ and $\dim(W) = l$, and let $\{\mathbf{v}_1, \ldots, \mathbf{v}_k\}$ be a basis for V and $\{\mathbf{w}_1, \ldots, \mathbf{w}_l\}$ be a basis for W. By Theorem (1.6.3) each vector $T(\mathbf{v}_j)$, $j = 1, \ldots, k$, can be expressed uniquely in terms of $\{\mathbf{w}_1, \ldots, \mathbf{w}_l\}$. For each vector $T(\mathbf{v}_j)$ there are l scalars:

$$a_{1j}, a_{2j}, \ldots, a_{lj} \in \mathbf{R}$$

such that

$$T(\mathbf{v}_j) = a_{1j}\mathbf{w}_1 + \cdots + a_{lj}\mathbf{w}_l$$

Since T is determined by its values on the basis $\{\mathbf{v}_1, \ldots, \mathbf{v}_k\}$ and each $T(\mathbf{v}_j)$ is determined by the l scalars $a_{1j}, \ldots, a_{lj}$, the transformation T is determined by the $l \cdot k$ scalars a_{ij} with $i = 1, \ldots, l$ and $j = 1, \ldots, k$. That is, if $\mathbf{v} = x_1\mathbf{v}_1 + \cdots + x_k\mathbf{v}_k$, then $T(\mathbf{v}) = T(x_1\mathbf{v}_1 + \cdots + x_k\mathbf{v}_k) = x_1T(\mathbf{v}_1) + \cdots + x_kT(\mathbf{v}_k) = x_1(a_{11}\mathbf{w}_1 + \cdots + a_{l1}\mathbf{w}_l) + \cdots + x_k(a_{1k}\mathbf{w}_1 + \cdots + a_{lk}\mathbf{w}_l)$, which specifies the mapping T completely.

(2.2.1) Proposition. Let $T: V \to W$ be a linear transformation between the finite-dimensional vector spaces V and W. If $\{\mathbf{v}_1, \ldots, \mathbf{v}_k\}$ is a basis for V and $\{\mathbf{w}_1, \ldots, \mathbf{w}_l\}$ is a basis for W, then $T: V \to W$ is uniquely determined by the $l \cdot k$ scalars used to express $T(\mathbf{v}_j)$, $j = 1, \ldots, k$, in terms of $\mathbf{w}_1, \ldots, \mathbf{w}_l$.

Proof: Exercise 1. ■

Here are three explicit examples of linear transformations specified in this way.

(2.2.2) Example. Let $V = W = \mathbf{R}^2$. Choose the standard basis: $\mathbf{e}_1 = (1, 0)$ and $\mathbf{e}_2 = (0, 1)$ for both V and W. Define T by $T(\mathbf{e}_1) = \mathbf{e}_1 + \mathbf{e}_2$ and $T(\mathbf{e}_2) = 2\mathbf{e}_1 - 2\mathbf{e}_2$. The four scalars $a_{11} = 1$, $a_{21} = 1$, $a_{12} = 2$, and $a_{22} = -2$ determine T.

(2.2.3) Example. Let $V = W = P_2(\mathbf{R})$. Choose the basis $p_0(x) = 1$, $p_1(x) = x$, and $p_2(x) = x^2$ for both V and W. Define T by $T(p(x)) = x\,\dfrac{d}{dx}\,p(x)$. Then $T(p_0(x)) = 0$, $T(p_1(x)) = p_1(x)$, and $T(p_2(x)) = 2p_2(x)$. Thus, T is determined by the scalars a_{ij}, $0 \leq i, j \leq 2$, with $a_{ij} = 0$ if $i \neq j$ and $a_{00} = 0$, $a_{11} = 1$, and $a_{22} = 2$.

(2.2.4) Example. Let $V = \mathrm{Span}(\{\mathbf{v}_1, \mathbf{v}_2\}) \subset \mathbf{R}^3$, where $\mathbf{v}_1 = (1, 1, 0)$ and $\mathbf{v}_2 = (0, 1, 1)$, and let $W = \mathbf{R}^3$ with the standard basis. Define T by $T(\mathbf{v}_1) = \mathbf{e}_1 + \mathbf{e}_2$ and $T(\mathbf{v}_2) = \mathbf{e}_2$. Then T is determined by the six scalars $a_{11} = 1$, $a_{21} = 1$, $a_{31} = 0$, $a_{12} = 0$, $a_{22} = 1$, and $a_{32} = 0$. Can you describe T geometrically?

In order to simplify the discussion of transformations once we have chosen bases in V and W, we now introduce the concept of a matrix of scalars. For the moment, this may be thought of as a way to organize the information given by Proposition (2.2.1). However, we will shortly see how matrices can be used to evaluate a linear transformation on a vector and perform other algebraic operations on linear transformations. The following definition is the same as the one in the exercises accompanying Section 1.2.

(2.2.5) Definition. Let a_{ij}, $1 \leq i \leq l$ and $1 \leq j \leq k$ be $l \cdot k$ scalars. The *matrix whose entries are the scalars* a_{ij} is the rectangular array of l rows and k columns:

$$\begin{bmatrix} a_{11} & a_{12} & a_{13} & \cdots & a_{1k} \\ a_{21} & a_{22} & a_{23} & \cdots & a_{2k} \\ & \vdots & & & \\ a_{l1} & a_{l2} & a_{l3} & \cdots & a_{lk} \end{bmatrix}$$

Thus, the scalar a_{ij} is the entry in the ith row and the jth column of the array. A matrix with l rows and k columns will be called an $l \times k$ matrix ("l by k matrix"). Matrices are usually denoted by capital letters, $A, B, C, \ldots$.

If we begin with a linear transformation between finite-dimensional vector spaces V and W as in the beginning of this section, we have shown that the transformation is determined by the choice of bases in V and W and a set of $l \cdot k$ scalars, where $k = \dim(V)$ and $l = \dim(W)$. If we write these scalars in the form of a matrix, we have a concise way to work with the transformation.

(2.2.6) Definition. Let $T: V \rightarrow W$ be a linear transformation between the finite-dimensional vector spaces V and W, and let $\alpha = \{\mathbf{v}_1, \ldots, \mathbf{v}_k\}$ and $\beta = \{\mathbf{w}_1, \ldots, \mathbf{w}_l\}$, respectively, be any bases for V and W. Let a_{ij}, $1 \leq i \leq l$ and $1 \leq j \leq k$ be the $l \cdot k$ scalars that determine T with respect to the bases α and β. The matrix whose entries are the scalars a_{ij}, $1 \leq i \leq l$ and $1 \leq j \leq k$, is called the *matrix of the linear transformation* T *with respect to the bases* α *for* V *and* β *for* W. This matrix is denoted by $[T]_\alpha^\beta$.

We see that if $T(\mathbf{v}_j) = a_{1j}\mathbf{w}_1 + \cdots + a_{lj}\mathbf{w}_l$, then the coefficients expressing $T(\mathbf{v}_j)$ in terms of the $\mathbf{w}_1, \ldots, \mathbf{w}_l$ form the jth column of $[T]_\alpha^\beta$. We use this observation in the following examples.

This observation about the columns of the matrix $[T]_\alpha^\beta$ also illustrates the fact that in forming the matrix of a linear transformation, we will always be assuming that the basis vectors in the domain and target spaces are written in some particular order. If we rearranged the vectors in the basis α for V in some way to form what must be viewed as another (ordered) basis α', then we would obtain a matrix $[T]_{\alpha'}^\beta$ whose columns were the corresponding rearrangement of the columns of $[T]_\alpha^\beta$. Similarly, the reader may check that if we rearranged the vectors in the basis β for W, the rows of the matrix would be permuted correspondingly.

(2.2.7) Example. Let $T: V \rightarrow V$ be the identity transformation of a finite-dimensional vector space to itself, $T = I$. Then with respect to any choice of basis α for V, the matrix of I is the $k \times k$ matrix with 1 in each diagonal position and 0 in each off-diagonal position, that is

$$[I]_\alpha^\alpha = \begin{bmatrix} 1 & & & & & \\ & 1 & & & & \\ & & \cdot & & & \\ & & & \cdot & & \\ & & & & \cdot & \\ & & & & & 1 \\ & & & & & & 1 \end{bmatrix}$$

(In writing matrices with large numbers of zero entries, we often omit the zeros as a matter of convenience.)

Using the notation of these definitions, we obtain the matrices of Examples (2.2.2), (2.2.3), and (2.2.4), respectively:

(2.2.2) $\begin{bmatrix} 1 & 2 \\ 1 & -2 \end{bmatrix}$

(2.2.3) $\begin{bmatrix} 0 & 0 & 0 \\ 0 & 1 & 0 \\ 0 & 0 & 2 \end{bmatrix}$

(2.2.4) $\begin{bmatrix} 1 & 0 \\ 1 & 1 \\ 0 & 0 \end{bmatrix}$

(2.2.8) Example. The matrix of a rotation. [See Example (2.1.12).] In this example $V = W = \mathbf{R}^2$, and we take both bases α and β to be the standard basis: $\mathbf{e}_1 = (1, 0)$ and $\mathbf{e}_2 = (0, 1)$. Let $T = R_\theta$ be rotation through an angle θ in the plane. Then for an arbitrary vector $\mathbf{v} = (v_1, v_2)$

$$R_\theta(\mathbf{v}) = (v_1 \cos(\theta) - v_2 \sin(\theta), v_1 \sin(\theta) + v_2 \cos(\theta))$$

Thus, $R_\theta(\mathbf{e}_1) = (\cos(\theta), \sin(\theta)) = \cos(\theta)\mathbf{e}_1 + \sin(\theta)\mathbf{e}_2$ and $R_\theta(\mathbf{e}_2) = -\sin(\theta)\mathbf{e}_1 + \cos(\theta)\mathbf{e}_2$. Therefore, the matrix of R_θ is

$$[R_\theta]_\alpha^\alpha = \begin{bmatrix} \cos(\theta) & -\sin(\theta) \\ \sin(\theta) & \cos(\theta) \end{bmatrix}$$

where α is the standard basis in $\mathbf{R}^2$.

(2.2.9) Example. The matrix of a projection in the plane. [See Example (2.1.13).] In this example $V = W = \mathbf{R}^2$ and both bases α and β are taken to be the standard basis. Let $\mathbf{a} = (a_1, a_2)$ be a fixed nonzero vector. Let $T = P_\mathbf{a}$ be the projection to the line spanned by the vector $\mathbf{a}$. Then for an arbitrary vector $\mathbf{v} = (v_1, v_2)$

$$P_\mathbf{a}(\mathbf{v}) = \|\mathbf{a}\|^{-2}(a_1^2 v_1 + a_1 a_2 v_2, a_1 a_2 v_1 + a_2^2 v_2)$$

Thus

$$P_{\mathbf{a}}(\mathbf{e}_1) = \|\mathbf{a}\|^{-2}a_1^2\mathbf{e}_1 + \|\mathbf{a}\|^{-2}a_1a_2\mathbf{e}_2$$

and

$$P_{\mathbf{a}}(\mathbf{e}_2) = \|\mathbf{a}\|^{-2}a_1a_2\mathbf{e}_1 + \|\mathbf{a}\|^{-2}a_2^2\mathbf{e}_2$$

Therefore, the matrix of $P_{\mathbf{a}}$ with respect to the standard basis is

$$[P_{\mathbf{a}}]_\alpha^\alpha = \|\mathbf{a}\|^{-2}\begin{bmatrix} a_1^2 & a_1a_2 \\ a_1a_2 & a_2^2 \end{bmatrix}$$

(The scalar $\|\mathbf{a}\|^{-2}$ multiplies each entry of the matrix. We use this notation on occasion to simplify writing matrices, all of whose entries contain some fixed scalar factor. Also see Exercise 10 in this section.)

If $\mathbf{v} = a_1\mathbf{v}_1 + \cdots + a_k\mathbf{v}_k$ and $\mathbf{w} = b_1\mathbf{w}_1 + \cdots + b_l\mathbf{w}_l$, we can express $\mathbf{v}$ and $\mathbf{w}$ in coordinates, respectively, as a $k \times 1$ matrix and as an $l \times 1$ matrix, that is, as *column vectors*. We call these the *coordinate vectors* of the given vectors with respect to the chosen bases. These coordinate vectors will be denoted by $[\mathbf{v}]_\alpha$ and $[\mathbf{w}]_\beta$, respectively. Thus

$$[\mathbf{v}]_\alpha = \begin{bmatrix} a_1 \\ \cdot \\ \cdot \\ \cdot \\ a_k \end{bmatrix} \quad \text{and} \quad [\mathbf{w}]_\beta = \begin{bmatrix} b_1 \\ \cdot \\ \cdot \\ \cdot \\ b_l \end{bmatrix}$$

In order to use the matrix $[T]_\alpha^\beta$ to express $T(\mathbf{v})$ in the coordinates of the basis β for W given the basis α for V, we need to introduce the notion of multiplication of a vector by a matrix.

(2.2.10) Definition. Let A be an $l \times k$ matrix, and let $\mathbf{x}$ be a column vector with k entries, then the *product of the vector* $\mathbf{x}$ *by the matrix* A is defined to be the column vector with l entries:

$$\begin{bmatrix} a_{11}x_1 + a_{12}x_2 + \cdots + a_{1k}x_k \\ \cdot \\ \cdot \\ \cdot \\ a_{l1}x_1 + a_{l2}x_2 + \cdots + a_{lk}x_k \end{bmatrix}$$

and is denoted by $A\mathbf{x}$. If we write out the entire matrix A and the vector $\mathbf{x}$, this becomes

$$\begin{bmatrix} a_{11} & a_{12} & a_{13} & \cdots & a_{1k} \\ a_{21} & a_{22} & a_{23} & \cdots & a_{2k} \\ & & \cdot & & \\ & & \cdot & & \\ & & \cdot & & \\ a_{l1} & a_{l2} & a_{l3} & \cdots & a_{lk} \end{bmatrix} \begin{bmatrix} x_1 \\ \cdot \\ \cdot \\ \cdot \\ x_k \end{bmatrix}$$

(2.2.11) Remark. The ith entry of the product $A\mathbf{x}$, $a_{i1}x_1 + \cdots + a_{ik}x_k$, can be thought of as the product of the ith row of A, considered as a $1 \times k$ matrix, with the column vector $\mathbf{x}$, using this same definition.

(2.2.12) Remark. The product of a $1 \times k$ matrix, which we can think of as a row vector, and a column vector generalizes the notion of the dot product in the plane. Recall if $\mathbf{x}$ and $\mathbf{y} \in \mathbf{R}^2$, $<\mathbf{x}, \mathbf{y}> = x_1y_1 + x_2y_2$. If we write $\mathbf{x}$ as a matrix $[x_1\ x_2]$, then $[x_1\ x_2]\begin{bmatrix} y_1 \\ y_2 \end{bmatrix} = <\mathbf{x}, \mathbf{y}>$.

(2.2.13) Remark. If the number of columns of the matrix A is not equal to the number of entries in the column vector $\mathbf{x}$, matrix multiplication $A\mathbf{x}$ is not defined.

(2.2.14) Example. The reader should verify the following calculations:

a) $$\begin{bmatrix} 1 & 0 & 3 & 2 \\ 2 & 5 & 0 & 1 \\ 2 & -1 & 1 & 9 \end{bmatrix}\begin{bmatrix} 1 \\ 0 \\ 2 \\ -1 \end{bmatrix} = \begin{bmatrix} 5 \\ 1 \\ -5 \end{bmatrix}$$

b) $$\begin{bmatrix} 1 & 0 \\ 0 & 2 \\ 1 & 1 \end{bmatrix}\begin{bmatrix} 3 \\ 2 \end{bmatrix} = \begin{bmatrix} 3 \\ 4 \\ 5 \end{bmatrix}$$

Matrix multiplication allows us to evaluate linear transformations between finite-dimensional vector spaces in a concise manner. To evaluate $T(\mathbf{v})$, we (1) compute the matrix of T with respect to given bases for V and W, (2) express the given vector $\mathbf{v}$ in terms of the basis for V, and (3) multiply the coordinate vector of $\mathbf{v}$ by the matrix of T to obtain a second coordinate vector. The result of these steps will be a coordinate vector with l coordinates that when interpreted as the coordinates of a vector in W with respect to the chosen basis of W, is precisely the coordinate vector of $T(\mathbf{v}) \in W$. The following proposition is a rigorous statement of this fact.

(2.2.15) Proposition. Let $T: V \rightarrow W$ be a linear transformation between vector spaces V of dimension k and W of dimension l. Let $\alpha = \{\mathbf{v}_1, \ldots, \mathbf{v}_k\}$ be a basis for V and $\beta = \{\mathbf{w}_1, \ldots, \mathbf{w}_l\}$ be a basis for W. Then for each $\mathbf{v} \in V$

$$[T(\mathbf{v})]_\beta = [T]_\alpha^\beta[\mathbf{v}]_\alpha$$

Proof: Let $\mathbf{v} = x_1\mathbf{v}_1 + \cdots + x_k\mathbf{v}_k \in V$. Then if $T(\mathbf{v}_j) = a_{1j}\mathbf{w}_1 + \cdots + a_{lj}\mathbf{w}_l$

$$\begin{aligned} T(\mathbf{v}) &= \sum_{j=1}^{k} x_j T(\mathbf{v}_j) \\ &= \sum_{j=1}^{k} x_j \left(\sum_{i=1}^{l} a_{ij}\mathbf{w}_i \right) \\ &= \sum_{i=1}^{l} \left(\sum_{j=1}^{k} x_j a_{ij} \right) \mathbf{w}_i \end{aligned}$$

Thus, the ith coefficient of $T(\mathbf{v})$ in terms of β is $\sum_{j=1}^{k} x_j a_{ij}$ and

$$[T(\mathbf{v})]_\beta = \begin{bmatrix} \sum_{j=1}^{k} x_j a_{1j} \\ \cdot \\ \cdot \\ \cdot \\ \sum_{j=1}^{k} x_j a_{lj} \end{bmatrix}$$

which is precisely $[T]_\alpha^\beta[\mathbf{v}]_\alpha$. ■

(2.2.16) Remark. If $\mathbf{v}_j$ is the jth member of the basis α of V, the coordinate vector of $\mathbf{v}_j$ with respect to the basis α is $(0, \ldots, 1, \ldots, 0)$, the coordinate vector that has a 1 in the jth position and 0 in the remaining positions. The coordinate vector of $T(\mathbf{v}_j)$, $[T(\mathbf{v}_j)]_\beta$, is

$$[T]_\alpha^\beta\,[\mathbf{v}_j]_\alpha = \begin{bmatrix} a_{11} & a_{12} & a_{13} & \cdots & a_{1k} \\ a_{21} & a_{22} & a_{23} & \cdots & a_{2k} \\ & & \vdots & & \\ a_{l1} & a_{l2} & a_{l3} & \cdots & a_{lk} \end{bmatrix} \begin{bmatrix} 0 \\ \vdots \\ 1 \\ \vdots \\ 0 \end{bmatrix} = \begin{bmatrix} a_{1j} \\ \vdots \\ a_{lj} \end{bmatrix}$$

which is the jth column of the matrix $[T]_\alpha^\beta$. This is another way of stating the description of the columns of $[T]_\alpha^\beta$ given after Definition (2.2.6).

(2.2.17) Example. If T is the transformation of Example (2.2.2) and $\mathbf{v} = x_1\mathbf{e}_1 + x_2\mathbf{e}_2$ is any vector in $\mathbf{R}^2$, then to evaluate $T(\mathbf{v})$ it suffices to perform the multiplication

$$\begin{bmatrix} 1 & 2 \\ 1 & -2 \end{bmatrix}\begin{bmatrix} x_1 \\ x_2 \end{bmatrix} = \begin{bmatrix} x_1 + 2x_2 \\ x_1 - 2x_2 \end{bmatrix}$$

For example, if $\mathbf{v} = 3\mathbf{e}_1 - 2\mathbf{e}_2$, $T(\mathbf{v})$ corresponds to the coordinate vector

$$\begin{bmatrix} 1 & 2 \\ 1 & -2 \end{bmatrix}\begin{bmatrix} 3 \\ -2 \end{bmatrix} = \begin{bmatrix} -1 \\ 7 \end{bmatrix}$$

The preceding procedure, which associates a matrix to a transformation $T: V \rightarrow W$ once a choice of bases in V and W has been made, can be reversed. That is, given a matrix of the correct size, $l \times k$, and choices of bases for V and W, we can associate a linear transformation to the matrix. However, in order to ensure that we obtain a linear transformation by this procedure, we must first verify that multiplication of a vector by a matrix is a linear function of the vector.

(2.2.18) Proposition. Let A be an $l \times k$ matrix and $\mathbf{u}$ and $\mathbf{v}$ be column vectors with k entries. Then for every pair of real numbers a and b

$$A(a\mathbf{u} + b\mathbf{v}) = aA\mathbf{u} + bA\mathbf{v}$$

Proof: Exercise 2. ■

It should be noted (as we did in our first discussion of linear transformations) that the addition and scalar multiplication that occur on the left-hand side of the preceding equation involve column vectors with k entries, whereas the addition and scalar multiplication that occur on the right-hand side of the preceding equation involve column vectors with l entries.

Now we can use a matrix to define a linear transformation. Let A be a fixed $l \times k$ matrix with entries a_{ij}, $1 \leq i \leq l$ and $1 \leq j \leq k$. Let $\alpha = \{\mathbf{v}_1, \ldots, \mathbf{v}_k\}$ be a basis for V and $\beta = \{\mathbf{w}_1, \ldots, \mathbf{w}_l\}$ be a basis for W. We define a function $T: V \rightarrow W$ as follows:

For each vector $\mathbf{v} = x_1\mathbf{v}_1 + \cdots + x_k\mathbf{v}_k$ define $T(\mathbf{v})$ to be the vector in W whose coordinate vector in the β coordinates is $A[\mathbf{v}]_\alpha$, that is, $[T(\mathbf{v})]_\beta = A[\mathbf{v}]_\alpha$ so that

$$T(\mathbf{v}) = \sum_{i=1}^{l} \left(\sum_{j=1}^{k} x_j a_{ij} \right) \mathbf{w}_i$$

Not only does this define a linear transformation, but if we start with the matrix of a linear transformation and perform this construction, we arrive back at the original transformation. Similarly, if we construct the matrix of a transformation that in turn was constructed from a matrix, we arrive back at the original matrix.

(2.2.19) Proposition. Let $\alpha = \{\mathbf{v}_1, \ldots, \mathbf{v}_k\}$ be a basis for V and $\beta = \{\mathbf{w}_1, \ldots, \mathbf{w}_l\}$ be a basis for W, and let $\mathbf{v} = x_1\mathbf{v}_1 + \cdots + x_k\mathbf{v}_k \in V$.

(i) If A is an $l \times k$ matrix, then the function

$$T(\mathbf{v}) = \mathbf{w}$$

where $[\mathbf{w}]_\beta = A[\mathbf{v}]_\alpha$ is a linear transformation.

(ii) If $A = [S]_\alpha^\beta$ is the matrix of a transformation $S: V \rightarrow W$, then the transformation T constructed from $[S]_\alpha^\beta$ is equal to S.

(iii) If T is the transformation of (i) constructed from A, then

$$[T]_\alpha^\beta = A.$$

Proof: (i) We apply (2.2.18). For a and $b \in \mathbf{R}$ and $\mathbf{u}$ and $\mathbf{v} \in V$, $T(a\mathbf{u} + b\mathbf{v}) = \mathbf{w}$, where $[\mathbf{w}]_\beta = A(a[\mathbf{u}]_\alpha + b[\mathbf{v}]_\alpha) = aA[\mathbf{u}]_\alpha + bA[\mathbf{v}]_\alpha$. Therefore, $\mathbf{w} = aT(\mathbf{u}) + bT(\mathbf{v})$, and we conclude that $T(a\mathbf{u} + b\mathbf{v}) = aT(\mathbf{u}) + bT(\mathbf{v})$.

(ii) It suffices to prove that $S(\mathbf{v})$ and $T(\mathbf{v})$ have the same coordinate vectors. We have that $[T(\mathbf{v})]_\beta = A[\mathbf{v}]_\alpha$ from the definition of T. But $A = [S]_\alpha^\beta$ so that $A[\mathbf{v}]_\alpha = [S]_\alpha^\beta [\mathbf{v}]_\alpha$, which by Proposition (2.2.15) is $[S(\mathbf{v})]_\beta$. Therefore, $[T(\mathbf{v})]_\beta = [S(\mathbf{v})]_\beta$, and we conclude that $T = S$.

(iii) The jth column of $[T]_\alpha^\beta$ is the coordinate vector whose entries are the coefficients of $T(\mathbf{v}_j)$ expressed in terms of β. But by definition $T(\mathbf{v}_j) =$

$a_{1j}\mathbf{w}_1 + \cdots + a_{lj}\mathbf{w}_l$, where $\begin{bmatrix} a_{1j} \\ \vdots \\ a_{lj} \end{bmatrix}$ is the jth column of A. Therefore, for each j, the jth columns of $[T]_\alpha^\beta$ and A are equal so that $[T]_\alpha^\beta = A$. ■

Proposition (2.2.19) can be rephrased as follows.

(2.2.20) Proposition. Let V and W be finite-dimensional vector spaces. Let α be a basis for V and β a basis for W. Then the assignment of a matrix to a linear transformation from V to W given by T goes to $[T]_\alpha^\beta$ is injective and surjective (see Appendix 1 for definitions of these terms). (The proof follows immediately from (2.2.19).)

(2.2.21) Remark. This assignment of a matrix to a transformation depends on the choice of the bases α and β. If the bases α and β are replaced by bases α' and β', then in general $[T]_\alpha^\beta \neq [T]_{\alpha'}^{\beta'}$. Section 2.7 explores the effect of changing the basis on the matrix of a transformation. Also see Exercise 9 at the end of this section for simple examples of this phenomenon.

EXERCISES

1. Prove Proposition (2.2.1).
2. Prove Proposition (2.2.18).
3. a) Let $V = \mathbf{R}^3$ and $W = \mathbf{R}^4$, and let $T: V \to W$ be defined by $T(\mathbf{x}) = (x_1 - x_2, x_2 - x_3, x_1 + x_2 - x_3, x_3 - x_1)$. What is the matrix of T with respect to the standard bases in V and W?
   b) Let $\mathbf{a} = (a_1, \ldots, a_n)$ be a fixed vector in $V = \mathbf{R}^n$, let $W = \mathbf{R}$, and define $T: V \to W$ by $T(\mathbf{v}) = \sum_{i=1}^{n} a_i v_i$. What is the matrix of T with respect to the standard bases in V and W?
   c) Let V be the vector subspace of $C(\mathbf{R})$ spanned by $\sin(x)$ and $\cos(x)$. Define $D: V \to V$ by $D(f(x)) = f'(x)$. What is the matrix of D with respect to the basis $\{\sin(x), \cos(x)\}$?
4. (Reflection in a line). Let $L = \text{Span}(\{\mathbf{a}\})$ for $\mathbf{a} \in \mathbf{R}^2$ and $\mathbf{a} \neq \mathbf{0}$. Define $R_\mathbf{a}(\mathbf{v}) = 2P_\mathbf{a}(\mathbf{v}) - \mathbf{v}$, where $P_\mathbf{a}$ is the projection to the line spanned by $\mathbf{a}$.
   a) Show that $R_\mathbf{a}$ is a linear transformation.
   b) What is the matrix of $R_\mathbf{a}$ with respect to the standard basis in $\mathbf{R}^2$?
   c) Give a geometric interpretation of $R_\mathbf{a}$ similar to that of $P_\mathbf{a}$.
5. Let $T: \mathbf{R}^3 \to \mathbf{R}^3$ be defined by

$$T(\mathbf{x}) = (x_1, x_1 + x_2, x_1 - x_2 + x_3)$$

a) Let β be the standard basis in $\mathbf{R}^3$. Find $[T]_\beta^\beta$.

b) Let $U \subset \mathbf{R}^3$ be the subspace spanned by $\alpha = \{(1, 1, 0), (0, 0, 1)\}$ and let $S: U \to \mathbf{R}^3$ be the restriction of T to U: that is, $S = T|_U$. (See Exercise 14 of Section 2.1.) Compute $[S]_\alpha^\beta$.

c) Is there any relationship between your answers to parts a and b of this exercise?

6. Let V be a vector space of dimension n and let $\alpha = \{\mathbf{v}_1, \ldots, \mathbf{v}_n\}$ be a basis for V.

a) Let $c \in \mathbf{R}$ be fixed. What is the matrix $[cI]_\alpha^\alpha$, where I is the identity transformation of V?

b) Let $T: V \to V$ be defined by

$$T(a_1\mathbf{v}_1 + \cdots + a_n\mathbf{v}_n) = a_1\mathbf{v}_1 + \cdots + a_k\mathbf{v}_k$$

for $k < n$ fixed. What is $[T]_\alpha^\alpha$?

7. Compute the following products:

a) $\begin{bmatrix} 2 & 0 & 1 \\ 1 & -1 & 2 \end{bmatrix} \begin{bmatrix} 1 \\ 0 \\ 3 \end{bmatrix}$

b) $\begin{bmatrix} 1 & 0 & 2 \end{bmatrix} \begin{bmatrix} 0 \\ 3 \\ -5 \end{bmatrix}$

c) $\begin{bmatrix} 1 & 0 & 1 & 0 \\ & 2 & 0 & 2 \\ & & 3 & 0 \\ & & & 4 \end{bmatrix} \begin{bmatrix} -1 \\ 5 \\ 3 \\ 2 \end{bmatrix}$

8. a) Let $T: \mathbf{R}^2 \to \mathbf{R}^2$ be rotation through an angle $\pi/3$. Compute $T((1, 1))$ using the matrix of T.

b) Let $T: \mathbf{R}^2 \to \mathbf{R}^2$ be projection to Span($\{(1, 2)\}$). Compute $T((1, 1))$ using the matrix of T.

9. a) Let α be the basis $\{(1, 2), (3, -4)\}$ for $\mathbf{R}^2$, and let β be the standard basis for $\mathbf{R}^2$. Compute the matrix $[I]_\alpha^\beta$ for the identity transformation $I: \mathbf{R}^2 \to \mathbf{R}^2$.

b) Let α be the basis $\{(a_{11}, a_{21}), (a_{12}, a_{22})\}$ for $\mathbf{R}^2$. Let β be the standard basis for $\mathbf{R}^2$. As before, compute $[I]_\alpha^\beta$.

c) Extend the results of part b of this exercise to $\mathbf{R}^n$.

Just as we defined addition and scalar multiplication in $\mathbf{R}^n$ by adding and multiplying in each coordinate, we can define addition and scalar multiplication of matrices. Let $M_{m \times n}(\mathbf{R})$ be the set of $m \times n$ matrices with real entries. (This definition was also used in the exercises for Chapter 1.)

Definition. Let $A, B \in M_{m \times n}(\mathbf{R})$ with entries a_{ij} and b_{ij}, respectively, $1 \leq i \leq m$, and $1 \leq j \leq n$. The sum $A + B$ is defined to be the $m \times n$ matrix whose entries are $a_{ij} + b_{ij}$. For $c \in \mathbf{R}$ the scalar product cA is defined to be the $m \times n$ matrix whose entries are ca_{ij}.

10. Prove that $M_{m \times n}(\mathbf{R})$ is a vector space with the operations of addition and scalar multiplication defined as in the previous definition. [*Hint:* The additive identity in $M_{m \times n}(\mathbf{R})$ is the $m \times n$ *zero matrix*—the $m \times n$ matrix all of whose entries are 0.]

11. Let S and T be linear transformations from V to W. Let $\alpha = \{\mathbf{x}_1, \ldots, \mathbf{x}_n\}$ and $\beta = \{\mathbf{w}_1, \ldots, \mathbf{w}_m\}$ be bases for V and W, respectively.
    a) Prove that $[S + T]_\alpha^\beta = [S]_\alpha^\beta + [T]_\alpha^\beta$.
    b) Prove that for $c \in \mathbf{R}$, $[cS]_\alpha^\beta = c[S]_\alpha^\beta$.
    c) Define a mapping Mat: $L(V, W) \to M_{m \times n}(\mathbf{R})$ by $\text{Mat}(S) = [S]_\alpha^\beta$ for each $S \in L(V, W)$ (i.e., Mat assigns to each linear mapping S its matrix with respect to the bases α and β). Conclude that Mat is a linear mapping from the vector space $L(V, W)$ to the vector space $M_{m \times n}(\mathbf{R})$. (See Exercise 12 of Section 2.1.)

12. Let $V = M_{m \times n}(\mathbf{R})$ and $W = M_{n \times m}(\mathbf{R})$. Define a function $T: V \to W$ as follows: If $\mathbf{A}$ is the $m \times n$ matrix with entries a_{ij}, then $B = T(A)$ is the $n \times m$ matrix with entries $b_{ji} = a_{ij}$. In other words, the rows of A are the columns of B, and vice versa. The matrix B is called the *transpose* of A, and is usually denoted by A^t. For example

$$\begin{bmatrix} 1 & 3 & 2 \\ -1 & 0 & 1 \end{bmatrix}^t = \begin{bmatrix} 1 & -1 \\ 3 & 0 \\ 2 & 1 \end{bmatrix}$$

    a) Prove that the transpose mapping $T: M_{m \times n}(\mathbf{R}) \to M_{n \times m}(\mathbf{R})$ is a linear transformation; that is, $T(cA + B) = (cA + B)^t = cA^t + B^t = cT(A) + T(B)$.
    b) Let $V = W = M_{n \times n}(\mathbf{R})$. If T again denotes the transpose linear transformation, prove that $\{A \in M_{n \times n}(\mathbf{R}) \mid T(A) = A\}$ is a subspace of $M_{n \times n}(\mathbf{R})$. (This subspace is called the space of *symmetric matrices*. Compare with Section 1.2, Exercise 14. Why are these matrices called symmetric?)
    c) Again, let $V = W = M_{n \times n}(\mathbf{R})$, and let T be the transpose mapping. Show that $\{A \in M_{n \times n}(\mathbf{R}) \mid T(A) = -A\}$ is a subspace of $M_{n \times n}(\mathbf{R})$. (This subspace is called the space of *skew-symmetric* matrices. Compare with Section 1.2, Exercise 15.)

13. (See Section 1.3, Exercise 12.) Let $V = M_{n \times n}(\mathbf{R})$, let W_+ be the space of symmetric matrices in V, and let W_- be the space of skew-symmetric matrices in V.
    a) Show that if $A \in V$, then $(1/2)(A + A^t) \in W_+$.
    b) Show that if $A \in V$, then $(1/2)(A - A^t) \in W_-$.

c) Show that $W_+ \cap W_- = \{0\}$ (where 0 denotes the $n \times n$ zero matrix).
d) Use parts a, b, and c of this exercise to show that $V = W_+ \oplus W_-$.

14. a) What is the dimension of the vector space $M_{m\times n}(\mathbf{R})$? (*Hint:* Find a basis.)
b) What are the dimensions of the subspaces W_+ and W_- of $M_{n\times n}(\mathbf{R})$ from Exercise 13? (*Hint:* Find bases for each.)

15. Let $V = M_{2\times 2}(\mathbf{R})$.
a) Using the basis for V you found in Exercise 14a, what is the matrix of the transpose mapping $T: V \to V$ $(T(A) = A^t)$?
b) Using the bases you found in Exercise 14b for the subspaces W_+ and W_- of V, what are the matrices of the restricted mappings $T|_{W_+}$ and $T|_{W_-}$?

§2.3. KERNEL AND IMAGE

Let $T: V \to W$ be a linear transformation. In this section we associate to T a subspace $\mathrm{Ker}(T) \subset V$ called the kernel of T and a subspace $\mathrm{Im}(T) \subset W$ called the image of T. These subspaces will prove to be important in determining the properties of T. In the case that V and W are finite-dimensional we use the elimination technique of Section 1.5 to find bases of these subspaces. The systems of equations we have to solve will be constructed using the matrix of T. In turn, understanding these subspaces will shed light on the properties of solutions of systems of linear equations. It should be noted, however, that the definitions of the kernel and image of a linear mapping make sense even if V and W are infinite-dimensional spaces. We begin with $\mathrm{Ker}(T)$.

(2.3.1) Definition. The *kernel* of T, denoted $\mathrm{Ker}(T)$, is the subset of V consisting of all vectors $\mathbf{v} \in V$ such that $T(\mathbf{v}) = \mathbf{0}$.

Of course, it must be shown that $\mathrm{Ker}(T)$ is, in fact, a subspace.

(2.3.2) Proposition. Let $T: V \to W$ be a linear transformation. $\mathrm{Ker}(T)$ is a subspace of V.

Proof: Since $\mathrm{Ker}(T) \subset V$, it suffices to prove that $\mathrm{Ker}(T)$ is closed under addition and scalar multiplication. Since T is linear, for all $\mathbf{u}$ and $\mathbf{v} \in \mathrm{Ker}(T)$ and $a \in \mathbf{R}$, $T(\mathbf{u} + a\mathbf{v}) = T(\mathbf{u}) + aT(\mathbf{v}) = \mathbf{0} + a\mathbf{0} = \mathbf{0}$; that is, $\mathbf{u} + a\mathbf{v} \in \mathrm{Ker}(T)$. ■

Let us consider several examples of linear transformations and try to determine their kernels.

(2.3.3) Example. [See Example (2.1.7) of this chapter.] Let $V = \mathbf{R}^3$ and $W = \mathbf{R}$ and define $T: V \to W$ by $T((x_1, x_2, x_3)) = 2x_1 - x_2 + 3x_3$. Then $\mathbf{x} \in \mathrm{Ker}(T)$ if and only if $\mathbf{x}$ satisfies the equation $2x_1 - x_2 + 3x_3 = 0$. We conclude that Ker(T) coincides with the subspace of $\mathbf{R}^3$ defined by the single linear equation $2x_1 - x_2 + 3x_3 = 0$.

In general, let $V = \mathbf{R}^n$ and $W = \mathbf{R}$, and let $\mathbf{a} = (a_1, \ldots, a_n)$ be a fixed vector in $\mathbf{R}^n$. If T is defined as in Example (2.1.7) by $T(\mathbf{x}) = a_1x_1 + \cdots + a_nx_n$, then Ker($T$) is equal to the subspace of $V = \mathbf{R}^n$ defined by the single linear equation $a_1x_1 + \cdots + a_nx_n = 0$.

(2.3.4) Example. Let $V = P_3(\mathbf{R})$. Define $T: V \to V$ by $T(p(x)) = \frac{d}{dx}p(x)$. A polynomial is in Ker(T) if its derivative is zero. From calculus, you should recall that the only polynomials with zero derivatives are the constant polynomials, $p(x) = a_0$. Therefore, Ker(T) is the subspace of V consisting of the constant polynomials.

(2.3.5) Example. Let $V = W = \mathbf{R}^2$, and let T be a rotation R_θ. [See Example (2.1.12).] From the geometric description of R_θ as rotation through an angle θ, which preserves the length of a vector, we can see that the only vector $\mathbf{v}$ for which $R_\theta(\mathbf{v}) = \mathbf{0}$ is the zero vector. Since $R_\theta(\mathbf{v})$ must have the same length as $\mathbf{v}$, if $R_\theta(\mathbf{v}) = \mathbf{0}$, its length is zero, in which case the length of $\mathbf{v}$ is also zero and $\mathbf{v}$ is the zero vector. Therefore, Ker(R_θ) $= \{\mathbf{0}\}$.

(2.3.6) Example. Let $V = W = \mathbf{R}^2$, and let $\mathbf{a} = (1, 1)$. Let T be the linear transformation which is projection to the line spanned by $\mathbf{a}$, that is, $T = P_\mathbf{a}$. Then from Example (2.2.9), the matrix of $P_\mathbf{a}$ with respect to the standard bases in V and W is

$$A = \begin{bmatrix} 1/2 & 1/2 \\ 1/2 & 1/2 \end{bmatrix}$$

From Proposition (2.2.15) it follows that if $P_a(\mathbf{x}) = \mathbf{0}$, then $A\mathbf{x} = \mathbf{0}$, that is

$$\begin{bmatrix} 1/2 & 1/2 \\ 1/2 & 1/2 \end{bmatrix}\begin{bmatrix} x_1 \\ x_2 \end{bmatrix} = \begin{bmatrix} 1/2x_1 + 1/2x_2 \\ 1/2x_1 + 1/2x_2 \end{bmatrix} = \begin{bmatrix} 0 \\ 0 \end{bmatrix}$$

By substitution into this equation we can see that any vector $\mathbf{x}$ of the form $t(1, -1) = (t, -t)$ satisfies $A\begin{bmatrix} t \\ -t \end{bmatrix} = 0$. Therefore, Ker($P_\mathbf{a}$) $=$ Span($\{(1, -1)\}$). Notice that the vector $(1, -1)$ is perpendicular to $(1, 1)$ (see Exercise 2).

Now let V and W be finite-dimensional and choose bases $\alpha = \{\mathbf{v}_1, \ldots, \mathbf{v}_k\}$ for V and $\beta = \{\mathbf{w}_1, \ldots, \mathbf{w}_l\}$ for W. Let $T: V \to W$ be a linear transformation. If $\mathbf{x} \in$ Ker(T), then $T(\mathbf{x}) = \mathbf{0}$ and the coordinate vector of $T(\mathbf{x})$ must have all its entries equal to zero. Proposition (2.2.15) says that the coordinate vector of $T(\mathbf{x})$ is $[T]_\alpha^\beta[\mathbf{x}]_\alpha$. Thus, if $\mathbf{x} = x_1\mathbf{v}_1 + \cdots + x_k\mathbf{v}_k \in$ Ker(T) and $[T]_\alpha^\beta$ has entries a_{ij}, $T(\mathbf{x}) = \mathbf{0}$ implies that

$$[T]_\alpha^\beta[\mathbf{x}]_\alpha = \begin{bmatrix} a_{11} & \cdots & a_{1k} \\ & \cdot & \\ & \cdot & \\ & \cdot & \\ a_{l1} & \cdots & a_{lk} \end{bmatrix} \begin{bmatrix} x_1 \\ \cdot \\ \cdot \\ \cdot \\ x_k \end{bmatrix}$$

$$= \begin{bmatrix} a_{11}x_1 + \cdots + a_{1k}x_k \\ \cdot \\ \cdot \\ \cdot \\ a_{l1}x_1 + \cdots + a_{lk}x_k \end{bmatrix} = \begin{bmatrix} 0 \\ \cdot \\ \cdot \\ \cdot \\ 0 \end{bmatrix}$$

The converse is also true: If the coordinate vector $[\mathbf{x}]_\alpha$ of $\mathbf{x} \in V$ satisfies $[T]_\alpha^\beta[\mathbf{x}]_\alpha = \mathbf{0}$, then $[T(\mathbf{x})]_\beta = \mathbf{0}$ and $T(\mathbf{x}) = \mathbf{0}$. Thus, we have shown Proposition (2.3.7).

(2.3.7) Proposition. Let $T: V \to W$ be a linear transformation of finite-dimensional vector spaces, and let α and β be bases for V and W, respectively. Then $\mathbf{x} \in \text{Ker}(T)$ if and only if the coordinate vector of $\mathbf{x}$, $[\mathbf{x}]_\alpha$, satisfies the system of equations

$$\begin{aligned} a_{11}x_1 + \cdots + a_{1k}x_k &= 0 \\ &\cdot \\ &\cdot \\ &\cdot \\ a_{l1}x_1 + \cdots + a_{lk}x_k &= 0 \end{aligned}$$

where the coefficients a_{ij} are the entries of the matrix $[T]_\alpha^\beta$.

This proposition will be our guide when dealing with the kernel of a transformation. For example, in order to find a nonzero vector in Ker(T), it suffices to find a nonzero solution to the corresponding system of equations. Further, in order to find a basis for Ker(T), it suffices to find a basis for the set of solutions of the corresponding system of equations. This last statement requires proof; that is, if the coordinate vectors of a collection of vectors $\mathbf{x}_1, \ldots, \mathbf{x}_m$ are linearly independent, are the vectors themselves linearly independent in V? The answer is, Yes. Although the answer can be phrased more easily in the general context of Section 2.6, the following proposition directly formulates the answer.

(2.3.8) Proposition. Let V be a finite-dimensional vector space, and let $\alpha = \{\mathbf{v}_1, \ldots, \mathbf{v}_k\}$ be a basis for V. Then the vectors $\mathbf{x}_1, \ldots, \mathbf{x}_m \in V$ are linearly independent if and only if their corresponding coordinate vectors $[\mathbf{x}_1]_\alpha, \ldots, [\mathbf{x}_m]_\alpha$ are linearly independent.

Proof: Assume $\mathbf{x}_1, \ldots, \mathbf{x}_m$ are linearly independent and

$$\mathbf{x}_i = a_{1i}\mathbf{v}_1 + \cdots + a_{ki}\mathbf{v}_k$$

If $b_1, \ldots, b_m$ is any m-tuple of scalars with

$$b_1[\mathbf{x}_1]_\alpha + \cdots + b_m[\mathbf{x}_m]_\alpha = b_1\begin{bmatrix} a_{11} \\ \cdot \\ \cdot \\ \cdot \\ a_{k1} \end{bmatrix} + b_2\begin{bmatrix} a_{12} \\ \cdot \\ \cdot \\ \cdot \\ a_{k2} \end{bmatrix} + \cdots + b_m\begin{bmatrix} a_{1m} \\ \cdot \\ \cdot \\ \cdot \\ a_{km} \end{bmatrix} = \begin{bmatrix} 0 \\ \cdot \\ \cdot \\ \cdot \\ 0 \end{bmatrix}$$

then equating each of the components to zero, we have $\sum_{i=1}^{m} b_i a_{ji} = 0$, for all j, $1 \leq j \leq k$. Thus

$$\left(\sum_{i=1}^{m} b_i a_{1i}\right)\mathbf{v}_1 + \cdots + \left(\sum_{i=1}^{m} b_i a_{ki}\right)\mathbf{v}_k = \mathbf{0}$$

Rearranging the terms yields

$$b_1\left(\sum_{j=1}^{k} a_{j1}\mathbf{v}_j\right) + \cdots + b_m\left(\sum_{j=1}^{k} a_{jm}\mathbf{v}_j\right) = \mathbf{0}$$

or

$$b_1\mathbf{x}_1 + \cdots + b_m\mathbf{x}_m = \mathbf{0}$$

Therefore, $b_1 = b_2 = \cdots = b_m = 0$ and the m coordinate vectors are also linearly independent. The converse is left as an exercise. ■

It is now possible to determine dim(Ker(T)) using the techniques developed in Chapter 1 for solving systems of equations. That is, given $T: V \to W$, choose bases α for V and β for W and construct $[T]_\alpha^\beta$. Use Proposition (2.3.7) to obtain the corresponding system of equations. Apply the techniques of Corollary (1.6.15) to find a basis for the set of solutions to the system of equations and thus determine the dimension. By Proposition (2.3.8), the corresponding vectors in Ker(T) must also be linearly independent, and they must form a maximal linearly independent set in Ker(T), hence a basis. Then the dimension of Ker(T) must be equal to the dimension of the set of solutions of the system of equations.

(2.3.9) Example

a) Let V be a vector space of dimension 4, and let W be a vector space of dimension 3. Let $\alpha = \{\mathbf{v}_1, \ldots, \mathbf{v}_4\}$ be a basis for V, $\beta = \{\mathbf{w}_1, \ldots, \mathbf{w}_3\}$ be a basis for W. Let T be the linear transformation such that

$$[T]_\alpha^\beta = \begin{bmatrix} 1 & 0 & 1 & 2 \\ 2 & 1 & 0 & 1 \\ 1 & -1 & 3 & 5 \end{bmatrix}$$

Let us find the dimension of Ker(T).

From the preceding discussion we must solve the system

$$\begin{aligned} x_1 + 0\cdot x_2 + \quad x_3 + 2x_4 &= 0 \\ 2x_1 + \quad x_2 + 0\cdot x_3 + \quad x_4 &= 0 \\ x_1 - \quad x_2 + \quad 3x_3 + 5x_4 &= 0 \end{aligned}$$

Elimination yields the following equivalent system:

$$\begin{aligned} x_1 + 0\cdot x_2 + x_3 + 2x_4 &= 0 \\ x_2 - 2x_3 - 3x_4 &= 0 \\ 0 &= 0 \end{aligned}$$

The free variables are x_3 and x_4. Substitution as in Example (1.6.16) yields two solutions $(-1, 2, 1, 0)$ and $(-2, 3, 0, 1)$. Therefore, $\dim(\text{Ker}(T)) = 2$ and a basis for Ker(T) is $\{-\mathbf{v}_1 + 2\mathbf{v}_2 + \mathbf{v}_3, -2\mathbf{v}_1 + 3\mathbf{v}_2 + \mathbf{v}_4\}$.

b) Let V be a vector space of dimension 2 and let W be a vector space of dimension 4. Let $\alpha = \{\mathbf{v}_1, \mathbf{v}_2\}$ be a basis for V and $\beta = \{\mathbf{w}_1, \ldots, \mathbf{w}_4\}$ be a basis for W. Let T be the linear transformation such that

$$[T]_\alpha^\beta = \begin{bmatrix} 1 & 2 \\ 2 & 1 \\ -1 & 1 \\ 1 & -1 \end{bmatrix}$$

Let us find the dimension of Ker(T). As in part a, we must solve the system

$$\begin{aligned} x_1 + 2x_2 &= 0 \\ 2x_1 + x_2 &= 0 \\ -x_1 + x_2 &= 0 \\ x_1 - x_2 &= 0 \end{aligned}$$

Elimination yields the following equivalent system:

$$\begin{aligned} x_1 \quad &= 0 \\ x_2 &= 0 \\ 0 &= 0 \\ 0 &= 0 \end{aligned}$$

The only solution to this system of equations is the trivial solution $\mathbf{x} = \mathbf{0}$. Therefore $\text{Ker}(T) = \{\mathbf{0}\}$ and $\dim(\text{Ker}(T)) = 0$.

The second subspace associated with each linear mapping $T: V \to W$ is the image of T, which is a subspace of W.

(2.3.10) Definition. The subset of W consisting of all vectors $\mathbf{w} \in W$ for which there exists a $\mathbf{v} \in V$ such that $T(\mathbf{v}) = \mathbf{w}$ is called the *image* of T and is denoted by $\text{Im}(T)$.

Again, we must show that $\text{Im}(T)$ actually is a subspace of W.

(2.3.11) Proposition. Let $T: V \to W$ be a linear transformation. The image of T is a subspace of W.

Proof: Let $\mathbf{w}_1$ and $\mathbf{w}_2 \in \text{Im}(T)$, and let $a \in \mathbf{R}$. Since $\mathbf{w}_1$ and $\mathbf{w}_2 \in \text{Im}(T)$, there exist vectors $\mathbf{v}_1$ and $\mathbf{v}_2 \in V$ with $T(\mathbf{v}_1) = \mathbf{w}_1$ and $T(\mathbf{v}_2) = \mathbf{w}_2$. Then we have $a\mathbf{w}_1 + \mathbf{w}_2 = aT(\mathbf{v}_1) + T(\mathbf{v}_2) = T(a\mathbf{v}_1 + \mathbf{v}_2)$, since T is linear. Therefore, $a\mathbf{w}_1 + \mathbf{w}_2 \in \text{Im}(T)$ and $\text{Im}(T)$ is a subspace of W. ■

If V is finite-dimensional, it is quite easy to obtain a finite spanning set for $\text{Im}(T)$.

(2.3.12) Proposition. If $\{\mathbf{v}_1, \ldots, \mathbf{v}_m\}$ is any set that spans V (in particular, it could be a basis of V), then $\{T(\mathbf{v}_1), \ldots, T(\mathbf{v}_m)\}$ spans $\text{Im}(T)$.

Proof: Let $\mathbf{w} \in \text{Im}(T)$, then there exists $\mathbf{v} \in V$ with $T(\mathbf{v}) = \mathbf{w}$. Since $\text{Span}\{\mathbf{v}_1, \ldots, \mathbf{v}_m\} = V$, there exist scalars $a_1, \ldots, a_m$ such that $a_1\mathbf{v}_1 + \cdots + a_m\mathbf{v}_m = \mathbf{v}$. Then

$$\begin{aligned}\mathbf{w} = T(\mathbf{v}) &= T(a_1\mathbf{v}_1 + \cdots + a_m\mathbf{v}_m) \\ &= a_1T(\mathbf{v}_1) + \cdots + a_mT(\mathbf{v}_m)\end{aligned}$$

since T is linear. Therefore, $\text{Im}(T)$ is contained in $\text{Span}\{T(\mathbf{v}_1), \ldots, T(\mathbf{v}_m)\}$. The opposite inclusion follows by reversing the steps of the preceding argument. ■

(2.3.13) Corollary. If $\alpha = \{\mathbf{v}_1, \ldots, \mathbf{v}_k\}$ is a basis for V and $\beta = \{\mathbf{w}_1, \ldots, \mathbf{w}_l\}$ is a basis for W, then the vectors in W whose coordinate vectors (in terms of β) are the columns of $[T]_\alpha^\beta$ span $\text{Im}(T)$.

Proof: By Remark (2.2.16) these vectors are the vectors $T(\mathbf{v}_j)$, $1 \leq j \leq k$, which, by the proposition, span $\text{Im}(T)$. ■

It now remains to construct a basis for $\text{Im}(T)$. We will give two procedures. The first involves the matrix $[T]_\alpha^\beta$ of T, whereas the second uses the construction of a basis for a subspace.

(2.3.14) Procedure 1: From Corollary (2.3.13), it suffices to determine a set of columns of $[T]_\alpha^\beta$ that is linearly independent and that spans the same subspace as the set of all the columns of the matrix. We claim that those columns of $[T]_\alpha^\beta$

that correspond to basic variables of the system of equations $[T]_\alpha^\beta[\mathbf{x}]_\alpha = \mathbf{0}$ are linearly independent.

To prove the claim we introduce some notation. Let $\mathbf{a}_j$ denote the jth column of $[T]_\alpha^\beta$. Thus, $\mathbf{a}_j$ is the coordinate vector of $T(\mathbf{v}_j)$, $\mathbf{a}_j = [T(\mathbf{v}_j)]_\beta$. Without loss of generality, assume that the columns that correspond to basic variables are the first p columns. If $b_1\mathbf{a}_1 + \cdots + b_p\mathbf{a}_p = \mathbf{0}$, then it follows that $b_1\mathbf{a}_1 + \cdots + b_p\mathbf{a}_p + 0 \cdot \mathbf{a}_{p+1} + \cdots + 0 \cdot \mathbf{a}_k = \mathbf{0}$. Therefore, $(b_1, \ldots ,b_p, 0, \ldots ,0)$ is a solution of the system of equations obtained by setting all the free variables equal to zero.

Since the first p columns correspond to the basic variables, the echelon form of this system of equations is

$$\begin{aligned}
x_1 \qquad\qquad + a'_{1,p+1}x_{p+1} + \cdots + a'_{1k}x_k &= 0\\
x_2 \qquad + a'_{2,p+1}x_{p+1} + \cdots + a'_{2k}x_k &= 0\\
&\vdots\\
x_p + a'_{p,p+1}x_{p+1} + \cdots + a'_{pk}x_k &= 0\\
0 &= 0\\
&\vdots\\
0 &= 0
\end{aligned}$$

where the coefficients a'_{ij} result from the elimination process. Substituting our solution into this system, we see that $b_1 = 0, \ldots ,b_p = 0$.

Thus, the first p columns are linearly independent. Now, note that if we include a later column corresponding to one of the free variables in our prospective basis, we obtain a nonzero solution by setting the free variable equal to 1 (i.e., those $p + 1$ columns of the matrix $[T]_\alpha^\beta$ are linearly dependent). We conclude that the columns of $[T]_\alpha^\beta$ corresponding to the basic variables form a basis for $\text{Im}(T)$.

(2.3.15) Procedure 2: Choose a basis $\{\mathbf{v}_1, \ldots ,\mathbf{v}_k\}$ for V so that $\mathbf{v}_1, \ldots ,\mathbf{v}_q$ (with $q \leq k$) is a basis of $\text{Ker}(T)$. Then $T(\mathbf{v}_1) = \cdots = T(\mathbf{v}_q) = \mathbf{0}$. We claim that $\{T(\mathbf{v}_{q+1}), \ldots ,T(\mathbf{v}_k)\}$ is a basis of $\text{Im}(T)$. By Proposition (2.3.12) it is a spanning set, so it suffices to show that it is linearly independent as well. To see this, suppose $b_{q+1}T(\mathbf{v}_{q+1}) + \cdots + b_kT(\mathbf{v}_k) = \mathbf{0}$. Then, using the linearity of T, we see that $T(b_{q+1}\mathbf{v}_{q+1} + \cdots + b_k\mathbf{v}_k) = \mathbf{0}$, and it follows that $b_{q+1}\mathbf{v}_{q+1} + \cdots + b_k\mathbf{v}_k \in \text{Ker}(T)$. Thus, $b_{q+1}\mathbf{v}_{q+1} + \cdots + b_k\mathbf{v}_k$ is a linear combination of $\mathbf{v}_1, \ldots ,\mathbf{v}_q$. That is, there exist $c_1, \ldots ,c_q$ such that $c_1\mathbf{v}_1 + \cdots + c_q\mathbf{v}_q = b_{q+1}\mathbf{v}_{q+1} + \cdots + b_k\mathbf{v}_k$, or equivalently, $c_1\mathbf{v}_1 + \cdots + c_q\mathbf{v}_q - b_{q+1}\mathbf{v}_{q+1}$

$- \cdots - b_k\mathbf{v}_k = \mathbf{0}$. Since $\{\mathbf{v}_1, \ldots, \mathbf{v}_k\}$ is a basis of V, all the coefficients in this linear combination must be zero. In particular, $b_{q+1} = \cdots = b_k = 0$, so $\{T(\mathbf{v}_{q+1}), \ldots, T(\mathbf{v}_k)\}$ forms a basis of $\mathrm{Im}(T)$.

We call the maximum number of linearly independent columns of a matrix A the *rank* of A. With this terminology, the rank of $[T]_\alpha^\beta$ is the dimension of the image of T.

Let us return to the elimination algorithm for a moment. Let A denote the matrix $[T]_\alpha^\beta$. If we apply elimination to the system of equations $A\mathbf{x} = \mathbf{0}$, we ultimately obtain a new system of equations $A'\mathbf{x} = \mathbf{0}$ in echelon form. If p is the number of basic variables, the matrix A' has only p rows which contain non-zero entries. Thus the span of these rows, considered as coordinate vectors in $\mathbf{R}^k$, has dimension at most p. In fact, these rows must also be linearly independent, so that this span has dimension equal to p. Thus we see that the maximum number of linearly independent rows of A' is p, which is equal to the maximum number of linearly independent columns of A'. (The basic variables of $A'\mathbf{x} = \mathbf{0}$ are the same as the basic variables of $A\mathbf{x} = \mathbf{0}$, so that the rank of A is also the rank of A'.) It is natural to ask if this is also true for the matrix A; that is, is the maximum number of linearly independent columns of A equal to the maximum number of linearly independent rows of A? The proof of this result is left to the reader in Exercise 10 of this section.

(2.3.16) Example

a) Find a basis for the image of T for the transformation T of Example (2.3.9a). From the elimination procedure carried out before, we see that x_3 and x_4 are the free variables and x_1 and x_2 are basic variables. Therefore, the first two columns of $[T]_\alpha^\beta$ are the coordinate vectors of a basis for $\mathrm{Im}(T)$. The basis is $\{\mathbf{w}_1 + 2\mathbf{w}_2 + \mathbf{w}_3, \mathbf{w}_2 - \mathbf{w}_3\}$.

b) Find a basis for the image of the transformation T of (2.3.9b). From the elimination procedure we see that there are no free variables and x_1 and x_2 are the basic variables. It follows that the columns of $[T]_\alpha^\beta$ form a basis for $\mathrm{Im}(T)$.

If we examine Procedure 2 [(2.3.15)] carefully, we see that it relates the dimension of $\mathrm{Ker}(T)$ to the dimension of $\mathrm{Im}(T)$.

(2.3.17) Theorem. If V is a finite-dimensional vector space and $T: V \to W$ is a linear transformation, then

$$\dim(\mathrm{Ker}(T)) + \dim(\mathrm{Im}(T)) = \dim(V)$$

Proof: We follow the construction of Procedure 2 in (2.3.15). Choose a basis $\{\mathbf{v}_1, \ldots, \mathbf{v}_k\}$ for V such that $\{\mathbf{v}_1, \ldots, \mathbf{v}_q\}$ is a basis for $\mathrm{Ker}(T)$, $q = \dim(\mathrm{Ker}(T))$. We showed that $\{T(\mathbf{v}_q), \ldots, T(\mathbf{v}_k)\}$ is a basis for $\mathrm{Im}(T)$. Therefore, $k - q = \dim(\mathrm{Im}(T))$, or equivalently

$$\dim(\mathrm{Ker}(T)) + \dim(\mathrm{Im}(T)) = \dim(V) \quad \blacksquare$$

In the following section, we explore the consequences of this theorem, which is known as the dimension theorem.

(2.3.18) Example. Let $T: V \to W$ be a linear transformation from a four-dimensional vector space V with basis $\alpha = \{\mathbf{v}_1, \ldots, \mathbf{v}_4\}$ to a two-dimensional vector space W with basis $\beta = \{\mathbf{w}_1, \mathbf{w}_2\}$ with matrix

$$[T]_\alpha^\beta = \begin{bmatrix} 1 & 2 & 0 & 1 \\ 1 & 2 & 1 & 0 \end{bmatrix}$$

We will find a basis for V so that the first dim(Ker(T)) members are in Ker(T).

From the preceding, a basis for Ker(T) can be found by elimination on the system of equations $[T]_\alpha^\beta[\mathbf{x}]_\alpha = \mathbf{0}$. Here we obtain the system

$$\begin{aligned} x_1 + 2x_2 \qquad + x_4 &= 0 \\ x_3 - x_4 &= 0 \end{aligned}$$

The free variables are x_2 and x_4 so that a basis $\{\mathbf{u}_1, \mathbf{u}_2\}$ for Ker(T) contains the vectors $\mathbf{u}_1 = -\mathbf{v}_1 + \mathbf{v}_3 + \mathbf{v}_4$ and $\mathbf{u}_2 = -2\mathbf{v}_1 + \mathbf{v}_2$. That is, in coordinates, $\mathbf{u}_1 = (-1, 0, 1, 1)$ and $\mathbf{u}_2 = (-2, 1, 0, 0)$. From the constructions of Chapter 1, we see that if we set $\mathbf{u}_3 = \mathbf{v}_1$ and $\mathbf{u}_4 = \mathbf{v}_4$, then the set $\{\mathbf{u}_1, \ldots, \mathbf{u}_4\}$ is a basis of V (verify this choice). Therefore, from the theorem, $T(\mathbf{u}_3) = T(\mathbf{v}_1) = \mathbf{w}_1 + \mathbf{w}_2$ and $T(\mathbf{u}_4) = T(\mathbf{v}_4) = \mathbf{w}_1$ must form a basis for Im(T). Note that in this case Im(T) $= W$.

EXERCISES

1. Find a basis for the vector space V so that the first dim(Ker(T)) vectors are a basis for Ker(T), $T: V \to W$ a linear transformation.
   a) $T: \mathbf{R}^3 \to \mathbf{R}^4$ whose matrix with respect to the standard basis is

   $$\begin{bmatrix} 2 & 0 & 1 \\ 1 & 2 & 2 \\ -1 & 2 & 1 \\ 3 & 2 & 3 \end{bmatrix}$$

   b) $T: \mathbf{R}^3 \to \mathbf{R}$, which is given by $T(\mathbf{x}) = x_1 + x_2 + x_3$.
   c) $T: \mathbf{R}^2 \to \mathbf{R}^2$ whose matrix with respect to the standard basis is

   $$\begin{bmatrix} 1 & 2 \\ 2 & 1 \end{bmatrix}$$

   d) $T: \mathbf{R}^4 \to P_2(\mathbf{R})$ defined by $T(a_1, \ldots, a_4) = (a_1 + a_2) + (a_2 + a_3)x + (a_3 + a_4)x^2$
   e) $T: \mathbf{R}^3 \to \mathbf{R}^2$ such that $T(\mathbf{e}_1) = \mathbf{e}_1$, $T(\mathbf{e}_2) = \mathbf{0}$, and $T(\mathbf{e}_3) = \mathbf{e}_2$.
   f) $T: P_n(\mathbf{R}) \to P_n(\mathbf{R})$, which is given by differentiation.

2. Find a basis for Ker($P_\mathbf{a}$), where $P_\mathbf{a}: \mathbf{R}^2 \to \mathbf{R}^2$ is the projection to the line spanned by the vector $\mathbf{a} \neq \mathbf{0}$.

3. For each of the following matrices, defining linear maps T between vector spaces of the appropriate dimensions, find bases for Ker(T) and Im(T).

a) $\begin{bmatrix} 1 & 2 \\ 2 & 2 \end{bmatrix}$ b) $\begin{bmatrix} -1 & 2 & 2 \\ 2 & -4 & 2 \end{bmatrix}$

c) $\begin{bmatrix} 1 & 0 & 1 & -1 & 0 & 1 \\ -1 & 1 & 2 & 1 & 1 & 0 \\ 0 & 1 & 3 & 2 & 2 & 0 \end{bmatrix}$ d) $\begin{bmatrix} 0 & 1 & 2 & 3 \\ 1 & 0 & 1 & 0 \\ 1 & 1 & 3 & 3 \\ 1 & 2 & 5 & 6 \end{bmatrix}$

4. For each of the transformations of Exercise 1, find a basis for Im(T) by extending the basis for Ker(T) to a basis for V as in (2.3.15), Procedure 2.

5. Let V be an n-dimensional vector space with basis $\alpha = \{\mathbf{v}_1, \ldots, \mathbf{v}_n\}$. Let T: $V \to V$ be a linear transformation such that $T(\mathbf{v}_i) = a_i\mathbf{v}_i$, where each $a_i \neq 0$. What are dim(Ker(T)) and dim(Im(T))?

6. Let $T: \mathbf{R}^3 \to \mathbf{R}^4$ be the transformation of Exercise 1a.
   a) Let $U \subset \mathbf{R}^3$ be the subspace spanned by the first two standard basis vectors. What is dim(Ker($T|_U$))? (See Exercise 14 of Section 2.1.)
   b) Let $U \subset \mathbf{R}^3$ be the subspace spanned by (1, 0, 0) and (0, 3, −4). What is dim(Ker($T|_U$))?

7. a) Let $\mathbf{v} = (1, 1)$ and $\mathbf{w} = (2, 1)$ in $\mathbf{R}^2$. Construct a linear mapping T: $\mathbf{R}^2 \to \mathbf{R}^2$ such that $\mathbf{v} \in$ Ker(T) and $\mathbf{w} \in$ Im(T). (Find the matrix of such a mapping with respect to the standard basis.)
   b) If $\{\mathbf{v}_1, \ldots, \mathbf{v}_k\}$ is a linearly independent subset of $\mathbf{R}^n$ and $\{\mathbf{w}_1, \ldots, \mathbf{w}_l\}$ is a linearly independent subset of $\mathbf{R}^m$, is there always a linear mapping $T: \mathbf{R}^n \to \mathbf{R}^m$ such that $\{\mathbf{v}_1, \ldots, \mathbf{v}_k\} \subset$ Ker(T) and $\{\mathbf{w}_1, \ldots, \mathbf{w}_l\} \subset$ Im(T)?
   c) Suppose there is a linear mapping T as in part b. How could you construct a matrix for T?

8. Let V be an n-dimensional vector space.
   a) Is it always possible to find a linear transformation T: $V \to V$ such that Ker(T) = Im(T)? Is it ever possible to find such a transformation?
   b) Can you find such a T for $V = \mathbf{R}^2$?

9. Let V and W be finite-dimensional vector spaces, and let α and β be fixed bases for V and W, respectively. In Exercise 11 of Section 2.2 we defined a linear transformation Mat: $L(V, W) \to M_{m \times n}(\mathbf{R})$ (m = dim(W), n = dim(V)) by the rule Mat(T) = $[T]_\alpha^\beta$ for each T: $V \to W$.
   a) What is the kernel of this transformation?
   b) What is the image of this transformation?
   [*Hint:* Recall Proposition (2.2.19).]

In this section we saw that finding the kernel of a linear mapping T between finite-dimensional vector spaces V and W could be accomplished by solving the system of linear equations $[T]_\alpha^\beta[\mathbf{x}]_\alpha = \mathbf{0}$. The elimination algorithm we learned for doing this in Section 1.5 relied on the *elementary operations* [see Proposition (1.5.3)] to simplify the form of the system of equations. Since the coefficients appearing in the equations defining Ker(T) are precisely the entries of the matrix $[T]_\alpha^\beta$ (each row corresponds to one equation), we may also define analogous operations (known as elementary row operations) directly *on matrices*. There are three types:

a) Adding a scalar multiple of one row of the matrix to another row, leaving all the remaining rows unchanged

b) Multiplying any one row of a matrix by a nonzero scalar, leaving the other rows unchanged

c) Interchanging two rows of the matrix, leaving the other rows unchanged

We can define echelon form matrices in a fashion similar to the way we defined echelon form systems of equations in section 1.5. By applying elementary row operations, we can always eliminate entries and reduce any matrix to echelon form.

10. a) Show that applying an elementary operation to the rows of a matrix does not change the span of the rows.
    b) Show that in an echelon form $m \times n$ matrix, the nonzero rows are linearly independent, considered as vectors in $\mathbf{R}^n$.
    c) Use the elimination procedure and parts a and b of this problem to prove that the dimension of the span of the rows of A is equal to the dimension of the span of the columns of A. The dimension of the span of the rows of a matrix is called the *row rank* of the matrix. The *rank* of the matrix (the dimension of the span of the columns) is also sometimes called the *column rank*. Thus, the result of this exercise shows that for every matrix

$$\text{row rank} = \text{column rank}$$

11. Since our elementary row operations may be applied to any matrices, we can consider them as *mappings* from $M_{m\times n}(\mathbf{R})$ to itself. Let T denote any one particular row operation defined on $m \times n$ matrices [e.g., the operation that to any matrix A assigns the matrix obtained from A by adding three times the first row to the third ($m \geqslant 3$)].
    a) Prove that $T: M_{m\times n}(\mathbf{R}) \to M_{m\times n}(\mathbf{R})$ is linear.
    b) Prove that $\dim(\text{Ker}(T)) = 0$.

12. Define a mapping Tr: $M_{n\times n}(\mathbf{R}) \to \mathbf{R}$ by the following rule. If $A \in M_{n\times n}(\mathbf{R})$ is the matrix whose entries are a_{ij}, let $\text{Tr}(A) = a_{11} + \cdots + a_{nn}$ (the sum of the diagonal entries). Tr(A) is called the *trace* of the matrix A.
    a) Show that Tr is a linear mapping; that is, for all $A, B \in M_{n\times n}(\mathbf{R})$ and all $c \in \mathbf{R}$, $\text{Tr}(cA + B) = c\text{Tr}(A) + \text{Tr}(B)$.
    b) Find the dimensions of Ker(Tr) and Im(Tr).
    c) Find a basis for Ker(Tr).

13. Let $V = M_{n\times n}(\mathbf{R})$, and let $T: V \to V$ be the mapping defined by $T(A) = (1/2)(A + A')$ for $A \in V$. (A' is the transpose of A. See Exercise 12 of Section 2.2.)
    a) Prove that Ker(T) is the subspace of all skew-symmetric matrices in V and Im(T) is the subspace of all symmetric matrices in V (see Exercise 13 of Section 2.2.)

b) Compute dim(Ker(T)).
c) Use part b and Theorem (2.3.17) to compute dim(Im(T)).

14. Let $C^\infty(\mathbf{R})$ be the vector space of functions $f: \mathbf{R} \to \mathbf{R}$ with derivatives of all orders.
a) Let $T: C^\infty(\mathbf{R}) \to C^\infty(\mathbf{R})$ be the linear mapping defined by $T(f) = f''$. Find a basis for Ker(T). Does the conclusion of Theorem (2.3.17) apply to T?
b) Find a set of two linearly independent functions in Ker(T) for $T: C^\infty(\mathbf{R}) \to C^\infty(\mathbf{R})$ defined by $T(f) = f - f''$.

§2.4. APPLICATIONS OF THE DIMENSION THEOREM

The dimension theorem (2.3.17) is, in a sense, the first major theorem of the text. The proof utilized all the key concepts we have developed so far concerning bases, dimensions of subspaces, and linear transformations. The strength of the theorem lies in its simplicity: The only condition on the linear transformation $T: V \to W$ is that its domain V is finite-dimensional. It remains for us to explore the consequences of this theorem. We begin this section with a coordinate free approach, that is, we avoid choosing bases, in order to determine conditions under which a transformation might or might not be surjective or injective. We then allow ourselves to choose coordinates and explore the consequences of the dimension theorem in the context of systems of equations.

Let us recall some basic definitions concerning functions (see Appendix 1 for more details). A function between sets $f: S_1 \to S_2$ is said to be *injective* if whenever $f(p_1) = f(p_2)$ for $p_1, p_2 \in S_1$, we have $p_1 = p_2$. The function f is said to be *surjective* if for each $q \in S_2$ there is some $p \in S_1$ with $f(p) = q$.

(2.4.1) Examples

a) Let $V = W = \mathbf{R}^2$, and let $T = R_\theta$, rotation through an angle θ [see Example (2.1.12)]. We can see geometrically that T is injective. If $\mathbf{w} = R_\theta(\mathbf{u}) = R_\theta(\mathbf{v})$, then $\mathbf{w}$ was obtained by rotating both $\mathbf{u}$ and $\mathbf{v}$ through an angle θ. Therefore, rotation of $\mathbf{w}$ through an angle $-\theta$ yields both $\mathbf{u}$ and $\mathbf{v}$. Thus it must be the case that $\mathbf{u} = \mathbf{v}$. Further, if $\mathbf{v}$ is obtained from $\mathbf{w}$ by rotating by $-\theta$, $R_\theta(\mathbf{v}) = \mathbf{w}$. This holds for all $\mathbf{w}$, so that R_θ is also surjective. Let us outline the algebraic method of showing that R_θ is injective and surjective. One can see that R_θ is injective by showing that there is at most one solution to the system of equations $[R_\theta]_\alpha^\alpha[\mathbf{u}]_\alpha = [\mathbf{w}]_\alpha$ for every choice of $\mathbf{w}$, where α is the standard basis for $\mathbf{R}^2$. To show that R_θ is surjective, one must show that there is at least one solution to the same system of equations for every choice of $\mathbf{w}$. We leave the algebraic calculations to the reader.

b) Again, let $V = W = \mathbf{R}^2$ and let $T = P_\mathbf{a}$, the projection to the line spanned by the vector $\mathbf{a} \neq \mathbf{0}$ [see Example (2.1.13)]. From the geometry we have developed we see that if $\mathbf{u}$ and $\mathbf{v}$ are both perpendicular to $\mathbf{a}$, then $P_\mathbf{a}(\mathbf{v}) = P_\mathbf{a}(\mathbf{u}) = \mathbf{0}$. Choosing $\mathbf{u} \neq \mathbf{v}$ we see that $P_\mathbf{a}$ is not injective. Clearly, we were able to choose these vectors because the kernel of $P_\mathbf{a}$ is not just the zero vector. Is $P_\mathbf{a}$ surjective?

Motivated by the examples, we relate injectivity to the dimension of the kernel.

(2.4.2) Proposition. A linear transformation $T: V \to W$ is injective if and only if $\dim(\text{Ker}(T)) = 0$.

Proof: If T is injective, then by definition, there is only one vector $\mathbf{v} \in V$ with $T(\mathbf{v}) = \mathbf{0}$. Since we know that $T(\mathbf{0}) = \mathbf{0}$ for all linear mappings, the zero vector is the unique vector $\mathbf{v}$ satisfying $T(\mathbf{v}) = \mathbf{0}$. Thus, the kernel of T consists of only the zero vector. Therefore, $\dim(\text{Ker}(T)) = 0$.

Conversely, assume that $\dim(\text{Ker}(T)) = 0$. Let $\mathbf{v}_1$ and $\mathbf{v}_2 \in V$ with $T(\mathbf{v}_1) = T(\mathbf{v}_2)$. We must show $\mathbf{v}_1 = \mathbf{v}_2$. Since $T(\mathbf{v}_1) = T(\mathbf{v}_2)$, $T(\mathbf{v}_1 - \mathbf{v}_2) = \mathbf{0}$, so $\mathbf{v}_1 - \mathbf{v}_2 \in \text{Ker}(T)$. But if $\dim(\text{Ker}(T)) = 0$, it follows that $\text{Ker}(T) = \{\mathbf{0}\}$. Hence, $\mathbf{v}_1 - \mathbf{v}_2 = \mathbf{0}$; that is, $\mathbf{v}_1 = \mathbf{v}_2$, and we conclude that T is injective. ■

If we apply Theorem (2.3.17) and Proposition (2.4.2) together, we immediately obtain Corollary (2.4.3).

(2.4.3) Corollary. A linear mapping $T: V \to W$ on a finite-dimensional vector space V is injective if and only if $\dim(\text{Im}(T)) = \dim(V)$.

Since the image of T is a subspace of W, if the dimension of W is strictly less than the dimension of V, then it follows that the dimension of the image of T is strictly less than the dimension of V; that is, $\dim(\text{Im}(T)) < \dim(V)$. If we apply Theorem (2.3.17) in this setting, we see that $\dim(\text{Ker}(T)) > 0$. Thus we obtain Corollary (2.4.4).

(2.4.4) Corollary. If $\dim(W) < \dim(V)$ and $T: V \to W$ is a linear mapping, then T is not injective.

This can be rephrased as Corollary (2.4.5).

(2.4.5) Corollary. If V and W are finite dimensional, then a linear mapping $T: V \to W$ can be injective only if $\dim(W) \geqslant \dim(V)$.

(2.4.6) Example. Let $V = P_3(\mathbf{R})$, and let $W = P_2(\mathbf{R})$. Since $\dim(V) = 4$ and $\dim(W) = 3$, any $T: V \to W$ is not injective. Thus, if $T = d/dx$, which we know is a linear transformation, T is not injective. Of course, we also know that the kernel of differentiation consists of the constant polynomials.

We now turn to the notion of surjectivity. If $T: V \to W$ is surjective, then for every $\mathbf{w} \in W$, there is a $\mathbf{v} \in V$ with $T(\mathbf{v}) = \mathbf{w}$. This is equivalent to saying that the image of T is all of W; that is, $\text{Im}(T) = W$. As a consequence, if W is finite-dimensional, $\dim(\text{Im}(T)) = \dim(W)$. The converse is also true. Assume that W is finite dimensional and $\dim(\text{Im}(T)) = \dim(W)$. Since $\text{Im}(T)$ is a subspace of W that has the same dimension as W, it must be all of W; that is, $W = \text{Im}(T)$. But if $W = \text{Im}(T)$, every vector $\mathbf{w} \in W$ is of the form $\mathbf{w} = T(\mathbf{v})$ for some $\mathbf{v} \in V$. Therefore T is surjective. Thus, we have proven Proposition (2.4.7).

(2.4.7) Proposition. If W is finite-dimensional, then a linear mapping $T: V \rightarrow W$ is surjective if and only if $\dim(\text{Im}(T)) = \dim(W)$.

Since $\dim(\text{Im}(T)) \leqslant \dim(V)$ by the theorem, if $\dim(V) < \dim(W)$, then we have immediately that $\dim(\text{Im}(T)) < \dim(W)$, and hence, T is not surjective.

(2.4.8) Corollary. If V and W are finite-dimensional, with $\dim(V) < \dim(W)$, then there is no surjective linear mapping $T: V \rightarrow W$.

This can be rephrased as Corollary (2.4.9).

(2.4.9) Corollary. A linear mapping $T: V \rightarrow W$ can be surjective only if $\dim(V) \geqslant \dim(W)$.

Both (2.4.4) and (2.4.8) deal with the case where $\dim(V) \neq \dim(W)$. If $\dim(V) = \dim(W)$, then we have Proposition (2.4.10).

(2.4.10) Proposition. Let $\dim(V) = \dim(W)$. A linear transformation $T: V \rightarrow W$ is injective if and only if it is surjective.

Proof: $\rightarrow$: If T is injective, then $\dim(\text{Ker}(T)) = 0$ by Proposition (2.4.2). By Theorem (2.3.17), then, $\dim(\text{Im}(T)) = \dim(V)$. Therefore, by Proposition (2.4.7), T is surjective.

$\leftarrow$: Conversely, if T is surjective, then by Proposition (2.4.7), $\dim(\text{Im}(T)) = \dim(W) = \dim(V)$. Therefore, by Theorem (2.3.17), $\dim(\text{Ker}(T)) = 0$. Hence, by Proposition (2.4.2), T is injective. ■

It should be emphasized that we can only apply Proposition (2.4.10) if V and W are finite-dimensional. If $\dim(V) > \dim(W)$, a linear transformation $T: V \rightarrow W$ may be surjective, but it cannot be injective. Conversely, if $\dim(V) < \dim(W)$, then $T: V \rightarrow W$ may be injective, but it cannot be surjective. In each of these cases we must compute $\dim(\text{Ker}(T))$ or $\dim(\text{Im}(T))$ using systems of equations as we have done previously.

In order to apply these ideas to systems of equations, we first pose a question about a linear transformation $T: V \rightarrow W$. Let $\mathbf{w} \in W$. We ask, How can we describe the set of all vectors $\mathbf{v}$ with $T(\mathbf{v}) = \mathbf{w}$? This set is denoted by $T^{-1}(\{\mathbf{w}\})$ and is called the *inverse image of* **w** *under the transformation* T. (The reader may wish to consult Appendix 1 for a discussion of inverse images of mappings in general.) We are not claiming that T has an inverse function here, we are simply considering the set $T^{-1}(\{\mathbf{w}\}) = \{\mathbf{v} \in V \mid T(\mathbf{v}) = \mathbf{w}\}$. Notice that if $\mathbf{w} = \mathbf{0}$, $T^{-1}(\{\mathbf{w}\}) = \text{Ker}(T)$. Clearly, if $\mathbf{w} \notin \text{Im}(T)$, then there do not exist any vectors $\mathbf{v}$ with $T(\mathbf{v}) = \mathbf{w}$. In that case we would have $T^{-1}(\{\mathbf{w}\}) = \phi$. Hence, to make things more interesting, let us assume that $\mathbf{w} \in \text{Im}(T)$.

If $\mathbf{v}_1 \in T^{-1}(\{\mathbf{w}\})$, then $T(\mathbf{v}_1) = \mathbf{w}$. If $\mathbf{v}_2$ is some other element of $T^{-1}(\{\mathbf{w}\})$, then $T(\mathbf{v}_2) = \mathbf{w}$ as well. But then, by subtracting these two equations, we see that $T(\mathbf{v}_1) - T(\mathbf{v}_2) = T(\mathbf{v}_1 - \mathbf{v}_2) = \mathbf{0}$. Thus, the difference, $\mathbf{v}_1 - \mathbf{v}_2 \in \text{Ker}(T)$. We can also turn this around. If $\mathbf{u}_1 = T^{-1}(\{\mathbf{w}\})$ and $\mathbf{u}_2 \in \text{Ker}(T)$, then $T(\mathbf{u}_1) = \mathbf{w}$ and $T(\mathbf{u}_2) = \mathbf{0}$. Therefore, $T(\mathbf{u}_1 + \mathbf{u}_2) = T(\mathbf{u}_1) + T(\mathbf{u}_2) = \mathbf{w} + \mathbf{0} = \mathbf{w}$, so $\mathbf{u}_1 + \mathbf{u}_2 \in T^{-1}(\{\mathbf{w}\})$ as well. We can put these two arguments together in the following proposition.

(2.4.11) Proposition. Let $T: V \to W$ be a linear transformation, and let $\mathbf{w} \in \text{Im}(T)$. Let $\mathbf{v}_1$ be any fixed vector with $T(\mathbf{v}_1) = \mathbf{w}$. Then every vector $\mathbf{v}_2 \in T^{-1}(\{\mathbf{w}\})$ can be written uniquely as $\mathbf{v}_2 = \mathbf{v}_1 + \mathbf{u}$, where $\mathbf{u} \in \text{Ker}(T)$.

Proof: If $T(\mathbf{v}_2) = \mathbf{w}$, we let $\mathbf{u} = \mathbf{v}_2 - \mathbf{v}_1$. Then as before, $T(\mathbf{u}) = T(\mathbf{v}_1 - \mathbf{v}_2) = \mathbf{0}$. We claim that this choice of $\mathbf{u}$ is unique. Suppose that $\mathbf{u}'$ is another vector in Ker(T) with $\mathbf{v}_2 = \mathbf{v}_1 + \mathbf{u}'$. Then we have $\mathbf{v}_1 + \mathbf{u} = \mathbf{v}_1 + \mathbf{u}'$, which implies that $\mathbf{u} = \mathbf{u}'$. ■

(2.4.12) Remark. Of course, if a different $\mathbf{v}_1$ were used, the corresponding $\mathbf{u}$'s would change too.

(2.4.13) Remark. In this situation $T^{-1}(\{\mathbf{w}\})$ is a subspace of V if and only if $\mathbf{w} = \mathbf{0}$. Can you prove this?

We illustrate the proposition with two examples.

(2.4.14) Examples

a) Let $V = W = \mathbf{R}^2$, and let $T = P_{\mathbf{a}}$ for $\mathbf{a} = (1, 1)$. Let $\mathbf{w} = (2, 2)$. Then $T((0, 4)) = (2, 2)$ and $T((1, 3)) = (2, 2)$. Note that $(0, 4) - (1, 3) = (-1, 1) \in \text{Ker}(T)$. (See Figure 2.4.) From the figure we can see that $P_{\mathbf{a}}^{-1}\{(2, 2)\}$ is a line in $\mathbf{R}^2$ parallel to the kernel of $P_{\mathbf{a}}$ passing through the point $(2, 2)$.

b) Let $V = \mathbf{R}^3$ and $W = \mathbf{R}$. Define $T: V \to W$ by $T((a_1, a_2, a_3)) = a_1 + a_2$. Let $\mathbf{w} = 1 \in \mathbf{R}$. Then we have that $T^{-1}(\{\mathbf{w}\}) = \{(a_1, a_2, a_3) | a_1 + a_2 = 1\}$. Furthermore, $\text{Ker}(T) = \text{Span}(\{(0, 0, 1), (1, -1, 0)\})$. Note that we have $T((1, 0, 0)) = 1$ and $T((1/2, 1/2, 1)) = 1$ as well. As a result, $(1, 0, 0) - (1/2, 1/2, 1) \in \text{Ker}(T)$. (See Figure 2.5.) As in a, we see from the figure that $T^{-1}(\{1\})$ is a plane in $\mathbf{R}^3$ parallel to Ker(T) and passing through $(1, 0, 0)$.)

Given the results of this chapter, the following corollary of Proposition (2.4.11) requires no proof.

(2.4.15) Corollary. Let $T: V \to W$ be a linear transformation of finite-dimensional vector spaces, and let $\mathbf{w} \in W$. Then there is a unique vector $\mathbf{v} \in V$ such that $T(\mathbf{v}) = \mathbf{w}$ if and only if

(i) $\mathbf{w} \in \text{Im}(T)$ and

(ii) $\dim(\text{Ker}(T)) = 0$.

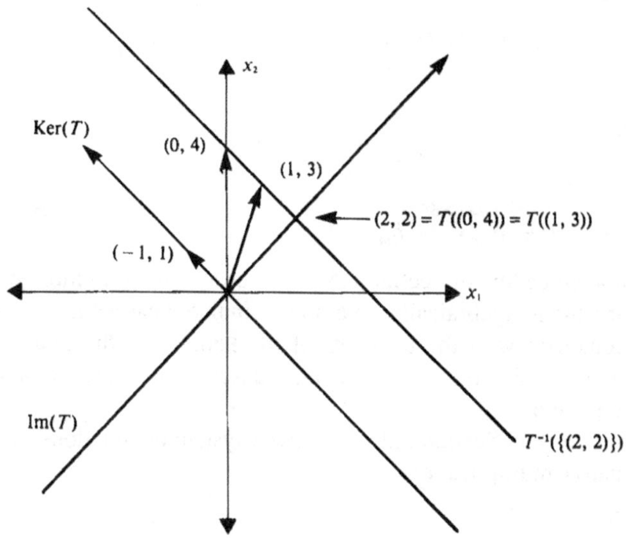

Figure 2.4

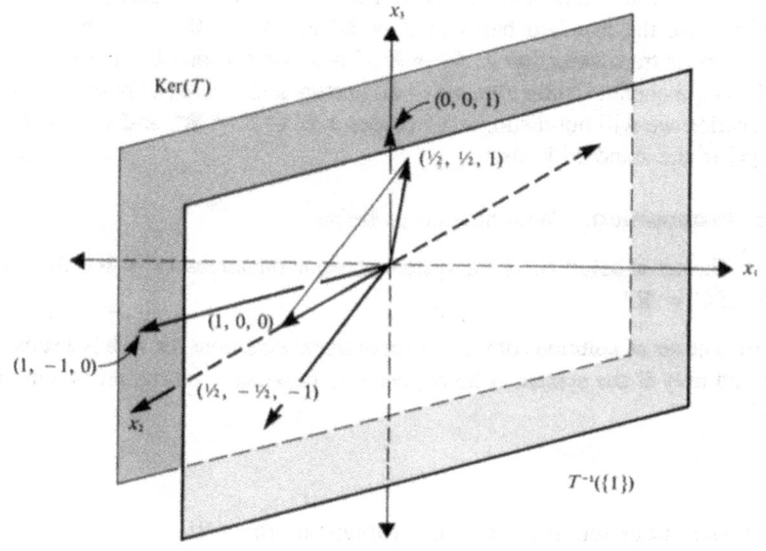

Figure 2.5

We now apply these ideas to systems of equations. We begin with a general system of linear equations:

$$a_{11}x_1 + \cdots + a_{mn}x_n = b_1$$
$$\vdots$$
$$a_{m1}x_1 + \cdots + a_{mn}x_n = b_m$$

The elimination procedure of Section 1.5 allows us to find all solutions of the system and to describe them algebraically. We now combine what we have learned about systems of equations with the concepts of the kernel and the image of a linear mapping in order to obtain a more geometric description of the set of solutions of a system of equations.

From our work in Section 2.2, we know a system of equations can be written in terms of matrix multiplication as

$$A\mathbf{x} = \mathbf{b}$$

where $A = [a_{ij}]$ is the matrix of coefficients, $\mathbf{x} = \begin{bmatrix} x_1 \\ \cdot \\ \cdot \\ \cdot \\ x_n \end{bmatrix}$ and $\mathbf{b} = \begin{bmatrix} b_1 \\ \cdot \\ \cdot \\ \cdot \\ b_m \end{bmatrix}$. Recall that the system is said to be homogeneous if $\mathbf{b} = \mathbf{0}$, and inhomogeneous otherwise.

If we use the standard bases in $V = \mathbf{R}^n$ and $W = \mathbf{R}^m$, then the matrix A defines a linear transformation $T: V \to W$. Thus, we see that solving the system $A\mathbf{x} = \mathbf{b}$ is equivalent to finding the set of all vectors $\mathbf{x} \in T^{-1}(\{\mathbf{b}\})$. (In the remainder of this section we will not distinguish between $\mathbf{x} \in V$, $V = \mathbf{R}^n$, and the coordinate vector $[\mathbf{x}]$ in the standard basis.)

(2.4.16) Proposition. With notation as before

(i) The set of solutions of the system of linear equations $A\mathbf{x} = \mathbf{b}$ is the subset $T^{-1}(\{\mathbf{b}\})$ of $V = \mathbf{R}^n$.

(ii) The set of solutions of the system of linear equations $A\mathbf{x} = \mathbf{b}$ is a subspace of V if and only if the system is homogeneous, in which case the set of solutions is $\text{Ker}(T)$.

Proof:

(i) The discussion of the previous paragraph proves (i).

(ii) Exercise. ■

Further, from our discussion of homogeneous systems in Chapter 1 and from Proposition (2.4.16), we obtain Corollary (2.4.17).

(2.4.17) Corollary

(i) The number of free variables in the homogeneous system $A\mathbf{x} = \mathbf{0}$ (or its echelon form equivalent) is equal to $\dim(\mathrm{Ker}(T))$.

(ii) The number of basic variables of the system is equal to $\dim(\mathrm{Im}(T))$.

Now, if the system is inhomogeneous, then from Proposition (2.4.16) we see that $A\mathbf{x} = \mathbf{b}$ has a solution if and only if $\mathbf{b} \in \mathrm{Im}(T)$. Let us assume then that $\mathbf{b} \in \mathrm{Im}(T)$. We have the following terminology.

(2.4.18) Definition. Given an inhomogeneous system of equations, $A\mathbf{x} = \mathbf{b}$, any single vector $\mathbf{x}$ satisfying the system (necessarily $\mathbf{x} \neq \mathbf{0}$) is called a *particular solution* of the system of equations.

In the context of systems of equations we can reword Proposition (2.4.11) and the Corollary, (2.4.15).

(2.4.19) Proposition. Let $\mathbf{x}_p$ be a particular solution of the system $A\mathbf{x} = \mathbf{b}$. Then every other solution to $A\mathbf{x} = \mathbf{b}$ is of the form $\mathbf{x} = \mathbf{x}_p + \mathbf{x}_h$, where $\mathbf{x}_h$ is a solution of the corresponding homogeneous system of equations $A\mathbf{x} = \mathbf{0}$. Furthermore, given $\mathbf{x}_p$ and $\mathbf{x}$, there is a unique $\mathbf{x}_h$ such that $\mathbf{x} = \mathbf{x}_p + \mathbf{x}_h$.

Proof: If $A\mathbf{x} = \mathbf{b}$ is a $k \times l$ system of equations, we may interpret A as the matrix of a linear transformation $T: \mathbf{R}^l \to \mathbf{R}^k$ with respect to the standard bases in $\mathbf{R}^l$ and $\mathbf{R}^k$. [See Proposition (2.2.19) (i).] The set of solutions to $A\mathbf{x} = \mathbf{b}$ is the same as the subset $T^{-1}(\{\mathbf{b}\}) \subset \mathbf{R}^l$. Thus a particular solution $\mathbf{x}_p$ to the system $A\mathbf{x} = \mathbf{b}$ is simply a vector in $T^{-1}(\{\mathbf{b}\})$. Apply Proposition (2.4.11) with $\mathbf{b}$ playing the role of $\mathbf{w}$ and $\mathbf{x}_p$ playing the role of $\mathbf{v}_1$. Then (2.4.11) says that every vector $\mathbf{x} \in T^{-1}\{(\mathbf{b})\}$ can be written uniquely in the form $\mathbf{x} = \mathbf{x}_p + \mathbf{x}_h$ where $\mathbf{x}_h \in \mathrm{Ker}(T)$. But we know that $\mathbf{x}_h \in \mathrm{Ker}(T)$ if and only if $A\mathbf{x}_h = \mathbf{0}$, thus we have proved the result. ■

(2.4.20) Corollary. The system $A\mathbf{x} = \mathbf{b}$ has a unique solution if and only if $\mathbf{b} \in \mathrm{Im}(T)$ and the only solution to $A\mathbf{x} = \mathbf{0}$ is the zero vector.

Geometrically, the result of Proposition (2.4.19) may be interpreted as follows. When $\mathbf{b} \in \mathrm{Im}(T)$, the set of solutions to the system $A\mathbf{x} = \mathbf{b}$ consists of all the vectors in $\mathrm{Ker}(T)$, a subspace of $V = \mathbf{R}^n$, *translated* by (any) particular solution $\mathbf{x}_p$. By a translation of a subspace $U \subset V$ by a vector $\mathbf{a}$, we mean the set $\{\mathbf{v} \in V \mid$ there exists $\mathbf{u} \in U$ such that $\mathbf{v} = \mathbf{a} + \mathbf{u}\} \subset V$.

(2.4.21) Example. Find a formula for the general solution of the system

$$\begin{bmatrix} 1 & 0 & 1 & 0 \\ 2 & 1 & 0 & 1 \\ 3 & 1 & 1 & 1 \end{bmatrix} \begin{bmatrix} x_1 \\ x_2 \\ x_3 \\ x_4 \end{bmatrix} = \begin{bmatrix} 2 \\ 3 \\ 5 \end{bmatrix}$$

According to Proposition (2.4.19), we will find a particular solution $\mathbf{x}_p$; next, we will find a basis of Ker(T); then, every solution will be expressed as the sum of $\mathbf{x}_p$ and some linear combination of the basis vectors for Ker(T).

To obtain $\mathbf{x}_p$, we reduce the system to echelon form. Since the operations performed in doing this actually only involve the entries of the matrix A and the vector $\mathbf{b}$ from the system, we perform our elementary operations on the rows of the so-called *augmented matrix* of the system

$$\left[\begin{array}{cccc|c} 1 & 0 & 1 & 0 & 2 \\ 2 & 1 & 0 & 1 & 3 \\ 3 & 1 & 1 & 1 & 5 \end{array}\right]$$

The rows of this matrix correspond to the equations of the system, and the last column just consists of the vector $\mathbf{b}$. Since this is the first time we have seen the reduction done in this way, we will display each step. (See Exercise 10 of Section 2.3 and the preceding comments for a detailed discussion of elimination applied directly to matrices.)

First, we subtract twice the first row from the second and three times the first row from the third, yielding

$$\left[\begin{array}{cccc|c} 1 & 0 & 1 & 0 & 2 \\ 0 & 1 & -2 & 1 & -1 \\ 0 & 1 & -2 & 1 & -1 \end{array}\right]$$

We now subtract the second row from the third (notice that operations we performed with equations previously correspond to similar operations on the rows of the matrix):

$$\left[\begin{array}{cccc|c} 1 & 0 & 1 & 0 & 2 \\ 0 & 1 & -2 & 1 & -1 \\ 0 & 0 & 0 & 0 & 0 \end{array}\right]$$

The matrix is now in echelon form.

To find a particular solution, we set all the free variables in the corresponding system of equations, in this case x_3 and x_4, equal to 0. When we do this, the system becomes

$$x_1 + 0 + 0 + 0 = 2$$

$$x_2 + 0 + 0 = -1$$

Therefore, we can take $\mathbf{x}_p = (2, -1, 0, 0)$. A basis for Ker($T$) ($T$ is the linear mapping $T\colon \mathbf{R}^4 \to \mathbf{R}^3$ defined by the matrix A) is obtained by setting $x_3 = 1$, $x_4 = 0$, then $x_3 = 0$, $x_4 = 1$ in the corresponding homogeneous system. We obtain the pair of independent vectors $(-1, 2, 1, 0)$ and $(0, -1, 0, 1)$. We conclude that the general solution of the system $A\mathbf{x} = \mathbf{b}$ is of the form $\mathbf{x} = (2, -1, 0, 0) + s(-1, 2, 1, 0) + t(0, -1, 0, 1)$, where s and t are any real numbers. Note that the set of solutions is simply the set of vectors obtained by translating each vector in Ker(T) by the fixed vector $\mathbf{x}_p$.

(2.4.22) Example. Let us consider a special case that will reappear in later sections. If $V = W$ is finite-dimensional and $T: V \to V$ is injective (and thus surjective also), then $\dim(\mathrm{Ker}(T)) = 0$ and there is a unique solution to $T(\mathbf{x}) = \mathbf{b}$ for any $\mathbf{b} \in V$. By Corollary (2.4.17) we see that the corresponding system has *no* free variables. Thus, after reducing the system to echelon form, the system will look like

$$\begin{aligned} x_1 \qquad\qquad &= c_1 \\ x_2 \qquad &= c_2 \\ &\vdots \\ x_n &= c_n \end{aligned}$$

for scalars $c_i \in \mathbf{R}$. Thus, $\mathbf{x} = \mathbf{c}$, the vector with these components, is the only solution. A particular case of this is as follows: Let $T: \mathbf{R}^3 \to \mathbf{R}^3$ be defined by the matrix

$$A = \begin{bmatrix} 1 & 0 & 1 \\ 2 & 1 & -1 \\ 0 & 1 & 1 \end{bmatrix}$$

Suppose we want to find a vector $\mathbf{x}$ with $T(\mathbf{x}) = (0, 1, 0)$. We begin with the augmented matrix

$$\left[\begin{array}{ccc|c} 1 & 0 & 1 & 0 \\ 2 & 1 & -1 & 1 \\ 0 & 1 & 1 & 0 \end{array}\right]$$

After reducing to echelon form (check this), we obtain

$$\left[\begin{array}{ccc|c} 1 & 0 & 0 & 1/4 \\ 0 & 1 & 0 & 1/4 \\ 0 & 0 & 1 & -1/4 \end{array}\right]$$

corresponding to the echelon form system of equations

$$\begin{aligned} x_1 \qquad &= 1/4 \\ x_2 \quad &= 1/4 \\ x_3 &= -1/4 \end{aligned}$$

The solution is $(1/4, 1/4, -1/4)$.

EXERCISES

1. Determine if the following linear transformations are injective, surjective, both, or neither:

a) $T: \mathbf{R}^3 \to \mathbf{R}^3$ defined by $T(x) = (x_1 + x_2, x_1 + x_2 + x_3, x_2 + x_3)$.
b) $T: \mathbf{R}^3 \to \mathbf{R}^4$ defined by $T(x) = (x_2, x_3, 0, 0)$.
c) $T: P_2(\mathbf{R}) \to \mathbf{R}^2$ defined by $T(p(x)) = (p(0), p'(0))$.
d) $T: M_{n\times n}(\mathbf{R}) \to M_{n\times n}(\mathbf{R})$ defined by $T(A) = A^t$. (See Exercise 12 of Section 2.2 and the preceding comments for the definition of A^t.)
e) $T: M_{n\times n}(\mathbf{R}) \to \mathbf{R}$ defined by $T(A) = a_{11} + \cdots + a_{nn}$.

2. In each case determine whether $T: \mathbf{R}^k \to \mathbf{R}^l$ is injective, surjective, both, or neither, where T is defined by the matrix:

a) $\begin{bmatrix} 1 & 1 \\ 1 & 2 \end{bmatrix}$

b) $\begin{bmatrix} 0 & 1 & 1 \\ -1 & 0 & 1 \\ -1 & -1 & 0 \end{bmatrix}$

c) $\begin{bmatrix} 1 & 2 & 0 \\ 0 & 1 & 1 \\ -1 & 3 & 1 \\ 1 & 0 & 1 \end{bmatrix}$

d) $\begin{bmatrix} 3 & -1 & 0 \\ 1 & 1 & 1 \end{bmatrix}$

e) $\begin{bmatrix} 1 & 1 & 1 & 1 \\ & 1 & 1 & 1 \\ & & 1 & 1 \\ & & & 1 \end{bmatrix}$

f) $\begin{bmatrix} 0 & 1 & & \\ 1 & 0 & & \\ & & 0 & 1 \\ & & 1 & 0 \end{bmatrix}$

3. Let T be the linear transformation $T: V \to W$, $\dim(V) = 4$ and $\dim(W) = 3$, whose matrix is

$$[T]_\alpha^\beta = \begin{bmatrix} 2 & -1 & 1 & 1 \\ 0 & 1 & 2 & -1 \\ 1 & -1 & 0 & 1 \end{bmatrix}$$

with respect to the bases $\alpha = \{\mathbf{v}_1, \ldots, \mathbf{v}_4\}$ for V and $\beta = \{\mathbf{w}_1, \mathbf{w}_2, \mathbf{w}_3\}$ for W. Determine $T^{-1}(\{\mathbf{w}\})$ for the following:
a) $\mathbf{w} = \mathbf{0}$
b) $\mathbf{w} = 4\mathbf{w}_1 + 2\mathbf{w}_2 + \mathbf{w}_3$
c) $\mathbf{w} = \mathbf{w}_3$

4. Find a formula for the general solution to the system of equations given in matrix form by the following:

a) $\begin{bmatrix} 1 & 0 & 2 & 1 \\ -3 & 2 & 1 & -1 \end{bmatrix} \mathbf{x} = \begin{bmatrix} 1 \\ 1 \end{bmatrix}$

b) $\begin{bmatrix} 2 & 0 & -1 \\ 1 & 2 & 0 \\ -1 & 2 & 1 \\ 1 & 2 & 0 \end{bmatrix} \mathbf{x} = \begin{bmatrix} -1 \\ 2 \\ 3 \\ 2 \end{bmatrix}$

c) $\begin{bmatrix} 1 & 2 \\ 0 & 1 \end{bmatrix} \mathbf{x} = \begin{bmatrix} 1 \\ 1 \end{bmatrix}$

d) $\begin{bmatrix} 0 & 1 & 2 \\ 1 & 1 & 1 \end{bmatrix} \mathbf{x} = \begin{bmatrix} -1 \\ 2 \end{bmatrix}$

e) $\begin{bmatrix} 0 & -1 \\ 2 & 1 \\ 1 & 2 \\ 1 & 0 \end{bmatrix} \mathbf{x} = \begin{bmatrix} -2 \\ 4 \\ 5 \\ 1 \end{bmatrix}$

f) $\begin{bmatrix} 1 & 3 & 1 & 0 & 0 \\ 0 & 1 & 0 & 0 & 0 \\ 1 & 3 & 1 & 0 & 0 \\ 0 & 0 & 0 & 0 & 0 \end{bmatrix} \mathbf{x} = \begin{bmatrix} 1 \\ 1 \\ 1 \\ 0 \end{bmatrix}$

5. Let $V = P(\mathbf{R})$, the vector space of polynomials of all degrees.
   a) Define $T: V \to V$ by $T(p(x)) = xp(x)$. Is T injective, surjective, both, or neither?
   b) Define $T: V \to V$ by $T(p(x)) = d/dx(p(x))$. Is T injective, surjective, both, or neither?
   c) Do your answers contradict Proposition (2.4.10)?
6. Let S and $T: U \to V$ be linear transformations.
   a) If S and T are injective (surjective), is $S + T$ necessarily injective (surjective)?
   b) If S is injective (surjective) and $a \neq 0$, is aS necessarily injective (surjective)?
7. For each fixed vector $\mathbf{a} \in \mathbf{R}^n$ define a linear transformation $D_{\mathbf{a}}$ by $D_{\mathbf{a}}(\mathbf{v}) = a_1v_1 + \cdots + a_nv_n$. This defines a function $*: \mathbf{R}^n \to L(\mathbf{R}^n, \mathbf{R}), *(\mathbf{a}) = D_{\mathbf{a}}$.
   a) Prove that $*$ is a linear transformation. (*Hint:* Show that $D_{\alpha\mathbf{a}+\beta\mathbf{b}} = \alpha D_{\mathbf{a}} + \beta D_{\mathbf{b}}$ as linear transformations.)
   b) Prove that $*$ is injective and surjective.
8. a) Prove that if $T: V \to W$ is injective and $U \subset V$ is a subspace, then $T|_U$ is also injective. [*Hint:* What is $\text{Ker}(T|_U)$?]
   b) Give an example to show that if $T: V \to W$ is surjective and $U \subset V$ is a subspace, then $T|_U$ may or may not be surjective.
9. Let $T: V \to W$ be a linear transformation.
   a) Let $U \subset V$ be a subspace of V such that $U \cap \text{Ker}(T) = \{\mathbf{0}\}$. Prove that $T|_U$ is injective. [*Hint:* What is $\text{Ker}(T|_U)$?]
   b) Assume further that T is surjective and that U satisfies $U + \text{Ker}(T) = V$. Prove that $T|_U$ is surjective.
10. Let $T: V \to W$ be a surjective linear transformation and let $X \subset W$ be a subspace. Assume that $\text{Ker}(T)$ and X are finite dimensional.
    a) Prove that $T^{-1}(X) = \{\mathbf{v} | T(\mathbf{v}) \in X\}$ is a subspace of V.
    b) Prove that $\text{Ker}(T) \subset T^{-1}(X)$.
    c) Prove that $\dim(T^{-1}(X)) = \dim(\text{Ker}(T)) + \dim(X)$. [*Hint:* Extend a basis for $\text{Ker}(T)$ to a basis for $T^{-1}(X)$.]
11. Let $V = U_1 \oplus U_2$ be a direct sum decomposition of the vector space V.
    a) Prove that if $T_i: U_i \to U_i$, $i = 1, 2$, are linear transformations, then $T: V \to V$ defined by $T(\mathbf{v}) = T_1(\mathbf{u}_1) + T_2(\mathbf{u}_2)$, where $\mathbf{v} = \mathbf{u}_1 + \mathbf{u}_2$, is a linear transformation.

b) Prove that if T_1 and T_2 are injective, then T is injective.
c) Prove that if T_1 and T_2 are surjective, then T is surjective.

12. a) Let V and W be two finite-dimensional vector spaces, and assume $\dim(V) \leqslant \dim(W)$. Show that there exists an injective linear mapping $T: V \to W$.
b) Now assume that $\dim(V) \geqslant \dim(W)$. Show that there exists a surjective linear mapping $T: V \to W$. (*Hint:* For both parts you will need to consider the effect of T on bases in V and W.)

§2.5. COMPOSITION OF LINEAR TRANSFORMATIONS

In this section we consider composition of linear transformations. This is a necessary prelude to our discussion of when a linear transformation has an inverse. Recall that if $f: S_1 \to S_2$ and $g: S_2 \to S_3$ are functions of three sets S_1, S_2, and S_3, their composition $g \circ f: S_1 \to S_3$ is defined by $g \circ f(s) = g(f(s))$. Notice that the composition only makes sense if the image of f is contained in the domain of g.

We obtain a simple method to find the matrix of a composition. We have already laid the groundwork for this in Section 2.2, in our discussion of the multiplication of a vector by a matrix. Here we show how to multiply two matrices together provided they are of the correct size. The matrix of the composition will be the product of the corresponding matrices. But first, we look at the composition in a coordinate free manner, that is, without reference to bases.

Let U, V, and W be vector spaces, and let $S: U \to V$ and $T: V \to W$ be linear transformations. The *composition of S and T* is denoted $TS: U \to W$ and is defined by

$$TS(\mathbf{v}) = T(S(\mathbf{v}))$$

Notice that this is well defined since the image of S is contained in V, which is the domain of T. Naturally, we need to know that TS is a linear transformation. This follows from the fact that both T and S are linear transformations.

(2.5.1). Proposition. Let $S: U \to V$ and $T: V \to W$ be linear transformations, then TS is a linear transformation.

Proof: Let a and $b \in \mathbf{R}$ and let $\mathbf{u}_1$ and $\mathbf{u}_2 \in U$. We must show that TS satisfies Proposition (2.1.2).

$$\begin{aligned} TS(a\mathbf{u}_1 + b\mathbf{u}_2) &= T(S(a\mathbf{u}_1 + b\mathbf{u}_2)), \text{ by the definition of } TS, \\ &= T(aS(\mathbf{u}_1) + bS(\mathbf{u}_2)), \text{ by the linearity of } S, \\ &= aT(S(\mathbf{u}_1)) + bT(S(\mathbf{u}_2)), \text{ by the linearity of } T, \\ &= aTS(\mathbf{u}_1) + bTS(\mathbf{u}_2). \quad \blacksquare \end{aligned}$$

For our first examples we return once again to rotations, projections, and the derivative.

(2.5.2) Examples

a) Let $U = V = W = \mathbf{R}^2$. Let $S = R_\theta$ be rotation through an angle θ, and let T be rotation through an angle φ. Then $TS = R_\varphi R_\theta = R_{\varphi+\theta}$, since, from the geometry, rotating first by an angle θ and then by an angle φ is equal to rotating by an angle $\varphi + \theta$. We obtain the same result for ST.

b) Let $U = V = W = \mathbf{R}^2$. Let $S = R_{\pi/2}$ and let $T = P_{(1,0)}$ (projection to the line spanned by (1,0), which is the x_1-axis). TS first rotates the plane by an angle $\pi/2$ and then projects to the x_1-axis. Let us, for example, compute the image of the x_1-axis by TS. If $\mathbf{x} = (x_1, 0)$, then $R_{\pi/2}(\mathbf{x}) = (0, x_1)$. $P_{(1,0)}(0, x_1) = \mathbf{0}$, since $(0, x_1)$ is perpendicular to the x_1-axis. Thus $TS(\mathbf{x}) = \mathbf{0}$ for $\mathbf{x} = (x_1, 0)$. That is, the x_1-axis is contained in the kernel of TS. What is the image of TS?

c) Let $U = P_4(\mathbf{R})$, $V = P_3(\mathbf{R})$, and $W = P_2(\mathbf{R})$, and let $S = d/dx$ and $T = d/dx$. Then the composition is $TS = d^2/dx^2$.

(2.5.3) Remark. In Example (2.5.2a) the composition is defined in either order, ST or TS, and as we noted is the same in either order. In general, as in Example (2.5.2b), ST is not equal to TS. In fact, if we consider the definition of composition carefully, ST will not make sense in the general case. We emphasize that the composition is well defined only if the image of the first transformation is contained in the domain of the second.

Composition of linear transformations satisfies the following basic properties whose proofs we leave as an exercise.

(2.5.4) Proposition

(i) Let $R: U \to V$, $S: V \to W$, and $T: W \to X$ be linear transformations of the vector spaces U, V, W, and X as indicated. Then

$$T(SR) = (TS)R \qquad \text{(associativity)}$$

(ii) Let $R:U \to V$, $S:U \to V$, and $T:V \to W$ be linear transformations of the vector spaces U, V, and W as indicated. Then

$$T(R + S) = TR + TS \qquad \text{(distributivity)}$$

(iii) Let $R: U \to V$, $S: V \to W$, and $T: V \to W$ be linear transformations of the vector spaces U, V, and W as indicated. Then

$$(T + S)R = TR + SR \qquad \text{(distributivity)}$$

Proof: We will prove (i) and leave (ii) and (iii) to the reader. (i) We must show that $T(SR)(\mathbf{u}) = (TS)R(\mathbf{u})$ for all $\mathbf{u} \in U$. $T(SR)(\mathbf{u})$ is the composition of T with SR, thus $T(SR)(\mathbf{u}) = T(SR(\mathbf{u}))$. Using the definition of the composition SR, we see that $T(SR(\mathbf{u})) = T(S(R(\mathbf{u})))$. By the definition of TS, this is $TS(R(\mathbf{u}))$. Finally by the definition of the composition of TS with R, we see that this is $(TS)R(\mathbf{u})$ as we desired. ■

(2.5.5) Remark. As a consequence of Proposition (2.5.4) (i), although composition is defined for only two transformations at a time, we may unambiguously write TSR and compute the composition in either order. Clearly, this extends to any finite number of transformations for which the composition is well defined.

If $S: U \to V$ and $T: V \to W$ are transformations of finite-dimensional vector spaces, we have shown how to determine the kernel and image of S and of T. Given this information, can we say anything about the kernel and image of TS? In general, we have Proposition (2.5.6).

(2.5.6) Proposition. Let $S: U \to V$ and $T: V \to W$ be linear transformations. Then

(i) $\text{Ker}(S) \subset \text{Ker}(TS)$

(ii) $\text{Im}(TS) \subset \text{Im}(T)$.

Proof: We prove (i) and leave the proof of (ii) to the reader. (i) If $\mathbf{u} \in \text{Ker}(S)$, $S(\mathbf{u}) = \mathbf{0}$. Then $TS(\mathbf{u}) = T(\mathbf{0}) = \mathbf{0}$. Therefore $\mathbf{u} \in \text{Ker}(TS)$ and we have proven (i). ■

The following corollary is an immediate consequence of the proposition.

(2.5.7) Corollary. Let $S: U \to V$ and $T: V \to W$ be linear transformations of finite-dimensional vector spaces. Then

(i) $\dim(\text{Ker}(S)) \leqslant \dim(\text{Ker}(TS))$

(ii) $\dim(\text{Im}(TS)) \leqslant \dim(\text{Im}(T))$.

(2.5.8) Example. Let $U = V = W = \mathbf{R}^2$, and let $S = R_\theta$ and $T = P_\mathbf{a}$ for $\mathbf{a} \neq \mathbf{0}$. First, consider $P_\mathbf{a}R_\theta$: $\dim(\text{Ker}(R_\theta)) = 0$ and $\dim(\text{Ker}(P_\mathbf{a}R_\theta)) = 1$ (what is the subspace $\text{Ker}(P_\mathbf{a}R_\theta)$?), whereas $\dim(\text{Im}(P_\mathbf{a})) = 1$ and $\dim(\text{Im}(P_\mathbf{a}R_\theta)) = 1$. Now, consider $R_\theta P_\mathbf{a}$: $\dim(\text{Ker}(P_\mathbf{a})) = 1$ and $\dim(\text{Ker}(R_\theta P_\mathbf{a})) = 1$, whereas $\dim(\text{Im}(R_\theta)) = 2$ and $\dim(\text{Im}(R_\theta P_\mathbf{a})) = 1$.

Assume now that $S: U \to V$ and $T: V \to W$ are linear transformations between finite-dimensional vector spaces. Choose bases $\alpha = \{\mathbf{u}_1, \ldots, \mathbf{u}_m\}$ for U, $\beta = \{\mathbf{v}_1, \ldots, \mathbf{v}_n\}$ for V and $\gamma = \{\mathbf{w}_1, \ldots, \mathbf{w}_p\}$ for W. Can we express $[TS]_\alpha^\gamma$ in terms of $[S]_\alpha^\beta$ and $[T]_\beta^\gamma$?

Let the entries of $[S]_\alpha^\beta$ be denoted by a_{ij}, where $1 \leqslant i \leqslant n$ and $1 \leqslant j \leqslant m$, and let the entries of $[T]_\beta^\gamma$ be denoted by b_{kl}, where $1 \leqslant k \leqslant p$ and $1 \leqslant l \leqslant n$.

From Remark (2.2.16) we have that the jth column of $[TS]_\alpha^\gamma$ is the coordinate vector of $TS(\mathbf{u}_j)$ in the coordinates of γ, $[TS(\mathbf{u}_j)]_\gamma$. Using (2.2.15), $[TS(\mathbf{u}_j)]_\gamma = [T]_\beta^\gamma[S(\mathbf{u}_j)]_\beta$, where again by (2.2.15) $[S(\mathbf{u}_j)]_\beta$ is the jth column of $[S]_\alpha^\beta$

$$\begin{bmatrix} a_{1j} \\ \cdot \\ \cdot \\ \cdot \\ a_{nj} \end{bmatrix}$$

Then $[TS(\mathbf{u}_j)]_\gamma$ is given by

$$[T]_\beta^\gamma \begin{bmatrix} a_{1j} \\ \cdot \\ \cdot \\ \cdot \\ a_{nj} \end{bmatrix} = \begin{bmatrix} b_{11} & \cdots & b_{1n} \\ & \cdot & \\ & \cdot & \\ & \cdot & \\ b_{p1} & \cdots & b_{pn} \end{bmatrix} \begin{bmatrix} a_{1j} \\ \cdot \\ \cdot \\ \cdot \\ a_{nj} \end{bmatrix} = \begin{bmatrix} \sum_{l=1}^{n} b_{1l}a_{lj} \\ \vdots \\ \sum_{l=1}^{n} b_{pl}a_{lj} \end{bmatrix}$$

Thus, we have shown Proposition (2.5.9).

(2.5.9) Proposition. If $[S]_\alpha^\beta$ has entries a_{ij}, $i = 1, \ldots, n$ and $j = 1, \ldots, m$ and $[T]_\beta^\gamma$ has entries b_{kl}, $k = 1, \ldots, p$ and $l = 1, \ldots, n$, then the entries of $[TS]_\alpha^\gamma$ are $\sum_{l=1}^{n} b_{kl}a_{lj}$.

If we analyze this formula carefully, it leads us to define an operation called multiplication of two matrices.

(2.5.10) Definition. Let A be an $n \times m$ matrix and B a $p \times n$ matrix, then the *matrix product BA* is defined to be the $p \times m$ matrix whose entries are $\sum_{l=1}^{n} b_{kl}a_{lj}$ for $k = 1, \ldots, p$ and $j = 1, \ldots, m$.

(2.5.11) Remark. The product of the matrices B and A can be obtained as follows: The entry in the kth row and jth column of BA is the product of the kth row of B (considered as a $1 \times n$ matrix) with the jth column of A considered as a column vector. We may write the matrix product as follows.

$$k\text{th row} \rightarrow \begin{bmatrix} b_{11} & \cdots & & b_{1n} \\ \cdot & & & \\ \cdot & & & \\ \cdot & & & \\ b_{k1} & b_{k2} & \cdots & b_{kn} \\ \cdot & & & \\ \cdot & & & \\ \cdot & & & \\ b_{p1} & \cdots & & b_{kn} \end{bmatrix} \begin{bmatrix} a_{11} & a_{1j} & a_{1m} \\ & a_{2j} & \\ & \cdot & \\ & \cdot & \\ & \cdot & \\ a_{n1} & a_{nj} & a_{mn} \\ & \uparrow & \\ & j\text{th column} & \end{bmatrix} = \begin{bmatrix} & \vdots & \\ \cdots & \sum_{l=1}^{n} b_{kl}a_{lj} & \cdots \\ & \vdots & \end{bmatrix}$$

$$\qquad\qquad\qquad\qquad\qquad\qquad\qquad\qquad\qquad\qquad j\text{th column}$$

We caution the reader that just as the composition of transformations TS is not defined if the range of S is not equal to the domain of T, the multiplication of matrices BA is not defined if the number of columns (or equivalently, the number of entries in each row) of B is not equal to the number of rows (or equivalently, the number of entries in each column) of A.

(2.5.12) Example. Calculate the matrix product

$$\begin{bmatrix} 1 & 2 & 0 \\ 1 & 3 & 1 \end{bmatrix} \begin{bmatrix} 1 & 0 & 1 & 0 \\ 0 & 1 & 2 & 1 \\ 2 & -1 & 3 & 1 \end{bmatrix}$$

The product of a 2×3 matrix with a 3×4 matrix is a 2×4 matrix. The entry in the first row and first column of the product is the product

$$[1 \quad 2 \quad 0] \begin{bmatrix} 1 \\ 0 \\ 2 \end{bmatrix} = 1 \cdot 1 + 2 \cdot 0 + 0 \cdot 2 = 1$$

The entry in the first row and second column of the product is the product

$$[1 \quad 2 \quad 0] \begin{bmatrix} 0 \\ 1 \\ -1 \end{bmatrix} = 1 \cdot 0 + 2 \cdot 1 + 0 \cdot -1 = 2$$

Continuing in this way, we find that the resulting 2×4 matrix is

$$\begin{bmatrix} 1 & 2 & 5 & 2 \\ 3 & 2 & 10 & 4 \end{bmatrix}$$

The following is a direct consequence of the definition of matrix multiplication and Proposition (2.5.9), and it shows how to compute the matrix of the composition of linear transformations.

(2.5.13) Proposition. Let $S: U \to V$ and $T: V \to W$ be linear transformations between finite-dimensional vector spaces. Let α, β, and γ be bases for U, V, and W, respectively. Then

$$[TS]_\alpha^\gamma = [T]_\beta^\gamma \, [S]_\alpha^\beta$$

In words, the matrix of the composition of two linear transformations is the product of the matrices of the transformations.

Proposition (2.5.4) can now be restated for multiplication of matrices and can be proved directly or by recourse to Propositions (2.5.4) and (2.5.13).

(2.5.14) Proposition

(i) Let A, B, and C be $m \times n$, $n \times p$, and $p \times r$ matrices, respectively, then

$$(AB)C = A(BC) \qquad \text{(associativity)}$$

(ii) Let A be an $m \times n$ matrix and B and C $n \times p$ matrices. Then

$$A(B + C) = AB + AC \qquad \text{(distributivity)}$$

(iii) Let A and B be $m \times n$ matrices, and let C be an $n \times p$ matrix, then

$$(A + B)C = AC + BC$$

EXERCISES

1. Compute the matrix of TS directly by evaluating TS on a basis for $\mathbf{R}^2$ for the following pairs of linear transformations S and T.
   a) $S = R_{\pi/2}$ and $T = P_{(1,1)}$
   b) $S = P_{(1,1)}$ and $T = R_{\pi/2}$
   c) $S = R_\theta$ and $T = R_\varphi$ (*Hint:* Use trigonometric identities to express your answer in terms of the angle $\varphi + \theta$.)
   d) $S = R_{(0,1)}$ and $T = R_{(1,0)}$, where $R_\mathbf{a}$ is a reflection (see Exercise 4 of Section 2.2). Do you recognize the matrix of TS?

2. Prove Proposition (2.5.4) (ii) and (iii).

3. Prove Proposition (2.5.6) (ii).

4. For part c of Example (2.5.2), compute the matrix product corresponding to the composition of transformations.

5. Compute the following matrix products if possible:

a) $\begin{bmatrix} 1 & 0 & 3 & 1 \\ 0 & 2 & 4 & 0 \\ 1 & -1 & 2 & 5 \end{bmatrix} \begin{bmatrix} 1 & -2 \\ -1 & 1 \\ 1 & -1 \\ 3 & 2 \end{bmatrix}$ b) $\begin{bmatrix} 1 & 5 \\ 6 & -1 \\ 0 & 2 \\ 3 & 1 \end{bmatrix} \begin{bmatrix} 1 & 0 & 1 \\ -1 & 2 & 0 \end{bmatrix}$

c) $\begin{bmatrix} 3 & 0 & 9 \end{bmatrix} \begin{bmatrix} 2 & 0 \\ -3 & -7 \\ 3 & 1 \end{bmatrix}$ d) $\begin{bmatrix} 1 & 9 \\ 2 & 13 \end{bmatrix} \begin{bmatrix} 2 & 3 & 0 \\ 7 & 8 & 0 \\ 0 & 0 & 0 \end{bmatrix}$

e) $\begin{bmatrix} 3 \\ 3 \\ 1 \\ 6 \end{bmatrix} \begin{bmatrix} 2 & -1 & 1 \end{bmatrix}$ f) $\begin{bmatrix} 2 & -1 & 1 \end{bmatrix} \begin{bmatrix} 3 \\ 3 \\ 1 \\ 6 \end{bmatrix}$

6. Let I be the identity linear transformation, $I: V \to V$.
   a) If $S: U \to V$ is a linear transformation, prove that $IS = S$.
   b) If $T: V \to W$ is a linear transformation, prove that $TI = T$.
   c) Repeat a and b of this exercise for the zero transformation. What are the compositions in this case?

7. Let I be the $n \times n$ identity matrix; that is, all the diagonal entries are 1 and all the off diagonal entries are zero.

a) Show that if A is an $n \times p$ matrix $IA = A$.
b) Show that if B is an $m \times n$ matrix $BI = B$.
c) Repeat a and b of this exercise for the zero matrix. What are the products in this case?

8. Let $V = P_3(\mathbf{R})$ and $W = P_4(\mathbf{R})$. Let $D: W \to V$ be the derivative mapping $D(p) = p'$, and let Int: $V \to W$ be the integration mapping $\text{Int}(p) = \int_0^x p(t)dt$ (i.e., Int(p) is the antiderivative of p such that $\text{Int}(p)(0) = 0$). Let $\alpha = \{1, x, x^2, x^3\}$ and $\beta = \{1, x, x^2, x^3, x^4\}$ be the "standard" bases in V and W.
   a) Compute $[D]_\beta^\alpha$ and $[\text{Int}]_\alpha^\beta$.
   b) Compute $[D\text{Int}]_\alpha^\alpha$ and $[\text{Int}D]_\beta^\beta$ using the appropriate matrix products. What theorem of calculus is reflected in your answer for $[D\text{Int}]_\alpha^\alpha$?

9. An $n \times n$ matrix A is called upper (lower) triangular if all its entries below (above) the diagonal are zero. That is, A is upper triangular if $a_{ij} = 0$ for all $i > j$, and lower triangular if $a_{ij} = 0$ for all $i < j$.
   a) Let A and B be $n \times n$ upper triangular matrices. Prove that AB is also upper triangular.
   b) Let A and B be $n \times n$ lower triangular matrices. Prove that AB is also lower triangular.

10. Let $T: V \to V$ be a linear transformation of a finite-dimensional vector space V. Let $\alpha = \{\mathbf{v}_1, \ldots, \mathbf{v}_n\}$ be a basis for V. Show that $[T]_\alpha^\alpha$ is upper triangular if and only if for each i, $1 \leq i \leq n$, $T(\mathbf{v}_i)$ is a linear combination of $\mathbf{v}_1, \ldots, \mathbf{v}_i$. What is the corresponding statement if $[T]_\alpha^\alpha$ is lower triangular?

11. Let A and B be matrices of the form

$$A = \left[\begin{array}{c|c} A_1 & A_2 \\ \hline 0 & A_3 \end{array}\right] \quad \text{and} \quad B = \left[\begin{array}{c|c} B_1 & B_2 \\ \hline 0 & B_3 \end{array}\right]$$

where A_1 and B_1 are $k \times k$ matrices, A_3 and B_3 are $l \times l$ matrices, A_2 and B_2 are $k \times l$ matrices, and 0 denotes an $l \times k$ block of zeros. Show that the matrix product AB has the same form, with an $l \times k$ block of zeros in the lower left corner.

12. Let $S: U \to V$ and $T: V \to W$ be linear transformations.
   a) If S and T are injective, is TS injective? If so, prove it; if not, find a counterexample.
   b) If S and T are surjective, is TS surjective? If so, prove it; if not, find a counterexample.

13. Give an example of linear transformations and vector spaces $S: U \to V$ and $T: V \to W$ such that TS is injective and surjective, but neither S nor T is both injective and surjective.

The elimination algorithm for solving systems of linear equations can be expressed in terms of matrix multiplication. That is, if A is an $m \times n$ matrix, each of the elementary operations used to reduce A to echelon form may be accomplished by multiplying A on the left by an appropriate matrix E. The underlying reason for this is that each of the three types of elementary operations is a *linear transformation* of each of the columns of A.

14. Let $E_{ij}(c)$ $(i \neq j)$ be the $m \times m$ matrix whose entries are

$$e_{kl} = \begin{cases} 1 & \text{if } k = l \\ c & \text{if } k = i \text{ and } l = j \\ 0 & \text{otherwise} \end{cases}$$

Show that for any $m \times n$ matrix A, $E_{ij}(c)A$ is the matrix obtained from A by adding c times the jth row to the ith row.

15. Let $F_i(c)$ $(c \neq 0)$ be the $m \times m$ matrix whose entries f_{kl} are

$$f_{kl} = \begin{cases} 1 & \text{if } k = l \neq i \\ c & \text{if } k = l = i \\ 0 & \text{otherwise} \end{cases}$$

Show that for any $m \times n$ matrix A, $F_i(c)A$ is the matrix obtained from A by multiplying the ith row of A by c.

16. Let G_{ij} be the $m \times m$ matrix whose entries g_{kl} are

$$g_{kl} = \begin{cases} 1 & \text{if } k = l \text{ but } k \neq i,j \\ 1 & \text{if } k = i \text{ and } l = j \text{ or } k = j \text{ and } l = i \\ 0 & \text{otherwise} \end{cases}$$

Show that for any $m \times n$ matrix A, $G_{ij}A$ is the matrix obtained from A by interchanging the ith row and the jth row.

The matrices of the forms $E_{ij}(c)$, $F_i(c)$, and G_{ij} are called *elementary matrices*. As a consequence of Exercises 14, 15, and 16, the elimination algorithm on a matrix A can be accomplished by multiplying A on the left by a product of elementary matrices. The first operation performed will correspond to the right-most factor in the product, and so on.

17. For each of the following matrices A, find the elementary matrices that will multiply A on the left to put A in echelon form.

a) $\begin{bmatrix} 2 & -1 \\ 1 & 2 \end{bmatrix}$ b) $\begin{bmatrix} 0 & 1 & -1 \\ 2 & 1 & 0 \end{bmatrix}$

c) $\begin{bmatrix} 1 & 2 & -1 \\ -1 & 0 & 1 \\ 1 & 1 & 0 \end{bmatrix}$ d) $\begin{bmatrix} 2 & -1 \\ 1 & 2 \\ 3 & -1 \end{bmatrix}$

18. Let A be an $n \times m$ matrix. What is the effect of multiplying A *on the right* by an $m \times m$ elementary matrix? That is, what are the matrices $AE_{ij}(c)$, $AF_i(c)$, and AG_{ij}?

19. Let $A, B \in M_{n\times n}(\mathbf{R})$. Show that $(AB)^t = B^tA^t$ gives the effect of taking the transpose of a product of two matrices. (See Exercise 12 of Section 2.2 of this chapter.)

§2.6. THE INVERSE OF A LINEAR TRANSFORMATION

In this section we use the special properties of linear transformations developed in Section 2.4 to determine when a linear transformation has an inverse. We begin by recalling the basic ideas about inverses of functions between any sets. If $f: S_1 \to S_2$ is a function from one set to another, we say that $g: S_2 \to S_1$ is the *inverse function of* f if for every $x \in S_1$, $g(f(x)) = x$ and for every $y \in S_2$, $f(g(y)) = y$.

In calculus, you have seen many examples of inverse functions. For example, let $S_1 = S_2 = \{x \in \mathbf{R} | x \geq 0\}$. Then $f(x) = x^2$ and $g(x) = \sqrt{x}$ are inverse functions. Another example is given by letting $S_1 = \{x \in \mathbf{R} | x > 0\}$, $S_2 = \mathbf{R}$, $f(x) = \ln(x)$ and $g(x) = e^x$.

If such a g exists, f must be both injective and surjective. To see this, notice that if $f(x_1) = f(x_2)$, then $x_1 = g(f(x_1)) = g(f(x_2)) = x_2$, so that f is injective. Further, if $y \in S_2$, then for $x = g(y)$, $f(x) = f(g(y)) = y$ so that f is surjective. Conversely, if f is both injective and surjective, g is defined by setting $g(y) = x$ if $f(x) = y$. Since f is surjective, such an x must exist. Since f is injective, this choice of x is unique. It follows that if an inverse function exists, it must be unique.

We now want to apply the same arguments to a linear transformation $T: V \to W$ to determine when it is possible to produce an inverse transformation $S: W \to V$ satisfying $ST(\mathbf{v}) = \mathbf{v}$ for all $\mathbf{v} \in V$ and $TS(\mathbf{w}) = \mathbf{w}$ for all $\mathbf{w} \in W$. From the arguments given before for functions between sets without any structure, we see that in order for S to exist, T must be injective and surjective. In this case the function S is defined as before, $S(\mathbf{w}) = \mathbf{v}$ for the unique $\mathbf{v}$ with $T(\mathbf{v}) = \mathbf{w}$. However, it remains to prove that S is a linear transformation.

(2.6.1) Proposition. If $T: V \to W$ is injective and surjective, then the inverse function $S: W \to V$ is a linear transformation.

Proof: Let $\mathbf{w}_1$ and $\mathbf{w}_2 \in W$ and a and $b \in \mathbf{R}$. By definition, $S(\mathbf{w}_1) = \mathbf{v}_1$ and $S(\mathbf{w}_2) = \mathbf{v}_2$ are the unique vectors $\mathbf{v}_1$ and $\mathbf{v}_2$ satisfying $T(\mathbf{v}_1) = \mathbf{w}_1$ and $T(\mathbf{v}_2) = \mathbf{w}_2$. By definition, $S(a\mathbf{w}_1 + b\mathbf{w}_2)$ is the unique vector $\mathbf{v}$ with $T(\mathbf{v}) = a\mathbf{w}_1 + b\mathbf{w}_2$, but $\mathbf{v} = a\mathbf{v}_1 + b\mathbf{v}_2$ satisfies $T(a\mathbf{v}_1 + b\mathbf{v}_2) = aT(\mathbf{v}_1) + bT(\mathbf{v}_2) = a\mathbf{w}_1 + b\mathbf{w}_2$. Thus, $S(a\mathbf{w}_1 + b\mathbf{w}_2) = a\mathbf{v}_1 + b\mathbf{v}_2 = aS(\mathbf{w}_1) + bS(\mathbf{w}_2)$ as we desired. ■

From the set theoretical argument, if a linear transformation S exists that is the inverse of T, then T must be injective and surjective. Thus, Proposition (2.6.1) can be extended as follows.

(2.6.2) Proposition. A linear transformation $T: V \to W$ has an inverse linear transformation S if and only if T is injective and surjective.

With these results in mind, we make the following two definitions.

(2.6.3) Definition. If $T: V \to W$ is a linear transformation that has an inverse transformation $S: W \to V$, we say that T is *invertible*, and we denote the inverse of T by T^{-1}.

(2.6.4) Definition. If $T: V \to W$ is an invertible linear transformation, T is called an *isomorphism*, and we say V and W are *isomorphic vector spaces*.

The word "isomorphism" is formed from Greek roots meaning "same form." Vector spaces that are isomorphic do have the same form in a certain sense, as we will see shortly.

We emphasize that the inverse transformation T^{-1} satisfies $T^{-1}T(\mathbf{v}) = \mathbf{v}$ for all $\mathbf{v} \in V$ and $TT^{-1}(\mathbf{w}) = \mathbf{w}$ for all $\mathbf{w} \in W$. Thus, $T^{-1}T$ is the identity linear transformation of V, $T^{-1}T = I_V$, and TT^{-1} is the identity linear transformation of W. $TT^{-1} = I_W$. If S is a linear transformation that is a candidate for the inverse, we need only verify that $ST = I_V$ and $TS = I_W$.

(2.6.5) Remark. We have used the notation T^{-1} in two different ways that can be distinguished by the contexts in which they appear. For any transformation T, which is not necessarily invertible, we have previously defined the inverse image of a vector $\mathbf{w}$ under the transformation T as a set, and it is denoted $T^{-1}(\{\mathbf{w}\})$, $T^{-1}(\{\mathbf{w}\}) = \{\mathbf{v} \in V | T(\mathbf{v}) = \mathbf{w}\}$. For a T that is not invertible, T^{-1} is only used in the symbol $T^{-1}(\{\mathbf{w}\})$, not as a well-defined function from W to V. If, as in Proposition (2.6.2), T is invertible, T^{-1} is a linear transformation and the symbol has meaning by itself. Thus, in Example (2.4.14) the transformations are not invertible, so there is no transformation T^{-1}, but it does make sense to consider the set $T^{-1}(\{\mathbf{w}\}) \subset V$.

(2.6.6) Examples

a) Let V be any vector space, and let $I: V \to V$ be the identity transformation. $I(\mathbf{v}) = \mathbf{v}$ for all $\mathbf{v} \in V$. I is an isomorphism and the inverse transformation is I itself, that is, $I(I(\mathbf{v})) = \mathbf{v}$ for all $\mathbf{v} \in V$. (Remember that the matrix of the identity, $[I]_\alpha^\alpha$, satisfies $[I]_\alpha^\alpha\,[I]_\alpha^\alpha = [I]_\alpha^\alpha$.)

b) Let $V = W = \mathbf{R}^2$, and let $T = R_\theta$. R_θ is injective and surjective. [See Example (2.4.1a).] The inverse transformation R_θ^{-1} is rotation through an angle $-\theta$, that is, $R_\theta^{-1} = R_{-\theta}$. To see this we need only verify that $R_\theta R_{-\theta} = R_{-\theta}R_\theta = I$. If we apply Example (2.5.2a), we see that $R_\theta R_{-\theta} = R_{-\theta}R_\theta = R_0 = I$.

c) Let $V = W = \mathbf{R}^2$, and let $T = P_{\mathbf{a}}$ be a projection to a vector $\mathbf{a} \neq \mathbf{0}$. Since $P_{\mathbf{a}}$ is neither surjective nor injective, it is not an isomorphism. [See Example (2.4.1b).]

If we now assume that V and W are finite-dimensional, the results of Section 2.4 give simple criteria for there to be an isomorphism $T: V \to W$.

(2.6.7) Proposition. If V and W are finite-dimensional vector spaces, then there is an isomorphism $T: V \to W$ if and only if $\dim(V) = \dim(W)$.

Proof: $\rightarrow$: If T is an isomorphism, T is injective and surjective so that $\dim(\mathrm{Ker}(T)) = 0$ and $\dim(\mathrm{Im}(T)) = \dim(W)$. By Theorem (2.3.17) $\dim(V) = \dim(W)$.

$\leftarrow$: If $\dim(V) = \dim(W)$, we must produce an isomorphism $T: V \to W$. Let

$\alpha = \{\mathbf{v}_1, \ldots, \mathbf{v}_n\}$ be a basis for V and $\beta = \{\mathbf{w}_1, \ldots, \mathbf{w}_n\}$ be a basis for W. Define T to be the linear transformation from V to W with $T(\mathbf{v}_i) = \mathbf{w}_i$, $i = 1, \ldots, n$. By Proposition (2.1.14), T is uniquely defined by this choice of values on α. To see that T is injective, notice that if $T(a_1\mathbf{v}_1 + \cdots + a_n\mathbf{v}_n) = \mathbf{0}$, then $a_1\mathbf{w}_1 + \cdots + a_n\mathbf{w}_n = \mathbf{0}$. Since the $\mathbf{w}$'s are a basis, $a_1 = \cdots = a_n = 0$. Then $\text{Ker}(T) = \{\mathbf{0}\}$ and T is injective. By Proposition (2.4.10) T is also surjective, and then by Proposition (2.6.2) it is an isomorphism. ■

Proposition (2.6.7) tells us that if two vector spaces have the same dimension, they are isomorphic. We can see from the proof that the isomorphism is not unique, however, since the bases α and β may be chosen in any way we like. Although certain vector spaces appear to have "natural" bases, for example, $\{\mathbf{e}_1, \ldots, \mathbf{e}_n\}$ in $\mathbf{R}^n$ or $\{1, x, x^2, \ldots, x^n\}$ in $P_n(\mathbf{R})$, which give rise to isomorphisms, natural has no meaning outside specific contexts. We will see in Chapter 4 that there are contexts in which there are very natural choices of basis for $\mathbf{R}^n$ that are not the standard basis. In Example (2.6.8a) we construct an isomorphism (as indicated in the proof of the proposition) between two subspaces of $\mathbf{R}^3$ of dimension two. In this example the choice of bases is arbitrary.

(2.6.8) Examples

a) Let $V \subset \mathbf{R}^3$ be defined by the single linear equation $x_1 - x_2 + x_3 = 0$, and let $W \subset \mathbf{R}^3$ be the subspace defined by the single linear equation $2x_1 + x_2 - x_3 = 0$. Since $\dim(V) = \dim(W) = 2$ it follows from Proposition (2.6.7) that V and W are isomorphic. To construct an isomorphism, we choose bases for V and W. A basis for V is given by $\mathbf{v}_1 = (1, 1, 0)$ and $\mathbf{v}_2 = (0, 1, 1)$. A basis for W is given by $\mathbf{w}_1 = (1, -1, 1)$ and $\mathbf{w}_2 = (-1/2, 1, 0)$. Define a linear transformation $T: V \to W$ by $T(\mathbf{v}_1) = \mathbf{w}_1$ and $T(\mathbf{v}_2) = \mathbf{w}_2$. From the proof of (2.6.7) it follows that T is an isomorphism, and that V and W are isomorphic vector spaces.

b) As we have seen, $\dim(\mathbf{R}^n) = n$, and $\dim(P_{n-1}(\mathbf{R})) = n$. Therefore, by Proposition (2.6.7), $\mathbf{R}^n$ and $P_{n-1}(\mathbf{R})$ are isomorphic vector spaces. An isomorphism $T: \mathbf{R}^n \to P_{n-1}(\mathbf{R})$ may be defined, for example, by $T(\mathbf{e}_i) = x^{i-1}$, for each i, $1 \leq i \leq n$.

c) The vector space $V = M_{m \times n}(\mathbf{R})$ (see Exercise 10 of Section 2.2) has dimension $m \cdot n$. One basis for V may be constructed as follows. Let E_{kl} be the $m \times n$ matrix whose entries e_{ij} are given by

$$e_{ij} = \begin{cases} 1 & \text{if } i = k \text{ and } j = l \\ 0 & \text{otherwise} \end{cases}$$

Thus, E_{kl} has one nonzero entry, a 1 in the kth row and the lth column. Then, $\alpha = \{E_{kl} | 1 \leq k \leq m, \text{ and } 1 \leq l \leq n\}$ forms a basis for V. By Proposition (2.6.7), V is isomorphic to the vector space $\mathbf{R}^{m\,n}$. As always, there are many different isomorphisms $T: M_{m\times n}(\mathbf{R}) \to \mathbf{R}^{m\cdot n}$.

The notion of isomorphism of vector spaces is an explanation and justification for the impression you have probably reached that certain pairs of vector spaces we have encountered are really "the same," even though the elements may be written in different ways. For instance, the isomorphism $T: \mathbf{R}^n \to P_{n-1}(\mathbf{R})$ given in Example (2.6.8b) shows that the vectors $(a_1, \ldots, a_n) \in \mathbf{R}^n$, and the polynomials $a_1 + a_2x + \cdots + a_nx^{n-1}$ really do behave in exactly the same way, with regard to vector addition and scalar multiplication. It is in this sense that isomorphic vector spaces have the same form.

If $T: V \to W$ is an isomorphism of finite-dimensional vector spaces, and we choose bases $\alpha = \{\mathbf{v}_1, \ldots, \mathbf{v}_n\}$ for V and $\beta = \{\mathbf{w}_1, \ldots, \mathbf{w}_n\}$ for W, T is represented by an $n \times n$ matrix $[T]_\alpha^\beta$. The inverse transformation is also represented by a matrix $[T^{-1}]_\beta^\alpha$. We will use the ideas of the previous section to construct $[T^{-1}]_\beta^\alpha$ if we already know $[T]_\alpha^\beta$. The algorithm we will use is called the *Gauss-Jordan method*. This is one of several methods for producing $[T^{-1}]_\beta^\alpha$. For n small, $n = 2, 3$, all the methods require roughly equal numbers of steps. However, for $n \geq 4$, the number of steps increases much less rapidly in the Gauss-Jordan method.

For simplicity, let A denote $[T]_\alpha^\beta$, and let B denote $[T^{-1}]_\beta^\alpha$ and denote the entries of A and B by a_{ij} and b_{ij}, respectively, where $i = 1, \ldots, n$ and $j = 1, \ldots, n$. Then $TT^{-1} = I_W$ and Proposition (2.5.13) yields the corresponding matrix equation; $AB = I$. If we view matrix multiplication as multiplying the jth column of B by A to obtain the jth column of I, we see that the jth column of B satisfies

$$\begin{bmatrix} a_{11} & \cdots & a_{1n} \\ \cdot & & \cdot \\ \cdot & & \cdot \\ \cdot & & \cdot \\ a_{n1} & \cdots & a_{nn} \end{bmatrix} \begin{bmatrix} b_{1j} \\ \cdot \\ \cdot \\ \cdot \\ b_{nj} \end{bmatrix} = \begin{bmatrix} 0 \\ \vdots \\ 1 \\ \vdots \\ 0 \end{bmatrix} \leftarrow j\text{th row}$$

It is clear then that to find the jth column of B, we need only solve the preceding system of equations. This we do by using elimination. Notice that since T is surjective we are guaranteed a solution, and since T is injective this solution must be unique. (This, in fact, gives another proof of the uniqueness of T^{-1}!)

To solve the system of equations, we begin by performing elimination on the augmented matrix of A [see Example (2.4.21)], which we write as

$$\left[\begin{array}{ccc|c} a_{11} & \cdots & a_{1n} & 0 \\ \cdot & & \cdot & \vdots \\ \cdot & & \cdot & 1 \\ \cdot & & \cdot & \vdots \\ a_{n1} & \cdots & a_{nn} & 0 \end{array}\right] \leftarrow j\text{th row}$$

Since T is an isomorphism, by Example (2.4.22), the resulting echelon form of the system is

$$\left[\begin{array}{cccccc|c} 1 & & & & & & c_1 \\ & 1 & & & & & \cdot \\ & & \cdot & & & & \cdot \\ & & & \cdot & & & \cdot \\ & & & & \cdot & & \\ & & & & & 1 & c_n \end{array}\right]$$

where $(c_1, \ldots, c_n)$ is the desired solution. That is, $c_1 = b_{1j}, \ldots, c_n = b_{nj}$. To determine each column of B, we must perform this procedure n times. If we begin with the $n \times 2n$ matrix

$$[A|I] = \left[\begin{array}{ccc|ccc} a_{11} & \cdots & a_{1n} & 1 & & \\ & & & & 1 & \\ \cdot & & \cdot & & & \cdot \\ \cdot & & \cdot & & & \cdot \\ \cdot & & \cdot & & & \cdot \\ a_{n1} & \cdots & a_{nn} & & & 1 \end{array}\right]$$

we can perform these n eliminations simultaneously. Perform all the row operations on A and I simultaneously until A is in echelon form, which, as we noted previously, is the identity. The resulting matrix on the right will be $B = [T^{-1}]_\beta^\alpha$.

(2.6.9) Example. Find the inverse transformation of $T: V \to W$, $\dim(V) = \dim(W) = 3$, defined in terms of $\alpha = \{\mathbf{v}_1, \mathbf{v}_2, \mathbf{v}_3\}$ and $\beta = \{\mathbf{w}_1, \mathbf{w}_2, \mathbf{w}_3\}$ by

$$\begin{aligned} T(\mathbf{v}_1) &= \mathbf{w}_1 + 2\mathbf{w}_2 \\ T(\mathbf{v}_2) &= \qquad\quad \mathbf{w}_2 + \mathbf{w}_3 \\ T(\mathbf{v}_3) &= \mathbf{w}_1 - \mathbf{w}_2 + \mathbf{w}_3 \end{aligned}$$

We have

$$A = [T]_\alpha^\beta = \begin{bmatrix} 1 & 0 & 1 \\ 2 & 1 & -1 \\ 0 & 1 & 1 \end{bmatrix}$$

To find $[T^{-1}]_\beta^\alpha$ and thus T^{-1}, we apply the Gauss-Jordan method to matrix A:

$$\left[\begin{array}{ccc|ccc} 1 & 0 & 1 & 1 & 0 & 0 \\ 2 & 1 & -1 & 0 & 1 & 0 \\ 0 & 1 & 1 & 0 & 0 & 1 \end{array}\right]$$

$$\left[\begin{array}{ccc|ccc} 1 & 0 & 1 & 1 & 0 & 0 \\ 0 & 1 & -3 & -2 & 1 & 0 \\ 0 & 1 & 1 & 0 & 0 & 1 \end{array}\right]$$

$$\left[\begin{array}{ccc|ccc} 1 & 0 & 1 & 1 & 0 & 0 \\ 0 & 1 & -3 & -2 & 1 & 0 \\ 0 & 0 & 4 & 2 & -1 & 1 \end{array}\right]$$

$$\left[\begin{array}{ccc|ccc} 1 & 0 & 1 & 1 & 0 & 0 \\ 0 & 1 & -3 & -2 & 1 & 0 \\ 0 & 0 & 1 & 1/2 & -1/4 & 1/4 \end{array}\right]$$

$$\left[\begin{array}{ccc|ccc} 1 & 0 & 1 & 1 & 0 & 0 \\ 0 & 1 & 0 & -1/2 & 1/4 & 3/4 \\ 0 & 0 & 1 & 1/2 & -1/4 & 1/4 \end{array}\right]$$

$$\left[\begin{array}{ccc|ccc} 1 & 0 & 0 & 1/2 & 1/4 & -1/4 \\ 0 & 1 & 0 & -1/2 & 1/4 & 3/4 \\ 0 & 0 & 1 & 1/2 & -1/4 & 1/4 \end{array}\right]$$

Therefore,

$$B = \begin{bmatrix} 1/2 & 1/4 & -1/4 \\ -1/2 & 1/4 & 3/4 \\ 1/2 & -1/4 & 1/4 \end{bmatrix}$$

As a consequence of the Gauss-Jordan construction, we see that if A is the matrix of an isomorphism $T: V \to W$—so that A is necessarily an $n \times n$ matrix—there is a matrix B so that $AB = I$. This leads us to make the following definition.

(2.6.10) Definition. An $n \times n$ matrix A is called *invertible* if there exists an $n \times n$ matrix B so that $AB = BA = I$. B is called the *inverse* of A and is denoted by A^{-1}.

Notice that we could rephrase our work in the previous example by saying that we produced a matrix inverse for the matrix A.

Just as the inverse of a transformation is unique, the inverse matrix is also unique. For if B and C satisfy $AB = BA = I$ and $AC = CA = I$, then $AB = AC$. If we multiply on the left by B (or C), we obtain

$$B = (BA)B = (BA)C = IC = C$$

(2.6.11) Proposition. Let $T: V \to W$ be an isomorphism of finite-dimensional vector spaces. Then for any choice of bases α for V and β for W

$$[T^{-1}]_\beta^\alpha = [T]_\alpha^{\beta\,-1}$$

Proof: It must be shown that $[T^{-1}]_\beta^\alpha$ is the matrix inverse of $[T]_\alpha^\beta$. We have $[T^{-1}]_\beta^\alpha\,[T]_\alpha^\beta = [T^{-1}T]_\alpha^\alpha$, by Proposition (2.5.13), which equals $[I]_\alpha^\alpha$, the $n \times n$ identity matrix. Since this also holds in the reverse order, $[T]_\alpha^\beta\,[T^{-1}]_\beta^\alpha = [I]_\beta^\beta$, the $n \times n$ identity matrix, the uniqueness argument given prior to this proposition gives the result. ■

The invertibility of a transformation or a matrix also plays a role in solving systems of equations. Let $T: V \to W$ be an isomorphism, or equivalently, an invertible linear transformation. Let a vector $\mathbf{w} \in W$ be given, and suppose we wish to find all $\mathbf{v} \in V$ that satisfy $T(\mathbf{v}) = \mathbf{w}$. Since an isomorphism is injective and surjective, there exists exactly one such $\mathbf{v}$. Since T is an isomorphism, T^{-1} exists and we may evaluate T^{-1} on both sides of $T(\mathbf{v}) = \mathbf{w}$. Thus, $\mathbf{v} = T^{-1}T(\mathbf{v}) = T^{-1}(\mathbf{w})$ and the solution is $T^{-1}(\mathbf{w})$.

In order to apply this to a system of equations $A\mathbf{x} = \mathbf{b}$, we need to know that the matrix A is the matrix of an isomorphism. This can occur only if A is a square matrix, say, $n \times n$, and if the echelon form of A is the identity matrix so that the Gauss-Jordan method can be carried successfully to completion. This is equivalent to saying that the columns of A are linearly independent. (There are n basic variables!) If this condition holds, A^{-1} exists and $A\mathbf{x} = \mathbf{b}$ is solved by $\mathbf{x} = A^{-1}\mathbf{b}$.

Before we get our hopes up, notice that to find A^{-1}, we must complete the elimination algorithm to reduce A to echelon form and perform the elimination steps on n columns of the matrix I. However, to find $\mathbf{x}$ directly by elimination we need only apply the elimination steps to one extra vector, $\mathbf{b}$. Thus, to solve $A\mathbf{x} = \mathbf{b}$ by writing $\mathbf{x} = A^{-1}\mathbf{b}$ is a clean theoretical tool; as a computational tool, it requires more calculations than applying elimination directly.

EXERCISES

1. Are the following transformations isomorphisms? If so find the inverse transformation:
   a) $T: P_3(\mathbf{R}) \to P_3(\mathbf{R})$ given by $T(p(x)) = x\,dp(x)/dx$.
   b) $T: \mathbf{R}^3 \to \mathbf{R}^3$ whose matrix with respect to the standard basis is
   $$\begin{bmatrix} 1 & 1 & 0 \\ & -1 & 1 \\ & & 1 \end{bmatrix}$$
   c) $T: P_2(\mathbf{R}) \to \mathbf{R}^3$ given by $T(p(x)) = (p(0), p(1), p(2))$.
   d) $T: V \to V$, $\dim(V) = 4$, and with respect to the basis $\{\mathbf{v}_1, \ldots, \mathbf{v}_4\}$, T is given by $T(\mathbf{v}_1) = \mathbf{v}_2$, $T(\mathbf{v}_2) = \mathbf{v}_1$, $T(\mathbf{v}_3) = \mathbf{v}_4$, and $T(\mathbf{v}_4) = \mathbf{v}_3$.

2. Which of the following pairs of vector spaces are isomorphic? If so, find an isomorphism $T: V \to W$. If not, say why not.
   a) $V = \mathbf{R}^4$, $W = \{p \in P_4(\mathbf{R}) | p(0) = 0\}$.
   b) $V = \{\mathbf{x} \in \mathbf{R}^2 | P_{\mathbf{a}}(\mathbf{x}) = \mathbf{0} \text{ for } \mathbf{a} \neq \mathbf{0}\}$, $\quad W = P_1(\mathbf{R})$.
   c) $V = P_5(\mathbf{R})$, $\quad W = M_{2\times 3}(\mathbf{R})$.
   d) $V = \mathbf{R}^3$, $\quad W = C^\infty(\mathbf{R})$.

3. Determine which of the following matrices are invertible and if so find the inverse.

   a) $\begin{bmatrix} 1 & 3 \\ 3 & 2 \end{bmatrix}$ $\qquad$ b) $\begin{bmatrix} 1 & 2 \\ 0 & 1 \end{bmatrix}$

   c) $\begin{bmatrix} 2 & 0 \\ 1 & 3 \end{bmatrix}$ $\qquad$ d) $\begin{bmatrix} \cos(\theta) & -\sin(\theta) \\ \sin(\theta) & \cos(\theta) \end{bmatrix}$

   e) $\begin{bmatrix} 1 & 0 & 1 \\ -1 & 2 & 1 \\ 1 & -1 & 1 \end{bmatrix}$ $\qquad$ f) $\begin{bmatrix} 1 & 1 & 0 \\ & 2 & 1 \\ & & 3 \end{bmatrix}$

   g) $\begin{bmatrix} 0 & 1 & 0 & 0 \\ 1 & 0 & 0 & 0 \\ 0 & 0 & 0 & 1 \\ 0 & 0 & 1 & 0 \end{bmatrix}$ $\qquad$ h) $\begin{bmatrix} 1 & -1 & 0 & 1 \\ 0 & 0 & 1 & 1 \\ 1 & -1 & 0 & 1 \\ 0 & 1 & 2 & 1 \end{bmatrix}$

4. Let A be an invertible 2×2 matrix. Find a general formula for A^{-1} in terms of the entries of A.

5. Let A and B be invertible $n \times n$ matrices. Prove that $(AB)^{-1} = B^{-1}A^{-1}$.

6. Prove that if $T: V \to W$ is an isomorphism, T^{-1} is also an isomorphism. What is the inverse of T^{-1}?

7. a) Prove that if $S: U \to V$ and $T: V \to W$ are isomorphisms, then TS is also an isomorphism.
   b) What is $(TS)^{-1}$?

8. Let $T: V \to W$ be an isomorphism. Let $\{\mathbf{v}_1, \ldots, \mathbf{v}_k\}$ be a subset of V. Prove that $\{\mathbf{v}_1, \ldots, \mathbf{v}_k\}$ is a linearly independent set if and only if $\{T(\mathbf{v}_1), \ldots, T(\mathbf{v}_k)\}$ is a linearly independent set.

9. Let $T: V \to W$ be an isomorphism of finite-dimensional vector spaces.
   a) Let U be a subspace of V, prove that $\dim(U) = \dim(T(U))$ and that $T|_U$ is an isomorphism between U and $T(U)$. (*Hint:* Use Exercise 8.)
   b) Let X be a subspace of W, prove that $\dim(T^{-1}(X)) = \dim(X)$ and that $T^{-1}|_X$ is an isomorphism between X and $T^{-1}(X)$. (*Hint:* Use Exercise 6 and part a of this exercise.)

10. Let $T: V \to W$ be an isomorphism of finite-dimensional vector spaces.
   a) Let $S: W \to X$ be a linear transformation. Show that $\dim(\text{Ker}(S)) = \dim(\text{Ker}(ST))$ and that $\dim(\text{Im}(S)) = \dim(\text{Im}(ST))$. (*Hint:* Apply Exercise 9.)
   b) Let $S: U \to V$ be a linear transformation and assume U is finite-dimensional. Show that $\dim(\text{Ker}(S)) = \dim(\text{Ker}(TS))$ and that $\dim(\text{Im}(S)) = \dim(\text{Im}(TS))$. (*Hint:* Apply Exercise 9.)

11. What will happen if we apply the Gauss-Jordan procedure discussed in this section to a matrix A that is not invertible?

12. Let $E_{ij}(c)$, $F_i(c)$, and G_{ij} denote the elementary matrices of Exercises 14, 15, and 16 of Section 2.5. Find $E_{ij}(c)^{-1}$, $F_i(c)^{-1}$ and G_{ij}^{-1} and express them in terms of elementary matrices.

As Exercises 14, 15, and 16 of Section 2.5 indicate, the elimination procedure can be expressed in terms of matrix multiplication. Exercise 12 shows that these matrices are invertible. Using these results, we can express the Gauss-Jordan procedure in terms of matrix multiplication as well.

13. Let A be an invertible $n \times n$ matrix. Let $E_1, \ldots, E_k$ be elementary matrices such that $E_k \cdots E_1 A$ is in echelon form. Show that $E_k \cdot \cdots \cdot E_1 = A^{-1}$.

14. Use the matrix version of the Gauss-Jordan procedure developed in Exercise 13 to find A^{-1} for the following matrices A:

a) $\begin{bmatrix} 2 & -1 \\ 3 & -1 \end{bmatrix}$ b) $\begin{bmatrix} 0 & 0 & 1 \\ 0 & 1 & 0 \\ 1 & 0 & 1 \end{bmatrix}$

c) $\begin{bmatrix} 1 & -1 & 2 \\ 1 & 2 & 1 \\ 3 & -1 & 0 \end{bmatrix}$ d) $\begin{bmatrix} 1 & a & b \\ 0 & 1 & c \\ 0 & 0 & 1 \end{bmatrix}$ $a, b, c \in \mathbf{R}$

15. Let A be an upper (lower) triangular matrix.
   a) Show that A is invertible if and only if the diagonal entries of A are all nonzero.
   b) Show that if A is invertible, then A^{-1} is also an upper (lower) triangular matrix.

16. Let A be an invertible $n \times n$ matrix of the form $A = \left[\begin{array}{c|c} A_1 & A_2 \\ \hline 0 & A_3 \end{array}\right]$ (see Exercise 11 of Section 2.5).
   a) Prove that A^{-1} is also of this form.
   b) Prove that A_1 and A_3 are invertible.
   c) Prove that if $A_2 = 0$, then $A^{-1} = \left[\begin{array}{c|c} A_1^{-1} & 0 \\ \hline 0 & A_3^{-1} \end{array}\right]$.

17. Show that isomorphism of vector spaces is an equivalence relation (see Appendix 1). That is, prove each of the following statements:
   a) Every vector space V is isomorphic to itself.
   b) If V and W are vector spaces, and V is isomorphic to W, then W is isomorphic to V.
   c) If U, V, and W are vector spaces and U is isomorphic to V and V is isomorphic to W, then U is isomorphic to W.
   (*Hint:* In each case the key point is to find the isomorphism in terms of the other information given.)

§2.7. CHANGE OF BASIS

In this chapter we have often begun the discussion of a linear transformation $T: V \to W$ between finite-dimensional vector spaces by choosing bases α for V and β for W. For example, in order to find a basis for Ker(T), we used the system of linear equations: $[T]_\alpha^\beta[\mathbf{x}]_\alpha = \mathbf{0}$. Or, to find the inverse transformation of an isomorphism, we applied elimination to $[T]_\alpha^\beta$ in the Gauss-Jordan procedure. In both cases dim(Ker(T)) and T^{-1} are actually defined without reference to a choice of bases and the accompanying systems of coordinates in V and W, but the calculations required the choice of a basis. It is, thus, a reasonable question to ask how these choices of bases affect our final answer. Certainly, we have no reason to believe that if α' and β' are different bases for V and W, respectively, then $[T]_\alpha^\beta = [T]_{\alpha'}^{\beta'}$. In this section we will see the precise (and elegant) way in which they are related.

We first address the question of how changing the basis in a vector space affects the coordinates of a vector.

(2.7.1) Examples

a) Let $V = \mathbf{R}^2$, and let β be the basis consisting of the vectors $\mathbf{u}_1 = (1, 1)$ and $\mathbf{u}_2 = (1, -1)$ and β' be the basis consisting of the vectors $\mathbf{u}_1' = (1, 2)$ and $\mathbf{u}_2' = (-2, 1)$. If $\mathbf{v}$ is any vector in V, then $\mathbf{v}$ can be expressed uniquely as $\mathbf{v} =$

$x_1\mathbf{u}_1 + x_2\mathbf{u}_2$, and also uniquely as $\mathbf{v} = x_1'\mathbf{u}_1' + x_2'\mathbf{u}_2'$. Given the coefficients (x_1, x_2), can we determine (x_1', x_2')?

Notice that $\mathbf{u}_1$ and $\mathbf{u}_2$ in particular can be expressed in terms of $\mathbf{u}_1'$ and $\mathbf{u}_2'$:

$$\mathbf{u}_1 = (3/5)\mathbf{u}_1' - (1/5)\mathbf{u}_2' \quad \text{and} \quad \mathbf{u}_2 = (-1/5)\mathbf{u}_1' - (3/5)\mathbf{u}_2'$$

Hence

$$\begin{aligned}\mathbf{v} = x_1\mathbf{u}_1 + x_2\mathbf{u}_2 &= x_1((3/5)\mathbf{u}_1' - (1/5)\mathbf{u}_2') + x_2((-1/5)\mathbf{u}_1' - (3/5)\mathbf{u}_2')\\ &= ((3/5)x_1 - (1/5)x_2)\mathbf{u}_1' + ((-1/5)x_1 - (3/5)x_2)\mathbf{u}_2'\end{aligned}$$

Since x_1' and x_2' are uniquely determined by $\mathbf{v} = x_1'\mathbf{u}_1' + x_2'\mathbf{u}_2'$, it follows that

$$x_1' = (3/5)x_1 - (1/5)x_2 \quad \text{and} \quad x_2' = (-1/5)x_1 - (3/5)x_2$$

From our description of systems of linear equations in terms of matrices, we see that this last pair of equations can be written in the form

$$\begin{bmatrix} 3/5 & -1/5 \\ -1/5 & -3/5 \end{bmatrix} \begin{bmatrix} x_1 \\ x_2 \end{bmatrix} = \begin{bmatrix} x_1' \\ x_2' \end{bmatrix} \tag{2.1}$$

Therefore, we can compute the coordinates of any vector $\mathbf{v}$ in the second basis β' by multiplying its vector of coordinates in the basis β by the matrix appearing in Eq. (2.1).

b) We now derive the expression of Eq. (2.1) in terms of linear transformations. Let $I: V \to V$ be the identity transformation from V to itself (i.e., $I(\mathbf{v}) = \mathbf{v}$, for all $\mathbf{v} \in V$). Let β and β' be as before. We calculate the matrix $[I]_\beta^{\beta'}$. To do this, we must express $I(\mathbf{u}_1) = \mathbf{u}_1$ and $I(\mathbf{u}_2) = \mathbf{u}_2$ in terms of $\mathbf{u}_1'$ and $\mathbf{u}_2'$, which is exactly the calculation we performed in part a of this example. From part a we find that

$$[I]_\beta^{\beta'} = \begin{bmatrix} 3/5 & -1/5 \\ -1/5 & -3/5 \end{bmatrix}$$

Thus, in this example we see that to express the coordinates of the vector $\mathbf{v}$ in terms of β', we need only determine the matrix $[I]_\beta^{\beta'}$ and use matrix multiplication.

(2.7.2) Remark. Notice that β and β' are *not* the same basis, so that $[I]_\beta^{\beta'}$ is *not* the identity matrix.

We are now ready to show that the pattern noticed in Example (2.7.1) holds in general, namely, Proposition (2.7.3).

(2.7.3) Proposition. Let V be a finite-dimensional vector space, and let α and α' be bases for V. Let $\mathbf{v} \in V$. Then the coordinate vector $[\mathbf{v}]_{\alpha'}$ of $\mathbf{v}$ in the basis α' is related to the coordinate vector $[\mathbf{v}]_\alpha$ of $\mathbf{v}$ in the basis α by

$$[I]_\alpha^{\alpha'}[\mathbf{v}]_\alpha = [\mathbf{v}]_{\alpha'}$$

Proof: By Proposition (2.2.15), $[I]_\alpha^{\alpha'}[\mathbf{v}]_\alpha$ is the coordinate vector of $I(\mathbf{v})$ in the basis α'. But $I(\mathbf{v}) = \mathbf{v}$, so the result follows immediately. ■

We will call the matrix $[I]_\alpha^{\alpha'}$ the *change of basis matrix* which changes from α coordinates to α' coordinates.

If we want to express $[\mathbf{v}]_\alpha$ in terms of $[\mathbf{v}]_{\alpha'}$, we can simply apply the proposition with the roles of α and α' reversed. This yields

$$[I]_{\alpha'}^{\alpha}[\mathbf{v}]_{\alpha'} = [\mathbf{v}]_\alpha$$

Notice that I is an invertible linear mapping, and $I^{-1} = I$, so that $[I^{-1}]_{\alpha'}^{\alpha} = [I]_{\alpha'}^{\alpha}$. Thus, another equation for $[\mathbf{v}]_\alpha$ can be obtained by multiplying the conclusion of the proposition by the matrix $([I]_\alpha^{\alpha'})^{-1}$ on the left. Hence, we find that $([I]_\alpha^{\alpha'})^{-1} = [I]_{\alpha'}^{\alpha}$.

(2.7.4) Example. There is a special case of Proposition (2.7.3) that will prove useful in calculations.

Let $V = \mathbf{R}^n$, and let $\alpha = \{\mathbf{u}_1, \ldots, \mathbf{u}_n\}$ be any basis for $\mathbf{R}^n$. If we let $\alpha' = \{\mathbf{e}_1, \ldots, \mathbf{e}_n\}$ be the standard basis of $\mathbf{R}^n$, then the change of basis matrix $[I]_\alpha^{\alpha'}$ is particularly easy to compute. If we write $\mathbf{u}_j = (a_{1j}, \ldots, a_{nj})$ for the coordinate vector of $\mathbf{u}_j$ with respect to the standard basis, then by definition, $\mathbf{u}_j = a_{1j}\mathbf{e}_1 + \cdots + a_{nj}\mathbf{e}_n$. Hence

$$[I]_\alpha^{\alpha'} = \begin{bmatrix} a_{11} & \cdots & a_{1n} \\ & \vdots & \\ a_{n1} & \cdots & a_{nn} \end{bmatrix}$$

In other words, the matrix that changes coordinates from the coordinates with respect to an arbitrary basis α for $\mathbf{R}^n$ to the coordinates with respect to the standard basis is just the matrix whose columns are the standard coordinate vectors of the elements of α.

We now move to the more general problem of determining the effect of a change of basis on the matrix of an arbitrary linear transformation $T: V \to W$ between finite-dimensional vector spaces.

(2.7.5) Theorem. Let $T: V \to W$ be a linear transformation between finite-dimensional vector spaces V and W. Let $I_V: V \to V$ and $I_W: W \to W$ be the respective identity transformations of V and W. Let α and α' be two bases for V, and let β and β' be two bases for W. Then

$$[T]_{\alpha'}^{\beta'} = [I_W]_\beta^{\beta'} \cdot [T]_\alpha^\beta \cdot [I_V]_{\alpha'}^{\alpha}$$

Proof: We apply Proposition (2.5.13). Since I_V and I_W are the respective identity transformations, $T = I_W T I_V$. Applying Proposition (2.5.13) once, we see that $[T]_{\alpha'}^{\beta'} = [I_W]_{\beta}^{\beta'}\ [TI_V]_{\alpha'}^{\beta}$. Applying (2.5.13) a second time, we have that $[TI_V]_{\alpha'}^{\beta} = [T]_{\alpha}^{\beta}[I_V]_{\alpha'}^{\alpha}$. Together these prove the theorem. ■

Before giving an example of this type of calculation, notice that (as we remarked before) $[I_W]_{\beta}^{\beta'} = ([I_W]_{\beta'}^{\beta})^{-1}$. Hence, the formula of the theorem can be expressed as

$$[T]_{\alpha'}^{\beta'} = ([I_W]_{\beta'}^{\beta})^{-1} \cdot [T]_{\alpha}^{\beta} \cdot [I_V]_{\alpha'}^{\alpha}$$

This formulation is particularly convenient when $V = W$ and $\alpha = \beta$ and $\alpha' = \beta'$. In this case we have

$$[T]_{\alpha'}^{\alpha'} = ([I_V]_{\alpha'}^{\alpha})^{-1} \cdot\ [T]_{\alpha}^{\alpha} \cdot [I_V]_{\alpha'}^{\alpha}$$

so that only one change of basis matrix $[I_V]_{\alpha'}^{\alpha}$ (together with its inverse) is involved. This leads to a definition that will be used extensively in Chapter 4.

(2.7.6) Definition. Let A, B be $n \times n$ matrices. A and B are said to be *similar* if there is an invertible $n \times n$ matrix Q such that

$$B = Q^{-1}AQ$$

Thus, our previous observation could be rephrased as follows: If $T: V \to V$ is a linear transformation and α and α' are two bases for V, then $A = [T]_{\alpha}^{\alpha}$ is similar to $B = [T]_{\alpha'}^{\alpha'}$ and the invertible matrix Q in the definition is the matrix $Q = [I_V]_{\alpha'}^{\alpha}$.

We finish this section with several examples. The first example is the most general.

(2.7.7) Examples

a) Let $V = \mathbf{R}^3$ and $W = \mathbf{R}^2$. Let $\alpha = \{(3, 0, 1), (3, 1, 1), (2, 1, 1)\}$, $\alpha' = \{(1, 1, 1), (1, 1, 0), (1, 0, 0)\}$, $\beta = \{(1, 1), (1, -1)\}$, and $\beta' = \{(1, 2), (-2, 1)\}$. Let $T: V \to W$ be the linear mapping with

$$[T]_{\alpha}^{\beta} = \begin{bmatrix} 1 & 2 & -1 \\ 0 & 1 & -1 \end{bmatrix}$$

What is the matrix of T with respect to the bases α' and β'?

From Theorem (2.7.5) we have

$$[T]_{\alpha'}^{\beta'} = [I_W]_{\beta}^{\beta'} \cdot\ [T]_{\alpha}^{\beta} \cdot\ [I_V]_{\alpha'}^{\alpha}$$

Now, the matrix $[I]_{\beta}^{\beta'}$ was calculated in Example (2.7.1). $[T]_{\alpha}^{\beta}$ was given previously. We need only calculate $[I]_{\alpha'}^{\alpha}$. As in Example (2.7.1), we need to express the members of α' in terms of the members of α. You should check that the resulting matrix is

$$[I]^{\alpha}_{\alpha'} = \begin{bmatrix} 0 & -1 & 0 \\ -1 & 2 & 1 \\ 2 & -1 & -1 \end{bmatrix}$$

Therefore

$$[T]^{\beta'}_{\alpha'} = \begin{bmatrix} 3/5 & -1/5 \\ -1/5 & -3/5 \end{bmatrix} \cdot \begin{bmatrix} 1 & 2 & -1 \\ 0 & 1 & -1 \end{bmatrix} \cdot \begin{bmatrix} 0 & -1 & 0 \\ -1 & 2 & 1 \\ 2 & -1 & -1 \end{bmatrix}$$

$$= \begin{bmatrix} -9/5 & 9/5 & 7/5 \\ 13/5 & -13/5 & -9/5 \end{bmatrix}$$

b) Let T: $\mathbf{R}^3 \to \mathbf{R}^3$ be the transformation whose matrix with respect to the basis α of part a is

$$[T]^{\alpha}_{\alpha} = \begin{bmatrix} 1 & 0 & 0 \\ 1 & 1 & 0 \\ 0 & 1 & 1 \end{bmatrix}$$

What is $[T]^{\alpha'}_{\alpha'}$ for the basis α' of part a? We must compute $[I]^{\alpha'}_{\alpha} = ([I]^{\alpha}_{\alpha'})^{-1}$. The reader should verify that

$$[I]^{\alpha'}_{\alpha} = \begin{bmatrix} 1 & 1 & 1 \\ -1 & 0 & 0 \\ 3 & 2 & 1 \end{bmatrix}$$

by applying the Gauss-Jordan procedure. Therefore

$$[T]^{\alpha'}_{\alpha'} = [I]^{\alpha'}_{\alpha} \cdot [T]^{\alpha}_{\alpha} \cdot [I]^{\alpha}_{\alpha'}$$

$$= \begin{bmatrix} 1 & 1 & 1 \\ -1 & 0 & 0 \\ 3 & 2 & 1 \end{bmatrix} \cdot \begin{bmatrix} 1 & 0 & 0 \\ 1 & 1 & 0 \\ 0 & 1 & 1 \end{bmatrix} \cdot \begin{bmatrix} 0 & -1 & 0 \\ -1 & 2 & 1 \\ 2 & -1 & -1 \end{bmatrix}$$

$$= \begin{bmatrix} 0 & 1 & 1 \\ 0 & 1 & 0 \\ -1 & 0 & 2 \end{bmatrix}$$

c) Let T: $\mathbf{R}^3 \to \mathbf{R}^3$ be the transformation of part b, and let α be the basis of part a. Let γ be the standard basis of $\mathbf{R}^3$. What is $[T]^{\gamma}_{\gamma}$? We know that $[T]^{\gamma}_{\gamma} = [I]^{\gamma}_{\alpha}[T]^{\alpha}_{\alpha}\,[I]^{\alpha}_{\gamma}$. From Example (2.7.4) we see that

$$[I]^{\gamma}_{\alpha} = \begin{bmatrix} 3 & 3 & 2 \\ 0 & 1 & 1 \\ 1 & 1 & 1 \end{bmatrix}$$

Then

$$[I]^{\alpha}_{\gamma} = ([I]^{\gamma}_{\alpha})^{-1} = \begin{bmatrix} 0 & -1 & 1 \\ 1 & 1 & -3 \\ -1 & 0 & 3 \end{bmatrix}$$

Therefore

$$[T]^{\gamma}_{\gamma} = [I]^{\gamma}_{\alpha} \cdot [T]^{\alpha}_{\alpha} \cdot [I]^{\alpha}_{\gamma}$$

$$= \begin{bmatrix} 3 & 3 & 2 \\ 0 & 1 & 1 \\ 1 & 1 & 1 \end{bmatrix} \cdot \begin{bmatrix} 1 & 0 & 0 \\ 1 & 1 & 0 \\ 0 & 1 & 1 \end{bmatrix} \cdot \begin{bmatrix} 0 & -1 & 1 \\ 1 & 1 & -3 \\ -1 & 0 & 3 \end{bmatrix}$$

$$= \begin{bmatrix} 3 & -1 & -3 \\ 1 & 1 & -2 \\ 1 & 0 & -1 \end{bmatrix}$$

(2.7.8) Example. If we are working in $\mathbf{R}^n$ and α and α' are bases for $\mathbf{R}^n$, we can write an explicit formula for $[I]^{\alpha'}_{\alpha}$ without solving for one basis in terms of the other. Let β be the standard basis in $\mathbf{R}^n$, then it is clear from our discussion that $[I]^{\alpha'}_{\alpha} = [I]^{\alpha'}_{\beta}\,[I]^{\beta}_{\alpha}$. Thus, if we know the coordinates of the basis vectors of α and α', we have, following Example (2.7.4), that

$$[I]^{\alpha'}_{\alpha} = \begin{bmatrix} a'_{11} & \cdots & a'_{1n} \\ & \vdots & \\ a'_{n1} & \cdots & a'_{1n} \end{bmatrix}^{-1} \begin{bmatrix} a_{11} & \cdots & a_{1n} \\ & \vdots & \\ a_{n1} & \cdots & a_{nn} \end{bmatrix}$$

where the ith member of α is $(a_{1i}, \ldots, a_{ni})$ in standard coordinates and the jth member of α' is $(a'_{1j}, \ldots, a'_{nj})$ in standard coordinates.

For example, to find the change of basis matrix from $\alpha = \{(1, 2), (3, 4)\}$ to $\alpha' = \{(0, 3), (1, 1)\}$ in $\mathbf{R}^2$, we can use the standard basis $\beta = \{(1, 0), (0, 1)\}$ in an intermediary role. We have

$$[I]^{\beta}_{\alpha} = \begin{bmatrix} 1 & 3 \\ 2 & 4 \end{bmatrix} \quad \text{and} \quad [I]^{\beta}_{\alpha'} = \begin{bmatrix} 0 & 1 \\ 3 & 1 \end{bmatrix}$$

Therefore

$$[I]^{\alpha'}_{\alpha} = [I]^{\alpha'}_{\beta}\,[I]^{\beta}_{\alpha} = ([I]^{\beta}_{\alpha'})^{-1}\,[I]^{\beta}_{\alpha}$$

$$= \begin{bmatrix} 0 & 1 \\ 3 & 1 \end{bmatrix}^{-1} \cdot \begin{bmatrix} 1 & 3 \\ 2 & 4 \end{bmatrix}$$

$$= \begin{bmatrix} -1/3 & 1/3 \\ 1 & 0 \end{bmatrix} \cdot \begin{bmatrix} 1 & 3 \\ 2 & 4 \end{bmatrix}$$

$$= \begin{bmatrix} 1/3 & 1/3 \\ 1 & 3 \end{bmatrix}$$

EXERCISES

1. Let $T: \mathbf{R}^2 \to \mathbf{R}^2$ be rotation through an angle θ.
   a) What is the matrix of T with respect to the basis $\{(2, 1), (1, -2)\}$?
   b) What is the matrix of T with respect to the basis $\{(1, 1), (1, 0)\}$?

2. Let $\alpha = \{(1, 1, 1), (1, 1, 0), (1, 0, 0)\}$ be a basis for $\mathbf{R}^3$. Let T satisfy $T((1, 1, 1)) = (2, 2, 2)$, $T((1, 1, 0)) = (3, 3, 0)$, and $T((1, 0, 0)) = (-1, 0, 0)$. Find $[T]_\beta^\beta$ for β the standard basis for $\mathbf{R}^3$.

3. Let $T: \mathbf{R}^2 \to \mathbf{R}^3$ be defined by $T(\mathbf{u}_1) = \mathbf{v}_1 + \mathbf{v}_2$ and $T(\mathbf{u}_2) = \mathbf{v}_2 + \mathbf{v}_3$ for the bases $\mathbf{u}_1 = (1, 1)$ and $\mathbf{u}_2 = (1, -1)$ for $\mathbf{R}^2$ and $\mathbf{v}_1 = (0, 0, 2)$, $\mathbf{v}_2 = (1, 1, 0)$ and $\mathbf{v}_3 = (0, 1, 0)$ for $\mathbf{R}^3$. What is the matrix of T with respect to the bases $\mathbf{u}_1' = (1, 0)$ and $\mathbf{u}_2' = (1, 1)$ for $\mathbf{R}^2$ and $\mathbf{v}_1' = (0, 1, 0)$, $\mathbf{v}_2' = (0, 0, 1)$, and $\mathbf{v}_3' = (1, 0, 0)$ for $\mathbf{R}^3$?

4. Each of the following matrices A is the matrix of a linear mapping $T: \mathbf{R}^n \to \mathbf{R}^n$ with respect to the standard basis. Compute $[T]_\alpha^\alpha$ for the given basis α.

a) $A = \begin{bmatrix} 1 & 3 \\ 3 & 1 \end{bmatrix}$, $\alpha = \{(1, 1), (1, -1)\}$

b) $A = \begin{bmatrix} 0 & 1 \\ 1 & 0 \end{bmatrix}$, $\alpha = \{(0, 1), (1, 0)\}$

c) $A = \begin{bmatrix} 1 & 0 & 1 \\ 2 & 2 & 2 \\ 3 & 0 & 0 \end{bmatrix}$, $\alpha = \{(1, 0, 1), (0, 1, 1), (1, 1, 0)\}$

d) $A = \begin{bmatrix} -1 & 0 & 0 & 0 \\ 1 & 2 & 0 & 0 \\ 0 & -1 & -1 & 0 \\ -2 & 1 & 0 & 1 \end{bmatrix}$, $\alpha = \{(1, 0, 0, 0,), (1, 1, 0, 0), (1, 1, 1, 0), (1, 1, 1, 1)\}$

5. (See Exercise 10 of Section 2.6.) Let V and W be finite-dimensional vector spaces. Let $P: V \to V$ be an isomorphism of V and $Q: W \to W$ be an isomorphism of W. Let $T: V \to W$ be a linear transformation. Prove that
   a) $\dim(\mathrm{Ker}(T)) = \dim(\mathrm{Ker}(QTP))$ and
   b) $\dim(\mathrm{Im}(T)) = \dim(\mathrm{Im}(QTP))$

6. a) Apply Exercise 5 to prove that similar matrices have the same rank.
   b) Show that if A and B are similar, then A is invertible if and only if B is invertible. What is the inverse of B in terms of the inverse of A?

7. Let $P_{\mathbf{a}}: \mathbf{R}^2 \to \mathbf{R}^2$ be the projection to the line spanned by $\mathbf{a} \neq \mathbf{0}$. Show that there is a basis α for $\mathbf{R}^2$ such that $[P_{\mathbf{a}}]_\alpha^\alpha = \begin{bmatrix} 1 & 0 \\ 0 & 0 \end{bmatrix}$.

8. Let A and B be $n \times n$ matrices. Prove that if A and B are similar, then there is a linear transformation $T: \mathbf{R}^n \to \mathbf{R}^n$ and bases α and β for $\mathbf{R}^n$ such that $[T]_\alpha^\alpha = A$ and $[T]_\beta^\beta = B$.

9. Let $S = M_{n\times n}(\mathbf{R})$ be the set of all $n \times n$ matrices with real entries. Prove that similarity of matrices defines an equivalence relation on S (see Appendix 1). That is, prove the following three statements.
   a) Every matrix $A \in S$ is similar to itself.
   b) If $A \in S$ is similar to $B \in S$, then B is similar to A.

c) If $A \in S$ is similar to $B \in S$ and B is similar to $C \in S$, then A is similar to C.

CHAPTER SUMMARY

In this chapter we began the study of linear transformations between vector spaces, the functions of importance in linear algebra. Aside from restrictions on the dimension of the vector spaces involved and at times requiring matrices or transformations to be invertible, our study was quite general. In subsequent chapters, especially in Chapters 4, 5, and 6, we will study individual transformations in greater detail.

A *linear transformation* $T: V \to W$ preserves vector addition and scalar multiplication in the domain and target spaces. That is [Definition (2.1.1)], T is linear if and only if for all $\mathbf{v}_1, \mathbf{v}_2 \in V$ and all $a_1, a_2 \in \mathbf{R}$, we have $T(a_1\mathbf{v}_1 + a_2\mathbf{v}_2) = a_1T(\mathbf{v}_1) + a_2T(\mathbf{v}_2)$. We saw several types of examples in Section 2.1. Among these, *rotations* [Example (2.1.12)] and *projections* [Example (2.1.13)] in $\mathbf{R}^2$ received special treatment, both because of their algebraic importance and because of the way they display the interaction between geometry and algebra in the vector space $\mathbf{R}^2$. In later chapters we will generalize both the algebraic and geometric aspects of these examples.

If $T: V \to W$ is a linear transformation and V is finite-dimensional, then, as we saw, the value of T on any vector $\mathbf{v} \in V$ is determined by the finite set of values $T(\mathbf{v}_1), \ldots, T(\mathbf{v}_k)$, for any basis $\alpha = \{\mathbf{v}_1, \ldots, \mathbf{v}_k\}$ of V [Proposition (2.1.14)]. If W is also finite-dimensional with basis $\beta = \{\mathbf{w}_1, \ldots, \mathbf{w}_l\}$, then T is determined by the $l \cdot k$ scalars used to express the vectors $T(\mathbf{v}_i)$ in terms of the basis β [Proposition (2.2.1)]. These $l \cdot k$ scalars form the entries of $[T]_\alpha^\beta$, the *matrix of T with respect to the bases α and β* [Definition (2.2.6)]. The matrix $[T]_\alpha^\beta$ is the primary calculational tool available to us in dealing with transformations between finite-dimensional vector spaces.

Given any linear mapping $T: V \to W$, we associated two subspaces to T, the *kernel* of T, $\text{Ker}(T) \subset V$, and the *image* of T, $\text{Im}(T) \subset W$ [Definitions (2.3.1) and (2.3.10)]. We were able to use the elimination algorithm as it applied to the system of equations $[T]_\alpha^\beta[\mathbf{v}]_\alpha = \mathbf{0}$ to calculate the dimensions of both $\text{Ker}(T)$ and $\text{Im}(T)$:

$$\dim(\text{Ker}(T)) = \text{the number of free variables in the system } [T]_\alpha^\beta[\mathbf{v}]_\alpha = \mathbf{0}$$

$$\dim(\text{Im}(T)) = \text{the number of basic variables in the same system}$$

Since the number of free variables plus the number of basic variables is the number of columns of $[T]_\alpha^\beta$, which is the dimension of V, this gives us one proof of the dimension theorem [Theorem (2.3.17)], under the assumption that W is finite-dimensional. This is the first major result in the theory of linear transformations.

Given any linear transformation $T: V \to W$ (V and W both finite-dimensional), we saw in Section 2.4 that we could check whether T is injective, surjective, or neither by checking $\dim(\text{Ker}(T))$ and $\dim(\text{Im}(T))$. The primary results in this direction were Propositions (2.4.2) and (2.4.7). T is injective if and only if $\dim(\mathrm{Ker}(T)) = 0$, and T is surjective if and only if $\dim(\mathrm{Im}(T)) = \dim(W)$. We used these results to characterize the set of solutions to systems of linear equations from the geometric point of view.

Using the algebra of matrix multiplication [Definition (2.5.10)], we exhibited a concise formula for the matrix of the composition of two linear mappings [Proposition (2.5.13)]. Furthermore, using the composition of transformations, we were able to say that a transformation is *invertible* if and only if it is both injective and surjective. We then applied a version of the elimination algorithm, called the *Gauss-Jordan procedure,* to calculate the matrix of the inverse of an invertible transformation.

Some of the proofs in this chapter appeared to rely on choices of bases in the domain and target spaces of the transformations in question. For example, one proof of the dimension theorem, (2.3.15), relied on a matrix representation of the transformation. In order to clear up any apparent gaps and to provide a convenient method for converting from one choice of bases to any other, we analyzed the effect of changing bases on the matrices of linear mappings in Theorem (2.7.5). This theorem is the justification for saying that anything that can be done with one choice of basis can be done with any other basis. As a result, we are free when dealing with a linear transformation to choose any bases that suit us depending on the context. This is a crucial point, which will be used repeatedly in the remainder of the text.

SUPPLEMENTARY EXERCISES

1. **True-False**. For each true statement, give a brief proof or reason. For each false statement, give a counterexample.
   a) If $T: \mathbf{R}^2 \to \mathbf{R}^2$ is a function such that $T(\mathbf{0}) = \mathbf{0}$, then T is a linear transformation.
   b) Every injective and surjective linear transformation $T: \mathbf{R}^2 \to \mathbf{R}^2$ is a rotation.
   c) If $T: V \to W$ is a linear transformation and $\dim(V) < \infty$, then it is always the case that $\dim(\mathrm{Im}(T)) < \infty$.
   d) If V and W are finite-dimensional, and $T: V \to W$ is a linear transformation, then $\dim(\mathrm{Ker}(T)) \leq \dim(\mathrm{Im}(T))$.
   e) If $T: \mathbf{R}^{13} \to \mathbf{R}^{17}$ and T is linear, then T is not surjective.
   f) Let S and T be linear transformations from V to W (both finite-dimensional). Let α be a basis of V, and β be a basis of W. Then $S = T$ if and only if $[S]_\alpha^\beta = [T]_\alpha^\beta$.
   g) If $S: U \to V$ and $T: V \to W$ are invertible linear transformations, then $(TS)^{-1} = T^{-1}S^{-1}$.
   h) If V and W are isomorphic vector spaces, there is a unique isomorphism $T: V \to W$.
   i) If the $n \times n$ matrices A and B are both similar to a third $n \times n$ matrix C, then A is similar to B.
   j) There are no linear transformations $T: V \to V$, V finite-dimensional, such that $[T]_\alpha^\alpha = [T]_\beta^\beta$ for all bases α, β of V.

2. Define $T: \mathbf{R}^3 \to \mathbf{R}^3$ by $T(\mathbf{e}_1) = \mathbf{e}_1 - \mathbf{e}_2$, $T(\mathbf{e}_2) = \mathbf{e}_2 - \mathbf{e}_3$, and $T(\mathbf{e}_3) = \mathbf{e}_3 - \mathbf{e}_1$. What is $T(\mathbf{v})$ for the following vectors $\mathbf{v}$?
   a) $\mathbf{v} = (1, 1, 1)$
   b) $\mathbf{v} = (1, 0, 2)$
   c) $\mathbf{v} = (-1, -1, 4)$

3. a) What is the matrix of the transformation T of Exercise 2 with respect to the standard basis in $\mathbf{R}^3$?
   b) What is the matrix of the transformation T of Exercise 2 with respect to the basis $\alpha = \{(1, 1, 1), (1, 0, 2,), (-1, -1, 4)\}$ in the domain and the basis $\beta = \{(1, -1, 0), (2, 0, 1), (0, 0, 1)\}$ in the target?

4. a) Find the dimensions of the kernel and the image of the transformation of Exercise 2.
   b) Find bases for the kernel and the image.

5. If $\dim(V) = 3$, and $T: V \to V$ has the matrix $[T]_\alpha^\alpha = \begin{bmatrix} 1 & -1 & 0 \\ 2 & 1 & -1 \\ 0 & 3 & -1 \end{bmatrix}$ with respect to some basis α for V, is T injective? surjective?

6. Determine whether the following matrices are invertible, and if so, find the inverses.

   a) $\begin{bmatrix} 2 & 4 \\ -1 & 2 \end{bmatrix}$ b) $\begin{bmatrix} -1 & 3 & 0 \\ -1 & 1 & 2 \\ 0 & -1 & 1 \end{bmatrix}$

   c) $\begin{bmatrix} 1 & 1 & 1 & 1 \\ 1 & -1 & 2 & -2 \\ 1 & 1 & 4 & 4 \\ 1 & -1 & 8 & -8 \end{bmatrix}$

7. Let A be an $n \times n$ matrix such that $A^k = I$ for some integer $k \geqslant 1$. Is A invertible? If so, what is its inverse? If not, how large is its kernel?

8. Show that if A and B are $n \times n$ matrices, then $(AB)^t = B^tA^t$, where A^t denotes the transpose of A (See Exercise 12 of Section 2.2.)

9. Let $T: V \to W$ be a linear transformation. Let $\alpha_1 = \{\mathbf{v}_1, \ldots, \mathbf{v}_k\}$ and $\beta_1 = \{\mathbf{w}_1, \ldots, \mathbf{w}_l\}$ be bases for V and W, respectively. Given the matrix $[T]_{\alpha_1}^{\beta_1}$, how could you find the matrix $[T]_{\alpha_2}^{\beta_2}$ where $\alpha_2 = \{\mathbf{v}_k, \mathbf{v}_{k-1}, \ldots, \mathbf{v}_1\}$ and $\beta_2 = \{\mathbf{w}_l, \mathbf{w}_{l-1}, \ldots, \mathbf{w}_1\}$?

10. Let $T: V \to W$ be a linear transformation. Prove that if $\{T(\mathbf{v}_1), \ldots, T(\mathbf{v}_k)\}$ is a linearly independent set in W, then $\{\mathbf{v}_1, \ldots, \mathbf{v}_k\}$ is a linearly independent set in V.

11. Let $T: V \to W$ be an invertible linear transformation of finite-dimensional vector spaces. Prove that if $U \subset V$ is a subspace, then $\dim(T(U)) = \dim(U)$.

12. Let $V = C^\infty(\mathbf{R})$, the vector space of functions $f: \mathbf{R} \to \mathbf{R}$ with derivatives of all orders. Let $S: V \to V$ be the mapping defined by $S(f) = x \cdot d^2f/dx^2 + x^2 f$, and let $T: V \to V$ be the mapping defined by $T(f) = df/dx + e^x f$.
   a) Show that S and T are linear mappings.
   b) Compute the composite mappings ST and TS. (Be careful—you will need to use the product rules for derivatives!)

13. Let A be a fixed 2×2 matrix and define a mapping mult_A: $M_{2\times 2}(\mathbf{R}) \to M_{2\times 2}(\mathbf{R})$ by $\text{mult}_A(B) = AB$.
   a) Prove that mult_A is a linear transformation of the vector space $M_{2\times 2}(\mathbf{R})$.
   b) Find the matrix of mult_A with respect to the basis α of $M_{2\times 2}(\mathbf{R})$ consisting of the matrices $E_{11} = \begin{bmatrix} 1 & 0 \\ 0 & 0 \end{bmatrix}$, $E_{12} = \begin{bmatrix} 0 & 1 \\ 0 & 0 \end{bmatrix}$, $E_{21} = \begin{bmatrix} 0 & 0 \\ 1 & 0 \end{bmatrix}$, and $E_{22} = \begin{bmatrix} 0 & 0 \\ 0 & 1 \end{bmatrix}$.

14. Let $R_{\mathbf{a}}: \mathbf{R}^2 \to \mathbf{R}^2$ denote reflection in the line spanned by $\mathbf{a}$ (see Exercise 4, Section 2.2).
   a) prove that if $\mathbf{b} = \alpha\mathbf{a}$, $\alpha \neq \mathbf{0}$, then $R_{\mathbf{b}} = R_{\mathbf{a}}$. (Thus a reflection may always be written in terms of a vector of unit length.)
   b) Let $\mathbf{a} = (\cos(\theta), \sin(\theta))$. What is the matrix of $R_{\mathbf{a}}$ with respect to the standard basis for $\mathbf{R}^2$.
   c) Let $\mathbf{a} = (\cos(\theta), \sin(\theta))$ and $\mathbf{b} = (\cos(\varphi), \sin(\varphi))$. Prove that the linear transformation $R_{\mathbf{a}}R_{\mathbf{b}}$ is a rotation. For which angle γ is this equal to R_γ? (*Hint*: Find the matrix of the composition with respect to the standard basis for $\mathbf{R}^2$ and then use trigonometric identities to find γ.)

15. Let $\mathbf{x} = (x_1, \ldots, x_n) \in \mathbf{R}^n$. Define $l_i: \mathbf{R}^n \to \mathbf{R}$ by $l_i(\mathbf{x}) = x_i$, $1 \leq i \leq n$.
   a) Prove that l_i is a linear transformation.
   b) Prove that $\{l_1, \ldots, l_n\}$ is a basis for $L(\mathbf{R}^n,\mathbf{R})$ (see Exercise 12, part c, of Section 2.1).
   c) Let $\mathbf{e}_1, \ldots, \mathbf{e}_n$ be the standard basis vectors for $\mathbf{R}^n$. Prove that $*(\mathbf{e}_i) = l_i$, where $*$ is the linear transformation defined in Exercise 7, Section 2.4.

CHAPTER 3

The Determinant Function

Introduction

This rather brief chapter is devoted to the study of one function of square matrices, the determinant function. This function assigns to each $n \times n$ matrix A a real number $\det(A)$. For us, the primary use of the determinant will be to determine if a matrix is invertible. The principal result is that a matrix is invertible if and only if its determinant is nonzero. This is an elegant result whose proof requires a fair amount of algebra in the general case, yet it also has an equally elegant geometric interpretation that we will see by considering the cases $n = 2$ and $n = 3$. We begin with the motivation and attempt to understand the result before we prove it.

There are several very nice applications of the determinant. As a consequence of our construction and particular choice of formula for the determinant (there are other equivalent formulas), we will be able to give a formula for the inverse of a matrix in terms of the determinant of the matrix. The multiplicative formula for the determinant of the product of matrices will allow us to compute the determinant of the inverse of a matrix and to define the determinant of a linear transformation.

The determinant will be used extensively in Chapters 4 and 6. It is somewhat curious to note that in this chapter we seem to want the determinant in order to show that matrices are invertible, whereas in the later chapters we will be more interested in finding when matrices are not invertible.

§3.1. THE DETERMINANT AS AREA

Our approach to the determinant is motivated by certain geometric constructions in $\mathbf{R}^2$. These arise in an effort to answer the question, if $T: V \to V$ is a linear transformation of a finite-dimensional vector space, is there a "quick" test to determine whether or not T is an isomorphism?

To begin, let $V = \mathbf{R}^2$, and let α be the standard basis. $[T]_\alpha^\alpha$ is a 2×2 matrix, which we denote by

$$A = \begin{bmatrix} a_{11} & a_{12} \\ a_{21} & a_{22} \end{bmatrix}$$

We will do all our calculations in terms of the rows of A, although we could just as well use the columns of A. Let $\mathbf{a}_1 = (a_{11}, a_{12})$ and $\mathbf{a}_2 = (a_{21}, a_{22})$ denote the rows of A.

In Chapter 2 we saw that if T is an isomorphism, then the number of linearly independent columns of A is the dimension of V, in our case this is two. Equivalently, the number of linearly independent rows is also equal to the dimension of V (see Exercise 10 of Section 2.3). Thus, to determine if T is an isomorphism, we need only check to see if the rows of A are linearly independent.

There is a simple geometric interpretation of the linear independence of two vectors in $\mathbf{R}^2$. Using $\mathbf{a}_1$ and $\mathbf{a}_2$, construct a parallelogram whose vertices are $\mathbf{0}$, $\mathbf{a}_1$, $\mathbf{a}_2$, and $\mathbf{a}_1 + \mathbf{a}_2$. See Figure 3.1.

If $\mathbf{a}_1$ and $\mathbf{a}_2$ are linearly independent, then $\mathbf{a}_1$ is not a scalar multiple of $\mathbf{a}_2$. Thus the sides of the parallelogram are not collinear and the area of the parallelogram is not zero. Conversely, if the area is not zero, $\mathbf{a}_1$ and $\mathbf{a}_2$ are linearly independent. Let us give a rigorous proof of this.

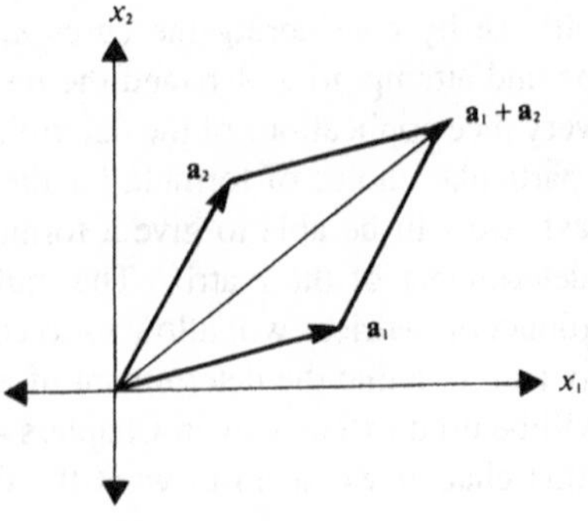

Figure 3.1

(3.1.1) Proposition

(i) The area of the parallelogram with vertices $\mathbf{0}$, $\mathbf{a}_1$, $\mathbf{a}_2$, and $\mathbf{a}_1 + \mathbf{a}_2$ is $\pm(a_{11}a_{22} - a_{12}a_{21})$.

(ii) The area is not zero if and only if the vectors $\mathbf{a}_1$ and $\mathbf{a}_2$ are linearly independent.

Proof:

(i) The area of the parallelogram is $b\|\mathbf{a}_1\|$, where b is the length of the perpendicular from $\mathbf{a}_2 = (a_{21}, a_{22}) \neq \mathbf{0}$ to the line spanned by $\mathbf{a}_1$. Elementary geometry shows that this is the length of $\mathbf{a}_2 - P_{\mathbf{a}_1}(\mathbf{a}_2)$, where $P_{\mathbf{a}_1}$ is the projection to the line spanned by $\mathbf{a}_1$. (Verify this!) That is

$$b = \|\mathbf{a}_2 - (\langle \mathbf{a}_1, \mathbf{a}_2\rangle / \langle \mathbf{a}_1, \mathbf{a}_1\rangle)\mathbf{a}_1\|$$

$$= \|\langle \mathbf{a}_1, \mathbf{a}_1\rangle\mathbf{a}_2 - \langle \mathbf{a}_1, \mathbf{a}_2\rangle\mathbf{a}_1\| / \|\mathbf{a}_1\|^2$$

Therefore, the area of the parallelogram is

$$\|\langle \mathbf{a}_1, \mathbf{a}_1\rangle\mathbf{a}_2 - \langle \mathbf{a}_1, \mathbf{a}_2\rangle\mathbf{a}_1\| / \|\mathbf{a}_1\|$$

If we expand inside $\| \ \|$ and take the square root, we obtain

$$\pm(a_{11}a_{22} - a_{12}a_{21})$$

(ii) We leave the proof to the reader as Exercise 1. ■

(3.1.2) Corollary. Let $V = \mathbf{R}^2$. $T: V \to V$ is an isomorphism if and only if the area of the parallelogram constructed previously is nonzero.

Proof: Since the independence of the columns of $[T]_\alpha^\alpha$ is equivalent to T being an isomorphism, the corollary follows immediately from Proposition (3.1.1) (ii). ■

The formula in the proposition determines the area up to a sign + or −, since our calculations involved taking a square root. We would like to remove this indeterminacy. To do this, we establish a convention that allows for both positive and negative areas. Negative area occurs for the same reason that negative area occurs in calculus, in that we must make a choice of ordering. In calculus it is the left to right ordering of the real numbers on a number line that is used to determine the sign of the area under a graph. That is, the area between the graph of $y = f(x)$ and the x-axis between a and b, $a < b$, is equal to the integral $\int_a^b f(x)\,dx$. (Note the integral is not defined to be the area.) See Figure 3.2. However, $\int_b^a f(x)\,dx = -\int_a^b f(x)\,dx$. Thus, we might say that the area from b to a under the graph is the negative of the area from a to b under the graph.

In linear algebra we use a similar ordering to measure angles. An angle will

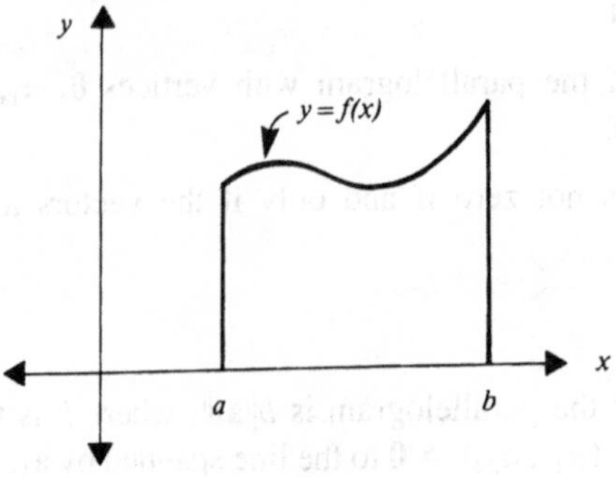

Figure 3.2

be positive if we measure the angle in the counterclockwise direction. To define the signed area of a parallelogram, we make a convention:

> If the angle from $\mathbf{a}_1$ to $\mathbf{a}_2$, which is $\leqslant \pi$ is traced counterclockwise (clockwise) from $\mathbf{a}_1$ to $\mathbf{a}_2$, the area is positive (negative). (See Figure 3.3.).

Returning to the formula Area $= \pm(a_{11}a_{22} - a_{12}a_{21})$, we see that if $\mathbf{a}_1 = (1, 0)$ and $\mathbf{a}_2 = (0, 1)$, the area must be positive. Thus, the correct sign is $a_{11}a_{22} - a_{12}a_{21}$. Henceforth, we will take the area to be $a_{11}a_{22} - a_{12}a_{21}$ and denote it by Area$(\mathbf{a}_1, \mathbf{a}_2)$.

We are interested in the algebraic properties of the area function that will be of general interest beyond $\mathbf{R}^2$.

(3.1.3) Proposition. The function Area$(\mathbf{a}_1, \mathbf{a}_2)$ has the following properties for $\mathbf{a}_1$, $\mathbf{a}_2$, $\mathbf{a}_1'$, and $\mathbf{a}_2' \in \mathbf{R}^2$:

(i) Area$(b\mathbf{a}_1 + c\mathbf{a}_1', \mathbf{a}_2) = b$ Area$(\mathbf{a}_1, \mathbf{a}_2) + c$ Area$(\mathbf{a}_1', \mathbf{a}_2)$ for $b, c \in \mathbf{R}$

(i′) Area$(\mathbf{a}_1, b\mathbf{a}_2 + c\mathbf{a}_2') = b$ Area$(\mathbf{a}_1, \mathbf{a}_2) + c$ Area$(\mathbf{a}_1, \mathbf{a}_2')$ for $b, c \in \mathbf{R}$

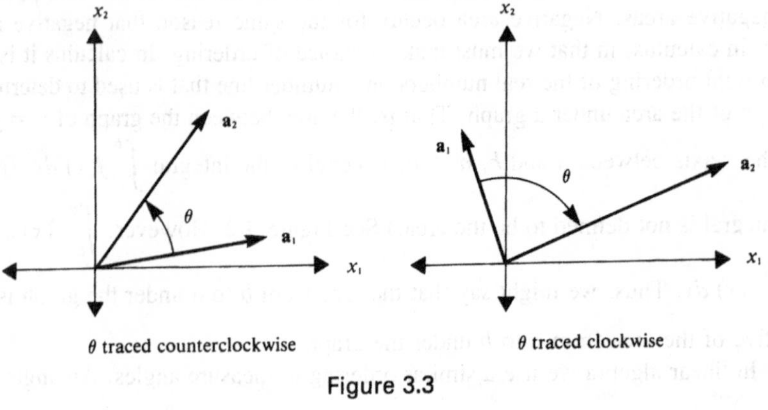

Figure 3.3

(ii) $\text{Area}(\mathbf{a}_1, \mathbf{a}_2) = -\text{Area}(\mathbf{a}_2, \mathbf{a}_1)$, and

(iii) $\text{Area}((1, 0), (0, 1)) = 1$.

Proof: Exercise 2. ■

Notice that (i) and (ii) imply (i′) so that we need only work with (i), (ii), and (iii).

Properties (i) and (i') tell us that Area is a linear function of each variable when the other is held fixed. We call this property *multilinearity*. Property (ii) is called the *alternating* property, and property (iii) is called the *normalization* property.

The three properties of the area function can also be derived in a purely geometric manner. Property (ii) follows from our sign convention, and Property (iii) simply says that the area of a unit square is one. Property (i) is somewhat more cumbersome to derive geometrically and this derivation does not readily generalize to higher dimensions. It is, however, revealing to consider the geometric constructions in dimension two. We separate (i) into two parts. We indicate the justification for the first part and leave the second as an exercise.

Property (i) is equivalent (why?) to

$$\text{Area}(b\mathbf{a}_1, \mathbf{a}_2) = b\,\text{Area}(\mathbf{a}_1, \mathbf{a}_2)$$

and

$$\text{Area}(\mathbf{a}_1 + \mathbf{a}_1', \mathbf{a}_2) = \text{Area}(\mathbf{a}_1, \mathbf{a}_2) + \text{Area}(\mathbf{a}_1', \mathbf{a}_2)$$

The first of these two properties says that if each of a pair of opposite sides of a parallelogram is increased by a factor of b, then the area increases by the same factor (see Figure 3.4). Exercise 11 asks you to give a geometric proof of the second of the two properties.

The properties in Proposition (3.1.3) are chosen from the many properties enjoyed by the area function because they are independent; that is, no two imply the third, and because they completely characterize the area function. This will be made precise in the following proposition. This is an instance of an important mathematical technique: the unique specification of a function by its properties

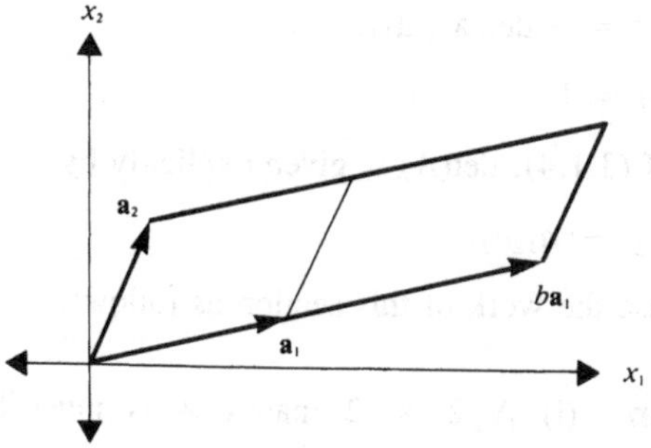

Figure 3.4

rather than by producing an explicit formula. Most likely you have already seen an example of this in calculus: The exponential function $u(x) = e^x$ is characterized by $du/dx = u$ and $u(0) = 1$.

(3.1.4) Proposition. If $B(\mathbf{a}_1, \mathbf{a}_2)$ is any real-valued function of $\mathbf{a}_1$ and $\mathbf{a}_2 \in \mathbf{R}^2$ that satisfies Properties (i), (ii), and (iii) of Proposition (3.1.3), then B is equal to the area function.

Proof: We apply Properties (i), (ii), and (iii) to $B(\mathbf{a}_1, \mathbf{a}_2)$ to simplify the value of the function:

$$\begin{aligned} B(\mathbf{a}_1, \mathbf{a}_2) &= B(a_{11}\mathbf{e}_1 + a_{12}\mathbf{e}_2, a_{21}\mathbf{e}_1 + a_{22}\mathbf{e}_2) \\ &= a_{11}B(\mathbf{e}_1, a_{21}\mathbf{e}_1 + a_{22}\mathbf{e}_2) + a_{12}B(\mathbf{e}_2, a_{21}\mathbf{e}_1 + a_{22}\mathbf{e}_2) \\ &= a_{11}a_{21}B(\mathbf{e}_1, \mathbf{e}_1) + a_{11}a_{22}B(\mathbf{e}_1, \mathbf{e}_2) \\ &\quad + a_{12}a_{21}B(\mathbf{e}_2, \mathbf{e}_1) + a_{12}a_{22}B(\mathbf{e}_2, \mathbf{e}_2) \end{aligned}$$

Property (ii) implies that $B(\mathbf{e}_2, \mathbf{e}_1) = -B(\mathbf{e}_1, \mathbf{e}_2)$. Thus, we may combine the second and third terms. Property (ii) also implies that $B(\mathbf{v}, \mathbf{v}) = -B(\mathbf{v}, \mathbf{v})$ so that for any vector $\mathbf{v}$, $B(\mathbf{v}, \mathbf{v}) = 0$. Therefore

$$B(\mathbf{a}_1, \mathbf{a}_2) = (a_{11}a_{22} - a_{12}a_{21})B(\mathbf{e}_1, \mathbf{e}_2)$$

By property (iii) $B(\mathbf{e}_1, \mathbf{e}_2) = 1$. Therefore

$$B(\mathbf{a}_1, \mathbf{a}_2) = a_{11}a_{22} - a_{12}a_{21} = \text{Area}(\mathbf{a}_1, \mathbf{a}_2)$$

Since this holds for all functions satisfying Properties (i), (ii), and (iii), all such functions are in fact the area function. ■

This leads to the definition of the determinant of a 2×2 matrix:

(3.1.5) Definition. The *determinant* of a 2×2 matrix A, denoted by $\det(A)$ or $\det(\mathbf{a}_1, \mathbf{a}_2)$, is the unique function of the rows of A satisfying

(i) $\det(b\mathbf{a}_1 + c\mathbf{a}_1', \mathbf{a}_2) = b \det(\mathbf{a}_1, \mathbf{a}_2) + c \det(\mathbf{a}_1', \mathbf{a}_2)$ for $b, c \in \mathbf{R}$.

(ii) $\det(\mathbf{a}_1, \mathbf{a}_2) = -\det(\mathbf{a}_2, \mathbf{a}_1)$, and

(iii) $\det(\mathbf{e}_1, \mathbf{e}_2) = 1$.

As a consequence of (3.1.4), $\det(A)$ is given explicitly by

$$\det(A) = a_{11}a_{22} - a_{12}a_{21}$$

We can rephrase the work of this section as follows.

(3.1.6) Proposition. (i) A 2×2 matrix A is invertible if and only if $\det(A) \neq 0$.

(ii) If $T: V \to V$ is a linear transformation of a two-dimensional vector space V, then T is an isomorphism if and only if $\det([T]_\alpha^\alpha) \neq 0$.

Proof: The proposition is a consequence of (3.1.1), (3.1.2), (3.1.4), and (3.1.5). ■

Notice that Proposition (3.1.6)(ii) does not specify which basis α we should use. It is quite reasonable to ask if $\det([T]_\alpha^\alpha)$ depends on the choice of α or only on T. In the last section of this chapter we will prove that $\det([T]_\alpha^\alpha) = \det([T]_\beta^\beta)$ for any choice of α and β. This will allow us to define the determinant of T by $\det(T) = \det([T]_\alpha^\alpha)$ for any choice of α.

Proposition (3.1.6) can be used to produce the inverse of a transformation. Suppose that $T: V \to V$, $\dim(V) = 2$, is an isomorphism. Let α be a basis for V, and let $A = [T]_\alpha^\alpha$. To find A^{-1}, we apply the Gauss-Jordan procedure to A.

Since T is an isomorphism, either $a_{11} \neq 0$ or $a_{21} \neq 0$. Assume for simplicity that $a_{11} \neq 0$. If $a_{11} = 0$ and $a_{21} \neq 0$, switch the first and second rows of A as the first step of the procedure. The Gauss-Jordan procedure yields

$$\left[\begin{array}{cc|cc} a_{11} & a_{12} & 1 & 0 \\ a_{21} & a_{22} & 0 & 1 \end{array}\right]$$

$$\to \left[\begin{array}{cc|cc} a_{11} & a_{12} & 1 & 0 \\ 0 & a_{22} - a_{21}a_{12}/a_{11} & -a_{21}/a_{11} & 1 \end{array}\right]$$

$$= \left[\begin{array}{cc|cc} a_{11} & a_{12} & 1 & 0 \\ 0 & \det(A)/a_{11} & -a_{21}/a_{11} & 1 \end{array}\right]$$

$$\to \left[\begin{array}{cc|cc} 1 & a_{12}/a_{11} & 1/a_{11} & 0 \\ 0 & 1 & -a_{21}/\det(A) & a_{11}/\det(A) \end{array}\right]$$

$$\to \left[\begin{array}{cc|cc} 1 & 0 & 1/a_{11} - (a_{12}/a_{11})(-a_{21}/\det(A)) & (-a_{12}/a_{11})(a_{11}/\det(A)) \\ 0 & 1 & -a_{21}/\det(A) & a_{11}/\det(A) \end{array}\right]$$

$$= \left[\begin{array}{cc|cc} 1 & 0 & a_{22}/\det(A) & -a_{12}/\det(A) \\ 0 & 1 & -a_{21}/\det(A) & a_{11}/\det(A) \end{array}\right]$$

Therefore

$$A^{-1} = \frac{1}{\det(A)} \begin{bmatrix} a_{22} & -a_{12} \\ -a_{21} & a_{11} \end{bmatrix}$$

It is easy to verify that $AA^{-1} = I$ holds for this matrix A, and it is clear that this formula is only valid if $\det(A) \neq 0$.

In closing, we note that the question we set out to answer has been answered. We have found a simple test to determine whether or not a transformation or matrix is invertible and, even better, we have given the relationship between the test and an explicit formula for the inverse.

EXERCISES

1. Prove Proposition (3.1.1) (ii).

2. Prove Proposition (3.1.3), using the formula Area($\mathbf{a}_1$, $\mathbf{a}_2$) $= a_{11}a_{22} - a_{12}a_{21}$.

3. Calculate the following 2 × 2 determinants.

   a) $\begin{bmatrix} 3 & -2 \\ 9 & 1 \end{bmatrix}$ b) $\begin{bmatrix} -1 & 2 \\ 0 & 4 \end{bmatrix}$ c) $\begin{bmatrix} 0 & 6 \\ -6 & 0 \end{bmatrix}$

4. Let A be the matrix of a rotation. Calculate det(A).

5. Let A be the matrix of a projection to a nonzero vector in the plane. Calculate det(A).

6. Let A be an invertible 2 × 2 matrix. Calculate det(A^{-1}).

7. Prove that any function f satisfying Properties (i) and (ii) of (3.1.3) satisfies $f(A) = \det(A)f(I)$.

8. Prove that det(A) is
   a) a linear function of the first column of A and
   b) an alternating function of the columns of A.

9. Prove that det(A) is the unique function of the columns of A satisfying Exercise 8, parts a and b, and det(I) = 1.

10. Prove that an upper (lower) triangular 2 × 2 matrix is invertible if and only if the diagonal entries are both nonzero.

11. Give a geometric proof of the following property of the area function: Area $(\mathbf{a}_1 + \mathbf{a}_1', \mathbf{a}_2)$ = Area$(\mathbf{a}_1, \mathbf{a}_2)$ + Area$(\mathbf{a}_1', \mathbf{a}_2)$. (*Hint:* Carefully sketch the three parallelograms in the formula, and by labeling regions of equal area in the diagram, verify the equality.)

12. Our geometric motivation for the determinant function was based on the area of the parallelogram whose vertices are $\mathbf{0}$, $\mathbf{a}_1$, $\mathbf{a}_2$, and $\mathbf{a}_1 + \mathbf{a}_2$. How would our work in this section change if we had used the area of the triangle with vertices $\mathbf{0}$, $\mathbf{a}_1$, and $\mathbf{a}_2$ instead?

§3.2. THE DETERMINANT OF AN $n \times n$ MATRIX

In the first section of this chapter we defined the determinant of a 2 × 2 matrix to be the unique function of the rows of a matrix that satisfies the multilinear, alternating, and normalization properties. The proof of uniqueness was purely algebraic, but the proof that the area function satisfies these three properties could proceed algebraically or geometrically.

In this section we would like to extend these constructions to define and compute the determinant of an $n \times n$ matrix. The geometric derivations very quickly run into trouble. For 2 × 2 matrices we considered areas of parallelograms in $\mathbf{R}^2$.

For a 3×3 matrix A with rows $\mathbf{a}_1$, $\mathbf{a}_2$, and $\mathbf{a}_3$ we could define the determinant by using the signed volume of the *parallelepiped* in $\mathbf{R}^3$ with vertices $\mathbf{0}$, $\mathbf{a}_1$, $\mathbf{a}_2$, $\mathbf{a}_3$, $\mathbf{a}_1 + \mathbf{a}_2$, $\mathbf{a}_1 + \mathbf{a}_3$, $\mathbf{a}_2 + \mathbf{a}_3$, and $\mathbf{a}_1 + \mathbf{a}_2 + \mathbf{a}_3$. This is the set of vectors $\{\mathbf{v} \mid \mathbf{v} = x_1\mathbf{a}_1 + x_2\mathbf{a}_2 + x_3\mathbf{a}_3, 0 \leq x_i \leq 1\}$. (See Figure 3.5.) With a bit more effort than was needed in the 2×2 case, we could verify that the volume function satisfies a similar list of properties. However, this geometric approach is quite difficult when we move to 4×4 matrices. How do we generalize a parallelepiped to $\mathbf{R}^4$, and how do we compute its volume? It is not even unreasonable to ask what "volume" means in $\mathbf{R}^4$.

Following the method of Section 3.1, we define the determinant function by the properties that it satisfies. Let A be an $n \times n$ matrix, and denote the rows of A by $\mathbf{a}_1, \ldots, \mathbf{a}_n$. We will be concerned with functions f that assign a real number $f(A)$ to each matrix A. Since we will be interested in the behavior of f as we change the rows in a specified manner, we often write $f(\mathbf{a}_1, \ldots, \mathbf{a}_n)$ for $f(A)$.

The three properties that will specify the determinant are modeled on the three properties of Proposition (3.1.3). In fact, we will still call them the multilinear, alternating, and normalization properties. The first two, which are equivalent to Properties (i) and (ii) of (3.1.3) are as follows.

(3.2.1) Definition. A function f of the rows of a matrix A is called *multilinear* if f is a linear function of each of its rows when the remaining rows are held fixed. That is, f is multilinear if for all b and $b' \in \mathbf{R}$

$$\begin{aligned} &f(\mathbf{a}_1, \ldots, b\mathbf{a}_i + b'\mathbf{a}'_i, \ldots, \mathbf{a}_n) \\ &\qquad = bf(\mathbf{a}_1, \ldots, \mathbf{a}_i, \ldots, \mathbf{a}_n) + b'f(\mathbf{a}_1, \ldots, \mathbf{a}'_i, \ldots, \mathbf{a}_n) \end{aligned}$$

(3.2.2) Definition. A function f of the rows of a matrix A is said to be *alternating* if whenever any two rows of A are interchanged f changes sign. That is, for all $i \neq j$, $1 \leq i, j \leq n$, we have

$$f(\mathbf{a}_1, \ldots, \mathbf{a}_i, \ldots, \mathbf{a}_j, \ldots, \mathbf{a}_n) = -f(\mathbf{a}_1, \ldots, \underset{\substack{\uparrow \\ i\text{th position}}}{\mathbf{a}_j}, \ldots, \underset{\substack{\uparrow \\ j\text{th position}}}{\mathbf{a}_i}, \ldots, \mathbf{a}_n)$$

The following lemma will prove to be very useful.

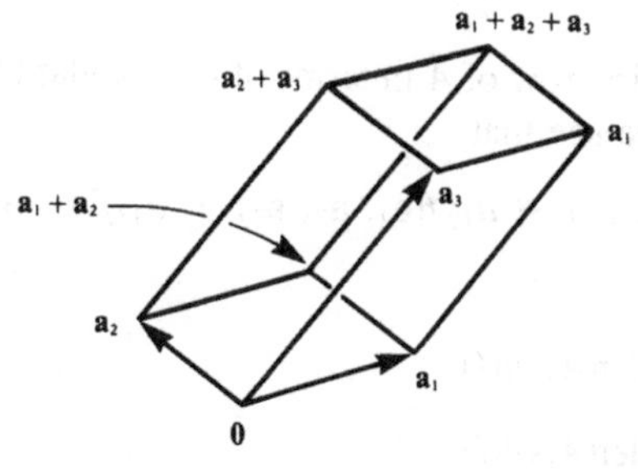

Figure 3.5

(3.2.3) Lemma. If f is an alternating real-valued function of the rows of an $n \times n$ matrix and two rows of the matrix A are identical, then $f(A) = 0$.

Proof: Assume $\mathbf{a}_i = \mathbf{a}_j$. Then $f(A) = f(\mathbf{a}_1, \ldots, \mathbf{a}_i, \ldots, \mathbf{a}_j, \ldots, \mathbf{a}_n) = -f(\mathbf{a}_1, \ldots, \mathbf{a}_j, \ldots, \mathbf{a}_i, \ldots, \mathbf{a}_n) = -f(A)$. Therefore, $f(A) = 0$. ■

Again, we note that this holds for a 2×2 matrix A:

$$\det\begin{bmatrix} a & b \\ a & b \end{bmatrix} = ab - ab = 0$$

Our goal is to show that any two multilinear alternating functions f and g that satisfy the normalization condition $f(\mathbf{e}_1, \ldots, \mathbf{e}_n) = g(\mathbf{e}_1, \ldots, \mathbf{e}_n) = 1$—the analog of condition (iii) of (3.1.3)—are equal. The techniques we will use are the same as those of Section 3.1. First, we introduce some terminology.

(3.2.4) Definition. Let A be an $n \times n$ matrix with entries a_{ij}, $i, j, = 1, \ldots, n$. The ijth *minor of* A is defined to be the $(n - 1) \times (n - 1)$ matrix obtained by deleting the ith row and jth column of A. The ijth minor is denoted by A_{ij}. Thus

$$A_{ij} = \begin{bmatrix} a_{11} & \cdots & a_{1,j-1} & a_{1,j+1} & \cdots & a_{1n} \\ & \vdots & & & \vdots & \\ a_{i-1,1} & \cdots & a_{i-1,j-1} & a_{i-1,j+1} & \cdots & a_{i-1,n} \\ a_{i+1,1} & \cdots & a_{i+1,j-1} & a_{i+1,j+1} & \cdots & a_{i+1,n} \\ & \vdots & & & \vdots & \\ a_{n1} & \cdots & a_{n,j-1} & a_{n,j+1} & \cdots & a_{nn} \end{bmatrix}$$

In the 3×3 case, for example

$$A_{12} = \begin{bmatrix} a_{21} & a_{23} \\ a_{31} & a_{33} \end{bmatrix}$$

In order to warm up for the $n \times n$ case, we will obtain a formula in the 3×3 case.

(3.2.5) Proposition. Let A be a 3×3 matrix, and let f be an alternating multilinear function. Then

$$f(A) = [a_{11} \det(A_{11}) - a_{12} \det(A_{12}) + a_{13} \det(A_{13})]f(I)$$

Proof: Expanding the first row of A in terms of the standard basis in $\mathbf{R}^3$ and using the multilinearity of f, we see that

$$f(A) = a_{11}f(\mathbf{e}_1, \mathbf{a}_2, \mathbf{a}_3) + a_{12}f(\mathbf{e}_2, \mathbf{a}_2, \mathbf{a}_3) + a_{13}f(\mathbf{e}_3, \mathbf{a}_2, \mathbf{a}_3)$$

We must now show that

$$f(\mathbf{e}_1, \mathbf{a}_2, \mathbf{a}_3) = \det(A_{11})f(I)$$

$$f(\mathbf{e}_2, \mathbf{a}_2, \mathbf{a}_3) = -\det(A_{12})f(I)$$

and

$$f(\mathbf{e}_3, \mathbf{a}_2, \mathbf{a}_3) = \det(A_{13})f(I)$$

We verify the first equality; the remaining two are similar. Expanding $\mathbf{a}_2$ in the same way, we obtain

$$\begin{aligned} f(\mathbf{e}_1, \mathbf{a}_2, \mathbf{a}_3) &= a_{21}f(\mathbf{e}_1, \mathbf{e}_1, \mathbf{a}_3) + a_{22}f(\mathbf{e}_1, \mathbf{e}_2, \mathbf{a}_3) + a_{23}f(\mathbf{e}_1, \mathbf{e}_3, \mathbf{a}_3) \\ &= a_{22}f(\mathbf{e}_1, \mathbf{e}_2, \mathbf{a}_3) + a_{23}f(\mathbf{e}_1, \mathbf{e}_3, \mathbf{a}_3) \end{aligned}$$

applying Lemma (3.2.3). Finally, expanding the third row yields

$$a_{22}a_{33}f(\mathbf{e}_1, \mathbf{e}_2, \mathbf{e}_3) + a_{23}a_{32}f(\mathbf{e}_1, \mathbf{e}_3, \mathbf{e}_2)$$

The other terms are zero by (3.2.3). Since f is alternating, we have $f(\mathbf{e}_1, \mathbf{e}_3, \mathbf{e}_2) = -f(\mathbf{e}_1, \mathbf{e}_2, \mathbf{e}_3)$, so the preceding expression equals

$$\det(A_{11})f(I)$$

This verifies the first equality. ■

If we impose the additional condition that $f(\mathbf{e}_1, \mathbf{e}_2, \mathbf{e}_3) = 1$, we see that

$$f(A) = a_{11}\det(A_{11}) - a_{12}\det(A_{12}) + a_{13}\det(A_{13})$$

It can be shown by a direct calculation that this formula defines an alternating multilinear function that satisfies $f(I) = 1$. We leave this as an exercise for the reader (see Exercise 5).

Thus, we have Corollary (3.2.6).

(3.2.6) Corollary. There exists exactly one multilinear alternating function f of the rows of a 3×3 matrix such that $f(I) = 1$.

(3.2.7) Definition. The determinant function of a 3×3 matrix is the unique alternating multilinear function f with $f(I) = 1$. This function will be denoted by $\det(A)$.

From Corollary (3.2.6) we see that $\det(A)$ may be rewritten in the form

$$\det(A) = (-1)^{1+1}a_{11}\det(A_{11}) + (-1)^{1+2}a_{12}\det(A_{12}) + (-1)^{1+3}a_{13}\det(A_{13})$$

Further, notice that there was nothing special in our choice of the first row of A in the proposition. So we see that the determinant is given by

$$\begin{aligned} \det(A) &= (-1)^{i+1}a_{i1}\det(A_{i1}) + (-1)^{i+2}a_{i2}\det(A_{i2}) + (-1)^{i+3}a_{i3}\det(A_{i3}) \\ &= \sum_{j=1}^{3} (-1)^{i+j}a_{ij}\det(A_{ij}) \end{aligned}$$

for $i = 1, 2,$ or 3.

Let us compute the determinant of a 3×3 matrix in two ways using this formula. Let A be the matrix

$$A = \begin{bmatrix} 2 & 1 & -3 \\ 3 & -1 & 0 \\ 2 & 1 & 4 \end{bmatrix}$$

Let us first compute the determinant by expanding along the first row:

$$\begin{aligned} \det(A) &= (-1)^2 \cdot 2((-1) \cdot 4 - 0 \cdot 1) + (-1)^3 \cdot 1(3 \cdot 4 - 0 \cdot 2) \\ &\qquad + (-1)^4 \cdot (-3)(3 \cdot 1 - (-1) \cdot 2) \\ &= -35. \end{aligned}$$

Now let us compute by expanding along the second row:

$$\begin{aligned} \det(A) &= (-1)^3 \cdot 3(1 \cdot 4 - (-3) \cdot 1) + (-1)^4 \\ &\qquad \cdot (-1)(2 \cdot 4 - (-3) \cdot 2) + (-1)^5 \cdot 0 \\ &= -35. \end{aligned}$$

Notice that by expanding along the second row, the fact that $a_{23} = 0$ simplified the calculation. When we calculate determinants by this method, we will usually expand along a row with the most zero entries in order to simplify the calculation.

We now want to extend the formula $\det(A) = \sum_{j=1}^{n} (-1)^{i+j} a_{ij} \det(A_{ij})$ to $n \times n$ matrices.

(3.2.8) Theorem. There exists exactly one alternating multilinear function $f\colon M_{n\times n}(\mathbf{R}) \to \mathbf{R}$ satisfying $f(I) = 1$, which is called the determinant function $f(A) = \det(A)$. Further, any alternating multilinear function f satisfies $f(A) = \det(A)f(I)$.

Proof: We prove existence by showing that

$$\det(A) = \sum_{j=1}^{n} (-1)^{i+j} a_{ij} \det(A_{ij}) \tag{3.1}$$

defines an alternating multilinear function that satisfies $\det(I) = 1$. We will prove uniqueness by showing that any alternating multilinear function f satisfies the formula

$$f(A) = \left[\sum_{j=1}^{n} (-1)^{i+j} a_{ij} \det(A_{ij}) \right] f(I) \tag{3.2}$$

The proof will be by induction on the size, $m \times m$, of the matrix, so that in Equations (3.1) and (3.2) $\det(A_{ij})$ refers to the unique alternating multilinear function of $(n - 1) \times (n - 1)$ matrices that takes the value 1 on the $(n - 1) \times (n - 1)$ identity matrix.

$m = 1$: If $A = [a_{11}]$ is a 1×1 matrix, $\det(A) = a_{11}$ defines an alternating multilinear function that satisfies $\det(I) = 1$. Both the summation formula and the alternating property are vacuous in this case and the other properties are trivially satisfied. The uniqueness of this function is also trivial.

$m = n$: We assume by the induction hypothesis that there is a unique alternating multilinear function det of $(n - 1) \times (n - 1)$ matrices that satisfies $\det(I_{(n-1)\times(n-1)}) = 1$. Further, we assume that det satisfies Equation (3.1). Notice that Equation (3.1) says that det may be expressed by expanding along any row of the matrix and that this expression is given in terms of the determinant of $(n - 2) \times (n - 2)$ matrices.

Existence: To prove existence we must prove that (3.1) defines an alternating multilinear function of the rows of an $n \times n$ matrix A and that when $A = I$ the formula yields the value 1. We prove this for $i = 1$ rather than for a general i, $1 \leq i \leq n$. This simplifies the use of indices in the proof. We can recover the general case by adjusting the indices.

First, we show the formula in Equation (3.1) gives an alternating function of the rows of A; that is, interchanging the kth and lth rows of A multiplies the determinant by a factor of -1. If we interchange the kth and lth rows of A, k and $l \neq 1$, the coefficients a_{1j} do not change but the $(k - 1)^{st}$ and $(l - 1)^{st}$ rows of each A_{1j} are interchanged. (Notice that the pth row of A contains the $(p - 1)^{st}$ row of A_{1j}. This is the first time that our use of the $i = 1$ simplifies the calculation.) By the induction hypothesis interchanging the rows of A_{1j} changes the sign of $\det(A_{1j})$. Thus each term of the sum in Equation (3.1) changes sign, and it follows that $\det(A)$ also changes sign. Suppose now that $k = 1$. In Equation (3.1) expand each $\det(A_{1j})$ along the pth row of A_{1j} where $p \neq l - 1$. In order to simplify the notation we assume that $l \neq 2$ and we choose $p = 2$. We obtain

$$\det(A) = \sum_{j=1}^{n} (-1)^{1+j} a_{1j} \left[\sum_{q=1}^{j-1} (-1)^{1+q} a_{2q} \det((A_{1j})_{1q}) + \sum_{q=j+1}^{n} (-1)^{1+(q-1)} a_{2q} \det((A_{1j})_{1,q-1}) \right]$$

(The difficulty in writing out this sum arises from the fact that the columns of A_{1j} are not numbered in the same way as the columns of A. That is, if $q > j$, the qth column of A contains the $(q - 1)^{st}$ column of A_{1j}.) If we rearrange the double sum we obtain

$$\sum_{j=1}^{n} \left[\sum_{q \neq j} (-1)^{1+j+q+1} a_{1j} a_{2q} \det((A_{1j})_{1q}) \right]$$

Notice that in this sum, for each $1 \leq q \leq n$, a_{2q} appears once multiplied by a_{1j} for each $j \neq q$. Thus if we sum first over q, $1 \leq q \leq n$, and then over j, the sum over j is for $j \neq q$. Writing this out we have

$$\sum_{q=1}^{n} \left[\sum_{j \neq q} (-1)^{1+j+q+1} a_{2q} a_{1j} \det((A_{1j})_{1q}) \right]$$

Reversing our calculations, we see that this is $-\det(A')$ where A' is the matrix obtained by interchanging the first and second rows of A. Thus we conclude that det is an alternating function of the rows of an $n \times n$ matrix.

Next we must show that the formula in Equation (3.1) defines a multilinear function of the rows of an $n \times n$ matrix. Let A denote the matrix whose kth row has entries of the form $a_{kj} = b'a'_{kj} + b''a''_{kj}$. Let A' denote the matrix whose entries are equal to the entries of A in all but the kth row, and in the kth row the entries are a'_{kj}. Similarly, let A'' denote the matrix whose entries are equal to the entries of A in all but the kth row, and in the kth row the entries are a''_{kj}. We must show that $\det(A) = b'\det(A') + b''\det(A'')$.

If $k = 1$, this follows directly from the formula in Equation (3.1).

$$\begin{aligned}
\det(A) &= \sum_{j=1}^{n} (-1)^{1+j} a_{1j} \det(A_{1j}) \\
&= \sum_{j=1}^{n} (-1)^{1+j} (b'a'_{1j} + b''a''_{1j}) \det(A_{1j}) \\
&= b'\left[\sum_{j=1}^{n} (-1)^{1+j} a'_{1j} \det(A_{1j})\right] + b''\left[\sum_{j=1}^{n} (-1)^{1+j} a''_{1j} \det(A_{1j})\right] \\
&= b'\det(A') + b''\det(A'')
\end{aligned}$$

since $A_{1j} = A'_{1j} = A''_{1j}$.

If $k \neq 1$, we must use the induction hypothesis. That is, we use the multilinearity of det for $(n-1) \times (n-1)$ matrices.

$$\begin{aligned}
\det(A) &= \sum_{j=1}^{n} (-1)^{1+j} a_{1j} \det(A_{1j}) \\
&= \sum_{j=1}^{n} (-1)^{1+j} a_{1j} \left[b'\det(A'_{1j}) + b''\det(A''_{1j}) \right] \\
&= b'\left[\sum_{j=1}^{n} (-1)^{1+j} a_{1j} \det(A'_{1j})\right] + b''\left[\sum_{j=1}^{n} (-1)^{1+j} a_{1j} \det(A''_{1j})\right] \\
&= b'\det(A') + b''\det(A'')
\end{aligned}$$

In order to finish the existence portion of the proof, we must show that $\det(I) = 1$. We consider Equation (3.1) for $A = I$. We obtain $\det(I) = 1 \cdot \det(I_{11})$ since the only nonzero entry in the first row of I occurs in the first column and this is one. The matrix I_{11} is the $(n - 1) \times (n - 1)$ identity matrix. By the induction hypothesis $\det(I_{(n-1)\times(n-1)}) = 1$. Therefore $\det(I) = 1$. This completes the proof of the existence of the determinant function for $n \times n$ matrices.

Uniqueness: We prove that Equation (3.2) holds for alternating multilinear functions f of $n \times n$ matrices assuming the previously stated induction hypothesis. We also use the existence proof which we have just completed.

Using the multilinearity of f, we have

$$f(A) = \sum_{j=1}^{n} a_{1j} f(\mathbf{e}_j, \mathbf{a}_2, \ldots, \mathbf{a}_n)$$

We must show that

$$f(\mathbf{e}_j, \mathbf{a}_2, \ldots, \mathbf{a}_n) = (-1)^{1+j} \det(A_{1j}) f(I) \tag{3.3}$$

In order to prove that Eq. (3.3) holds for a fixed j, let $\mathbf{a}_i' = \mathbf{a}_i - a_{ij}\mathbf{e}_j$, for $2 \leq i \leq n$, so that $\mathbf{a}_i = \mathbf{a}_i' + a_{ij}\mathbf{e}_j$. Then $\mathbf{a}_i'$ is a vector that agrees with $\mathbf{a}_i$ in each coordinate except the jth, and its jth coordinate is zero. We claim first that

$$f(\mathbf{e}_j, \mathbf{a}_2, \ldots, \mathbf{a}_n) = f(\mathbf{e}_j, \mathbf{a}_2', \ldots, \mathbf{a}_n') \tag{3.4}$$

We prove the claim by another induction argument, showing that $f(\mathbf{e}_j, \mathbf{a}_2', \ldots, \mathbf{a}_l', \mathbf{a}_{l+1}, \ldots, \mathbf{a}_n) = f(\mathbf{e}_j, \mathbf{a}_2, \ldots, \mathbf{a}_n)$ for all $l \leq n$. The initial case, $l = 1$ is trivially true. Assume that our claim holds for l, that is

$$f(\mathbf{e}_j, \mathbf{a}_2, \ldots, \mathbf{a}_n) = f(\mathbf{e}_j, \mathbf{a}_2', \ldots, \mathbf{a}_l', \mathbf{a}_{l+1}, \ldots, \mathbf{a}_n)$$

Then

$$\begin{aligned}
&f(\mathbf{e}_j, \mathbf{a}_2', \ldots, \mathbf{a}_l', \mathbf{a}_{l+1}, \ldots, \mathbf{a}_n)\\
&= f(\mathbf{e}_j, \mathbf{a}_2, \ldots, \mathbf{a}_l', \mathbf{a}_{l+1}' + a_{l+1,j}\mathbf{e}_j, \ldots, \mathbf{a}_n)\\
&= f(\mathbf{e}_j, \mathbf{a}_2', \ldots, \mathbf{a}_l', \mathbf{a}_{l+1}', \ldots, \mathbf{a}_n) + a_{l+1,j} f(\mathbf{e}_j, \mathbf{a}_2', \ldots, \mathbf{a}_l',\\
&\quad \mathbf{e}_j, \ldots, \mathbf{a}_n)\\
&= f(\mathbf{e}_j, \mathbf{a}_2', \ldots, \mathbf{a}_{l+1}', \mathbf{a}_{l+2}, \ldots, \mathbf{a}_n)
\end{aligned}$$

since f is alternating. This completes the induction proof of the claim and Eq. (3.4) holds for all j.

We now prove that

$$f(\mathbf{e}_j, \mathbf{a}_2', \ldots, \mathbf{a}_n') = \det(A_{1j}) f(\mathbf{e}_j, \mathbf{e}_1, \ldots, \mathbf{e}_{j-1}, \mathbf{e}_{j+1}, \ldots, \mathbf{e}_n) \tag{3.5}$$

which is almost what we need. Once we have proven Eq. (3.5) the fact that f is alternating allows us to rewrite Eq. (3.5) as

$$(-1)^{j-1} \det(A_{1j}) f(\mathbf{e}_1, \ldots, \mathbf{e}_n)$$

since we must exchange first $\mathbf{e}_j$ with $\mathbf{e}_1$, then $\mathbf{e}_j$ with $\mathbf{e}_2$, and so on for a total of $j - 1$ exchanges to reach the usual ordering of $\mathbf{e}_1, \ldots, \mathbf{e}_n$. Since $(-1)^{j-1} = (-1)^{j+1}$, we have the desired form.

We now apply the induction hypothesis to prove Eq. (3.5). Since f is alternating and multilinear, $f(\mathbf{e}_k, \mathbf{a}_2', \ldots, \mathbf{a}_n')$ is an alternating and multilinear function of the rows of A_{1j}. By the induction hypothesis

$$f(\mathbf{e}_j, \mathbf{a}_2', \ldots, \mathbf{a}_n') = \det(A_{1j}) f(\mathbf{e}_j, \mathbf{e}_1, \ldots, \mathbf{e}_{j-1}, \mathbf{e}_{j+1}, \ldots, \mathbf{e}_n)$$

since the matrix B with rows $\mathbf{e}_j, \mathbf{e}_1, \ldots, \mathbf{e}_{j-1}, \mathbf{e}_{j+1}, \ldots, \mathbf{e}_n$ has $B_{1j} = I_{(n-1)\times(n-1)}$. Thus, Eq. (3.5) holds and we have proven Eq. (3.2).

We use Eq. (3.1) for the determinant function to rewrite Eq. (3.2) as $f(A) = \det(A) \cdot f(I)$. Therefore if $f(I) = 1$, then the alternating multilinear function f is equal to the determinant function. Thus the determinant function given by Eq. (3.1) is the only alternating multilinear function of $n \times n$ matrices which takes the value one on the identity matrix. ■

Now let us do an explicit calculation.

(3.2.9) Example. Compute the determinant of

$$A = \begin{bmatrix} 1 & 2 & 0 & -1 \\ 0 & 2 & 1 & 4 \\ 1 & -1 & -1 & 2 \\ 2 & 1 & 4 & 7 \end{bmatrix}$$

We expand along the first row:

$$\begin{aligned} \det(A) &= 1 \cdot \det(A_{11}) - 2 \cdot \det(A_{12}) + 0 \cdot \det(A_{13}) - (-1)\det(A_{14}) \\ &= \{2(-1 \cdot 7 - 2 \cdot 4) - 1(-1 \cdot 7 - 2 \cdot 1) \\ &\quad + 4(-1 \cdot 4 - (-1) \cdot 1)\} \\ &\quad - 2\{0(-1 \cdot 7 - 2 \cdot 4) - 1(1 \cdot 7 - 2 \cdot 2) \\ &\quad + 4(1 \cdot 4 - (-1) \cdot 2)\} \\ &\quad + \{0(-1 \cdot 4 - (-1) \cdot 1) - 2(1 \cdot 4 - (-1) \cdot 2) \\ &\quad + 1(1 \cdot 1 - (-1) \cdot 2)\} \\ &= -33 - 2(21) + (-9) = -84. \end{aligned}$$

The determinant function of an $n \times n$ matrix should play the same role as the determinant of a 2×2 matrix. The determinant should be nonzero if and only if the matrix is invertible (or equivalently, if and only if the matrix is the matrix of an isomorphism).

Part of this relationship is straightforward.

(3.2.10) Proposition. If an $n \times n$ matrix A is not invertible, then $\det(A) = 0$.

Proof: If A is not invertible, then by Exercise 10 of Section 2.3 the span of the rows of A has dimension $\leq n - 1$. Therefore, there is a linear dependence among the rows of A, $x_1\mathbf{a}_1 + \cdots + x_n\mathbf{a}_n = \mathbf{0}$ with some $x_i \neq 0$. Without loss of generality, assume $x_1 \neq 0$. Then $\mathbf{a}_1 = -1/x_1(x_2\mathbf{a}_2 + \cdots + x_n\mathbf{a}_n)$ and

$$\begin{aligned} \det(A) &= \det(\mathbf{a}_1, \ldots, \mathbf{a}_n) \\ &= \det\left(\sum_{i=2}^{n} (-x_i/x_1)\mathbf{a}_i, \mathbf{a}_2, \ldots, \mathbf{a}_n\right) \end{aligned}$$

$$= \sum_{i=2}^{n} (-x_i/x_1) \det(\mathbf{a}_i, \mathbf{a}_2, \ldots, \mathbf{a}_n)$$

$$= 0$$

since each term in the final sum is the determinant of a matrix with repeated rows. ■

If A is invertible, we now want to show that $\det(A) \neq 0$. This result is somewhat more involved and requires an additional proposition. Essentially, we will use elimination to put the invertible matrix A into echelon form, which is necessarily a diagonal matrix with nonzero entries. Thus, we will need to know how the elimination operations affect the determinant of a matrix and then to compute the determinant of a diagonal matrix. First, we show that the operation of adding a multiple of one row to another does not change the value of the determinant.

(3.2.11) Proposition. $\det(\mathbf{a}_1, \ldots, \mathbf{a}_n) = \det(\mathbf{a}_1, \ldots, \underset{\substack{\uparrow \\ i\text{th position}}}{\mathbf{a}_i + b\mathbf{a}_j}, \ldots, \mathbf{a}_n)$

Proof:
$$\begin{aligned}\det(\mathbf{a}_1, \ldots, \mathbf{a}_i + b\mathbf{a}_j, \ldots, \mathbf{a}_n) &= \det(\mathbf{a}_1, \ldots, \mathbf{a}_i, \ldots, \mathbf{a}_n) \\ &\quad + b \det(\mathbf{a}_1, \ldots, \\ &\quad \mathbf{a}_j, \ldots, \mathbf{a}_j, \ldots, \mathbf{a}_n) \\ &= \det(\mathbf{a}_1, \ldots, \mathbf{a}_n)\end{aligned}$$

since $\det(\mathbf{a}_1, \ldots, \mathbf{a}_j, \ldots, \mathbf{a}_j, \ldots, \mathbf{a}_n) = 0$. ■

We will need to apply (3.2.11) repeatedly as we execute the elimination procedure to produce a diagonal matrix with nonzero diagonal entries. The following lemma shows that it is trivial to compute the determinant of a diagonal matrix.

(3.2.12) Lemma. If A is an $n \times n$ diagonal matrix, then $\det(A) = a_{11}a_{22} \cdots a_{nn}$.

Proof: Exercise 6a. ■

The proof of the following proposition is actually an algorithm for computing the determinant of an $n \times n$ matrix. In low dimensions, $n \leqslant 3$, this algorithm is not more efficient than the inductive formula given in Theorem (3.2.8). However, for $n > 3$, this algorithm is more efficient and, in fact, the larger the value of n is, the greater is the number of steps it saves. (Count the number of steps in each method!)

Note that we can give a more elegant proof of this result if we proceed by induction on the size of the matrix. (See Exercise 10.)

(3.2.13) Proposition. If A is invertible, then $\det(A) \neq 0$.

Proof: From Proposition (3.2.11) we see that elementary row operations of type (a)—adding multiples of one row to another—do not change the determinant of a matrix. Further, from the basic properties of the determinant, row operations of type (c)—interchanging two rows of a matrix—change the sign of the determinant of a matrix and thus do not affect whether or not the determinant is zero.

If the elimination algorithm is applied using only operations of type (a) and (c), the resulting matrix will differ from the echelon form only in that the leading nonzero entries of each row need not be equal to 1. Since the matrix A is invertible, it has n linearly independent columns, and thus, by Exercise 10 of Section 3 of Chapter 2 the matrix also has n linearly independent rows. Also, by Exercise 10 we see that the matrix resulting from the modified elimination algorithm has n linearly independent rows. It must, therefore, be a diagonal matrix with nonzero diagonal entries. By the lemma the determinant is nonzero.

From the initial remarks of the proof, we conclude that the determinant of the diagonal matrix obtained from the modified elimination algorithm is either the determinant of the original matrix A or its negative. If the modified elimination algorithm uses k elementary row operations of type (c), then the determinant of the resulting diagonal matrix is $(-1)^k$ times the determinant of the matrix A. Therefore, the determinant of the matrix A is nonzero. ■

Combining (3.2.10) and (3.2.13) we have the major result, Theorem (3.2.14), of this chapter.

(3.2.14) Theorem. Let A be an $n \times n$ matrix. A is invertible if and only if $\det(A) \neq 0$.

Thus, we see in general that the determinant plays the same role for $n \times n$ matrices that the area of a parallelogram plays in $\mathbf{R}^2$. It is worth noting that there is also a well-developed theory of volumes in vector spaces of a dimension higher than three and that the determinant as we have defined it plays a central role in that theory.

(3.2.15) Example. The modified version of the elimination algorithm used in the proof of Proposition (3.2.13) can be used to compute the determinant of a matrix. As an example, we compute the determinant of the matrix A of Example (3.2.9) by this method. We begin with A and proceed step by step through the algorithm until we obtain an upper triangular matrix.

$$\begin{bmatrix} 1 & 2 & 0 & -1 \\ 0 & 2 & 1 & 4 \\ 1 & -1 & -1 & 2 \\ 2 & 1 & 4 & 7 \end{bmatrix} \rightarrow \begin{bmatrix} 1 & 2 & 0 & -1 \\ 0 & 2 & 1 & 4 \\ 0 & -3 & -1 & 3 \\ 0 & -3 & 4 & 9 \end{bmatrix}$$

$$\rightarrow \begin{bmatrix} 1 & 2 & 0 & -1 \\ 0 & 2 & 1 & 4 \\ 0 & 0 & 1/2 & 9 \\ 0 & 0 & 11/2 & 15 \end{bmatrix} \rightarrow \begin{bmatrix} 1 & 2 & 0 & -1 \\ 0 & 2 & 1 & 4 \\ 0 & 0 & 1/2 & 9 \\ 0 & 0 & 0 & -84 \end{bmatrix}$$

At this point the proof indicates that we should proceed with the algorithm until we obtain a diagonal matrix. However, as a result of Exercise 6 of this section, we can already calculate the determinant at this stage by computing the product of the diagonal entries. Thus the determinant of the preceding upper triangular matrix is $1 \cdot 2 \cdot (1/2) \cdot (-84) = -84$. As a consequence of the proof of (3.2.13) -84 is also the determinant of the original matrix A.

EXERCISES

1. Compute the determinant of each of the following matrices in two ways using Equation (3.1) of Theorem (3.2.8), first by expanding along the first row and second by expanding along the third row.

a) $\begin{bmatrix} 1 & 0 & 2 \\ 0 & 2 & 0 \\ 1 & 0 & 3 \end{bmatrix}$ b) $\begin{bmatrix} -1 & 1 & 2 \\ 1 & 0 & 1 \\ -1 & 1 & 0 \end{bmatrix}$

c) $\begin{bmatrix} 1 & 1 & 1 \\ 3 & 1 & 1 \\ 0 & 2 & 1 \end{bmatrix}$ d) $\begin{bmatrix} 1 & 2 & -1 \\ 1 & 3 & 0 \\ 2 & 0 & 1 \end{bmatrix}$

e) $\begin{bmatrix} 1 & 1 & 3 & 2 \\ & 2 & -1 & 0 \\ & & 4 & 1 \\ & & & 7 \end{bmatrix}$ f) $\begin{bmatrix} 1 & -1 & 0 & 2 \\ 0 & 1 & 0 & 1 \\ 3 & -1 & 2 & 0 \\ 1 & 1 & 1 & 1 \end{bmatrix}$

2. Use the version of the elimination algorithm developed in Example (3.2.15) to calculate the determinant of each of the following matrices.

a) $\begin{bmatrix} 1 & 2 & 1 \\ -1 & 0 & 1 \\ 2 & 3 & 1 \end{bmatrix}$ b) $\begin{bmatrix} 1 & 2 & 0 \\ 1 & 3 & 4 \\ 3 & 2 & 1 \end{bmatrix}$

c) $\begin{bmatrix} 1 & 3 & 2 & 1 \\ 0 & 2 & 1 & 0 \\ 1 & 0 & 3 & 4 \\ 1 & 1 & 0 & 4 \end{bmatrix}$ d) $\begin{bmatrix} 1 & 1 & & & \\ 1 & 1 & 1 & & \\ & 1 & 1 & 1 & \\ & & 1 & 1 & 1 \\ & & & 1 & 1 \end{bmatrix}$

3. Calculate the determinants of the following matrices by applying one or more of the properties of the determinant—do not use either Equation (3.1) of Theorem (3.2.8) or the elimination algorithm of Example (3.2.15).

a) $\begin{bmatrix} 1 & 3 & 2 \\ 1 & 0 & 1 \\ 0 & 0 & 0 \end{bmatrix}$ b) $\begin{bmatrix} 0 & 0 & 1 \\ 0 & 1 & 0 \\ 1 & 0 & 0 \end{bmatrix}$

c) $\begin{bmatrix} 1 & 2 & 3 \\ 3 & 2 & 1 \\ 1 & 2 & 3 \end{bmatrix}$ d) $\begin{bmatrix} 1 & -1 & 2 \\ 2 & -2 & 4 \\ 3 & 2 & 1 \end{bmatrix}$

4. For what values of the scalar a is each of the following matrices invertible?

a) $\begin{bmatrix} a & 1 \\ -1 & a \end{bmatrix}$ b) $\begin{bmatrix} -1-a & 2 \\ 3 & -a \end{bmatrix}$

c) $\begin{bmatrix} a & & \\ & a & \\ & & a \end{bmatrix}$ d) $\begin{bmatrix} a & 1 & 1 \\ a & 1 & \\ a & & \end{bmatrix}$

5. Let A be a 3×3 matrix. Expanding the formula for the determinant of A, we have $\det(A) = a_{11}a_{22}a_{33} - a_{11}a_{23}a_{32} - a_{12}a_{21}a_{33} + a_{12}a_{23}a_{31} + a_{13}a_{21}a_{32} - a_{13}a_{22}a_{31}$. Use this formula to prove that $\det(A)$ is an alternating multilinear function of the rows of A satisfying $\det(I) = 1$.

6. a) Prove that the determinant of a diagonal matrix (all of its off-diagonal entries are zero) is the product of its diagonal entries. (*Hint:* Prove by induction on the size of the matrix.)

   b) Prove that the determinant of an upper (lower) triangular matrix is the product of its diagonal entries. (*Hint:* Prove by induction on the size of the matrix.)

7. Let A be a $k \times k$ matrix, and let B be an $l \times l$ matrix. Prove that

$$\det\begin{bmatrix} A & 0 \\ 0 & B \end{bmatrix} = \det(A)\det(B)$$

[*Hint:* Use the technique of the proof of Proposition (3.2.13).]

8. Let A be a $k \times k$ matrix, let B be an $l \times l$ matrix, and let C be a $k \times l$ matrix. Prove that

$$\det\begin{bmatrix} A & C \\ 0 & B \end{bmatrix} = \det(A)\det(B)$$

9. Let A be an $n \times n$ matrix. Rewrite the existence portion of Theorem (3.2.8) for a general i, $1 \leqslant i \leqslant n$; that is, prove for a general i that Eq. (3.1) defines an alternating multilinear function with $\det(I) = 1$.

10. Give an inductive proof of Proposition (3.2.13). (What is the induction hypothesis?)

11. Use the formula for the volume of a parallelepiped in $\mathbf{R}^3$ to prove that the volume of the parallelepiped with vertices $\mathbf{0}$, $\mathbf{a}_1$, $\mathbf{a}_2$, $\mathbf{a}_3$, $\mathbf{a}_1 + \mathbf{a}_2$, $\mathbf{a}_1 + \mathbf{a}_3$, $\mathbf{a}_2 + \mathbf{a}_3$, and $\mathbf{a}_1 + \mathbf{a}_2 + \mathbf{a}_3$ is equal to $\det(\mathbf{a}_1, \mathbf{a}_2, \mathbf{a}_3)$. (*Hint:* Assume that $\mathbf{a}_1$, $\mathbf{a}_2$ lie in the $x_1 - x_2$ plane.)

12. Prove the multilinearity property of the determinant of a 3×3 matrix directly from a geometric construction as in Exercise 11 of Section 3.1.

The Cross Product in $\mathbf{R}^3$

From Theorem (3.2.14) we see that the determinant function can be used to determine if a set of n vectors in $\mathbf{R}^n$ is linearly independent or linearly dependent. In this sequence of problems we will show that determinants of 2×2 matrices can be used to show two vectors in $\mathbf{R}^3$ are linearly independent. Let $\mathbf{x} = (x_1, x_2, x_3)$ and $\mathbf{y} = (y_1, y_2, y_3)$ be vectors in $\mathbf{R}^3$. Define the *cross product* of $\mathbf{x}$ and $\mathbf{y}$, $\mathbf{x} \times \mathbf{y}$, to be the vector

$$\mathbf{x} \times \mathbf{y} = (x_2y_3 - y_2x_3, -(x_1y_3 - y_1x_3), x_1y_2 - y_1x_2)$$

(Where are the 2×2 determinants?)

13. Prove that the cross product in $\mathbf{R}^3$ satisfies the following properties for all $\mathbf{x}$, $\mathbf{y}$, and $\mathbf{z} \in \mathbf{R}^3$ and $a, b \in \mathbf{R}$:
    a) $(a\mathbf{x} + b\mathbf{y}) \times \mathbf{z} = a(\mathbf{x} \times \mathbf{z}) + b(\mathbf{y} \times \mathbf{z})$
    b) $\mathbf{x} \times \mathbf{y} = -\mathbf{y} \times \mathbf{x}$
    c) $\mathbf{e}_i \times \mathbf{e}_j = \pm\mathbf{e}_k$ for $\mathbf{e}_i$, $\mathbf{e}_j$, and $\mathbf{e}_k$ three distinct standard basis vectors for $\mathbf{R}^3$ taken in any order.

14. Prove that $\mathbf{x}$ and $\mathbf{y}$ are linearly independent in $\mathbf{R}^3$ if and only if $\mathbf{x} \times \mathbf{y} \neq \mathbf{0}$.

15. Let $\mathbf{x}$ and $\mathbf{y}$ be linearly independent vectors in $\mathbf{R}^3$. Prove that $\mathbf{x}$, $\mathbf{y}$, and $\mathbf{x} \times \mathbf{y}$ form a basis for $\mathbf{R}^3$.

16. Compute the cross products of the following pairs of vectors:
    a) $(1, -1, 2)$ $(3, -1, 0)$
    b) $(1, -1, 0)$ $(2, 1, 0)$
    c) $(0, 2, 2)$ $(0, -1, -1)$

§3.3. FURTHER PROPERTIES OF THE DETERMINANT

Our discussion of the determinant function was motivated by a desire to find an efficient theoretical means of determining if a transformation $T: V \to V$ or a square matrix A is invertible. Theorem (3.2.14) provided a solution to this problem. However, we have yet to see a direct relationship between the nonzero determinant of A and the inverse of A. The first goal of this section is to give an alternate construction of A^{-1} that utilizes the determinant of A. The second goal of this section is to prove the product rule for determinants, which, among other applications, will allow us to compute $\det(A^{-1})$.

Begin with an invertible matrix A so that $\det(A) \neq 0$. If we fix a row of A, say, the ith row, we have by Theorem (3.2.8)

$$\det(A) = \sum_{j=1}^{n} (-1)^{i+j} a_{ij} \det(A_{ij})$$

Consider the matrix B_{kl} whose rows are

$$\mathbf{a}_1, \mathbf{a}_2, \ldots, \mathbf{a}_k, \ldots, \underset{\substack{\uparrow \\ l\text{th position}}}{\mathbf{a}_k}, \ldots, \mathbf{a}_n$$

That is, B_{kl} is equal to A except that we have replaced the lth row of A by the kth row of A. If $k \neq l$, B_{kl} has a repeated row so that $\det(B_{kl}) = 0$. If $k = l$, $B_{kk} = A$ so that $\det(B_{kl}) = \det(A)$. If we use the formula to compute $\det(B_{kl})$ for $k \neq l$ and expand along the lth row, we see that

$$0 = \det(B_{kl}) = \sum_{j=1}^{n} (-1)^{j+l} a_{kj} \det(A_{lj}) \tag{3.5}$$

These facts will enable us to construct A^{-1} directly in terms of the entries of A.

Let A' be the matrix whose entries a'_{ij} are the scalars $(-1)^{i+j} \det(A_{ji})$. The quantity a'_{ij} is called the jith *cofactor* of A. (Notice that the indices are interchanged in the minor–it is A_{ji} and not A_{ij}.)

(3.3.1) Proposition. $AA' = \det(A)I$.

Proof: The klth entry of AA' is

$$\sum_{j=1}^{n} a_{kj}a'_{jl} = \sum_{j=1}^{n} a_{kj}(-1)^{j+l} \det(A_{lj}) = \det(B_{kl})$$

From the preceding calculations we see that this is zero if $k \neq l$ and is the determinant of A if $k = l$. Therefore, $AA' = \det(A)I$. ■

If $\det(A) \neq 0$ we may divide by $\det(A)$ in (3.3.1) to obtain Corollary (3.3.2).

(3.3.2) Corollary. If A is an invertible $n \times n$ matrix, then A^{-1} is the matrix whose ijth entry is $(-1)^{i+j} \det(A_{ji})/\det(A)$.

(3.3.3) Example. We will use Corollary (3.3.2) to compute the inverse of the matrix

$$A = \begin{bmatrix} 1 & 0 & 3 \\ 2 & -1 & 4 \\ 5 & 7 & -2 \end{bmatrix}$$

First we compute A' the matrix of cofactors of A

$$A' = \begin{bmatrix} (-1)^2(-26) & (-1)^3(-21) & (-1)^4(3) \\ (-1)^3(-24) & (-1)^4(-17) & (-1)^5(-2) \\ (-1)^4(19) & (-1)^5(7) & (-1)^6(-1) \end{bmatrix}$$

$$= \begin{bmatrix} -26 & 21 & 3 \\ 24 & -17 & 2 \\ 19 & -7 & -1 \end{bmatrix}$$

Second we compute the determinant of A by expanding along the first row.

$$\det(A) = (-1)^2 \cdot 1 \cdot (-26) + (-1)^3 \cdot 0 + (-1)^4 \cdot 3 \cdot 19 = 31$$

Therefore

$$A^{-1} = (1/31)\begin{bmatrix} -26 & 21 & 3 \\ 24 & -17 & 2 \\ 19 & -7 & -1 \end{bmatrix}$$

An unexpected consequence of (3.3.1) and (3.3.2) is that we can calculate $\det(A)$ by expanding along columns instead of rows if we choose to do so.

(3.3.4) Proposition. For any fixed j, $1 \leq j \leq n$

$$\det(A) = \sum_{i=1}^{n} (-1)^{i+j} a_{ij} \det(A_{ij})$$

Proof: If A is invertible, then $A^{-1}A = I$ implies that $A'A = \det(A)I$. The diagonal entries of $A'A$ are

$$\det(A) = \sum_{i=1}^{n} a'_{ji} a_{ij}$$

$$= \sum_{i=1}^{n} (-1)^{i+j} \det(A_{ij}) a_{ij}$$

$$= \sum_{i=1}^{n} (-1)^{i+j} a_{ij} \det(A_{ij})$$

which is the formula we wished to establish.

If A is not invertible, then $\det(A) = 0$ and $AA' = 0$. However, this does not immediately imply that $A'A = 0$. A direct proof in this case is somewhat more complicated, so we leave this fact to Exercise 8, which presents an alternate proof of Proposition (3.3.4) using the uniqueness of the determinant. ■

It should be emphasized that this formula for A^{-1} is efficient only for small n, that is, $n \leq 3$. In general, it is more efficient to use the Gauss-Jordan method.

(3.3.5) Remark. In general, if $\mathbf{b}$ is a vector in $\mathbf{R}^n$, $A'\mathbf{b}$ is a vector whose ith entry is $\sum_{j=1}^{n} a'_{ij} b_j = \sum_{j=1}^{n} b_j (-1)^{i+j} \det(A_{ji})$. This is the determinant of the matrix whose *columns* are $\mathbf{a}_1, \ldots, \mathbf{a}_{i-1}, \mathbf{b}, \mathbf{a}_{i+1}, \ldots, \mathbf{a}_n$, where $\mathbf{a}_j$, $1 \leq j \leq n$, is the jth column of A. The determinant is expanded along the ith column. This fact will be used in the discussion of Cramer's rule, which appears later in this section.

The next example motivates the product rule for determinants.

(3.3.6) Example

a) If A and B are 2×2 rotation matrices, we have shown that AB is also a rotation matrix. Further we know that every rotation matrix has a determinant equal to 1, since $\det(R_\theta) = \cos^2(\theta) + \sin^2(\theta) = 1$. Thus, $\det(AB) = \det(A)\det(B)$, since each determinant in the equation is equal to one.

b) Now replace B by the 2×2 matrix of a projection to a nonzero vector. Then B is not invertible and $\det(B) = 0$. The product AB is also not invertible for A, a rotation matrix. We can see this directly from the geometry or we can appeal to Corollary (2.5.7). Thus $\det(AB) = 0$. Then $\det(AB) = \det(B) = 0$ implies that $\det(AB) = \det(A)\det(B)$.

c) Let $A = \begin{bmatrix} 1 & 2 \\ 1 & 1 \end{bmatrix}$ and $B = \begin{bmatrix} 2 & 0 \\ 1 & 3 \end{bmatrix}$. Then $\det(A)\det(B) = -1 \cdot 6 = -6$. On the other hand, $AB = \begin{bmatrix} 4 & 6 \\ 3 & 3 \end{bmatrix}$, so that $\det(AB) = -6$ also. Here, too, we have $\det(AB) = \det(A)\det(B)$.

These examples lead us to expect that the determinant satisfies the general rule $\det(AB) = \det(A)\det(B)$.

(3.3.7) Proposition. If A and B are $n \times n$ matrices, then

a) $\det(AB) = \det(A)\det(B)$.

b) If A is invertible, then $\det(A^{-1}) = 1/\det(A)$.

Proof:

a) First, consider the case when B is not invertible. Then $\det(B) = 0$ so that the right-hand side is zero. By Corollary (2.5.7) (i), AB also fails to be invertible. Therefore, $\det(AB) = 0$ and the equation holds.

Now assume that B is invertible and fix B. The function of matrices A defined by

$$f(A) = \det(AB)$$

is an alternating multilinear function of the rows of A. (See Exercise 3.) Since $f(I) = \det(IB) = \det(B)$, it follows from Theorem (3.2.8) that

$$f(A) = \det(A)\det(B)$$

as we desired. Statement b follows from a, and its proof is left to the reader as Exercise 4. ■

This result has a nice geometric interpretation if we utilize Proposition (3.3.4). We will restrict ourselves to $\mathbf{R}^3$. From (3.3.4) we can view the determinant as a function of the columns of a matrix. If we let $\mathbf{b}_1$, $\mathbf{b}_2$, and $\mathbf{b}_3$ denote the columns of B, then we may interpret $\det(B)$ as the volume of the parallelepiped defined by these vectors. Then we may view the matrix product as the matrix whose columns are $A\mathbf{b}_1$, $A\mathbf{b}_2$, and $A\mathbf{b}_3$. Thus, $\det(AB)$ is the volume of the parallelepiped defined by these three vectors. If we view A as the matrix of a linear transformation, then (3.3.7) says that the transformation increases (or decreases) the volume of parallelepipeds by a factor of det(A). [In multivariable calculus, this fact is a key part of the change of the variable formula for multiple integrals, which involves the determinant of the derivative matrix of a (usually nonlinear) mapping $T: \mathbf{R}^n \to \mathbf{R}^n$.]

As a consequence of Proposition (3.3.7), we have Corollary (3.3.8).

(3.3.8) Corollary. If $T: V \to V$ is a linear transformation, $\dim(V) = n$, then

$$\det([T]_\alpha^\alpha) = \det([T]_\beta^\beta)$$

for all choices of bases α and β for V.

Proof: Exercise 5. ■

This corollary allows us to define the determinant of a linear transformation $T: V \to V$ if $\dim(V) < \infty$.

(3.3.9) Definition. The *determinant* of a linear transformation $T: V \to V$ of a finite-dimensional vector space is the determinant of $[T]_\alpha^\alpha$ for any choice of α. We denote this by det(T).

(3.3.10) Example. Let $V = \text{Span}(\{\cos(x), \sin(x)\})$ considered as a subspace of the vector space of differentiable functions on $\mathbf{R}$. Define $T: V \to V$ by $T(f) = f'$, the derivative of f. It is an easy calculation to show that the determinant of T is 1, since the matrix of T with respect to the basis $\{\cos(x), \sin(x)\}$ is $\begin{bmatrix} 0 & -1 \\ 1 & 0 \end{bmatrix}$.

Of course, now we have Proposition (3.3.11).

(3.3.11) Proposition. A linear transformation $T: V \to V$ of a finite-dimensional vector space is an isomorphism if and only if $\det(T) \neq 0$.

Proof: This follows immediately since T is an isomorphism if and only if $A = [T]_\alpha^\alpha$ is invertible. ■

(3.3.12) Proposition. Let $S: V \to V$ and $T: V \to V$ be linear transformations of a finite-dimensional vector space, then

(i) $\det(ST) = \det(S)\det(T)$ and

(ii) if T is an isomorphism $\det(T^{-1}) = \det(T)^{-1}$.

Proof: Exercise 6. ■

We will consider one final application of the formula for the inverse of a matrix known as Cramer's rule. Consider an $n \times n$ system of equations $A\mathbf{x} = \mathbf{b}$, where A is invertible. From the preceding we see that

$$\mathbf{x} = A^{-1}\mathbf{b} = \frac{1}{\det(A)} A'\mathbf{b}$$

If we apply Remark (3.3.5), we see that the jth entry of $A'\mathbf{b}$ is the determinant of the matrix obtained from A by replacing the jth column of A by the vector $\mathbf{b}$. Thus, we have proven Proposition (3.3.13).

(3.3.13) Proposition. (Cramer's rule) Let A be an invertible $n \times n$ matrix. The solution $\mathbf{x}$ to the system of equations $A\mathbf{x} = \mathbf{b}$ is the vector whose jth entry is the quotient

$$\det(B_j)/\det(A)$$

where B_j is the matrix obtained from A by replacing the jth column of A by the vector $\mathbf{b}$.

We emphasize that this is not a practical method for solving systems of equations for large n; elimination is far more efficient.

(3.3.14) Example. We use Cramer's rule to solve

$$\begin{bmatrix} 2 & 1 \\ -1 & 3 \end{bmatrix} \mathbf{x} = \begin{bmatrix} 0 \\ 1 \end{bmatrix}$$

We obtain

$$x_1 = \frac{\det \begin{bmatrix} 0 & 1 \\ 1 & 3 \end{bmatrix}}{\det \begin{bmatrix} 2 & 1 \\ -1 & 3 \end{bmatrix}} = -1/7 \quad \text{and} \quad x_2 = \frac{\det \begin{bmatrix} 2 & 0 \\ -1 & 1 \end{bmatrix}}{\det \begin{bmatrix} 2 & 1 \\ -1 & 3 \end{bmatrix}} = 2/7$$

Therefore, $\mathbf{x} = (-1/7, 2/7)$.

EXERCISES

1. Compute the inverses of the following matrices:

a) $\begin{bmatrix} 3 & 2 \\ 1 & 0 \end{bmatrix}$ b) $\begin{bmatrix} 1 & 0 & -1 \\ 2 & 1 & -3 \\ 0 & 1 & 4 \end{bmatrix}$ c) $\begin{bmatrix} 1 & 1 & 0 & 0 \\ 1 & 2 & 1 & 0 \\ 0 & 1 & 3 & 1 \\ 0 & 0 & 1 & 4 \end{bmatrix}$

using the formula involving the determinant. Check your answer. (How?)

2. Verify that the determinants of the matrices of Exercise 1 are the same whether we expand along the first row or first column of the matrix.

3. Let B be an invertible matrix. Prove that the function

$$f(A) = \det(AB)$$

of $n \times n$ matrices A is

a) a multilinear function of the rows of A.
b) an alternating function of the rows of A.

4. Prove that if A is an invertible matrix $\det(A^{-1}) = \det(A)^{-1}$.

5. Prove Corollary (3.3.8).

6. Prove Proposition (3.3.12).

7. Use Cramer's rule to solve for $\mathbf{x}$:

a) $\begin{bmatrix} 1 & 3 \\ -1 & 2 \end{bmatrix} \mathbf{x} = \begin{bmatrix} 2 \\ 2 \end{bmatrix}$ b) $\begin{bmatrix} 1 & 0 & -1 \\ 2 & 1 & -3 \\ 0 & 1 & 4 \end{bmatrix} \mathbf{x} = \begin{bmatrix} 1 \\ 1 \\ 1 \end{bmatrix}$

c) $\begin{bmatrix} 1 & 1 & 0 & 0 \\ 1 & 2 & 1 & 0 \\ 0 & 1 & 3 & 1 \\ 0 & 0 & 1 & 4 \end{bmatrix} \mathbf{x} = \begin{bmatrix} 1 \\ 2 \\ 4 \\ 6 \end{bmatrix}$

8. Let A be a $n \times n$ matrix. This problem will show that the determinant of A can be calculated using the columns of A whether or not A is invertible. Prove that for each fixed j $(1 \leqslant j \leqslant n)$

$$\det(A) = \sum_{i=1}^{n} (-1)^{i+j} a_{ij} \det(A_{ij})$$

by showing that this formula defines an alternating multilinear function of the rows of the matrix A [notice that when either rows are exchanged or a multiple of one row is added to another, both the coefficients a_{ij} and the determinants $\det(A_{ij})$ may change] and showing that this formula yields the value 1 when $A = I$.

9. Prove that for an arbitrary $n \times n$ matrix A, $\det(A) = \det(A^t)$, where A^t is the transpose of A, which was defined in Chapter 2, Section 2.2, Exercise 12. (See Exercise 8.)

10. Show that for any $n \times n$ matrix A and any scalar c, $\det(cA) = c^n \cdot \det(A)$.

11. Let A be an $n \times n$ skew-symmetric matrix, that is, $A^t = -A$ (see Section 2.2, Exercise 12). Prove that if n is an odd integer, then A is not invertible. (*Hint:* Use Exercise 10 of this section.)

12. This problem will give another proof of Proposition (3.2.13) based on Proposition (3.3.7). Let $E_{ij}(c)$, $F_i(c)$, and G_{ij} denote the elementary matrices used in the elimination procedure (see Section 2.5, Exercises 14, 15, and 16).
a) Prove that $\det(E_{ij}(c)) = 1$, $\det(F_i(c)) = c$, and $\det(G_{ij}) = -1$ if $i \neq j$.
b) Let A be an $n \times n$ matrix. Prove that if the matrix B is obtained from A by elimination in such a way that no matrix $F_i(c)$ is used, then $\det(B) = \pm\det(A)$.
c) Let A be an $n \times n$ matrix. Let B be any matrix obtained from A by any sequence of elimination steps. What is the relationship between $\det(B)$ and $\det(A)$?

CHAPTER SUMMARY

The determinant is defined to be the unique alternating multilinear function of the rows of an $n \times n$ matrix with values in real numbers such that $\det(I) = 1$ [Theorem (3.2.8)]. In order to make this definition, we had to show that these three properties—multilinearity [Definition (3.2.1)], the alternating property [Definition (3.2.2)], and normalization [$f(I) = 1$]—were sufficient to define uniquely a function of the rows of a square matrix. Our proof was based on induction on the size of the matrix. Using the fact that the $(n - 1) \times (n - 1)$ determinant existed and was unique, we were able to show that the $n \times n$ determinant existed and was unique. The method of proof gave us the formula

$$\det(A) = \sum_{j=1}^{n} (-1)^{i+j} a_{ij} \det(A_{ij})$$

where A_{ij} is the ijth minor of A [Definition (3.2.4)].

This was motivated by our consideration of the area of a parallelogram in $\mathbf{R}^2$. In turn, the determinant can be interpreted as giving the volume of the parallelepiped determined by the rows of an $n \times n$ matrix in $\mathbf{R}^n$, generalizing the cases $n = 2$ and 3.

The first use of the determinant function is as a test to determine if a matrix A is invertible [Theorem (3.2.14)]. We also gave an explicit formula for A^{-1} in terms of $\det(A)$ [Corollary (3.3.2)]. These same calculations yielded the fact that the determinant could also be expressed as a function of the columns of a matrix instead of the rows [Proposition (3.3.4)].

Using the multiplicative formula for the determinant [Proposition (3.3.7)], we were able to define the determinant of a linear transformation [Corollary (3.3.8)]. Thus, although the determinant of a transformation would appear to depend on the particular matrix chosen to represent the transformation, it does not.

Finally, if $\det(A) \neq 0$, Cramer's rule [Proposition (3.3.13)] provided an alternate method of solving $A\mathbf{x} = \mathbf{b}$ for any $\mathbf{b}$.

SUPPLEMENTARY EXERCISES

1. **True-False**. For each true statement, give a brief proof or reason. For each false statement give a counterexample.
   a) If $\mathbf{a}_1$, $\mathbf{a}_2$, and $\mathbf{a}_3$ lie in a plane in $\mathbf{R}^3$, then the determinant of the matrix whose rows are these vectors is zero.
   b) If A is an $n \times n$ matrix with rows $\mathbf{v}_1, \ldots, \mathbf{v}_n$ and B is the $n \times n$ matrix with rows $a_1\mathbf{v}_1, \ldots, a_n\mathbf{v}_n$, then $\det(B) = a_1 \cdots a_n \det(A)$.
   c) Every isomorphism $T: V \to V$ of a finite-dimensional vector space has a determinant equal to 1.
   d) If two matrices have the same rows, although possibly in different orders, then they have the same determinant.
   e) If a matrix has repeated rows, then its determinant is zero.
   f) Cramer's rule applies to every square matrix A to solve $A\mathbf{x} = \mathbf{b}$.
   g) Similar matrices have the same determinant.

h) If $A = \begin{bmatrix} 0 & \cdots & 0 & 0 & a_{1n} \\ 0 & & 0 & a_{2,n-1} & a_{2n} \\ & & \vdots & & \\ 0 & a_{n-1,2} & & \cdots & a_{n-1,n} \\ a_{n1} & & & \cdots & a_{nn} \end{bmatrix}$

then $\det(A) = a_{1n}\, a_{2,n-1} \cdots a_{n1}$.

i) If A is an $n \times n$ matrix, then $\det(aA) = a \det(A)$ for $a \in \mathbf{R}$.

j) If A and B are $n \times n$ matrices, then $\det(A + B) = \det(A) + \det(B)$.

2. What is the area of the parallelogram in $\mathbf{R}^2$ whose vertices are
   a) $\mathbf{0}$, (1, 2), (2, 2), and (3, 4)?
   b) $\mathbf{0}$, (−1, 1), (2, 4), and (1, 5)?

3. What is the determinant of each of the following 2 × 2 matrices?

   a) $\begin{bmatrix} 3 & -1 \\ 2 & 1 \end{bmatrix}$ b) any rotation matrix

   c) any projection matrix

4. Compute the determinants of the following matrices. Which are invertible?

   a) $\begin{bmatrix} 3 & -1 & 2 \\ 1 & 0 & 1 \\ 2 & -1 & 1 \end{bmatrix}$ b) $\begin{bmatrix} 1 & 2 & 3 & 4 & 5 \\ 0 & 2 & 3 & 4 & 5 \\ 0 & 0 & 3 & 4 & 5 \\ 0 & 0 & 0 & 4 & 5 \\ 0 & 0 & 0 & 0 & 5 \end{bmatrix}$

5. Are there multilinear functions of the rows of a matrix that are not alternating or alternating functions that are not multilinear? If so, give an example, if not, why not?

6. Let A and B be $n \times n$ matrices. What is $\det(A^k B^l)$?

7. Prove that if A is an $n \times n$ matrix (with real entries) and $\det(A^k) = 1$ for some integer k, then $\det(A) = \pm 1$.

8. Prove that if $\det(AB) = 0$, then either A or B is not invertible.

9. a) Show that the determinant of the 3 × 3 matrix

   $$V = \begin{bmatrix} 1 & 1 & 1 \\ a & b & c \\ a^2 & b^2 & c^2 \end{bmatrix}$$

   is $\det(V) = (c - b)(c - a)(b - a)$. Conclude that V is invertible if and only if a, b, and c are distinct.

   b) What are the analogous $n \times n$ matrices? Prove a similar formula for their determinants. (These matrices are sometimes called *Vandermonde matrices*.)

CHAPTER 4

Eigenvalues, Eigenvectors, Diagonalization, and the Spectral Theorem in $\mathbf{R}^n$

Introduction

In this chapter we begin by developing some results on the fine structure of individual linear mappings $T: V \to V$. Our main goal is to be able to understand the geometry of a linear mapping in more detail. Apart from our study of some special classes of mappings such as projections and rotations, we have not taken this approach previously, because we were more interested in investigating the properties shared by all linear mappings. However, in using linear algebra, understanding the special behavior of an individual mapping is often a necessary part of solving problems. Furthermore, by studying the different ways an individual mapping can be special, we gain even more insight into the structure of the space of all mappings $T: V \to V$.

The technique we will use to analyze linear mappings will be to try to find special vectors in the space V on which T acts in a particularly simple way—vectors that are mapped to scalar multiples of themselves. We call such vectors eigenvectors of the linear mapping in question.

If we can find a basis of V consisting of eigenvectors for T, then we are in the fortunate position of being able to describe T completely in a very direct way. T simply acts by stretching by different scalar factors (which are called eigenvalues) in the different coordinate directions determined by the basis of eigenvectors. We will derive precise criteria that we can use to determine whether such a basis exists.

In the remainder of the chapter, we consider more of the geometry of the vector space $\mathbf{R}^n$—facts about lengths of vectors, angles between vectors, and so forth, which can all be expressed in terms of the standard inner product (also known as the dot product) on $\mathbf{R}^n$. We will then use the geometry of $\mathbf{R}^n$ to study a special, and extremely important, class of linear mappings $T: \mathbf{R}^n \to \mathbf{R}^n$ known as symmetric mappings. We will show that for every symmetric mapping, there are bases of $\mathbf{R}^n$ consisting of eigenvectors of T, and that furthermore, we can find such a basis in which each pair of basis vectors is perpendicular (or, as we will say, orthogonal). This is part of the spectral theorem for symmetric mappings, a theorem with many applications in all branches of mathematics. We will apply the spectral theorem to study the geometry of conic sections and quadric surfaces.

§4.1. EIGENVALUES AND EIGENVECTORS

Let us begin with an example to illustrate some of the ideas we will be considering.

(4.1.1) Example. Consider the linear mapping $T: \mathbf{R}^2 \to \mathbf{R}^2$ defined by the matrix $A = \begin{bmatrix} 2 & 1 \\ 1 & 2 \end{bmatrix}$. We ask, given a vector $\mathbf{x} \in \mathbf{R}^2$, how are $\mathbf{x}$ and its image under T related geometrically? As we know, to compute values of T, we just apply matrix multiplication. For example, the images under T of the standard basis vectors are

$$\begin{bmatrix} 2 & 1 \\ 1 & 2 \end{bmatrix}\begin{bmatrix} 1 \\ 0 \end{bmatrix} = \begin{bmatrix} 2 \\ 1 \end{bmatrix} \quad \text{and} \quad \begin{bmatrix} 2 & 1 \\ 1 & 2 \end{bmatrix}\begin{bmatrix} 0 \\ 1 \end{bmatrix} = \begin{bmatrix} 1 \\ 2 \end{bmatrix}$$

Thus, T acts on the standard basis vectors by both "stretching" and rotating them (see Figure 4.1).

To understand what T does in more detail, we need to understand how the stretching and rotation in going from $\mathbf{x}$ to $T(\mathbf{x})$ vary with the vector $\mathbf{x}$. In particular,

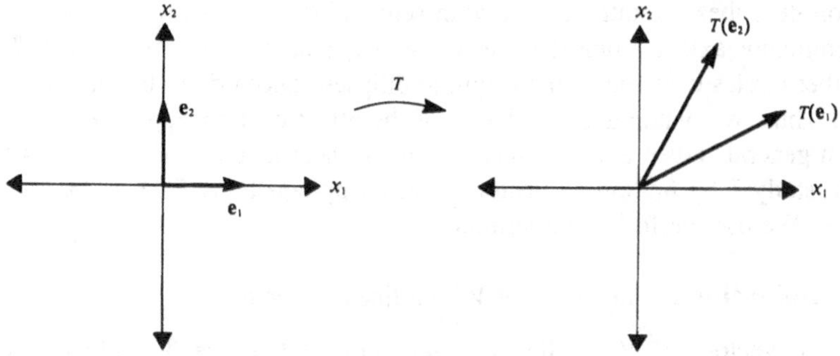

Figure 4.1

are there any vectors that undergo only stretching (or shrinking) and no rotation under T? To see if there are any such vectors, we ask, Are there any vectors $\mathbf{x} = \begin{bmatrix} x_1 \\ x_2 \end{bmatrix}$ (other than $\begin{bmatrix} 0 \\ 0 \end{bmatrix}$) such that $A\mathbf{x} = \lambda\mathbf{x}$ for some scalar λ?

The components of such a vector must satisfy the equations

$$\begin{aligned} 2x_1 + x_2 &= \lambda x_1 \\ x_1 + 2x_2 &= \lambda x_2 \end{aligned}$$

or

$$\begin{aligned} (2 - \lambda)x_1 + \quad\quad x_2 &= 0 \\ x_1 + (2 - \lambda)x_2 &= 0 \end{aligned} \tag{4.1}$$

As we know (by the results of Section 3.2), this homogeneous system of equations has a nontrivial solution if and only if

$$0 = \det\begin{bmatrix} (2 - \lambda) & 1 \\ 1 & (2 - \lambda) \end{bmatrix} = \lambda^2 - 4\lambda + 3 = (\lambda - 1)(\lambda - 3)$$

There are two roots of this equation, namely, $\lambda = 1$ and $\lambda = 3$. From Eq. (4.1) we find that for $\lambda = 1$, $\begin{bmatrix} x_1 \\ x_2 \end{bmatrix} = \begin{bmatrix} 1 \\ -1 \end{bmatrix}$ (or any scalar multiple) satisfies the corresponding equations. Similarly, for $\lambda = 3$, $\begin{bmatrix} x_1 \\ x_2 \end{bmatrix} = \begin{bmatrix} 1 \\ 1 \end{bmatrix}$ (or any scalar multiple) satisfies that system of equations.

Hence, we can form a picture of the action of T as follows. T leaves all the vectors on the line Span $\{\begin{bmatrix} 1 \\ -1 \end{bmatrix}\}$ fixed, and stretches every vector on the line Span $\{\begin{bmatrix} 1 \\ 1 \end{bmatrix}\}$ by a factor of 3. Other vectors are both stretched and rotated, but note that we can say, for example, that every $\begin{bmatrix} x_1 \\ x_2 \end{bmatrix}$ on the unit circle C (the locus of the equation $x_1^2 + x_2^2 = 1$) is mapped to a point on the locus $5x_1^2 - 8x_1x_2 + 5x_2^2 = 9$. [The reader should check that the components of the image vector $T(\mathbf{x}) = A\mathbf{x} = \begin{bmatrix} 2x_1 + x_2 \\ x_1 + 2x_2 \end{bmatrix}$ satisfy this equation if $x_1^2 + x_2^2 = 1$.] This equation describes a rotated ellipse with semimajor axis 3 along the line $x_1 = x_2$ and semiminor axis 1 along the line $x_1 = -x_2$ [see $T(C)$ in Figure 4.2]. T will take other circles centered in the origin to ellipses concentric to the one described earlier. Thus, we obtain a good picture of the effect of the mapping T.

In general, given a vector space V and a linear mapping $T: V \to V$, we will want to study T by finding all vectors $\mathbf{x}$ satisfying equations $T(\mathbf{x}) = \lambda\mathbf{x}$ for some scalar λ. We use the following terminology.

(4.1.2) Definitions. Let $T: V \to V$ be a linear mapping.

a) A vector $\mathbf{x} \in V$ is called an *eigenvector* of T if $\mathbf{x} \neq \mathbf{0}$ and there exists a scalar $\lambda \in \mathbf{R}$ such that $T(\mathbf{x}) = \lambda\mathbf{x}$.

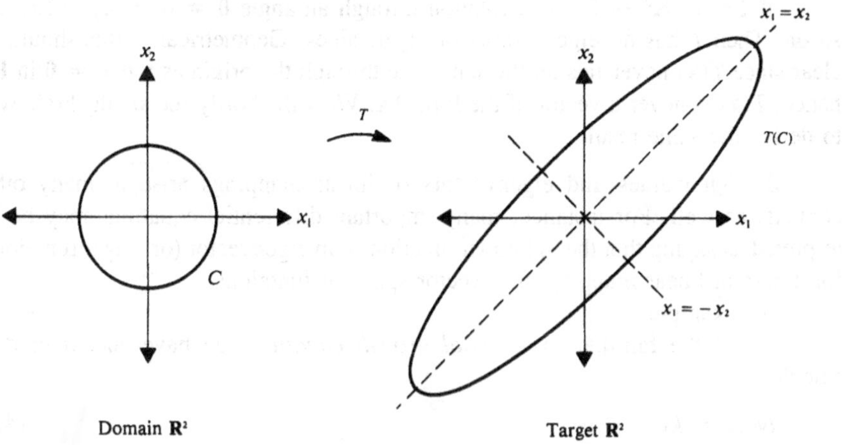

Figure 4.2

b) If $\mathbf{x}$ is an eigenvector of T and $T(\mathbf{x}) = \lambda\mathbf{x}$, the scalar λ is called the *eigenvalue* of T corresponding to $\mathbf{x}$.

In other words, an eigenvalue of T is a scalar for which there exists a corresponding eigenvector—that is, a nonzero vector $\mathbf{x}$ such that $T(\mathbf{x}) = \lambda\mathbf{x}$.

(4.1.3) Remark. A comment regarding these strange-looking terms "eigenvector" and "eigenvalue" is probably in order. These words are both combinations involving the German prefix "eigen-," which means *characteristic* or *proper* (in the sense of *belonging to*). The terminology is supposed to suggest the way eigenvalues and eigenvectors characterize the behavior of a linear mapping. In some linear algebra books, the authors, wishing to avoid these hybrid terms, use the alternate names *characteristic value* and *characteristic vector* instead of eigenvalue and eigenvector.

(4.1.4) Examples

a) In Example (4.1.1), the vector $\begin{bmatrix} 1 \\ 1 \end{bmatrix}$ is an eigenvector of T with eigenvalue $\lambda = 3$, whereas the vector $\begin{bmatrix} 1 \\ -1 \end{bmatrix}$ is an eigenvector with eigenvalue $\lambda = 1$. Any nonzero scalar multiple of either of these two vectors is also an eigenvector for the mapping T.

b) Let $T: \mathbf{R}^2 \to \mathbf{R}^2$ be the orthogonal projection on the x_1-axis [see Example (2.1.13) for the definition]. Then every vector of the form $\begin{bmatrix} a \\ 0 \end{bmatrix}$ $(a \neq 0)$ is an eigenvector of T with eigenvalue $\lambda = 1$ (why?). Vectors of the form $\begin{bmatrix} 0 \\ b \end{bmatrix}$ (again, with $b \neq 0$) are eigenvectors of T with eigenvalue $\lambda = 0$.

c) Let $T\colon \mathbf{R}^2 \rightarrow \mathbf{R}^2$ be a rotation through an angle $\theta \neq 0, \pm\pi, \pm 2\pi$, and so on. Then T has no eigenvectors or eigenvalues. Geometrically, this should be clear since $T(\mathbf{x})$ never lies on the same line through the origin as $\mathbf{x}$ if $\mathbf{x} \neq \mathbf{0}$ in $\mathbf{R}^2$; hence, $T(\mathbf{x})$ is never a vector of the form $\lambda\mathbf{x}$. We will shortly see an algebraic way to derive the same result.

d) Eigenvalues and eigenvectors of linear mappings arise in many other contexts as well. For instance, many important differential equations may be interpreted as saying that the unknown function is an eigenvector (or "eigenfunction") for a certain linear mapping on a vector space of functions.

For example,

i) the familiar *exponential growth equation* you have encountered in calculus:

$$dy/dx = ky \tag{4.2}$$

may be viewed in this way. If we restrict our attention to functions $y(x) \in C^\infty(\mathbf{R})$ (the vector space of functions with derivatives of all orders), then Eq. (4.2) simply says that y is an eigenvector of the linear mapping $T = d/dx : C^\infty(\mathbf{R}) \rightarrow C^\infty(\mathbf{R})$, with eigenvalue $\lambda = k$. Solving the equation means finding these eigenvectors. By separation of variables and integration, we find that $y = ce^{kx}$ for some constant c. (Recall that the otherwise arbitrary constant c may be determined if we know the value of y at one x, for example, at $x = 0$.)

Other important examples of this type include the following.

ii) The *simple harmonic motion equation*

$$d^2y/dx^2 = -k^2y$$

(here any solution is an eigenvector of $T = d^2/dx^2$ with eigenvalue $-k^2$), and

iii) *Bessel's equation:*

$$x^2d^2y/dx^2 + x\,dy/dx + x^2y = p^2y$$

(here any solution is an eigenvector of the mapping $T = x^2d^2/dx^2 + xd/dx + x^2I$ with eigenvalue p^2). Bessel's equation appears in many problems in applied mathematics and physics.

We will return to examples of this type in Chapter 7.

In general, if $\mathbf{x}$ is an eigenvector of a linear mapping $T\colon V \rightarrow V$ with eigenvalue λ, we have $T(\mathbf{x}) = \lambda\mathbf{x}$, so that $(T - \lambda I)(\mathbf{x}) = \mathbf{0}$. Conversely, if $\mathbf{x} \in \text{Ker}(T - \lambda I)$ and $\mathbf{x} \neq \mathbf{0}$, then $(T - \lambda I)(\mathbf{x}) = \mathbf{0}$, so $T(\mathbf{x}) = \lambda\mathbf{x}$, and by definition $\mathbf{x}$ is an eigenvector of T with eigenvalue λ. In sum, we have proved Proposition (4.1.5).

(4.1.5) Proposition. A vector $\mathbf{x}$ is an eigenvector of T with eigenvalue λ if and only if $\mathbf{x} \neq \mathbf{0}$ and $\mathbf{x} \in \text{Ker}(T - \lambda I)$.

As it turns out, most of the basic general facts about eigenvalues and eigenvectors are consequences of the observations we have just made. For example, we make the following definition.

(4.1.6) Definition. Let $T: V \to V$ be a linear mapping, and let $\lambda \in \mathbf{R}$. The λ-*eigenspace* of T, denoted E_λ, is the set

$$E_\lambda = \{\mathbf{x} \in V \mid T(\mathbf{x}) = \lambda\mathbf{x}\}$$

That is, E_λ is the set containing all the eigenvectors of T with eigenvalue λ, together with the vector $\mathbf{0}$. (If λ is not an eigenvalue of T, then we have $E_\lambda = \{\mathbf{0}\}$.)

(4.1.7) Proposition. E_λ is a subspace of V for all λ.

Proof: By Proposition (4.1.5), $E_\lambda = \text{Ker}(T - \lambda I)$. The kernel of any linear mapping is a subspace of the domain by Proposition (2.3.2). ■

(4.1.8) Examples

a) In Example (4.1.1) the linear mapping T has two nontrivial eigenspaces, E_1 and E_3. As we saw there, both of these are lines through the origin in $\mathbf{R}^2$, hence subspaces of $\mathbf{R}^2$.

b) For a given real number k the corresponding eigenspace of the operator d^2/dx^2 in the simple harmonic motion equation $d^2y/dx^2 = -k^2y$ from Example (4.1.4d) is $E = \text{Span}(\{\sin(kx), \cos(kx)\})$. It is easy to see (and the reader should check) that any linear combination of these functions solves the equation. Showing that these are the only functions that do is somewhat more difficult, but a proof can be constructed using the mean value theorem.

We will be concerned mainly with the problem of finding eigenvectors and eigenvalues in the case of linear mappings $T: V \to V$, where V is *finite-dimensional*. As we know [by Definition (2.2.6)], once we choose a basis α in V, such a linear mapping is uniquely represented by a matrix $A = [T]_\alpha^\alpha$ [an n by n matrix, if $\dim(V)$ is n]. In other words, choosing a basis in V sets up isomorphisms between V and $\mathbf{R}^n$ and between $L(V, V)$ and $M_{n\times n}(\mathbf{R})$. (Recall, we proved this in Section 2.3, Exercise 9.) In practice, choosing a basis and making this reduction would almost always be the first step in finding the eigenvalues of a linear mapping on a finite-dimensional space. Therefore, for the remainder of this section we consider the case $V = \mathbf{R}^n$ with the standard basis and treat matrices $A \in M_{n\times n}(\mathbf{R})$ and the corresponding linear mappings $T: \mathbf{R}^n \to \mathbf{R}^n$ defined by $T(\mathbf{x}) = A\mathbf{x}$ interchangeably.

As a practical matter, to compute the eigenvectors of a matrix A, we find it easiest to begin by computing the eigenvalues of A, using the following result.

(4.1.9) Proposition. Let $A \in M_{n\times n}(\mathbf{R})$. Then $\lambda \in \mathbf{R}$ is an eigenvalue of A if and only if $\det(A - \lambda I) = 0$. (Here I denotes the $n \times n$ identity matrix.)

Proof: By Proposition (4.1.5), an eigenvector of A with eigenvalue λ is a *non-trivial* solution of the homogeneous system $(A - \lambda I)\mathbf{x} = \mathbf{0}$.

$\to$: If such a solution exists, then by Section 2.6 we know that the matrix

$A - \lambda I$ cannot be invertible. Hence, by Theorem (3.2.14), we have that its determinant must vanish.

$\leftarrow$: If $\det(A - \lambda I) = 0$, then by Theorem (3.2.14) the matrix $A - \lambda I$ is not invertible. Hence, there is some vector $\mathbf{x} \neq \mathbf{0}$ such that $(A - \lambda I)\mathbf{x} = \mathbf{0}$, so that $A\mathbf{x} = \lambda\mathbf{x}$. But this implies that $\mathbf{x}$ is an eigenvector of A with eigenvalue λ. ■

(4.1.10) Examples

a) Let $A = \begin{bmatrix} 1 & 0 & 0 \\ 1 & 0 & -1 \\ -1 & -1 & 1 \end{bmatrix}$. Then we have

$$\det(A - \lambda I) = \det\left[\begin{bmatrix} 1 & 0 & 0 \\ 1 & 0 & -1 \\ -1 & -1 & 1 \end{bmatrix} - \lambda \begin{bmatrix} 1 & 0 & 0 \\ 0 & 1 & 0 \\ 0 & 0 & 1 \end{bmatrix}\right]$$

$$= \det\begin{bmatrix} 1-\lambda & 0 & 0 \\ 1 & -\lambda & -1 \\ -1 & -1 & 1-\lambda \end{bmatrix}$$

$$= (1 - \lambda)(\lambda^2 - \lambda - 1)$$

(expanding along the first row). The eigenvalues of A are the roots of the equation $0 = (\lambda - 1)(\lambda^2 - \lambda - 1)$, which are $\lambda = 1$ and $\lambda = (1/2)(1 \pm \sqrt{5})$. (The last two come from applying the quadratic formula to the second factor.)

b) Let $A = \begin{bmatrix} 0 & 1 \\ -1 & 0 \end{bmatrix}$. Then $\det(A - \lambda I) = \det\begin{bmatrix} -\lambda & 1 \\ -1 & -\lambda \end{bmatrix} = \lambda^2 + 1$. There are no real roots of the equation $\lambda^2 + 1 = 0$, so A has no (real) eigenvalues.

In general, as the preceding examples should make clear, when we expand out a determinant of the form $\det(A - \lambda I)$, we obtain a *polynomial* in λ whose coefficients depend on the entries in the matrix A.

(4.1.11) Definition. Let $A \in M_{n \times n}(\mathbf{R})$. The polynomial $\det(A - \lambda I)$ is called the *characteristic polynomial* of A.

The alert reader is probably wondering why this is not called the "eigenpolynomial" of A to match the other terms we have introduced, eigenvalue and eigenvector. That would be consistent, and even logical in its own way, but the fact is that no one uses the word "eigenpolynomial." Mathematical terminology is sometimes arbitrary.

One interesting property of the characteristic polynomial is that it really only depends on the linear mapping defined by the matrix A and not on the matrix itself. By this, we mean that if we changed to another basis in $\mathbf{R}^n$ and expressed our linear mapping in terms of that new basis, the characteristic polynomial of the matrix of the mapping in the new basis would be the same polynomial as the characteristic polynomial of the original matrix. Of course, in a sense this is not unexpected, since the eigenvalues and eigenvectors of a linear mapping should reflect the geometric behavior of the mapping, and this is independent of the coordinates we use to describe it. To see why this is true, recall that changing basis amounts to passing from our original matrix A to a *similar matrix* $B = Q^{-1}AQ$. The invariance of the characteristic polynomial then results from the following proposition.

(4.1.12) Proposition. Similar matrices have equal characteristic polynomials.

Proof: Suppose A and B are two similar matrices, so that $B = Q^{-1}AQ$ for some invertible matrix Q. Then we have

$$\begin{aligned}\det(B - \lambda I) &= \det(Q^{-1}AQ - \lambda I)\\ &= \det(Q^{-1}AQ - Q^{-1}\lambda IQ) \qquad \text{(why?)}\\ &= \det(Q^{-1}(A - \lambda I)Q)\\ &= \det(Q^{-1})\det(A - \lambda I)\det(Q)\end{aligned}$$

Then, since $\det(Q^{-1}) = 1/\det(Q)$, we see that the outer two factors cancel, leaving $\det(A - \lambda I)$. Hence, $\det(B - \lambda I) = \det(A - \lambda I)$, as claimed. ■

By Proposition (4.1.9), the (real) roots of the equation obtained by setting the characteristic polynomial equal to zero are precisely the eigenvalues of the matrix. There are several general patterns and special cases that arise in computing characteristic polynomials, some of which are illustrated in the following examples.

(4.1.13) Examples

a) For a general 2×2 matrix $A = \begin{bmatrix} a & b \\ c & d \end{bmatrix}$ we have

$$\det\begin{bmatrix} a - \lambda & b \\ c & d - \lambda \end{bmatrix} = \lambda^2 - (a + d)\lambda + (ad - bc)$$

The constant term in this polynomial is just the determinant of the matrix A. The coefficient of λ is another important scalar, called the *trace* of the matrix A, and denoted by $\mathrm{Tr}(A)$. In general, for any square matrix, the trace is defined to be the sum of the diagonal entries. Thus, the characteristic polynomial of our 2×2 matrix may be written as

$$\det(A - \lambda I) = \lambda^2 - \mathrm{Tr}(A)\lambda + \det(A)$$

b) Similarly, for a general 3×3 matrix $A = \begin{bmatrix} a & b & c \\ d & e & f \\ g & h & i \end{bmatrix}$ we have

$$\det\begin{bmatrix} a - \lambda & b & c \\ d & e - \lambda & f \\ g & h & i - \lambda \end{bmatrix} = -\lambda^3 + (a + e + i)\lambda^2 - ((ae - bd) + (ai - cg) + (ei - fh))\lambda$$

$$+ (a(ei - fh) - b(di - fg) + c(dh - eg))$$
$$= -\lambda^3 + \text{Tr}(A)\lambda^2 - ((ae - bd) + (ai - cg) + (ei - fh))\lambda + \det(A)$$

c) In general, it is true (as you will prove in Exercises 5 and 6) that for any $n \times n$ matrix A, the characteristic polynomial has the form

$$(-1)^n\lambda^n + (-1)^{n-1}\,\text{Tr}(A)\lambda^{n-1} + c_{n-2}\lambda^{n-2} + \cdots + c_1\lambda + \det(A)$$

where the c_i are other polynomial expressions in the entries of the matrix A.

Question: Can you find the general pattern for the coefficients c_i? How are they formed from the entries of the matrix A? Try the 4×4 case to see if your guess is correct.

d) Let A be an $n \times n$ diagonal matrix: $A = \begin{bmatrix} a_1 & & & & 0 \\ & a_2 & & & \\ & & \cdot & & \\ & & & \cdot & \\ 0 & & & & a_n \end{bmatrix}$. Then the characteristic polynomial of A is $(a_1 - \lambda)(a_2 - \lambda) \cdots (a_n - \lambda)$. In this case the eigenvalues of the matrix are simply the diagonal entries.

The fact that the degree of the characteristic polynomial of an $n \times n$ matrix is exactly n in λ—part of the general pattern noted in Example (4.1.13c)—implies the following result on the number of eigenvalues of a matrix.

(4.1.14) Corollary. Let $A \in M_{n \times n}(\mathbf{R})$. Then A has no more than n distinct eigenvalues. In addition, if $\lambda_1, \ldots, \lambda_k$ are the distinct eigenvalues of A and λ_i is an m_i-fold root of the characteristic polynomial, then $m_1 + \cdots + m_k \leq n$.

Proof: By high school algebra, a polynomial equation of degree n has no more than n real roots. [Recall, this is true because $\lambda = a \in \mathbf{R}$ is a root of the equation $p(\lambda) = 0$ if and only if the degree one polynomial $\lambda - a$ divides $p(\lambda)$, that is, $p(\lambda) = (\lambda - a)q(\lambda)$ for some polynomial q. The degree of q is one less than the degree of p, so there can be at most n roots in all.]

Now, if for each i $(1 \leq i \leq k)$ we have that $\lambda = \lambda_i$ is an m_i-fold root of $p(\lambda) = \det(A - \lambda I) = 0$ (i.e., $(\lambda - \lambda_i)^{m_i}$ divides $p(\lambda)$ but no higher power does), then we have

$$p(\lambda) = (\lambda - \lambda_1)^{m_1}(\lambda - \lambda_2)^{m_2} \cdots (\lambda - \lambda_k)^{m_k}q(\lambda)$$

for some polynomial $q(\lambda)$. Comparing degrees gives $m_1 + \cdots + m_k \leq$ degree of $p = n$. ■

Once the eigenvalues of a given matrix have been found, the corresponding eigenvectors may be found by solving the systems $(A - \lambda_i I)\mathbf{x} = \mathbf{0}$ for each eigenvalue λ_i.

(4.1.15) Examples

a) Consider the matrix $A = \begin{bmatrix} 1 & 0 & 0 \\ 1 & 0 & -1 \\ -1 & -1 & 1 \end{bmatrix}$ from Example (4.1.10a).

We saw there that the eigenvalues of A are $\lambda = 1$, and $\lambda = (1/2)(1 \pm \sqrt{5})$. For $\lambda = 1$ the eigenvectors will be the nontrivial solutions of $(A - 1I)\mathbf{x} = \mathbf{0}$, or

$$\begin{bmatrix} 0 & 0 & 0 \\ 1 & -1 & -1 \\ -1 & -1 & 0 \end{bmatrix} \begin{bmatrix} x_1 \\ x_2 \\ x_3 \end{bmatrix} = \begin{bmatrix} 0 \\ 0 \\ 0 \end{bmatrix}$$

The set of solutions (i.e., the eigenspace E_1) is $\text{Span}\{\begin{bmatrix} 1/2 \\ -1/2 \\ 1 \end{bmatrix}\}$. (Check this by reducing the augmented matrix of the system to echelon form.)

For $\lambda = (1/2)(1 + \sqrt{5})$ we must solve the system $(A - \lambda I)\mathbf{x} = \mathbf{0}$, or

$$\begin{bmatrix} (1/2)(1-\sqrt{5}) & 0 & 0 \\ 1 & (1/2)(-1 - \sqrt{5}) & -1 \\ -1 & -1 & (1/2)(1 - \sqrt{5}) \end{bmatrix} \begin{bmatrix} x_1 \\ x_2 \\ x_3 \end{bmatrix} = \begin{bmatrix} 0 \\ 0 \\ 0 \end{bmatrix}$$

The set of solutions is $\text{Span}\ \{\begin{bmatrix} 0 \\ (1/2)(1 - \sqrt{5}\,) \\ 1 \end{bmatrix}\}$.

Similarly, for $\lambda = (1/2)(1 - \sqrt{5})$ we solve the appropriate system of equations and find that the set of solutions is $\text{Span}\{\begin{bmatrix} 0 \\ (1/2)(1+\sqrt{5}) \\ 1 \end{bmatrix}\}$. In this example we see that the geometric behavior of the mapping defined by the matrix A is very much like that of the mapping in Example (4.1.1). It is easy to check that the three eigenvectors found in the preceding calculations form a linearly independent set, hence a basis of $\mathbf{R}^3$. A acts by stretching by the different scalar factors 1, $(1/2)(1 \pm \sqrt{5})$ in the different coordinate directions.

b) Consider the 2×2 matrix $A = \begin{bmatrix} 2 & 5 \\ 0 & 2 \end{bmatrix}$. The characteristic polynomial is $\det(A - \lambda I) = (2 - \lambda)^2$. The only eigenvalue is $\lambda = 2$, and it appears as a double root. In this case the eigenspace E_2 is the set of solutions of $(A - 2I)\mathbf{x} = \mathbf{0}$. Hence, $E_2 = \text{Span}\ \{\begin{bmatrix} 1 \\ 0 \end{bmatrix}\}$, since there is no condition on the first component.

(4.1.16) Remark. The reader should be aware that finding the roots of the characteristic polynomial (the starting point for finding the eigenvectors of a matrix A) is not always so easy as it was in these examples. For 2×2 matrices we can always find the eigenvalues, because we have the quadratic formula to compute the roots of the (quadratic) characteristic polynomial. There are analogous, but much more complicated, general formulas for the roots of the third- and fourth-degree polynomials (which arise from 3×3 and 4×4 matrices). (They are so complicated, in fact, that they are rarely used!) However, it is known that no general formula, involving only a finite sequence of the algebraic operations of addition, subtraction, multiplication, division, and extraction of roots, exists for solving polynomial equations of degree 5 or higher, given the coefficients. For this reason, approximate numerical techniques (implemented using computers) are frequently used when eigenvalue problems arise in real-world applications. We will not consider such methods here. However, their study is an important part of the branch of mathematics known as numerical analysis.

In the examples we encounter it will always be possible to find the roots of the characteristic polynomial by using the elementary techniques for factoring polynomials you saw in high school algebra.

We close this section with another observation regarding the characteristic polynomial of a matrix. Because we can add and multiply matrices, given a polynomial, $p(t) = a_m t^m + \cdots + a_0$, and a matrix A, it makes sense to substitute the matrix A for t in the polynomial, provided that we interpret the constant term a_0 as the matrix $a_0 I$. We write

$$p(A) = a_m A^m + \cdots + a_1 A + a_0 I$$

(4.1.17) Example. Consider a general 2×2 matrix $A = \begin{bmatrix} a & b \\ c & d \end{bmatrix}$, and let $p(t)$ be the characteristic polynomial of A, which we can write as

$$p(t) = t^2 - (a + d)t + (ad - bc)$$

as in Example (4.1.13a). If we substitute A into its own characteristic polynomial, something interesting happens:

$$\begin{aligned} p(A) &= A^2 - (a + d)A + (ad - bc)I \\ &= \begin{bmatrix} a^2 + bc & ab + bd \\ ca + db & cb + d^2 \end{bmatrix} - \begin{bmatrix} a^2 + ad & ab + bd \\ ac + bd & ad + d^2 \end{bmatrix} \\ &\quad + \begin{bmatrix} ad - bc & 0 \\ 0 & ad - bc \end{bmatrix} \\ &= \begin{bmatrix} 0 & 0 \\ 0 & 0 \end{bmatrix} \end{aligned}$$

A "satisfies its own characteristic polynomial equation." In Exercise 8 you will show that the same is true for 3×3 matrices. These calculations prove the 2×2 and 3×3 cases of the celebrated Cayley-Hamilton theorem:

(4.1.18) Theorem. Let $A \in M_{n \times n}(\mathbf{R})$, and let $p(t) = \det(A - tI)$ be its characteristic polynomial. Then $p(A) = 0$ (the $n \times n$ zero matrix).

Proof: Deferred until Chapter 6. ■

EXERCISES

1. Verify that the given vector is an eigenvector of the given mapping, and find the eigenvalue.

   a) $\mathbf{x} = \begin{bmatrix} -3 \\ 4 \end{bmatrix} \in \mathbf{R}^2$, $T: \mathbf{R}^2 \to \mathbf{R}^2$ defined by $A = \begin{bmatrix} -1 & 3 \\ 0 & -5 \end{bmatrix}$

   b) $\mathbf{x} = \begin{bmatrix} 1 \\ 1 \\ 1 \end{bmatrix} \in \mathbf{R}^3$, $T: \mathbf{R}^3 \to \mathbf{R}^3$ defined by $A = \begin{bmatrix} 1 & -2 & 1 \\ 1 & -2 & 1 \\ 1 & -2 & 1 \end{bmatrix}$

   c) $p = x^3 \in P_3(\mathbf{R})$, $T: P_3(\mathbf{R}) \to P_3(\mathbf{R})$ defined by $T(p) = xp' - 4p$

   d) $f = e^{4x} \in C^\infty(\mathbf{R})$, $T: C^\infty(\mathbf{R}) \to C^\infty(\mathbf{R})$ defined by $T(f) = f'' + f$

2. Compute the characteristic polynomials of the mappings from parts a, b, and c of Exercise 1. (Why is part d not included?)

3. For each of the following linear mappings:
   (i) Find all the eigenvalues, and
   (ii) For each eigenvalue λ, find a basis of the eigenspace E_λ.

   a) $T: \mathbf{R}^2 \to \mathbf{R}^2$ defined by $A = \begin{bmatrix} 1 & 4 \\ 3 & 2 \end{bmatrix}$

   b) $T: \mathbf{R}^2 \to \mathbf{R}^2$ defined by $A = \begin{bmatrix} -1 & 1 \\ 0 & -1 \end{bmatrix}$

   c) $T: \mathbf{R}^3 \to \mathbf{R}^3$ defined by $A = \begin{bmatrix} 1 & 0 & 1 \\ 0 & 1 & 0 \\ 1 & 0 & 1 \end{bmatrix}$

   d) $T: \mathbf{R}^3 \to \mathbf{R}^3$ defined by $A = \begin{bmatrix} 1 & -1 & 3 \\ 0 & 1 & 0 \\ 0 & 4 & 1 \end{bmatrix}$

   e) $T: \mathbf{R}^4 \to \mathbf{R}^4$ defined by

   $$A = \begin{bmatrix} 1 & 2 & -3 & 1 \\ 2 & 1 & 7 & 2 \\ 0 & 0 & 2 & 1 \\ 0 & 0 & 1 & 2 \end{bmatrix}$$ (See Exercise 8 in Section 3.2)

   f) $T: P_3(\mathbf{R}) \to P_3(\mathbf{R})$ defined by $T(p)(x) = p'(x) + 2p(x)$.

   g) $V = \text{Span}\{e^x, xe^x, e^{2x}, xe^{2x}\}$, $T: V \to V$ defined by $T(f) = f''$.

4. Show that if $T: V \to V$ is any linear mapping, then the eigenspace E_0 in V is equal to $\text{Ker}(T)$.

5. a) Show that if A is any $n \times n$ matrix, the constant term in the characteristic polynomial of A is $\det(A)$.
   b) Deduce that A is invertible if and only if $\lambda = 0$ is not an eigenvalue of A.

6. a) Show that if A is an $n \times n$ matrix, with entries a_{ij}, then

$$\det(A - \lambda I) = (a_{11} - \lambda)(a_{22} - \lambda) \cdots (a_{nn} - \lambda) + (\text{terms of degree } \leq n - 2 \text{ in } \lambda)$$

(*Hint:* Use induction.)

   b) From part a deduce that the coefficient of λ^{n-1} in $\det(A - \lambda I)$ is $(-1)^{n-1}\,\text{Tr}(A)$.

7. Let $A \in M_{n\times n}(\mathbf{R})$, and assume that $\det(A - \lambda I)$ factors completely into $(\lambda_1 - \lambda)(\lambda_2 - \lambda) \cdots (\lambda_n - \lambda)$, a product of linear factors.
   a) Show that $\text{Tr}(A) = \lambda_1 + \cdots + \lambda_n$.
   b) Show that $\det(A) = \lambda_1 \cdots \lambda_n$.

8. Prove the 3×3 case of the Cayley-Hamilton theorem: If $A \in M_{3\times 3}(\mathbf{R})$ and $p(t)$ is its characteristic polynomial, then $p(A) = 0$ (the 3×3 zero matrix.)

9. a) Show that a matrix $A \in M_{2\times 2}(\mathbf{R})$ has distinct real eigenvalues if and only if $(\text{Tr}(A))^2 > 4 \cdot \det(A)$.
   b) Show that A has a real eigenvalue of multiplicity 2 if and only if $(\text{Tr}(A))^2 = 4 \cdot \det(A)$.
   c) What happens if $(\text{Tr}(A))^2 < 4 \cdot \det(A)$?

10. a) Show that a matrix $A \in M_{3\times 3}(\mathbf{R})$ always has at least one real eigenvalue. Give an example of such a matrix that has only one real eigenvalue, with multiplicity 1.
    b) How can this result be generalized?

11. a) Show that if $A = \begin{bmatrix} a_{11} & \cdots & & a_{1n} \\ 0 & a_{22} & \cdots & \\ \vdots & & & \\ 0 & \cdots & 0 & a_{nn} \end{bmatrix}$ is an "upper-triangular" $n \times n$ matrix (i.e., $a_{ij} = 0$ for all pairs $i > j$), then the eigenvalues of A are the diagonal entries $\lambda = a_{ii}$.
    b) Show that the eigenvalues of a "lower-triangular" matrix (i.e., $a_{ij} = 0$ for all pairs $i < j$) are the diagonal entries a_{ii}.

12. Let $A \in M_{n\times n}(\mathbf{R})$, and suppose A has the block form $\begin{bmatrix} B & C \\ 0 & D \end{bmatrix}$, where B is some $k \times k$ matrix, D is some $(n - k) \times (n - k)$ matrix, and there is an $(n - k) \times k$ block of zeros below the diagonal. Show that $\det(A - \lambda I_n) = \det(B - \lambda I_k) \cdot \det(D - \lambda I_{n-k})$. What does this imply about the eigenvalues of A?

13. a) Let $A \in M_{n\times n}(\mathbf{R})$, and let λ be an eigenvalue of A. Show that λ^m is an eigenvalue of A^m for all integers $m \geqslant 1$.
    b) Show more generally that if $p(t)$ is any polynomial and λ is an eigenvalue of A, then $p(\lambda)$ is an eigenvalue of $p(A)$.

14. A matrix $A \in M_{n\times n}(\mathbf{R})$ is said to be *nilpotent* if $A^k = 0$ for some integer $k \geqslant 1$.
    a) Give examples of (nonzero) 2×2 and 3×3 nilpotent matrices.
    b) Use Exercise 13 to show that if A is nilpotent, then the only eigenvalue of A is $\lambda = 0$.

15. Let V be a vector space, and let $T: V \to V$ be a linear mapping such that $T^2 = I$. (Such mappings are sometimes called involutions.)
    a) Show that if λ is an eigenvalue of T, then $\lambda = +1$ or $\lambda = -1$.
    b) Show that the eigenspaces satisfy $E_{+1} \cap E_{-1} = \{\mathbf{0}\}$.
    c) Show that $V = E_{+1} + E_{-1}$. Hence, $V = E_{+1} \oplus E_{-1}$. (*Hint*: For any $\mathbf{x} \in V$, $\mathbf{x} = (1/2)(\mathbf{x} + T(\mathbf{x})) + (1/2)(\mathbf{x} - T(\mathbf{x}))$.)

16. Let $T: M_{n\times n}(\mathbf{R}) \to M_{n\times n}(\mathbf{R})$ be the transpose mapping: $T(A) = A^t$.
    a) Show that T satisfies $T^2 = I$. (Hence, Exercise 15 applies.)
    b) What are the eigenspaces E_{+1} and E_{-1} in this case?

17. Let V be a vector space, and let $T: V \to V$ be a linear mapping with $T^2 = T$.
    a) Give an example of such a mapping other than $T = I$.
    b) Show that if λ is an eigenvalue of such a mapping, then $\lambda = 1$, or $\lambda = 0$.
    c) Show that the eigenspaces satisfy $E_1 \cap E_0 = \{\mathbf{0}\}$.
    d) Show that $V = E_1 + E_0$. Hence, $V = E_1 \oplus E_0$.

18. Let $p(t) = (-1)^n(t^n + a_{n-1}t^{n-1} + \cdots + a_1 t + a_0)$ be a polynomial with real coefficients. Show that the characteristic polynomial of the $n \times n$ matrix:

$$C_p = \begin{bmatrix} 0 & 0 & \cdots & 0 & -a_0 \\ 1 & 0 & \cdots & 0 & -a_1 \\ 0 & 1 & \cdots & 0 & -a_2 \\ 0 & 0 & \cdots & \cdots & \\ \cdot & \cdot & \cdots & \cdot & \cdot \\ \cdot & \cdot & \cdots & \cdot & \cdot \\ \cdot & \cdot & \cdots & \cdot & \cdot \\ 0 & 0 \cdots 1 & & 0 & -a_{n-2} \\ 0 & 0 \cdots 0 & & 1 & -a_{n-1} \end{bmatrix}$$

    is the polynomial $p(t)$. C_p is called the *companion matrix* of p. (*Hint:* Expand $\det(C_p - tI)$ along the first row and use induction.)

§4.2. DIAGONALIZABILITY

In Section 4.1 we saw several examples where, by finding the eigenvalues and eigenvectors of a linear mapping, we were able to analyze the behavior of the mapping completely. The special feature of the mappings in examples such as Example (4.1.1) [and Example (4.1.15a), which is similar] that allowed us to do this is that in each of those cases it was possible to find a basis of the domain vector space, each of whose elements was an eigenvector of the mapping in question. This leads us to focus on the following property of linear mappings.

(4.2.1) Definition. Let V be a finite-dimensional vector space, and let $T: V \to V$ be a linear mapping. T is said to be *diagonalizable* if there exists a basis of V, all of whose vectors are eigenvectors of T.

The reason for this terminology becomes clear if we interpret the condition in the definition in terms of matrices. We choose any basis α for V and represent the linear mapping T by its matrix with respect to α, say, $A = [T]_\alpha^\alpha$. Now, if T is diagonalizable, when we change to a basis, say, β, consisting of eigenvectors, then the matrix of T with respect to β takes on a special form. If $\beta = \{\mathbf{x}_1, \ldots, \mathbf{x}_n\}$, then for each $\mathbf{x}_i$, we have $T(\mathbf{x}_i) = \lambda_i\mathbf{x}_i$ for some eigenvalue λ_i. Hence, $[T]_\beta^\beta$ is a diagonal matrix with diagonal entries λ_i:

$$[T]_\beta^\beta = \begin{bmatrix} \lambda_1 & & 0 \\ & \ddots & \\ 0 & & \lambda_n \end{bmatrix} = D$$

If P is the change of basis matrix from β to α, then by Chapter 2, Theorem (2.7.6) we have a matrix equation $P^{-1}AP = D$. Thus, A is diagonalizable in the sense that A can be brought to diagonal form by passing from A to $P^{-1}AP$. In other words, we have shown half of Proposition (4.2.2).

(4.2.2) Proposition. $T: V \to V$ is diagonalizable if and only if, for any basis α of V, the matrix $[T]_\alpha^\alpha$ is similar to a diagonal matrix.

Proof: $\to$: done previously.

$\leftarrow$: If $A = [T]_\alpha^\alpha$ is similar to a diagonal matrix D, then by definition, there is some invertible matrix P such that $P^{-1}AP = D$. Thus, we have a matrix equation $AP = PD$. Since D is a diagonal matrix, the columns of the matrix PD are just scalar multiples of the columns of P. Hence, the columns of P are all eigenvectors for the matrix A, or equivalently, the columns of P are the coordinate vectors of eigenvectors for the mapping T. On the other hand, since P is invertible, we know its columns are linearly independent, hence those columns are the coordinate vectors for a basis of V. Therefore, by definition, T is diagonalizable. Note that the eigenvalues of T are the diagonal entries of D. ■

In order for a linear mapping or a matrix to be diagonalizable, it must have *enough linearly independent eigenvectors to form a basis of V.*

(4.2.3) Examples

a) The matrix A considered in Example (4.1.15a) is diagonalizable, since (as the reader will verify) the eigenvectors for the three different eigenvalues are linearly independent.

b) For another kind of example, consider the mapping of $V = \mathbf{R}^3$ defined by the matrix $A = \begin{bmatrix} 2 & 2 & 0 \\ 0 & 1 & 0 \\ 0 & 1 & 2 \end{bmatrix}$. The characteristic polynomial of A is $(2 - \lambda)^2(1 - \lambda)$ [expand $\det(A - \lambda I)$ along the second row, for example]. Hence the eigenvalues of A are $\lambda = 1$ and $\lambda = 2$. $\lambda = 2$ is a double root, so we say the eigenvalue 2 has multiplicity 2. Now we compute the eigenvectors of A. First, for $\lambda = 1$ we find the solutions of $(A - I)\mathbf{x} = \mathbf{0}$,

or $\begin{bmatrix} 1 & 2 & 0 \\ 0 & 0 & 0 \\ 0 & 1 & 1 \end{bmatrix} \begin{bmatrix} x_1 \\ x_2 \\ x_3 \end{bmatrix} = \begin{bmatrix} 0 \\ 0 \\ 0 \end{bmatrix}$ are all multiples of the vector $x = \begin{bmatrix} 2 \\ -1 \\ 1 \end{bmatrix}$.

(Check this.) Next, for $\lambda = 2$, the eigenvectors are the solutions of $(A - 2I)\mathbf{x} = \mathbf{0}$,

or $\begin{bmatrix} 0 & 2 & 0 \\ 0 & -1 & 0 \\ 0 & 1 & 0 \end{bmatrix} \begin{bmatrix} x_1 \\ x_2 \\ x_3 \end{bmatrix} = \begin{bmatrix} 0 \\ 0 \\ 0 \end{bmatrix}$. The rank of the coefficient matrix is 1, so we expect a two-dimensional space of solutions. Indeed, $E_2 = \text{Span}\{\begin{bmatrix} 1 \\ 0 \\ 0 \end{bmatrix}, \begin{bmatrix} 0 \\ 0 \\ 1 \end{bmatrix}\}$. It is easy to verify that these two vectors, together with the eigenvector for $\lambda = 1$ found previously, form a linearly independent set, so we do have a basis of $\mathbf{R}^3$ consisting of eigenvectors of A. If we form the change of basis matrix $Q = \begin{bmatrix} 1 & 0 & 2 \\ 0 & 0 & -1 \\ 0 & 1 & 1 \end{bmatrix}$ (whose columns are the three eigenvectors found before), then

it is easy to see that $Q^{-1}AQ = \begin{bmatrix} 2 & 0 & 0 \\ 0 & 2 & 0 \\ 0 & 0 & 1 \end{bmatrix}$, which diagonalizes A.

c) Now consider the matrix $A = \begin{bmatrix} 2 & 1 & 0 \\ 0 & 2 & 0 \\ 0 & 0 & 1 \end{bmatrix}$. Here the characteristic polynomial is $\det(A - \lambda I) = (2 - \lambda)^2(1 - \lambda)$ as in part b. We find that in this example, though, $E_1 = \text{Span}\{\begin{bmatrix} 0 \\ 0 \\ 1 \end{bmatrix}\}$, and $E_2 = \text{Span}\{\begin{bmatrix} 1 \\ 0 \\ 0 \end{bmatrix}\}$. In particular,

$\dim(E_2) = 1$. There is no way to get a basis of $\mathbf{R}^3$ consisting of eigenvectors for this matrix A, since the largest possible linearly independent set of eigenvectors will contain only two vectors. Hence this matrix is not diagonalizable.

d) Matrices such as the 2×2 rotation matrix $R_\theta = \begin{bmatrix} \cos(\theta) & -\sin(\theta) \\ \sin(\theta) & \cos(\theta) \end{bmatrix}$, where θ is not an integer multiple of π, which have no real eigenvalues (and hence no eigenvectors) are not diagonalizable either.

Our major goal in this section is to find some convenient criterion for determining whether a given linear mapping or matrix is diagonalizable over $\mathbf{R}$. If we consider the behavior seen in the Examples (4.2.3), we can start to get some intuition about what must be true. First, by considering Example (4.2.3d), for an $n \times n$ matrix A to be diagonalizable, we might expect that the roots of the characteristic polynomial of A must all be real. Second, by considering Example (4.2.3b and c), we might expect that for A to be diagonalizable, if some eigenvalue λ is a multiple root of the characteristic polynomial, then there must be as many linearly independent eigenvectors for that λ as the number of times that λ appears as a root.

We will see that these two conditions do, in fact, characterize diagonalizable mappings. The basic idea of our proof will be to show that if a mapping satisfies these two conditions, then we can make a basis of $\mathbf{R}^n$ by taking the union of bases of the eigenspaces corresponding to the different eigenvalues. To do this, we need to know the following facts about eigenvectors of a mapping, corresponding to distinct eigenvalues.

(4.2.4) Proposition. Let $\mathbf{x}_i$ $(1 \leq i \leq k)$ be eigenvectors of a linear mapping $T: V \to V$ corresponding to distinct eigenvalues λ_i. Then $\{\mathbf{x}_1, \ldots, \mathbf{x}_k\}$ is a linearly independent subset of V.

Proof: (by induction) If $k = 1$, then a set containing one eigenvector is automatically linearly independent, since eigenvectors are nonzero by definition. Now assume the result is true for sets of $k = m$ vectors, and consider a potential linear dependence

$$a_1\mathbf{x}_1 + \cdots + a_m\mathbf{x}_m + a_{m+1}\mathbf{x}_{m+1} = \mathbf{0} \tag{4.3}$$

If we apply T to both sides of this equation, we obtain

$$T(a_1\mathbf{x}_1 + \cdots + a_m\mathbf{x}_m + a_{m+1}\mathbf{x}_{m+1}) = \mathbf{0}$$

so

$$a_1T(\mathbf{x}_1) + \cdots + a_mT(\mathbf{x}_m) + a_{m+1}T(\mathbf{x}_{m+1}) = \mathbf{0}$$

Now each $\mathbf{x}_i$ is an eigenvector of T with corresponding eigenvalue λ_i, so this equation becomes

$$a_1\lambda_1\mathbf{x}_1 + \cdots + a_m\lambda_m\mathbf{x}_m + a_{m+1}\lambda_{m+1}\mathbf{x}_{m+1} = \mathbf{0}$$

If we take Eq. (4.3), multiply both sides by λ_{m+1}, and subtract from this last equation, we can cancel the $\mathbf{x}_{m+1}$ term to obtain

$$a_1(\lambda_1 - \lambda_{m+1})\mathbf{x}_1 + \cdots + a_m(\lambda_m - \lambda_{m+1})\mathbf{x}_m = \mathbf{0}$$

By the induction hypothesis we deduce that $a_i(\lambda_i - \lambda_{m+1}) = 0$, for all i. Furthermore, since the λ_i are distinct, $\lambda_i - \lambda_{m+1} \neq 0$, so it must be true that $a_i = 0$ for all i, $1 \leq i \leq m$. Finally, it follows from Eq. (4.3) that $a_{m+1} = 0$ as well. Hence, $\{\mathbf{x}_1, \ldots, \mathbf{x}_{m+1}\}$ is linearly independent. ■

A useful corollary of the proposition follows.

(4.2.5) Corollary. For each i $(1 \leq i \leq k)$, let $\{\mathbf{x}_{i,1}, \ldots, \mathbf{x}_{i,n_i}\}$ be a linearly independent set of eigenvectors of T all with eigenvalue λ_i and suppose the λ_i are distinct. Then $S = \{\mathbf{x}_{1,1}, \ldots, \mathbf{x}_{1,n_1}\} \cup \cdots \cup \{\mathbf{x}_{k,1}, \ldots, \mathbf{x}_{k,n_k}\}$ is linearly independent.

Proof: Consider a possible linear dependence

$$a_{1,1}\mathbf{x}_{1,1} + \cdots + a_{1,n_1}\mathbf{x}_{1,n_1} + \cdots + a_{k,1}\mathbf{x}_{k,1} + \cdots + a_{k,n_k}\mathbf{x}_{k,n_k} = \mathbf{0}$$

For each i the sum $a_{i,1}\mathbf{x}_{i,1} + \cdots + a_{i,n_i}\mathbf{x}_{i,n_i}$ is an element of the eigenspace E_{λ_i}, and hence either an eigenvector of T or $\mathbf{0}$. If any of these sums were nonzero, then we would have a contradiction to Proposition (4.2.4). It follows that for each i we have

$$a_{i,1}\mathbf{x}_{i,1} + \cdots + a_{i,n_i}\mathbf{x}_{i,n_i} = \mathbf{0}$$

But the sets of vectors $\{\mathbf{x}_{i,1}, \ldots, \mathbf{x}_{i,n_i}\}$ were assumed to be linearly independent individually. Hence, all the coefficients $a_{i,j} = 0$, and as a result, the whole set S is linearly independent. ■

Our next result will give upper and lower bounds on the number of linearly independent eigenvectors we can expect to find for each eigenvalue.

(4.2.6) Proposition. Let V be finite-dimensional, and let $T: V \to V$ be linear. Let λ be an eigenvalue of T, and assume that λ is an m-fold root of the characteristic polynomial of T. Then we have

$$1 \leq \dim(E_\lambda) \leq m$$

Proof: On the one hand, the inequality $1 \leq \dim(E_\lambda)$ follows directly from the Definition (4.1.2)—every eigenvalue has at least one (nonzero) eigenvector.

On the other hand, to see that the second inequality also holds, let $d = \dim(E_\lambda)$, and let $\{\mathbf{x}_1, \ldots, \mathbf{x}_d\}$ be a basis of E_λ. Extend to a basis $\beta = \{\mathbf{x}_1, \ldots, \mathbf{x}_d,$

$\mathbf{x}_{d+1}, \ldots, \mathbf{x}_n\}$ for V. Since each of the first d vectors in the basis β is an eigenvector for T with eigenvalue λ, we have

$$[T]_\beta^\beta = \left[\begin{array}{ccc|c} \lambda & & 0 & \\ & \ddots & & A \\ 0 & & \lambda & \\ \hline & 0 & & B \end{array}\right]$$

where A is some $d \times n - d$ submatrix, and B is some $n - d \times n - d$ submatrix. Because of this block decomposition, the characteristic polynomial of T factors as $(\lambda - t)^d \det(B - tI_{n-d})$. (We are using Exercise 12 of Section 4.1 here.) Now, since $(\lambda - t)^d$ divides the characteristic polynomial, while by assumption $(\lambda - t)^m$ is the largest power of $(\lambda - t)$ dividing it, we must have $d = \dim(E_\lambda) \leq m$, as claimed. ■

We are now ready to state and prove our characterization of diagonalizable linear mappings.

(4.2.7) Theorem. Let $T: V \to V$ be a linear mapping on a finite-dimensional vector space V, and let $\lambda_1, \ldots, \lambda_k$ be its distinct eigenvalues. Let m_i be the multiplicity of λ_i as a root of the characteristic polynomial of T. Then T is diagonalizable if and only if

(i) $m_1 + \cdots + m_k = n = \dim(V)$, and

(ii) for each i, $\dim(E_{\lambda_i}) = m_i$.

Remarks: The first condition in the theorem may be thought of in the following way. *Counting with multiplicities,* T must have n real eigenvalues. The second condition says that for each of the eigenvalues, the maximum possible dimension for the eigenspace E_{λ_i} [i.e., m_i, by Proposition (4.2.6)] must be attained.

Proof: $\to$: We leave this implication as an exercise for the reader.

$\leftarrow$: Assume that conditions (i) and (ii) of the theorem are satisfied for $T: V \to V$. Using condition (ii), for each i let $\beta_i = \{\mathbf{x}_{i,1}, \ldots, \mathbf{x}_{i,m_i}\}$ be a basis of E_{λ_i}. (Note that since all of these vectors must be nonzero, they are all eigenvectors of T with eigenvalue λ_i.) Now, form the set $\beta = \bigcup_{i=1}^{n} \beta_i$. By Corollary (4.2.5), we see that β is linearly independent. But then, by condition (i), we have that β contains $\sum_{i=1}^{k} m_i = n$ vectors in all. Since β is linearly independent and β contains $n = \dim(V)$ vectors, β is a basis for V, and it follows that T is diagonalizable, since all the vectors in β are eigenvectors of T. ■

We close this section with some corollaries of Theorem (4.2.7). One special case in which T is always diagonalizable is given in the following corollary.

(4.2.8) Corollary. Let $T: V \to V$ be a linear mapping on a finite-dimensional space V, and assume that T has $n = \dim(V)$ distinct real eigenvalues. Then T is diagonalizable.

Proof: Since $\dim(V) = n$, the characteristic polynomial of T has degree n. If this polynomial has n distinct roots, then each of them has multiplicity $m = 1$. From Proposition (4.2.6), we see that in this case $1 = \dim(E_\lambda) =$ (multiplicity of λ) for every eigenvalue. Hence, by the theorem, T is diagonalizable. ■

The reader is cautioned not to read too much into this statement! There are matrices with multiple eigenvalues that are also diagonalizable. We saw one such matrix in Example (4.2.3b). Any diagonal matrix with some repeated diagonal entries, such as $A = \left[\begin{array}{cc|ccc} 2 & 0 & 0 & 0 & 0 \\ 0 & 2 & 0 & 0 & 0 \\ \hline 0 & 0 & 3 & 0 & 0 \\ 0 & 0 & 0 & 3 & 0 \\ 0 & 0 & 0 & 0 & 3 \end{array}\right]$, gives another example (as does any matrix similar to one of this form).

Finally, we give some conditions that are equivalent to those presented in Theorem (4.2.7), which also imply diagonalizability.

(4.2.9) Corollary. A linear mapping $T: V \to V$ on a finite-dimensional space V is diagonalizable if and only if the sum of the multiplicities of the real eigenvalues is $n = \dim(V)$, and either

(i) We have $\sum_{i=1}^{k} \dim(E_{\lambda_i}) = n$, where the λ_i are the distinct eigenvalues of T, or

(ii) We have $\sum_{i=1}^{k} (n - \dim(\text{Im}(T - \lambda_i I))) = n$, where again λ_i are the distinct eigenvalues.

Proof: Exercise. ■

One consequence of Theorem (4.2.7) is that there are many linear mappings that are *not* diagonalizable. In the exercises we will see that even though diagonalizable mappings are the simplest ones to understand, it is also possible to analyze nondiagonalizable mappings geometrically on an individual basis. Nevertheless, one natural question to ask is whether there is some "next-best" standard form for the matrix of a general linear mapping that is not much more complicated than the diagonal form for diagonalizable mappings, and that would make the geometric behavior of nondiagonalizable mappings easier to understand. We will return to such questions in Chapter 6.

EXERCISES

1. Determine whether the given linear mappings are diagonalizable, and if so, find a basis of the appropriate vector space consisting of eigenvectors.

a) $T: \mathbf{R}^2 \to \mathbf{R}^2$ defined by $A = \begin{bmatrix} 3 & 1 \\ -1 & 1 \end{bmatrix}$

b) $T: \mathbf{R}^3 \to \mathbf{R}^3$ defined by $A = \begin{bmatrix} 2 & 0 & 0 \\ 2 & 1 & 3 \\ 5 & 0 & 1 \end{bmatrix}$

c) $T: \mathbf{R}^3 \to \mathbf{R}^3$ defined by $A = \begin{bmatrix} 0 & 3 & 0 \\ 1 & 0 & -1 \\ 0 & 2 & 0 \end{bmatrix}$

d) $T: \mathbf{R}^3 \to \mathbf{R}^3$ defined by $A = \begin{bmatrix} 1 & 2 & 1 \\ 2 & -1 & 2 \\ 1 & 2 & 1 \end{bmatrix}$

e) $T: \mathbf{R}^4 \to \mathbf{R}^4$ defined by $A = \begin{bmatrix} 4 & 2 & 7 & -1 \\ -1 & 1 & 3 & 2 \\ 0 & 0 & 1 & -1 \\ 0 & 0 & 1 & 1 \end{bmatrix}$

f) $T: P_3(\mathbf{R}) \to P_3(\mathbf{R})$ defined by $T(p)(x) = 3 \cdot p(x) + x \cdot p''(x)$.

g) $V = \text{Span}\{\sin(x), \cos(x), e^x, e^{-x}, 1\}$ in $C^\infty(\mathbf{R})$, $T: V \to V$ defined by $T(f) = f'' - 4f' + 3f$.

2. Show that the matrix $A = \begin{bmatrix} a & b \\ 0 & a \end{bmatrix} \in M_{2\times 2}(\mathbf{R})$ is diagonalizable if and only if $b = 0$.

3. Find necessary and sufficient conditions on the real numbers a, b, c for the matrix:

$$\begin{bmatrix} 1 & a & b \\ 0 & 1 & c \\ 0 & 0 & 2 \end{bmatrix}$$

to be diagonalizable.

4. Prove the "only if" part of Theorem (4.2.7). That is, show that if $T: V \to V$ is diagonalizable, $\lambda_1, \ldots, \lambda_k$ are its distinct eigenvalues, and the multiplicity of λ_i is m_i, then $\sum_{i=1}^{k} m_i = n = \dim(V)$, and for each i $\dim(E_{\lambda_i}) = m_i$.

5. Prove Corollary (4.2.9).

6. a) Show that if A is diagonalizable, then A^k is diagonalizable for all $k \geq 2$.
 b) Is the converse of the statement in part a true?

7. a) Show that if A is an invertible $n \times n$ matrix and $\mathbf{x}$ is an eigenvector of A with eigenvalue λ, then $\mathbf{x}$ is also an eigenvector of A^{-1}, with eigenvalue λ^{-1}.

b) From part a, deduce that if A is an invertible $n \times n$ matrix, then A is diagonalizable if and only if A^{-1} is diagonalizable.

8. Let V be finite-dimensional. Show that if $T: V \to V$ is linear, and $T^2 = I$, then T is diagonalizable. (See Exercise 15 of Section 4.1)

9. Let V be finite-dimensional. Show that if $T: V \to V$ is linear, and $T^2 = T$, then T is diagonalizable. (See Exercise 17 of Section 4.1.)

10. Let V be finite-dimensional. Show that if $T: V \to V$ is a linear mapping and $(T - aI)(T - bI) = 0$ where $a \neq b$, then T is diagonalizable. (*Hint:* What can the eigenvalues of T be? Then proceed as in Exercises 8 and 9.)

11. Describe the nondiagonalizable mapping T: $\mathbf{R}^2 \to \mathbf{R}^2$ defined by $A = \begin{bmatrix} a & 1 \\ 0 & a \end{bmatrix}$ geometrically. Begin by computing the image of the unit square spanned by the standard basis vectors in $\mathbf{R}^2$.

12. Describe the nondiagonalizable mapping defined by $A = \begin{bmatrix} 1/2 & -\sqrt{3}/2 & 0 \\ \sqrt{3}/2 & 1/2 & 0 \\ 0 & 0 & 5 \end{bmatrix}$ geometrically. Begin by determining if any lines are mapped into themselves. What happens to other vectors?

13. Let V be finite-dimensional, and let $T: V \to V$ be a linear mapping with distinct eigenvalues $\lambda_1, \ldots, \lambda_k$.
a) Show that for each i, $E_{\lambda_i} \cap (\sum_{j \neq i} E_{\lambda_j}) = \{\mathbf{0}\}$.
b) Show that T is diagonalizable if and only if $V = E_{\lambda_1} + \cdots + E_{\lambda_k}$.

14. Let $A, B \in M_{n \times n}(\mathbf{R})$ be two matrices with $AB = BA$.
a) Show that if $\mathbf{x} \in \mathbf{R}^n$ is an eigenvector of A with eigenvalue λ, then $B\mathbf{x}$ is also an eigenvector of A with eigenvalue λ.
b) Show that if $T: V \to V$ is a linear mapping, W is a subspace of V with $T(W) \subseteq W$, and λ is an eigenvalue of T, then the λ-eigenspace for $T|_W$ is equal to $E_\lambda \cap W$, where E_λ is the λ-eigenspace for T on all of V.
c) Show that if $T: V \to V$ is a diagonalizable linear mapping, and W is a subspace of V with $T(W) \subseteq W$, then the restricted mapping $T|_W$ is also diagonalizable.
d) Conclude from parts a and c that if A and B are each diagonalizable, then there is a basis β of $\mathbf{R}^n$ consisting of vectors that are eigenvectors of both A and B. (In other words, A and B can be diagonalized *simultaneously*.) [*Hint:* By part b of Exercise 13, and part b of this exercise, it suffices to show that if the distinct eigenvalues of T on V are $\lambda_1, \ldots, \lambda_k$, then $W = (E_{\lambda_1} \cap W) + \cdots + (E_{\lambda_k} \cap W)$. To prove this, show that if $\mathbf{x} \in W$, and $\mathbf{x} = \mathbf{x}_1 + \cdots + \mathbf{x}_k$, where $\mathbf{x}_i \in E_{\lambda_i}$ for each i, then $\mathbf{x}_i \in W$ as well. One way to do this is to consider the effect of the mapping $(T - \lambda_1 I) \cdot \cdots \cdot (T - \lambda_{i-1} I) \cdot (T - \lambda_{i+1} I) \cdot \cdots \cdot (T - \lambda_k I)$ (λ_i omitted) on $\mathbf{x}$.]

15. Let $A \in M_{n\times n}(\mathbf{R})$ be any matrix. Show that A is diagonalizable if and only if A^t is diagonalizable. [*Hint:* Use the criterion of Corollary (4.2.9ii) and the general fact that $\dim(\text{Im}(M^t)) = \dim(\text{Im}(M))$ for all matrices M that follows from Exercise 10 of Section 2.3.]

§4.3. GEOMETRY IN $\mathbf{R}^n$

In this section we lay some of the groundwork for the study of a special class of linear mappings that will follow in later sections of this chapter. Recall that in Section 2.1, when we introduced rotations and projections in $\mathbf{R}^2$, we showed that both distances between points and angles between vectors (and hence all of the basic geometry of $\mathbf{R}^2$) could be expressed in terms of the *dot product* on pairs of vectors in $\mathbf{R}^2$. It is possible to generalize these notions to the case of vectors in $\mathbf{R}^n$ as follows.

(4.3.1) Definition. The *standard inner product* (or dot product) on $\mathbf{R}^n$ is the function $\langle\ ,\ \rangle : \mathbf{R}^n \times \mathbf{R}^n \to \mathbf{R}$ defined by the following rule. If $\mathbf{x} = (x_1, \ldots, x_n)$ and $\mathbf{y} = (y_1, \ldots, y_n)$ in standard coordinates, then $\langle \mathbf{x}, \mathbf{y}\rangle$ is the scalar: $\langle \mathbf{x}, \mathbf{y}\rangle = x_1y_1 + \cdots + x_ny_n$.

We have chosen the notation $\langle \mathbf{x}, \mathbf{y}\rangle$ rather than the common alternate notation $\mathbf{x} \cdot \mathbf{y}$ for the "dot product" to emphasize the fact that we will be viewing this operation as a mapping that takes a pair of vectors and returns a scalar.

The standard inner product has the following properties (generalizing the properties of the dot product in $\mathbf{R}^2$ that we saw in Section 2.1, Exercise 6).

(4.3.2) Proposition

a) For all $\mathbf{x}, \mathbf{y}, \mathbf{z} \in \mathbf{R}^n$ and all $c \in \mathbf{R}$, $\langle c\mathbf{x} + \mathbf{y}, \mathbf{z}\rangle = c\langle \mathbf{x}, \mathbf{z}\rangle + \langle \mathbf{y}, \mathbf{z}\rangle$. (In other words, as a function of the first variable, the inner product is a linear mapping.)

b) For all $\mathbf{x}, \mathbf{y} \in \mathbf{R}^n$, $\langle \mathbf{x}, \mathbf{y}\rangle = \langle \mathbf{y}, \mathbf{x}\rangle$ (i.e., the inner product is "symmetric").

c) For all $\mathbf{x} \in \mathbf{R}^n$, $\langle \mathbf{x}, \mathbf{x}\rangle \geqslant 0$, and $\langle \mathbf{x}, \mathbf{x}\rangle = 0$ if and only if $\mathbf{x} = \mathbf{0}$. (We say the standard inner product is "positive-definite.")

Proof: Many of the simpler facts about the standard inner product may be derived by expanding out the expressions involved, using Definition (4.3.1).

a) Write $\mathbf{x} = (x_1, \ldots, x_n)$, $\mathbf{y} = (y_1, \ldots, y_n)$, and $\mathbf{z} = (z_1, \ldots, z_n)$. Then $c\mathbf{x} + \mathbf{y} = (cx_1 + y_1, \ldots, cx_n + y_n)$, so we have

$$\langle c\mathbf{x} + \mathbf{y}, \mathbf{z}\rangle = (cx_1 + y_1)z_1 + \cdots + (cx_n + y_n)z_n$$

After multiplying this out and rearranging the terms, we obtain

$$\begin{aligned}\langle c\mathbf{x} + \mathbf{y}, \mathbf{z}\rangle &= c(x_1z_1 + \cdots + x_nz_n) + \cdots + (y_1z_1 + \cdots + y_nz_n)\\ &= c\langle \mathbf{x}, \mathbf{z}\rangle + \langle \mathbf{y}, \mathbf{z}\rangle\end{aligned}$$

b) We have:

$$\begin{aligned}\langle \mathbf{x}, \mathbf{y}\rangle &= x_1y_1 + \cdots + x_ny_n \\ &= y_1x_1 + \cdots + y_nx_n \\ &= \langle \mathbf{y}, \mathbf{x}\rangle.\end{aligned}$$

c) By the definition, $\langle \mathbf{x}, \mathbf{x}\rangle = x_1^2 + \cdots + x_n^2$. Since the components of $\mathbf{x}$ are all real numbers, each $x_i^2 \geq 0$, so $\langle \mathbf{x}, \mathbf{x}\rangle \geq 0$. Furthermore, if $\langle \mathbf{x}, \mathbf{x}\rangle = 0$, then it is clear that $x_i = 0$ for all i, or in other words, that $\mathbf{x} = \mathbf{0}$. ■

Note that if we combine parts a and b of the proposition, we have

$$\begin{aligned}\langle \mathbf{z}, c\mathbf{x} + \mathbf{y}\rangle &= \langle c\mathbf{x} + \mathbf{y}, \mathbf{z}\rangle \quad \text{(by b)} \\ &= c\langle \mathbf{x}, \mathbf{z}\rangle + \langle \mathbf{y}, \mathbf{z}\rangle \quad \text{(by a)} \\ &= c\langle \mathbf{z}, \mathbf{x}\rangle + \langle \mathbf{z}, \mathbf{y}\rangle \quad \text{(by b again)}\end{aligned}$$

so that the inner product is linear in the second variable as well.

As in $\mathbf{R}^2$, we define lengths of vectors in terms of the inner product.

(4.3.3) Definitions

a) The *length* (or *norm*) of $\mathbf{x} \in \mathbf{R}^n$ is the scalar

$$\|\mathbf{x}\| = \sqrt{\langle \mathbf{x}, \mathbf{x}\rangle}$$

b) $\mathbf{x}$ is called a *unit vector* if $\|\mathbf{x}\| = 1$.

Note that part a of the definition makes sense, since $\langle \mathbf{x}, \mathbf{x}\rangle \geq 0$ for all vectors $\mathbf{x}$. For example, $\|(1, -4, 3)\| = \sqrt{(1)^2 + (-4)^2 + (3)^2} = \sqrt{26}$. Note also that the scalar multiple $(1/\sqrt{26}, -4\sqrt{26}, 3/\sqrt{26})$ is a unit vector. In general, if $\mathbf{x} \neq \mathbf{0}$, then $(1/\|\mathbf{x}\|)\mathbf{x}$ will be a unit vector.

Lengths of vectors, defined in this way, satisfy the same geometric properties that are familiar from $\mathbf{R}^2$. For example, we have Proposition (4.3.4).

(4.3.4) Proposition

a) (The triangle inequality) For all vectors $\mathbf{x}, \mathbf{y} \in \mathbf{R}^n$, $\|\mathbf{x} + \mathbf{y}\| \leq \|\mathbf{x}\| + \|\mathbf{y}\|$. (Geometrically, in a triangle, the length of any side is less than or equal to the sum of the lengths of the other two sides. See Figure 4.3.)

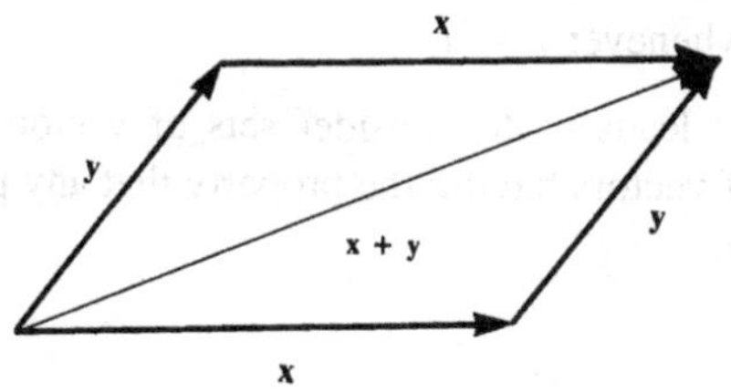

Figure 4.3

b) (The Cauchy-Schwarz inequality) For all vectors $\mathbf{x}, \mathbf{y} \in \mathbf{R}^n$,

$$|\langle \mathbf{x}, \mathbf{y} \rangle| \leq \|\mathbf{x}\| \cdot \|\mathbf{y}\|$$

The proofs of these inequalities will be left as exercises for the reader. In fact, the triangle inequality may be deduced from the Cauchy-Schwarz inequality, as you will see.

Note that by part b of Proposition (4.3.4), we have that for all nonzero vectors $\mathbf{x}, \mathbf{y} \in \mathbf{R}^n$, $|\langle \mathbf{x}, \mathbf{y} \rangle/(\|\mathbf{x}\| \cdot \|\mathbf{y}\|)| \leq 1$. Hence, there is a unique real number θ in the range $0 \leq \theta < \pi$ with

$$\cos(\theta) = \langle \mathbf{x}, \mathbf{y} \rangle/(\|\mathbf{x}\| \cdot \|\mathbf{y}\|)$$

This leads to the definition of the angle between two vectors in $\mathbf{R}^n$.

(4.3.5) Definition. The *angle*, θ, *between two nonzero vectors* $\mathbf{x}, \mathbf{y} \in \mathbf{R}^n$ is defined to be

$$\theta = \cos^{-1}(\langle \mathbf{x}, \mathbf{y} \rangle/(\|\mathbf{x}\| \cdot \|\mathbf{y}\|))$$

(4.3.6) Example. If $\mathbf{x} = (1, 2, 1)$ and $\mathbf{y} = (-1, 1, 1)$ in $\mathbf{R}^3$, then the angle between them is

$$\begin{aligned} \theta &= \cos^{-1}(\langle \mathbf{x}, \mathbf{y} \rangle/(\|\mathbf{x}\| \cdot \|\mathbf{y}\|)) \\ &= \cos^{-1}(2/(\sqrt{6} \cdot \sqrt{3})) \\ &= \cos^{-1}(\sqrt{2}/3) \\ &\approx 1.1 \text{ radians} \end{aligned}$$

By the properties of the cosine function, we know that for θ in the range $0 \leq \theta < \pi$, $\cos(\theta) = 0$ has the unique solution $\theta = \pi/2$. This justifies the following definition.

(4.3.7) Definition. Two vectors $\mathbf{x}, \mathbf{y} \in \mathbf{R}^n$ are said to be *orthogonal* (or perpendicular) if $\langle \mathbf{x}, \mathbf{y} \rangle = 0$.

(4.3.8) Examples

a) The vectors $\mathbf{x} = (-1, 1, -1)$ and $\mathbf{y} = (1, -1, -2)$ are orthogonal in $\mathbf{R}^3$.

b) In $\mathbf{R}^n$, the standard basis vectors $\mathbf{e}_i = (0, \ldots, 1, \ldots, 0)$ are orthogonal in pairs: $\langle \mathbf{e}_i, \mathbf{e}_j \rangle = 0$ whenever $i \neq j$.

Example (4.3.8b) leads us to consider sets of vectors, and we introduce a new term for any set of vectors having the property that any pair of distinct vectors in the set is orthogonal.

(4.3.9) Definitions

a) A set of vectors $S \subset \mathbf{R}^n$ is said to be *orthogonal* if for every pair of vectors $\mathbf{x}, \mathbf{y} \in S$ with $\mathbf{x} \neq \mathbf{y}$, we have $\langle \mathbf{x}, \mathbf{y} \rangle = 0$.

b) A set of vectors $S \subset \mathbf{R}^n$ is said to be *orthonormal* if S is orthogonal and, in addition, every vector in S is a unit vector.

Now, if a set of nonzero vectors is orthogonal, those vectors point in mutually perpendicular directions in $\mathbf{R}^n$, and because of this, we might expect that such sets are linearly independent. This is indeed the case, as we will now see. A first step in this direction is Proposition (4.3.10)

(4.3.10) Proposition. If $\mathbf{x}, \mathbf{y} \in \mathbf{R}^n$ are orthogonal, nonzero vectors, then $\{\mathbf{x}, \mathbf{y}\}$ is linearly independent.

Proof: Suppose $a\mathbf{x} + b\mathbf{y} = \mathbf{0}$ for some scalars a, b. Then we have

$$\begin{aligned}\langle a\mathbf{x} + b\mathbf{y}, \mathbf{x} \rangle &= a\langle \mathbf{x}, \mathbf{x}\rangle + b\langle \mathbf{y}, \mathbf{x}\rangle \qquad \text{by (4.3.2a)}\\ &= a\langle \mathbf{x}, \mathbf{x}\rangle \qquad \text{since } \langle \mathbf{x}, \mathbf{y}\rangle = 0\end{aligned}$$

On the other hand, $a\mathbf{x} + b\mathbf{y} = \mathbf{0}$, so $\langle a\mathbf{x} + b\mathbf{y}, \mathbf{x}\rangle = \langle \mathbf{0}, \mathbf{x}\rangle = 0$. By (4.3.2c), since $\mathbf{x} \neq \mathbf{0}$, $\langle \mathbf{x}, \mathbf{x}\rangle \neq 0$, so we must have $a = 0$. If we now repeat the argument, taking the inner product of $a\mathbf{x} + b\mathbf{y}$ with $\mathbf{y}$, the same reasoning shows that $b = 0$ as well. Hence, $\mathbf{x}$ and $\mathbf{y}$ are linearly independent. ■

Now, a simple extension of the argument given here shows in fact that any orthogonal set of nonzero vectors is linearly independent. You will show this in an exercise.

(4.3.11) Remark. An orthogonal set of vectors can contain the zero vector (by definition, the zero vector is orthogonal to every vector in $\mathbf{R}^n$). However, in an orthonormal set, since every vector has length 1, every vector is nonzero. By the extension of the result of Proposition (4.3.10) mentioned previously, it then follows that orthonormal sets are automatically linearly independent. Hence, for example, if S is a subset of $\mathbf{R}^n$ that is orthonormal, and S contains n vectors, then S is a basis of $\mathbf{R}^n$, a so-called *orthonormal basis*.

(4.3.12) Example. The set $\{(1/2)(1, 1, -1, 1), (1/\sqrt{2})(1, -1, 0, 0), (1/\sqrt{2})(0, 0, 1, 1), (1/2)(1, 1, 1, -1)\}$ is an orthonormal basis for $\mathbf{R}^4$. (You should check that all pairs of distinct vectors here are, in fact, orthogonal.)

EXERCISES

1. Let $\mathbf{x} = (1, 4, 3)$, $\mathbf{y} = (-1, 7, 2)$ in $\mathbf{R}^3$.
   a) Compute $\langle \mathbf{x}, \mathbf{y}\rangle$ and $\langle \mathbf{y}, \mathbf{x}\rangle$.
   b) Compute $\|\mathbf{x}\|$ and $\|\mathbf{y}\|$.
   c) Verify that the Cauchy-Schwarz and triangle inequalities hold for the vectors $\mathbf{x}$ and $\mathbf{y}$.
   d) Find the angle between $\mathbf{x}$ and $\mathbf{y}$.
2. Repeat Exercise 1 with $\mathbf{x} = (1, 1, 1, 1)$ and $\mathbf{y} = (1, 2, -4, 1)$ in $\mathbf{R}^4$.

3. a) Show that for any vectors $\mathbf{x}, \mathbf{y} \in \mathbf{R}^n$ and any scalar c,

$$\langle \mathbf{x} - c\mathbf{y}, \mathbf{x} - c\mathbf{y} \rangle = \|\mathbf{x}\|^2 - 2c\langle \mathbf{x}, \mathbf{y} \rangle + c^2\|\mathbf{y}\|^2 \geqslant 0$$

b) Set $c = \langle \mathbf{x}, \mathbf{y} \rangle / \|\mathbf{y}\|^2$ and deduce the Cauchy-Schwarz inequality from part a.
c) Now, use part a to deduce that $\langle \mathbf{x} + \mathbf{y}, \mathbf{x} + \mathbf{y} \rangle = \|\mathbf{x}\|^2 + 2\langle \mathbf{x}, \mathbf{y} \rangle + \|\mathbf{y}\|^2 \geqslant 0$.
d) Use parts b and c to deduce the triangle inequality.

4. Show that if $\mathbf{x}$ and $\mathbf{y}$ are orthogonal vectors in $\mathbf{R}^n$, then

$$\|\mathbf{x}\|^2 + \|\mathbf{y}\|^2 = \|\mathbf{x} - \mathbf{y}\|^2$$

What does this result mean geometrically?

5. Show that for any vectors $\mathbf{x}, \mathbf{y} \in \mathbf{R}^n$,

$$\|\mathbf{x} + \mathbf{y}\|^2 + \|\mathbf{x} - \mathbf{y}\|^2 = 2(\|\mathbf{x}\|^2 + \|\mathbf{y}\|^2)$$

What does this result mean geometrically?

6. Show that for any vectors $\mathbf{x}, \mathbf{y} \in \mathbf{R}^n$, $\langle \mathbf{x}, \mathbf{y} \rangle = (1/4)(\|\mathbf{x} + \mathbf{y}\|^2 - \|\mathbf{x} - \mathbf{y}\|^2)$. (This is called the *polarization identity*. It shows that we can recover the inner product on $\mathbf{R}^n$ just by measuring lengths of vectors.)

7. Determine whether or not the following sets of vectors are orthogonal:
a) $\{(1, 4, 2, 0), (-2, 0, 1, 7), (3, 7, 6, 2)\}$
b) $\{(0, 0, 0, 1), (0, 1, 1, 0), (1, 1, -1, 0)\}$

8. Prove the following extension of the result of Proposition (4.3.10). Let $S = \{\mathbf{x}_1, \ldots, \mathbf{x}_k\}$ be any orthogonal set of nonzero vectors in $\mathbf{R}^n$. Show that S is linearly independent.

9. a) Find an orthonormal basis of $\mathbf{R}^3$ other than the standard basis.
b) Find an orthonormal basis of $\mathbf{R}^3$ containing the vector $(1/\sqrt{3})(1, 1, 1)$.

In this section we dealt with the standard inner product on $\mathbf{R}^n$. This mapping is an example of a more general notion.

Definition. Let V be a vector space. An *inner product* on V is any mapping $\langle\ ,\ \rangle : V \times V \to \mathbf{R}$ with the following three properties:

(i) For all vectors $\mathbf{x}, \mathbf{y}, \mathbf{z} \in V$ and all $c \in \mathbf{R}$, $\langle c\mathbf{x} + \mathbf{y}, \mathbf{z} \rangle = c\langle \mathbf{x}, \mathbf{z} \rangle + \langle \mathbf{y}, \mathbf{z} \rangle$.

(ii) For all $\mathbf{x}, \mathbf{y} \in V$, $\langle \mathbf{x}, \mathbf{y} \rangle = \langle \mathbf{y}, \mathbf{x} \rangle$.

(iii) For all $\mathbf{x} \in V$, $\langle \mathbf{x}, \mathbf{x} \rangle \geqslant 0$, and $\langle \mathbf{x}, \mathbf{x} \rangle = 0$ only if $\mathbf{x} = \mathbf{0}$.

10. a) Show that the mapping $\langle\ ,\ \rangle : \mathbf{R}^2 \times \mathbf{R}^2 \to \mathbf{R}$ defined by

$$\langle (x_1, x_2), (y_1, y_2) \rangle = [x_1\ x_2] \begin{bmatrix} 2 & 1 \\ 1 & 1 \end{bmatrix} \begin{bmatrix} y_1 \\ y_2 \end{bmatrix}$$

(matrix product) is an inner product on $V = \mathbf{R}^2$.
b) Let $V = C[0, 1]$ (the vector space of continuous functions on the closed

interval [0, 1]). Show that $\langle f, g\rangle = \int_0^1 f(t)g(t)\,dt$ is an inner product on V.

c) Let $V = M_{n\times n}(\mathbf{R})$. Show that $\langle A, B\rangle = \mathrm{Tr}(A^tB)$ is an inner product on V.

Since $\langle \mathbf{x}, \mathbf{x}\rangle \geq 0$ for all $\mathbf{x} \in V$, we can define the *norm* of $\mathbf{x}$: $\|\mathbf{x}\| = \sqrt{\langle \mathbf{x}, \mathbf{x}\rangle}$ as we did in $\mathbf{R}^n$, and we interpret $\|\mathbf{x}\|$ as the length of the vector $\mathbf{x}$ with respect to the given inner product on V.

11. Show that the Cauchy-Schwarz and triangle inequalities are valid for the norm coming from any inner product.

12. Show that the polarization identity (see Exercise 6) is also valid. We may also define the angle between two vectors, and orthogonal sets of vectors exactly as in $\mathbf{R}^n$.

13. a) Show that $(-4, 5)$ and $(1, 3)$ are orthogonal with respect to the inner product on $\mathbf{R}^2$ defined in Exercise 10a.

    b) Show that $\sin(\pi x)$ and $\cos(\pi x)$ are orthogonal with respect to the inner product on $C[0, 1]$ defined in Exercise 10b.

    c) Show that $\begin{bmatrix} 1 & 2 \\ 2 & 3 \end{bmatrix}$ and $\begin{bmatrix} 0 & 5 \\ -5 & 0 \end{bmatrix}$ are orthogonal with respect to the inner product on $M_{2\times 2}(\mathbf{R})$ defined in Exercise 10c.

The general inner products introduced in these exercises are linear in each variable, just as the standard inner product on $\mathbf{R}^n$ is. Of course, this property is closely related to the property of *multilinearity* of functions from $M_{n\times n}(\mathbf{R})$ to $\mathbf{R}$ that we saw in developing the determinant function in Chapter 3. In the next exercises we will explore this connection.

Definition. A mapping $B: V \times V \to \mathbf{R}$ is said to be *bilinear* if B is linear in each variable, or more precisely if

i) $B(c\mathbf{x} + \mathbf{y}, \mathbf{z}) = cB(\mathbf{x}, \mathbf{z}) + B(\mathbf{y}, \mathbf{z})$, and

ii) $B(\mathbf{x}, c\mathbf{y} + \mathbf{z}) = cB(\mathbf{x}, \mathbf{y}) + B(\mathbf{x}, \mathbf{z})$ for all $\mathbf{x}, \mathbf{y}, \mathbf{z} \in V$ and all $c \in \mathbf{R}$.

14. Let $A \in M_{n\times n}(\mathbf{R})$ be any matrix and define $B: \mathbf{R}^n \times \mathbf{R}^n \to \mathbf{R}$ by $B(\mathbf{x}, \mathbf{y}) = \mathbf{x}^tA\mathbf{y}$ (matrix product, thinking of $\mathbf{x}$ and $\mathbf{y}$ as column vectors or $n \times 1$ matrices). Show that B is bilinear.

15. Let $\alpha = \{\mathbf{x}_1, \ldots, \mathbf{x}_n\}$ be a basis in $\mathbf{R}^n$, and let B be a bilinear mapping. Define $[B]_\alpha$ to be the $n \times n$ matrix $[B]_\alpha = (B(\mathbf{x}_i, \mathbf{x}_j))$. That is, the (i, j) entry of $[B]_\alpha$ is the scalar $B(\mathbf{x}_i, \mathbf{x}_j)$. Show that for all $\mathbf{x}, \mathbf{y} \in \mathbf{R}^n$

    $$B(\mathbf{x}, \mathbf{y}) = [\mathbf{x}]_\alpha^t \cdot [B]_\alpha \cdot [\mathbf{y}]_\alpha$$

    (This shows that every bilinear mapping on $\mathbf{R}^n$ is of the kind we saw in Exercise 14.) If B is the standard inner product, and α is the standard basis in $\mathbf{R}^n$, what is $[B]_\alpha$?

16. We say a bilinear mapping on a vector space V is *symmetric* if $B(\mathbf{x}, \mathbf{y}) = B(\mathbf{y}, \mathbf{x})$ for all $\mathbf{x}, \mathbf{y} \in V$. Show that B on $\mathbf{R}^n$ is symmetric if and only if the matrix $[B]_\alpha$ is symmetric.

17. Show that if we change basis from α to a second basis β for $\mathbf{R}^n$, letting $Q = [I]_\alpha^\beta$, we have

$$[B]_\alpha = Q^t \cdot [B]_\beta \cdot Q$$

§4.4. ORTHOGONAL PROJECTIONS AND THE GRAM-SCHMIDT PROCESS

In this section we consider some further geometric questions in $\mathbf{R}^n$ related to the concept of orthogonality introduced in Section 4.3. First, let $W \subset \mathbf{R}^n$ be any subspace. We will need to consider the set of all vectors orthogonal to every vector in W.

(4.4.1) Definition. The *orthogonal complement* of W, denoted $W^\perp$, is the set

$$W^\perp = \{\mathbf{v} \in \mathbf{R}^n \mid \langle \mathbf{v}, \mathbf{w} \rangle = 0 \quad \text{for all } \mathbf{w} \in W\}$$

(This is sometimes called "W perp" in informal speech.)

To gain a better understanding of what is involved in this definition, we consider how $W^\perp$ may be described in a more explicit fashion. If we choose a basis $\{\mathbf{w}_1, \ldots, \mathbf{w}_k\}$ for W, where $\mathbf{w}_i = (a_{i1}, \ldots, a_{in})$ in standard coordinates, then $\mathbf{v} = (x_1, \ldots, x_n) \in W^\perp$ if and only if

$$\begin{aligned}\langle \mathbf{w}_1, \mathbf{v} \rangle &= a_{11}x_1 + \cdots + a_{1n}x_n = 0\\ &\vdots\\ \langle \mathbf{w}_k, \mathbf{v} \rangle &= a_{k1}x_1 + \cdots + a_{kn}x_n = 0\end{aligned}$$

(Why is it enough to check that the basis vectors for W are orthogonal to $\mathbf{v}$?) Thus, $W^\perp$ may be viewed as the set of solutions of a system of homogeneous linear equations, and we are back in familiar territory.

(4.4.2) Examples

a) If we take $W = \{\mathbf{0}\}$, then the preceding system of equations reduces to the trivial system $0 = 0$. Every vector is a solution, so $\{\mathbf{0}\}^\perp = \mathbf{R}^n$.

b) If $W = \mathbf{R}^n$ on the other hand, then (e.g., using the standard basis) we see that $\mathbf{v} \in W^\perp$ only if all the components of $\mathbf{v}$ are 0. In other words, $(\mathbf{R}^n)^\perp = \{\mathbf{0}\}$.

c) If $W = \text{Span}(\{(1, 0, 3), (0, 1, -7)\}$ in $\mathbf{R}^3$, then $W^\perp$ is the set of solutions of the homogeneous system

$$\begin{aligned} x_1 + \qquad 3x_3 &= 0\\ x_2 - 7x_3 &= 0\end{aligned}$$

The system is already in echelon form, and we can read off a basis for the space of solutions by setting the free variable $x_3 = 1$. We obtain that $W^\perp = \text{Span}\{(-3,7,1)\}$.

From the definition, the description of $W^\perp$ as the set of solutions of a system of homogeneous linear equations, and general properties of subspaces, the following properties may be deduced immediately.

(4.4.3) Proposition

a) For every subspace W of $\mathbf{R}^n$, $W^\perp$ is also a subspace of $\mathbf{R}^n$.

b) We have $\dim(W) + \dim(W^\perp) = \dim(\mathbf{R}^n) = n$.

c) For all subspaces W of $\mathbf{R}^n$, $W \cap W^\perp = \{\mathbf{0}\}$.

d) Given a subspace W of $\mathbf{R}^n$, every vector $\mathbf{x} \in \mathbf{R}^n$ can be written uniquely as $\mathbf{x} = \mathbf{x}_1 + \mathbf{x}_2$, where $\mathbf{x}_1 \in W$ and $\mathbf{x}_2 \in W^\perp$. In other words, $\mathbf{R}^n = W \oplus W^\perp$.

The proofs are left as exercises for the reader. We will see an explicit way to produce the vectors $\mathbf{x}_1$ and $\mathbf{x}_2$ described in part d as a result of our next topic of discussion.

Recall that in Section 2.1 we defined projection mappings $P_\mathbf{a}: \mathbf{R}^2 \to \mathbf{R}^2$, which were constructed as follows. Given a one-dimensional subspace $W = \text{Span}(\{\mathbf{a}\})$ of $\mathbf{R}^2$ (in other words, a line through the origin) and a vector $\mathbf{v} \notin W$, we defined $P_\mathbf{a}(\mathbf{v})$ to be the intersection point of W and the line through $\mathbf{v}$ perpendicular to W. It is easy to check that this mapping actually depends only on the line W and not on the specific vector $\mathbf{a}$, so we could just as well have called the mapping P_W.

Given any subspace W of $\mathbf{R}^n$, we can define a mapping, P_W, called the *orthogonal projection* of $\mathbf{R}^n$ onto W by a similar process. We will describe this mapping algebraically first. By part d of Proposition (4.4.3), every vector $\mathbf{x}$ in $\mathbf{R}^n$ can be written uniquely as $\mathbf{x} = \mathbf{x}_1 + \mathbf{x}_2$, where $\mathbf{x}_1 \in W$ and $\mathbf{x}_2 \in W^\perp$. Hence, we can define a mapping $P_W: \mathbf{R}^n \to \mathbf{R}^n$ by $P_W(\mathbf{x}) = \mathbf{x}_1$.

Geometrically, this mapping may be viewed as follows (see also Exercise 15). Given W, and a vector $\mathbf{v} \in \mathbf{R}^n$, $\mathbf{v} \notin W$, there is a unique line l passing through $\mathbf{v}$ and orthogonal to W (i.e., l meets W at a right angle). We see that $P_W(\mathbf{v})$ is the vector $l \cap W$ (see Figure 4.4). If $\mathbf{v} \in W$ already, then by the definition $P_W(\mathbf{v}) = \mathbf{v}$.

(4.4.4) Example. Let $W = \text{Span}\{\begin{bmatrix} 1 \\ 1 \\ 0 \end{bmatrix}\}$ in $\mathbf{R}^3$. The reader will verify that

$\beta = \{\begin{bmatrix} 1 \\ 1 \\ 0 \end{bmatrix}, \begin{bmatrix} 1 \\ -1 \\ 0 \end{bmatrix}, \begin{bmatrix} 0 \\ 0 \\ 1 \end{bmatrix}\}$ is a basis of $\mathbf{R}^3$ and that the second two vectors are

orthogonal themselves, and lie in the subspace $W^\perp$. Let $\mathbf{x} = \begin{bmatrix} x_1 \\ x_2 \\ x_3 \end{bmatrix}$ be any vector

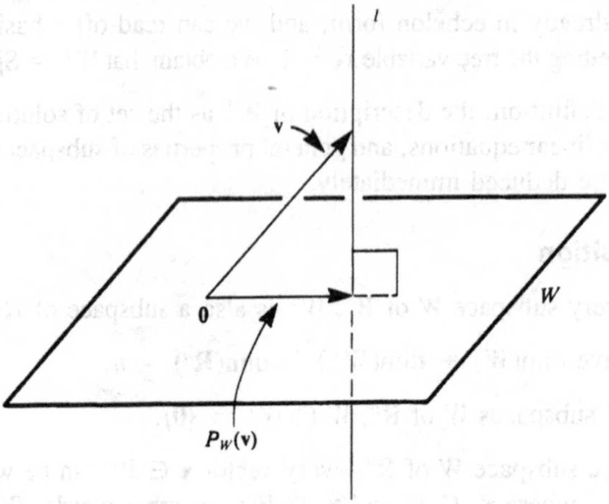

Figure 4.4

in $\mathbf{R}^3$. If we wish to apply the definition directly to produce the mapping P_W, then we must write $\mathbf{x}$ explicitly as the sum of a vector in W, and one in $W^\perp$. To do this, we can proceed as follows. [We will see a more systematic way to construct the mapping P_W later, in Proposition (4.4.6).] Note that our desired decomposition of the vector $\mathbf{x}$ will have the form

$$\begin{bmatrix} x_1 \\ x_2 \\ x_3 \end{bmatrix} = a_1 \begin{bmatrix} 1 \\ 1 \\ 0 \end{bmatrix} + (a_2 \begin{bmatrix} 1 \\ -1 \\ 0 \end{bmatrix} + a_3 \begin{bmatrix} 0 \\ 0 \\ 1 \end{bmatrix}) \tag{4.3}$$

for some scalars a_1, a_2, and a_3 and our projection will simply take $\begin{bmatrix} x_1 \\ x_2 \\ x_3 \end{bmatrix}$ to the vector $a_1 \begin{bmatrix} 1 \\ 1 \\ 0 \end{bmatrix}$, the scalar a_1 depending on the particular x_i appearing in the vector $\mathbf{x}$. The vector equation (4.3) yields a system of three equations in the unknowns a_1, a_2, and a_3, whose solutions are easily seen to be

$$a_1 = (1/2)(x_1 + x_2)$$

$$a_2 = (1/2)(x_1 - x_2)$$

$$a_3 = x_3$$

Hence, the mapping P_W is defined by $P_W(x) = (1/2)(x_1 + x_2)\begin{bmatrix} 1 \\ 1 \\ 0 \end{bmatrix}$. In other words,

P_W is the mapping from $\mathbf{R}^3$ to $\mathbf{R}^3$ whose matrix with respect to the standard basis is $\begin{bmatrix} 1/2 & 1/2 & 0 \\ 1/2 & 1/2 & 0 \\ 0 & 0 & 0 \end{bmatrix}$.

The major properties of the mappings P_W are given in the following proposition.

(4.4.5) Proposition

a) P_W is a linear mapping.

b) $\text{Im}(P_W) = W$, and if $\mathbf{w} \in W$, then $P_W(\mathbf{w}) = \mathbf{w}$.

c) $\text{Ker}(P_W) = W^{\perp}$.

Proof:

a) If $\mathbf{x}, \mathbf{y} \in \mathbf{R}^n$, then we have unique decompositions $\mathbf{x} = \mathbf{x}_1 + \mathbf{x}_2$, and $\mathbf{y} = \mathbf{y}_1 + \mathbf{y}_2$, where $\mathbf{x}_1$ and $\mathbf{y}_1 \in W$, and $\mathbf{x}_2$ and $\mathbf{y}_2 \in W^{\perp}$. Thus, for all $c \in \mathbf{R}$, $P_W(c\mathbf{x} + \mathbf{y}) = P_W((c\mathbf{x}_1 + \mathbf{y}_1) + (c\mathbf{x}_2 + \mathbf{y}_2))$. By the uniqueness of the vectors in the decomposition, this implies $P_W(c\mathbf{x} + \mathbf{y}) = c\mathbf{x}_1 + \mathbf{y}_1 = cP_W(\mathbf{x}) + P_W(\mathbf{y})$. Hence, P_W is linear.

b) Since every vector in $\text{Im}(P_W)$ is contained in W, it is clear that $\text{Im}(P_W) \subseteq W$. On the other hand, if we compute $P_W(\mathbf{w})$, where $\mathbf{w}$ is any vector in W, the decomposition given by Proposition (4.4.3d) in this case is $\mathbf{w} = \mathbf{w} + \mathbf{0}$, and we find that $P_W(\mathbf{w}) = \mathbf{w}$. This shows that $\text{Im}(P_W) = W$.

c) If $\mathbf{x} \in$ and $P_W(\mathbf{x}) = \mathbf{0}$, then since the decomposition of $\mathbf{x}$ given by Proposition (4.4.3d) is unique, it follows that $\mathbf{x} \in W^{\perp}$. Conversely, if $\mathbf{x} \in W^{\perp}$, it is clear that $P_W(\mathbf{x}) = \mathbf{0}$. Hence, $\text{Ker}(P_W) = W^{\perp}$. ■

Let $\{\mathbf{w}_1, \ldots, \mathbf{w}_k\}$ be an orthonormal basis for W. Then each vector in W and the orthogonal projection P_W may be written in a particularly simple form using these vectors.

(4.4.6) Proposition. Let $\{\mathbf{w}_1, \ldots, \mathbf{w}_k\}$ be an orthonormal basis for the subspace $W \subseteq \mathbf{R}^n$.

a) For each $\mathbf{w} \in W$, we have

$$\mathbf{w} = \sum_{i=1}^{k} \langle \mathbf{w}, \mathbf{w}_i \rangle \mathbf{w}_i.$$

b) For all $\mathbf{x} \in \mathbf{R}^n$, we have

$$P_W(\mathbf{x}) = \sum_{i=1}^{k} \langle \mathbf{x}, \mathbf{w}_i \rangle \mathbf{w}_i.$$

Proof: In each case, the real meaning of the statement is that we can use the inner product to compute the scalars needed to express the relevant vector in W ($\mathbf{w}$ in part a, and $P_W(\mathbf{x})$ in part b) as a linear combination of the basis vectors $\mathbf{w}_i$.

a) Since $\mathbf{w} \in W = \text{Span}\{\mathbf{w}_1, \ldots, \mathbf{w}_k\}$, we have $\mathbf{w} = \sum_{i=1}^{k} a_i\mathbf{w}_i$ for some scalars a_i. By the properties of the standard inner product, we have

$$\begin{aligned}\langle \mathbf{w},\mathbf{w}_j\rangle &= \langle \sum_{i=1}^{k} a_i\mathbf{w}_i, \mathbf{w}_j\rangle \\ &= \sum_{i=1}^{k} a_i \langle \mathbf{w}_i,\mathbf{w}_j\rangle\end{aligned}$$

However, since this is an orthonormal basis for W, we have

$$\langle \mathbf{w}_i,\mathbf{w}_j\rangle = \begin{cases} 1 \text{ if } i = j \\ 0 \text{ if } i \neq j \end{cases}$$

Hence, $\langle \mathbf{w},\mathbf{w}_j\rangle = a_j$ for all j, and our linear combination for $\mathbf{w}$ is

$$\mathbf{w} = \sum_{i=1}^{k} \langle \mathbf{w},\mathbf{w}_i\rangle\mathbf{w}_i$$

as claimed.

b) For each $\mathbf{x} \in V$ we have a unique decomposition $\mathbf{x} = \mathbf{w} + \mathbf{w}'$, where $\mathbf{w} \in W$ and $\mathbf{w}' \in W^{\perp}$. Therefore, for each i, we have $\langle \mathbf{x},\mathbf{w}_i\rangle = \langle \mathbf{w} + \mathbf{w}', \mathbf{w}_i\rangle = \langle \mathbf{w},\mathbf{w}_i\rangle + \langle \mathbf{w}',\mathbf{w}_i\rangle = \langle \mathbf{w},\mathbf{w}_i\rangle$, since $\langle \mathbf{w}', \mathbf{w}_i\rangle = 0$. Thus

$$\begin{aligned}\sum_{i=1}^{k} \langle \mathbf{x},\mathbf{w}_i\rangle\mathbf{w}_i &= \sum_{i=1}^{k} \langle \mathbf{w},\mathbf{w}_i\rangle\mathbf{w}_i \\ &= \mathbf{w} \text{ (by part a)} \\ &= P_W(\mathbf{x}) \text{ (by definition).} \quad \blacksquare\end{aligned}$$

(4.4.7) Examples

a) To begin, let us repeat the computation done in Example (4.4.4), using the new technique given by Proposition (4.4.6). We have $W = \text{Span}\{(1,1,0)\}$ in $\mathbf{R}^3$, so on orthonormal basis of W is the set containing the one vector $\mathbf{v} = (1/\sqrt{2})(1,1,0)$. By Proposition (4.4.6), given $\mathbf{x} = (x_1,x_2,x_3) \in \mathbf{R}^3$, we have

$$\begin{aligned}P_W(\mathbf{x}) &= \langle \mathbf{x},\mathbf{v}\rangle\mathbf{v} \\ &= (1/\sqrt{2})(x_1 + x_2)\cdot(1/\sqrt{2}, 1/\sqrt{2}, 0) \\ &= ((1/2)(x_1 + x_2), (1/2)(x_1 + x_2), 0)\end{aligned}$$

as we saw before. We remark, however, that this method is certainly easier.

b) For spaces of higher dimension, the idea is the same. For instance, let W be the $x_1 - x_3$ plane in $\mathbf{R}^3$. The set $\{\mathbf{e}_1, \mathbf{e}_3\}$ is an orthonormal basis for W. By the proposition, for any vector $\mathbf{x} = (x_1, x_2, x_3) \in \mathbf{R}^3$, we have

$$\begin{aligned} P_W(\mathbf{x}) &= \langle \mathbf{x}, \mathbf{e}_1 \rangle \mathbf{e}_1 + \langle \mathbf{x}, \mathbf{e}_3 \rangle \mathbf{e}_3 \\ &= (x_1, 0, x_3) \end{aligned}$$

P_W is the linear mapping whose matrix with respect to the standard basis is

$$\begin{bmatrix} 1 & 0 & 0 \\ 0 & 0 & 0 \\ 0 & 0 & 1 \end{bmatrix}$$

From examples like these, it should be fairly clear that orthonormal bases are, in a sense, the best bases to work with—at least for questions involving the inner product. We next address the question whether there are always orthonormal bases for subspaces of $\mathbf{R}^n$. To see that the answer is, in fact, yes, we consider the following procedure. Suppose we are given vectors $\{\mathbf{u}_1, \ldots, \mathbf{u}_k\}$ that are linearly independent, but not necessarily orthogonal, and we want to construct an orthogonal set of vectors $\{\mathbf{v}_1, \ldots, \mathbf{v}_k\}$ with the property that $\text{Span}(\{\mathbf{u}_1, \ldots, \mathbf{u}_k\}) = \text{Span}(\{\mathbf{v}_1, \ldots, \mathbf{v}_k\})$. Our strategy will be to produce the vectors $\mathbf{v}_i$ one at a time. We can start by defining $\mathbf{v}_1 = \mathbf{u}_1$. Then, to produce $\mathbf{v}_2$, we note that if $W_1 = \text{Span}(\{\mathbf{u}_1\})$, then by the definition of P_{W_1} the vector $\mathbf{u}_2 - P_{W_1}(\mathbf{u}_2)$ is orthogonal to W_1. We define $\mathbf{v}_2 = \mathbf{u}_2 - P_{W_1}(\mathbf{u}_2)$. For a picture of the situation at this stage of the process, see Figure 4.5.

At this point, we have an orthogonal set $\{\mathbf{v}_1, \mathbf{v}_2\}$, and it is easy to see that $\text{Span}(\{\mathbf{v}_1, \mathbf{v}_2\}) = \text{Span}(\{\mathbf{u}_1, \mathbf{u}_2\})$. Now, we continue. Let $W_2 = \text{Span}\{\mathbf{v}_1, \mathbf{v}_2\}$. Again, by the definition of the orthogonal projection mapping, the vector $\mathbf{u}_3 - P_{W_2}(\mathbf{u}_3)$ is orthogonal to W_2. Hence, we define $\mathbf{v}_3 = \mathbf{u}_3 - P_{W_2}(\mathbf{u}_3)$. (See Figure 4.6.)

Hence we have an orthogonal set $\{\mathbf{v}_1, \mathbf{v}_2, \mathbf{v}_3\}$, and by the way we produced the vector $\mathbf{v}_3$, $\text{Span}(\{\mathbf{u}_1, \mathbf{u}_2, \mathbf{u}_3\}) = \text{Span}(\{\mathbf{v}_1, \mathbf{v}_2, \mathbf{v}_3\})$.

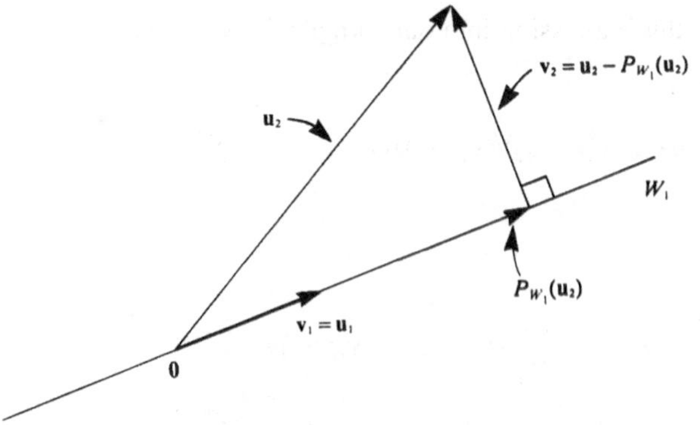

Figure 4.5

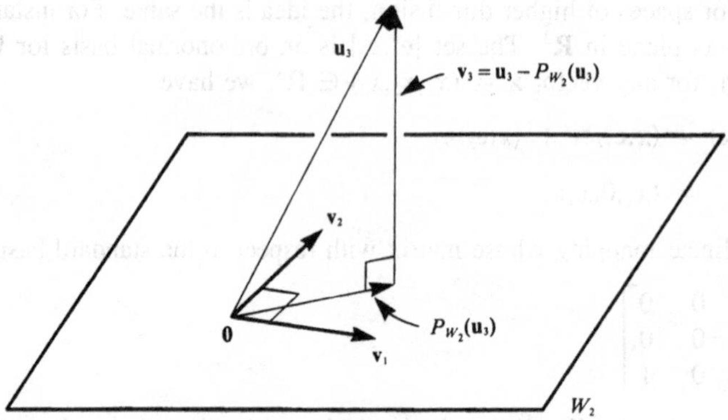

Figure 4.6

If we continue in this fashion, letting $W_i = \text{Span}(\{\mathbf{v}_1, \ldots, \mathbf{v}_i\})$ and defining $\mathbf{v}_{i+1} = \mathbf{u}_{i+1} - P_{W_i}(\mathbf{u}_{i+1})$ at each step, we obtain an orthogonal set of vectors $\{\mathbf{v}_1, \ldots, \mathbf{v}_k\}$ with the same span as the set of vectors with which we started.

If desired we can also modify the process slightly so that the vectors $\mathbf{v}_i$ that we produce are unit vectors. The idea is simple—just divide each $\mathbf{v}_i$ by its norm. If we do this, the modified set $\{\mathbf{v}_1, \ldots, \mathbf{v}_k\}$ will be an orthonormal set.

(4.4.8) The process we have just outlined is called the **Gram-Schmidt orthogonalization process.** Explicit formulas for the vectors $\mathbf{v}_i$ may be obtained by using the observation made in Proposition (4.4.6). By induction, suppose that we have already constructed the vectors $\{\mathbf{v}_1, \ldots, \mathbf{v}_k\}$. Note that the vectors $\{\mathbf{v}_1/\|\mathbf{v}_1\|, \ldots, \mathbf{v}_j/\|\mathbf{v}_j\|\}$ form an orthonormal basis for the subspace W_j. Then we have

$$P_{W_j}(\mathbf{v}) = \sum_{i=1}^{j} (\langle \mathbf{v}_i, \mathbf{v}\rangle / \langle \mathbf{v}_i, \mathbf{v}_i\rangle)\mathbf{v}_i$$

Substituting this expression into our formula for $\mathbf{v}_{j+1}$, we obtain

$$\mathbf{v}_1 = \mathbf{u}_1$$

$$\mathbf{v}_2 = \mathbf{u}_2 - (\langle \mathbf{v}_1, \mathbf{u}_2\rangle / \langle \mathbf{v}_1, \mathbf{v}_1\rangle)\mathbf{v}_1$$

$$\vdots$$

$$\mathbf{v}_{j+1} = \mathbf{u}_{j+1} - \sum_{i=1}^{j} (\langle \mathbf{v}_i, \mathbf{u}_{j+1}\rangle / \langle \mathbf{v}_i, \mathbf{v}_i\rangle)\mathbf{v}_i$$

We conclude by noting that if we apply this Gram-Schmidt process to a basis of a subspace W of $\mathbf{R}^n$, we obtain an orthonormal basis of W; hence Theorem (4.4.9).

(4.4.9) Theorem. Let W be a subspace of $\mathbf{R}^n$. Then there exists an orthonormal basis of W.

(4.4.10) Example. To illustrate how Gram-Schmidt may be used to produce orthonormal bases for subspaces, we consider the subspace W of $\mathbf{R}^5$ defined as the set of solutions of the system of homogeneous equations

$$\begin{aligned} x_1 + 2x_2 + + 2x_4 + 4x_5 &= 0 \\ -x_1 + x_3 &= 0 \end{aligned}$$

When we reduce the system to echelon form and apply the method of Section 1.6 we find that x_3, x_4, and x_5 are the free variables and $\{(1, -1/2, 1, 0, 0), (0, -1, 0, 1, 0), (0, -2, 0, 0, 1)\}$ is the corresponding (nonorthogonal) basis of W.

Now we apply Gram-Schmidt. We begin by letting $\mathbf{v}_1 = (1, -1/2, 1, 0, 0)$. Then the next vector in the orthogonalized basis may be computed by the formulas of (4.4.8):

$$\begin{aligned} \mathbf{v}_2 &= \mathbf{u}_2 - (\langle \mathbf{v}_1, \mathbf{v}_2\rangle / \langle \mathbf{v}_1, \mathbf{v}_1\rangle)\mathbf{v}_1 \\ &= (0, -1, 0, 1, 0) - (1/2)/(9/4)(1, -1/2, 1, 0, 0) \\ &= (-2/9, -8/9, -2/9, 1, 0) \end{aligned}$$

(The reader should check that, in fact, $\langle \mathbf{v}_1, \mathbf{v}_2\rangle = 0$.) Finally, the remaining vector $\mathbf{v}_3$ may be computed in a similar fashion:

$$\begin{aligned} \mathbf{v}_3 &= \mathbf{u}_3 - (\langle \mathbf{v}_1, \mathbf{u}_3\rangle / \langle \mathbf{v}_1, \mathbf{v}_1\rangle)\mathbf{v}_1 - (\langle \mathbf{v}_2, \mathbf{u}_3\rangle / \langle \mathbf{v}_2, \mathbf{v}_2\rangle)\mathbf{v}_2 \\ &= (0, -2, 0, 0, 1) - (1/(9/4))(1, -1/2, 1, 0, 0) \\ &\quad - ((16/9)/(153/81))(-2/9, -8/9, -2/9, 1, 0) \\ &= (-4/17, -16/17, -4/17, -16/17, 1) \end{aligned}$$

The set $\{\mathbf{v}_1, \mathbf{v}_2, \mathbf{v}_3\}$ is orthogonal. The corresponding orthonormal set is $\{\mathbf{v}_1 / \| \mathbf{v}_1 \|, \mathbf{v}_2 / \| \mathbf{v}_2 \|, \mathbf{v}_3 / \| \mathbf{v}_3 \|\}$. We have

$$\begin{aligned} \mathbf{v}_1 / \| \mathbf{v}_1 \| &= (2/3, -1/3, 2/3, 0, 0) \\ \mathbf{v}_2 / \| \mathbf{v}_2 \| &= (1/3\sqrt{17})(-2, -8, -2, 9, 0) \\ \mathbf{v}_3 / \| \mathbf{v}_3 \| &= (1/7\sqrt{17})(-4, -16, -4, -16, 17) \end{aligned}$$

As you might guess from this example, computations with the Gram-Schmidt process may become rather messy in practice. It is a simple matter to program a computer to do this calculation, however.

EXERCISES

1. a) Let $W \subset \mathbf{R}^4$ be the subspace $W = \text{Span}\{(0, 1, 1, -1), (3, 1, 4, 2)\}$. Find a basis for the orthogonal complement, $W^\perp$.
   b) Same question for $W = \text{Span}\{(1, -1, 3, 1), (2, 1, 1, 0), (0, 3, -5, -2)\}$.

2. Show that if $W_1 \subset W_2$ are subspaces of $\mathbf{R}^n$, then $W_1^\perp \supset W_2^\perp$.

3. Let W_1 and W_2 be subspaces of $\mathbf{R}^n$.
   a) Show that $(W_1 + W_2)^\perp = W_1^\perp \cap W_2^\perp$.
   b) Show that $(W_1 \cap W_2)^\perp = W_1^\perp + W_2^\perp$.
   c) Show that $(W_1^\perp)^\perp = W_1$.

4. Using Proposition (4.4.6), find the matrix of the orthogonal projection onto the subspace of $\mathbf{R}^4$ spanned by the given orthonormal set.
   a) $\{(1/\sqrt{5})(2,0,0,1)\}$
   b) $\{(1/\sqrt{2})(1,0,1,0), (1/\sqrt{2})(0,1,0,1)\}$
   c) $\{(1/\sqrt{2})\ (1,0,0,1), (0,0,1,0), (0,1,0,0)\}$

5. For each of the following sets of vectors, use the Gram-Schmidt process to find an orthonormal basis of the subspace they span.
   a) $\{(1, 2, 1), (-1, 4, 0)\}$ in $\mathbf{R}^3$.
   b) $\{(1, 1, -1), (2, 1, 3), (1, 4, 1)\}$ in $\mathbf{R}^3$.

6. Find an orthonormal basis of the subspace W in $\mathbf{R}^5$ defined by

$$\begin{aligned} x_1 + 2x_2 - x_3 + \ x_5 &= 0 \\ 3x_1 + x_3 + 3x_4 &= 0 \end{aligned}$$

7. Prove Proposition (4.4.3). [*Hints:* For part b consider $W^\perp$ as the set of solutions of a system of homogeneous equations and look at the rank of the system. For part d, use part c to compute $\dim(W_1 + W_2)$.]

8. What happens if we apply the Gram-Schmidt process to a set of vectors that is already orthonormal? Prove your assertion.

9. a) In Chapter 1, we saw that the set of solutions of a homogeneous linear equation $a_1x_1 + \cdots + a_nx_n = 0$ is a hyperplane in $\mathbf{R}^n$—a subspace of dimension $n - 1$. Let $\mathbf{N} = (a_1, \ldots, a_n) \in \mathbf{R}^n$, and let $W = \text{Span}\{\mathbf{N}\}$. Show that the set of solutions is the set of vectors satisfying $\langle \mathbf{N}, \mathbf{x} \rangle = 0$, which is the same as $W^\perp$. ($\mathbf{N}$ is called a *normal vector* to the hyperplane.)
   b) Show that the set of solutions of the inhomogeneous linear equation $a_1x_1 + \cdots + a_nx_n = b$ is equal to the set of vectors $\mathbf{x}$ satisfying $\langle \mathbf{N}, (\mathbf{x} - \mathbf{p}) \rangle = 0$, where $\mathbf{N}$ is the normal vector from part a and $\mathbf{p}$ is any particular solution vector.

10. a) Let $V = \text{Span}\{\mathbf{v}_1, \mathbf{v}_2\}$ be a plane containing the origin in $\mathbf{R}^3$. Show that the vector $\mathbf{v}_1 \times \mathbf{v}_2$ (cross product—see Exercise 13 of Section 3.2) is a normal vector to V, as in Exercise 9.
    b) Let $V = \text{Span}\{\mathbf{v}_1, \ldots, \mathbf{v}_{n-1}\}$ be a hyperplane containing the origin in $\mathbf{R}^n$ $(n \geq 2)$. For each i, $1 \leq i \leq n$, let $D_i = (-1)^{i+1}$ times the determinant of the $(n - 1) \times (n - 1)$ matrix whose rows are the vectors $\mathbf{v}_i$, but with the ith coordinates deleted. Show that $\mathbf{N} = (D_1, \ldots, D_n)$ is a normal vector to V. (*Hint:* The inner product $\langle \mathbf{N}, \mathbf{v}_i \rangle$ can be expressed as the determinant of a certain $n \times n$ matrix.)

11. Use the result of Exercise 10 to find normal vectors for each of the following hyperplanes.
   a) $\text{Span}\{(1, 2, 1), (0, 2, 2)\}$ in $\mathbf{R}^3$
   b) $\text{Span}\{(2, 2, 1), (-1, 7, 2)\}$ in $\mathbf{R}^3$
   c) $\text{Span}\{(2, 2, 2, 2), (-1, 0, 3, 0), (3, 3, 1, 1)\}$ in $\mathbf{R}^4$

12. Extend the results of Exercise 9 to systems of more than one linear equation.

13. Let $W \subset \mathbf{R}^n$ be a subspace, and let P_W be the orthogonal projection on W.
   a) Show that $P_W^2 = P_W$. Hence, the results of Exercise 17 of Section 4.1 apply.
   b) What is E_1 for P_W?
   c) What is E_0 for P_W?
   d) Show that P_W is diagonalizable.

14. As we did in $\mathbf{R}^2$, in Exercise 4 of Section 2.2, we can also define *reflection mappings* across subspaces in $\mathbf{R}^n$. Given $W \subset \mathbf{R}^n$ and $\mathbf{x} \in \mathbf{R}^n$, define $R_W(\mathbf{x}) = 2P_W(\mathbf{x}) - \mathbf{x}$.
   a) Show that R_W is a linear mapping on $\mathbf{R}^n$.
   b) Show that $R_W^2 = I$. Hence, the results of Exercise 15 of Section 4.1 apply.
   c) What is E_{+1} for R_W?
   d) What is E_{-1} for R_W?
   e) Show that R_W is diagonalizable.

15. This problem gives another geometric interpretation of the orthogonal projection mappings studied in this section. Let $W \subset \mathbf{R}^n$ be a subspace, and let $\mathbf{x} \in \mathbf{R}^n$. Show that for all $\mathbf{y} \in W$, $\| P_W(\mathbf{x}) - \mathbf{x} \| \leq \| \mathbf{y} - \mathbf{x} \|$, and if $\mathbf{y} \neq P_W(\mathbf{x})$, the inequality is strict. In other words, $P_W(\mathbf{x})$ is the *closest* vector to $\mathbf{x}$, which lies in the subspace W.

16. Show that an orthonormal basis for a subspace of $\mathbf{R}^n$ may always be extended to an *orthonormal* basis of $\mathbf{R}^n$.

17. Let $S = \{\mathbf{x}_1, \ldots, \mathbf{x}_k\}$ be an orthonormal set in $\mathbf{R}^n$, and let $W = \text{Span}(S)$.
   a) Show that if $\mathbf{w} \in W$, then $\| \mathbf{w} \|^2 = \sum_{i=1}^{k} (\langle \mathbf{w}, \mathbf{x}_i \rangle)^2$
   b) Show that if $\mathbf{x} \in \mathbf{R}^n$ is any vector, then $\| P_W(\mathbf{x}) \|^2 = \sum_{i=1}^{k} (\langle \mathbf{x}, \mathbf{x}_i \rangle)^2$.

18. Let $\{\mathbf{x}_1, \ldots, \mathbf{x}_k\}$ be an orthonormal set in $\mathbf{R}^n$.
   a) Prove *Bessel's inequality*: For all $\mathbf{x} \in \mathbf{R}^n$

$$\sum_{i=1}^{k} (\langle \mathbf{x}, \mathbf{x}_i \rangle)^2 \leq \| \mathbf{x} \|^2$$

   b) When is Bessel's inequality an equality? Prove your assertion.

19. a) Show that for every linear mapping $l: \mathbf{R}^n \to \mathbf{R}$ there is a unique vector $\mathbf{y} \in \mathbf{R}^n$ such that $l(\mathbf{x}) = \langle \mathbf{x}, \mathbf{y} \rangle$ for all $\mathbf{x} \in \mathbf{R}^n$. (*Hint:* Let $\{\mathbf{x}_1, \ldots, \mathbf{x}_n\}$ be an orthonormal basis of $\mathbf{R}^n$ and consider the real numbers $l(\mathbf{x}_i)$.)

b) Show that the mapping that takes l to the corresponding vector $\mathbf{y}$ is itself a linear mapping $D: L(\mathbf{R}^n, \mathbf{R}) \to \mathbf{R}^n$.

c) Conclude that $L(\mathbf{R}^n, \mathbf{R})$ and $\mathbf{R}^n$ are isomorphic vector spaces and that the mapping D from part b is an isomorphism between them. [The vector space $L(\mathbf{R}^n, \mathbf{R})$ is called the *dual space* of $\mathbf{R}^n$. The standard inner product thus sets up an isomorphism between $\mathbf{R}^n$ and its dual space.]

§4.5. SYMMETRIC MATRICES

In Section 4.2 we saw necessary and sufficient conditions for a linear mapping $T: V \to V$ to be diagonalizable. In particular, we found that not every mapping has this property. Nevertheless, there is a particular fairly large and quite important class of mappings all of which are diagonalizable, and which we now begin to study. In what follows, we consider the vector space $V = \mathbf{R}^n$, and we almost always identify linear mappings T with their matrices with respect to the standard basis. (Indeed, we frequently deal with the matrices themselves, without reference to linear mappings, because the results we prove will be valid for matrices produced in other ways as well.)

To introduce the class of matrices we will consider, we recall the following definition.

(4.5.1) Definition. A square matrix A is said to be *symmetric* if $A = A^t$, where A^t denotes the transpose of A as defined in Exercise 12 of Section 2.2.

For example, the 3×3 matrix

$$A = \begin{bmatrix} 4 & 1 & -7 \\ 1 & 0 & 2 \\ -7 & 2 & 3 \end{bmatrix}$$

is a symmetric matrix. By the definition of the transpose of a matrix, A is symmetric if and only if $a_{ij} = a_{ji}$ for all pairs i,j.

Our major motivation for considering symmetric matrices is the fact that these matrices arise frequently in many branches of mathematics. For example, recall from analytic geometry that every conic section in the plane (a circle, ellipse, parabola, or hyperbola) is defined by an equation of the form

$$Ax_1^2 + 2Bx_1x_2 + Cx_2^2 + Dx_1 + Ex_2 + F = 0$$

If we take only the terms of total degree 2 in x_1 and x_2, we see that

$$Ax_1^2 + 2Bx_1x_2 + Cx_2^2 = [x_1 \; x_2] \begin{bmatrix} A & B \\ B & C \end{bmatrix} \begin{bmatrix} x_1 \\ x_2 \end{bmatrix}$$

where the right-hand side is a matrix product. Note that the 2×2 matrix that appears here, containing the coefficients from the equation of the conic section, is symmetric. Geometrically, we know that some linear terms $Dx_1 + Ex_2 + F$ appear in the equation of the conic if we translate its center away from the origin, while the quadratic terms are the ones that determine the "shape" of the curve. Thus, the shape of the curve will depend on the properties of the symmetric matrix

$$\begin{bmatrix} A & B \\ B & C \end{bmatrix}.$$

For another example, if $f: \mathbf{R}^2 \to \mathbf{R}$ is a twice-differentiable function, we can construct the so-called "Hessian matrix" of f:

$$H_f = \begin{bmatrix} \partial^2 f/\partial x_1^2 & \partial^2 f/\partial x_1 \partial x_2 \\ \partial^2 f/\partial x_2 \partial x_1 & \partial^2 f/\partial x_2^2 \end{bmatrix}$$

using the second-order partial derivatives of f. In the calculus of functions of several variables it is shown that if the entries of the Hessian matrix are continuous functions of x_1 and x_2, then $\partial^2 f/\partial x_1\, \partial x_2 = \partial^2 f/\partial x_2\, \partial x_1$. If this is the case, then H_f is a symmetric matrix (of functions). If we substitute values for (x_1, x_2), we obtain symmetric matrices of scalars. The Hessian matrix is used at points where $\partial f/\partial x_1 = \partial f/\partial x_2 = 0$ to determine if f has a local maximum, a local minimum, or neither at that point—a second derivative test for functions of two variables.

There is also a strong connection between symmetric matrices and the geometry of $\mathbf{R}^n$ (the standard inner product), which is revealed by Proposition (4.5.2a).

(4.5.2a) Proposition. Let $A \in M_{n \times n}(\mathbf{R})$.

1) For all $\mathbf{x}, \mathbf{y} \in \mathbf{R}^n$, $\langle A\mathbf{x}, \mathbf{y}\rangle = \langle \mathbf{x}, A^t\mathbf{y}\rangle$.

2) A is symmetric if and only if $\langle A\mathbf{x}, \mathbf{y}\rangle = \langle \mathbf{x}, A\mathbf{y}\rangle$ for all vectors $\mathbf{x}$, $\mathbf{y} \in \mathbf{R}^n$.

Proof:

1) Since we know that the inner product is linear in each variable, it suffices to check that $\langle A\mathbf{e}_i, \mathbf{e}_j\rangle = \langle \mathbf{e}_i, A^t\mathbf{e}_j\rangle$ for all pairs of standard basis vectors $\mathbf{e}_i$ and $\mathbf{e}_j$. However, if $A = (a_{ij})$, then since the $\mathbf{e}_k$'s form an orthonormal basis of $\mathbf{R}^n$

$$\langle A\mathbf{e}_i, \mathbf{e}_j\rangle = a_{ji}$$

whereas

$$\langle \mathbf{e}_i, A^t\mathbf{e}_j\rangle = a_{ji}$$

for all i and j. It follows that $\langle A\mathbf{e}_i, \mathbf{e}_j\rangle = \langle \mathbf{e}_i, A^t\mathbf{e}_j\rangle$ for all i,j.

2) By the same argument, $\langle A\mathbf{e}_i, \mathbf{e}_j\rangle = \langle \mathbf{e}_i, A\mathbf{e}_j\rangle$ for all i and j if and only if $a_{ij} = a_{ji}$ for all i,j, that is, if and only if A is symmetric. ■

The proof of the proposition actually shows a bit more, and the following corollary will be useful to us as well.

(4.5.2b) Corollary. Let V be any subspace of $\mathbf{R}^n$, let $T: V \to V$ be any linear mapping, and let $\alpha = \{\mathbf{x}_1, \ldots, \mathbf{x}_k\}$ be any orthonormal basis of V. Then $[T]_\alpha^\alpha$ is a symmetric matrix if and only if $\langle T(\mathbf{x}), \mathbf{y}\rangle = \langle \mathbf{x}, T(\mathbf{y})\rangle$ for all vectors $\mathbf{x}, \mathbf{y} \in V$.

Proof: As in the proof of the proposition, if we let $A = [T]_\alpha^\alpha$, and $\mathbf{f}_j$ be the coordinate vectors of the $\mathbf{x}_j$ with respect to the basis α (these are just the standard basis vectors in $\mathbf{R}^k$), then it is enough to show that A is symmetric if and only if $\langle A\mathbf{f}_i, \mathbf{f}_j\rangle = \langle \mathbf{f}_i, A\mathbf{f}_j\rangle$ for all pairs of vectors in the basis of V. The proof of the proposition now goes through virtually unchanged. The key property of the standard basis that made everything work, so to speak, was that it is orthonormal with respect to the standard inner product, and that is also true of our basis for V by assumption. ■

The corollary explains the following terminology, which will be used in the next section.

(4.5.3) Definition. Let V be a subspace of $\mathbf{R}^n$. A linear mapping $T: V \to V$ is said to be *symmetric* if $\langle T(\mathbf{x}), \mathbf{y}\rangle = \langle \mathbf{x}, T(\mathbf{y})\rangle$ for all vectors $\mathbf{x}, \mathbf{y} \in \mathbf{V}$.

(4.5.4) Example. An important class of symmetric mappings is the class of orthogonal projections. Let P_W be the orthogonal projection on the subspace $W \subset \mathbf{R}^n$. As we know from Section 4.4, every vector $\mathbf{x} \in \mathbf{R}^n$ can be written as $\mathbf{x} = \mathbf{x}_1 + \mathbf{x}_2$, where $\mathbf{x}_1 = P_W(\mathbf{x}) \in W$, and $\mathbf{x}_2 \in \text{Ker}(P_W) = W^\perp$. If $\mathbf{y} = \mathbf{y}_1 + \mathbf{y}_2$ is the analogous decomposition of a vector $\mathbf{y} \in \mathbf{R}^n$, then on the one hand, we have

$$\begin{aligned}\langle P_W(\mathbf{x}), \mathbf{y}\rangle &= \langle \mathbf{x}_1, \mathbf{y}_1 + \mathbf{y}_2\rangle \\ &= \langle \mathbf{x}_1, \mathbf{y}_1\rangle + \langle \mathbf{x}_1, \mathbf{y}_2\rangle \\ &= \langle \mathbf{x}_1, \mathbf{y}_1\rangle \qquad (\text{since } \mathbf{y}_2 \in W^\perp)\end{aligned}$$

On the other hand

$$\begin{aligned}\langle \mathbf{x}, P_W(\mathbf{y})\rangle &= \langle \mathbf{x}_1 + \mathbf{x}_2, \mathbf{y}_1\rangle \\ &= \langle \mathbf{x}_1, \mathbf{y}_1\rangle + \langle \mathbf{x}_2, \mathbf{y}_1\rangle \\ &= \langle \mathbf{x}_1, \mathbf{y}_1\rangle\end{aligned}$$

Hence since $\langle P_W(\mathbf{x}), \mathbf{y}\rangle = \langle \mathbf{x}, P_W(\mathbf{y})\rangle$ for all vectors $\mathbf{x}, \mathbf{y} \in \mathbf{R}^n$, P_W is a symmetric mapping. From the corollary, it follows that if α is any orthonormal basis of $\mathbf{R}^n$, $[P_W]_\alpha^\alpha$ is a symmetric matrix.

Indeed, if we choose an orthonormal basis of V and extend it to an orthonormal basis α of $\mathbf{R}^n$ (this is always possible by Exercise 16 of Section 4.4), then the matrix of P_W will be one of the simplest possible matrices:

$$[P_W]_\alpha^\alpha = \left[\begin{array}{ccc|c} 1 & & 0 & \\ & \ddots & & \\ 0 & & 1 & 0 \\ \hline & 0 & & 0 \end{array}\right]$$

where there are $\dim(W)$ 1's in all along the diagonal, and zeros in every other entry. Note that as a by-product of this computation, we see a direct proof that orthogonal projections are *diagonalizable,* a fact that was proved by a different method in the exercises.

Caution: The basis α must be an orthonormal basis chosen in this way for the matrix of an orthogonal projection to have this form. It will not even be true that the matrix of an orthogonal projection is symmetric if an arbitrary basis is used.

We now consider some of the special properties of eigenvalues and eigenvectors of symmetric matrices. We begin with an example.

(4.5.5) Example. Consider the symmetric 2×2 matrix $A = \begin{bmatrix} 4 & 1 \\ 1 & 2 \end{bmatrix}$. The eigenvalues of A are the roots of the quadratic equation $\lambda^2 - 6\lambda + 7 = 0$, so $\lambda = 3 \pm \sqrt{2}$ by the quadratic formula. (Note that the roots of the characteristic polynomial are real in this case, although that would not be true for every quadratic polynomial.) Now, for $\lambda = 3 + \sqrt{2}$ we find that $\text{Ker}(A - (3 + \sqrt{2})I) = \text{Span}\left\{\begin{bmatrix} 1 \\ -1 + \sqrt{2} \end{bmatrix}\right\}$. For $\lambda = 3 - \sqrt{2}$, we find $\text{Ker}(A - (3 - \sqrt{2})I) = \text{Span}\left\{\begin{bmatrix} 1 \\ -1 - \sqrt{2} \end{bmatrix}\right\}$. You should check this by solving the appropriate systems of equations. If we compute the inner product of the two eigenvectors, we find that

$$\left\langle \begin{bmatrix} 1 \\ -1 + \sqrt{2} \end{bmatrix}, \begin{bmatrix} 1 \\ -1 - \sqrt{2} \end{bmatrix} \right\rangle = 1 \cdot 1 + (-1 + \sqrt{2}) \cdot (-1 - \sqrt{2}) = 0$$

It turns out that the eigenvectors of the matrix A are orthogonal.

In fact, it is true that for any symmetric matrix: (1) all the roots of the characteristic polynomial are real and (2) eigenvectors corresponding to distinct eigenvalues are orthogonal. For 2×2 matrices both of these assertions may be proved quite explicitly by using the quadratic formula to find the eigenvalues, then solving for the corresponding eigenvectors and checking that the desired orthogonality properties hold. However, for $n \times n$ matrices, to see the underlying reason why all the roots of the characteristic polynomial of a symmetric matrix are real, we must introduce matrices and polynomials over the field of complex numbers. Because this requires an extensive discussion of several new topics, we postpone this until Chapter 5. Here, we will give only an *ad hoc* argument to show that the result is true.

(4.5.6) Theorem. Let $A \in M_{n \times n}(\mathbf{R})$ be a symmetric matrix. Then all the roots of the characteristic polynomial of A are real. In other words, the characteristic polynomial has n roots in $\mathbf{R}$ (counted with multiplicities).

Proof: We first appeal to the so-called fundamental theorem of algebra, which says that every polynomial of degree n with real (or complex) coefficients has n roots (not necessarily distinct) in the set of complex numbers. Hence, if λ is any root of the characteristic polynomial of our symmetric matrix A, then we can write $\lambda = \alpha + i\beta$, where $\alpha, \beta \in \mathbf{R}$ and $i^2 = -1$.

Consider the real $n \times n$ matrix $(A - (\alpha - i\beta)I)(A - (\alpha + i\beta)I) = A^2 - 2\alpha A + (\alpha^2 + \beta^2)I = (A - \alpha I)^2 + \beta^2 I$. Since $\lambda = \alpha + i\beta$ is a root of the characteristic polynomial of A, the (complex) matrix $A - (\alpha + i\beta)I$ is not invertible. Since the product of two square matrices is not invertible if one of the factors is not, it follows that the real matrix $(A - \alpha I)^2 + \beta^2 I$ is not invertible either.

Let $\mathbf{x} \in \mathbf{R}^n$ be any nonzero vector satisfying $((A - \alpha I)^2 + \beta^2 I)\mathbf{x} = \mathbf{0}$. Then we have

$$\begin{aligned} 0 &= \langle (A - \alpha I)^2\mathbf{x} + \beta^2\mathbf{x}, \mathbf{x}\rangle \\ &= \langle (A - \alpha I)^2\mathbf{x}, \mathbf{x}\rangle + \beta^2 \langle \mathbf{x}, \mathbf{x}\rangle \qquad \text{by Proposition (4.3.3a)} \\ &= \langle (A - \alpha I)\mathbf{x}, (A - \alpha I)\mathbf{x}\rangle + \beta^2\langle \mathbf{x}, \mathbf{x}\rangle \end{aligned}$$

since A, and hence $A - \alpha I$ as well, is symmetric.

From this last equation we see that $(A - \alpha I)\mathbf{x} = \mathbf{0}$ and $\beta = 0$, since otherwise one or both terms on the right-hand side would be strictly positive. Hence, $\lambda = \alpha$ is real (and $\mathbf{x}$ is an eigenvector of A with eigenvalue α). ■

Showing that eigenvectors corresponding to distinct eigenvalues of a symmetric matrix are orthogonal is somewhat easier. Note first that once we know that all the eigenvalues of such a matrix are real, it follows immediately that we can find corresponding eigenvectors in $\mathbf{R}^n$ for each of them.

(4.5.7) Theorem. Let $A \in M_{n\times n}(\mathbf{R})$ be a symmetric matrix, let $\mathbf{x}_1$ be an eigenvector of A with eigenvalue λ_1, and let $\mathbf{x}_2$ be an eigenvector of A with eigenvalue λ_2, where $\lambda_1 \neq \lambda_2$. Then $\mathbf{x}_1$ and $\mathbf{x}_2$ are orthogonal vectors in $\mathbf{R}^n$.

Proof: Consider the inner product $\langle A\mathbf{x}_1, \mathbf{x}_2\rangle$. First, we can compute this by using the fact that $\mathbf{x}_1$ is an eigenvector

$$\langle A\mathbf{x}_1, \mathbf{x}_2\rangle = \langle \lambda_1\mathbf{x}_1, \mathbf{x}_2\rangle = \lambda_1\langle \mathbf{x}_1, \mathbf{x}_2\rangle$$

Now we can also use the property of symmetric matrices proved in Proposition (4.5.2a) and the fact that $\mathbf{x}_2$ is also an eigenvector of A to compute this a second way:

$$\langle A\mathbf{x}_1, \mathbf{x}_2\rangle = \langle \mathbf{x}_1, A\mathbf{x}_2\rangle = \langle \mathbf{x}_1, \lambda_2\mathbf{x}_2\rangle = \lambda_2\langle \mathbf{x}_1, \mathbf{x}_2\rangle$$

Hence, subtracting the final results of the two computations, we have

$$(\lambda_1 - \lambda_2)\langle \mathbf{x}_1, \mathbf{x}_2\rangle = 0$$

Since we assumed that $\lambda_1 \neq \lambda_2$, this implies that $\langle \mathbf{x}_1, \mathbf{x}_2\rangle = 0$, so the two eigenvectors are orthogonal by definition. ■

EXERCISES

1. a) Show that if A and B are symmetric matrices, then $A + B$ is symmetric.
   b) Show that if A is any $n \times n$ matrix, then $A + A^t$ and AA^t are symmetric. (*Hint:* For AA^t, see Section 2.5, Exercise 19.)

2. Find all the eigenvalues and a basis of each eigenspace for each of the following symmetric matrices. In each case show that the eigenvectors corresponding to distinct eigenvalues are orthogonal.

a) $A = \begin{bmatrix} 1 & -3 \\ -3 & 4 \end{bmatrix}$ b) $A = \begin{bmatrix} 1 & 0 & 1 \\ 0 & 1 & 0 \\ 1 & 0 & 0 \end{bmatrix}$ c) $A = \begin{bmatrix} 2 & 1 & 0 \\ 1 & 2 & 1 \\ 0 & 1 & 2 \end{bmatrix}$

3. Consider the basis $\beta = \{(1, 1, 0), (1, 0, -1), (2, 1, 0)\}$ for $\mathbf{R}^3$. Which of the following matrices $A = [T]_\beta^\beta$ define symmetric mappings of $\mathbf{R}^3$?

a) $\begin{bmatrix} 1 & 1 & 0 \\ 1 & 1 & 0 \\ 0 & 0 & 1 \end{bmatrix}$ b) $\begin{bmatrix} 2 & 0 & 0 \\ 0 & 2 & 0 \\ 0 & 0 & 2 \end{bmatrix}$ c) $\begin{bmatrix} -1 & 1 & 2 \\ 1 & 4 & 0 \\ 2 & 0 & 1 \end{bmatrix}$.

d) Let T be the symmetric mapping whose matrix with respect to the standard basis of $\mathbf{R}^3$ is $A = \begin{bmatrix} 2 & 3 & -1 \\ 3 & 0 & 1 \\ -1 & 1 & 1 \end{bmatrix}$. Compute $[T]_\beta^\beta$. Is your result symmetric? Why or why not?

4. Show that if $A \in M_{n \times n}(\mathbf{R})$ is symmetric and $\lambda \neq \mu$ are eigenvalues of A, then $E_\mu \subset (E_\lambda)^\perp$.

5. Let $A = \begin{bmatrix} a & b \\ b & c \end{bmatrix}$ be a general 2×2 symmetric matrix. Show directly that the eigenvalues of A are real and that there is always an orthonormal basis of $\mathbf{R}^2$ consisting of eigenvectors of A. (Do not appeal to the general theorems of this section.) To what diagonal matrix is A similar?

6. (Continuation of Exercise 5.) To what diagonal matrix is the special symmetric matrix $\begin{bmatrix} a & b \\ b & a \end{bmatrix}$ similar?

7. A matrix $A \in M_{n \times n}(\mathbf{R})$ is said to be *normal* if $AA^t = A^tA$.
   a) Show that if A is symmetric, then A is normal. (Thus, in a sense, normal matrices are generalizations of symmetric ones. We now see that they have some of the same properties with regard to eigenvalues and eigenvectors as well.)
   b) Find an example of a 2×2 normal matrix that is not symmetric.
   c) Show that if A is normal, then $\|A\mathbf{x}\| = \|A^t\mathbf{x}\|$ for all $\mathbf{x} \in \mathbf{R}^n$.
   d) Show that if A is normal, then $A - cI$ is normal for all $c \in \mathbf{R}$.
   e) Show that if $\lambda \in \mathbf{R}$ is an eigenvalue of a normal matrix, then λ is also an eigenvalue of A^t. (*Caution:* The roots of the characteristic polynomial of a normal matrix need not all be real, however.)
   f) Show that if λ_1 and λ_2 are distinct real eigenvalues of a normal matrix, then the corresponding eigenvectors are orthogonal.

In any vector space with an inner product (see the exercises accompanying Section 4.3), we may define symmetric mappings as in $\mathbf{R}^n$: A linear mapping $T: V \rightarrow V$ is symmetric if $\langle T(\mathbf{x}), \mathbf{y}\rangle = \langle \mathbf{x}, T(\mathbf{y})\rangle$ for all $\mathbf{x}, \mathbf{y} \in V$.

8. a) Show that the linear mapping $T: \mathbf{R}^2 \to \mathbf{R}^2$ defined by the matrix $A = \begin{bmatrix} 1 & 1 \\ 2 & 1 \end{bmatrix}$ is symmetric with respect to the nonstandard inner product on $\mathbf{R}^2$ defined in Section 4.3, Exercise 10a.
   b) Let $W \subset C[0, 1]$ be the subspace $W = \{f \in C[0, 1] \mid f'' \text{ exists}, f'' \in C[0, 1] \text{ and } f(0) = f(1) = 0\}$. Let $T: W \to V$ be the linear mapping $T(f) = f''$. Show that T is symmetric with respect to the inner product on $C[0, 1]$ defined in Section 4.3, Exercise 10b. (*Hint:* Integrate by parts.)
   c) Is the transpose mapping $T: M_{n\times n}(\mathbf{R}) \to M_{n\times n}(\mathbf{R})$ defined by $T(A) = A^t$ symmetric with respect to the inner product on $M_{n\times n}(\mathbf{R})$ defined in Section 4.3 Exercise 10c?

9. Show that if $\lambda_1, \ldots, \lambda_n$ are the eigenvalues of a symmetric matrix $A \in M_{n\times n}(\mathbf{R})$, then $\det(A) = \lambda_1 \cdots \lambda_n$ and $\text{Tr}(A) = \lambda_1 + \cdots + \lambda_n$.

10. a) Consider the standard form of the equation of an ellipse in the plane centered at the origin: $(x_1)^2/a^2 + (x_2)^2/b^2 = 1$. Write this equation in the form

$$[x_1\ x_2] \begin{bmatrix} A & B \\ B & C \end{bmatrix} \begin{bmatrix} x_1 \\ x_2 \end{bmatrix} = 1 \text{ for some symmetric matrix.}$$

   b) Do the same for the standard form of the equation of a hyperbola:

$$(x_1)^2/a^2 - (x_2)^2/b^2 = \pm 1$$

§4.6 THE SPECTRAL THEOREM

In this section we put everything together to prove a very important theorem known as the spectral theorem (in $\mathbf{R}^n$). As is the case with most major theorems, the spectral theorem may be interpreted in several ways. Algebraically, the spectral theorem says that every symmetric linear mapping $T: \mathbf{R}^n \to \mathbf{R}^n$ is diagonalizable (indeed, there is an *orthonormal* basis of $\mathbf{R}^n$ consisting of eigenvectors of T). Geometrically, this statement is equivalent to saying that a symmetric linear mapping can be "decomposed" into a linear combination of orthogonal projections (onto its eigenspaces). We also consider some applications of these results to the theory of conic sections and quadric surfaces.

Before starting, we would like to interject a few words about the name of this theorem. The name "spectral theorem" comes from the word spectrum—the connection being that the set of eigenvalues of a linear mapping is frequently referred to as the *spectrum* of the mapping. The reason for this terminology seems to be as follows.

In the nineteenth century it was discovered that when atoms of a substance (e.g., gases such as hydrogen) are excited, they give off light of certain precisely defined wavelengths (colors). This is what happens, for example, in "neon" signs. An electric current is passed through a glass tube containing a gas. The excited atoms of the gas give off light of the desired color. In fact, the wavelengths (or "spectral lines") of the light emitted by a glowing object are so characteristic of the elements present that they can be used to determine the composition of the object. There was no systematic explanation for this behavior available at the time it was first noticed, however.

Toward the end of the nineteenth century, mathematicians studying analogs of our symmetric linear mappings on function spaces noted a similarity between the discrete spectral lines of atoms and the pattern of distribution of eigenvalues of these linear mappings. Perhaps motivated by this analogy, the German mathematician Hilbert, who proved the first general version of the theorem in question (but in the function space setup), called his result the "spectral theorem."

In the 1920s and 1930s the new physical theory of quantum mechanics was developed to explain phenomena such as the spectral emission lines mentioned before, and in a rather amazing vindication of the earlier mathematicians' intuition, it was found that the wavelengths of the emission spectrum of an atom were indeed related to the eigenvalues of a certain linear operator. This was the Schrödinger operator that appears in the basic differential equation of quantum mechanics. Although we will not be able to deal with such questions here, quantum mechanics is one of the areas of modern physics where the mathematical theory of linear algebra and inner product spaces is used.

We begin with the algebraic version of the spectral theorem. We have already seen several examples of symmetric matrices that turned out to be diagonalizable. See Example (4.1.1) for one such matrix.

(4.6.1) Theorem. Let $T: \mathbf{R}^n \to \mathbf{R}^n$ be a symmetric linear mapping. Then there is an orthonormal basis of $\mathbf{R}^n$ consisting of eigenvectors of T. In particular, T is diagonalizable.

Proof: We will prove this by induction on n, the dimension of the space. The idea of the proof is to "split off" one unit eigenvector $\mathbf{x}_1$ of T, and write $\mathbf{R}^n = W \oplus W^\perp$, where W is the one-dimensional subspace Span$\{\mathbf{x}_1\}$. Then we will show that there is an orthonormal basis of $W^\perp$ consisting of eigenvectors of the restricted mapping $T|_{W^\perp}$. These vectors, together with $\mathbf{x}_1$ will form our orthonormal basis of $\mathbf{R}^n$.

If $n = 1$, then every linear mapping is symmetric and diagonalizable (why?) so there is nothing to prove. Now, assume the theorem has been proved for mappings from $\mathbf{R}^k$ to $\mathbf{R}^k$ and consider $T: \mathbf{R}^{k+1} \to \mathbf{R}^{k+1}$. By Theorem (4.5.6) we know that all the eigenvalues of T are real numbers. Let λ be any one of the eigenvalues, and let $\mathbf{x}_1$ be any unit eigenvector with eigenvalue λ. Let $W = \text{Span}(\{\mathbf{x}_1\})$. Note that $W^\perp$ is a k-dimensional subspace of $\mathbf{R}^{k+1}$, so $W^\perp$ is isomorphic to $\mathbf{R}^k$, and we can try to apply our induction hypothesis to T restricted to $W^\perp$. To do this, we must show two things: first, T restricted to $W^\perp$ takes values in $W^\perp$, and second, the restricted mapping is also symmetric, with respect to the restriction of the standard inner product to $W^\perp$.

To see that T takes vectors in $W^\perp$ to vectors in $W^\perp$, note that if $\mathbf{y} \in W^\perp$, then

$$\begin{aligned}\langle \mathbf{x}_1, T(\mathbf{y})\rangle &= \langle T(\mathbf{x}_1), \mathbf{y}\rangle && \text{since } T \text{ is symmetric}\\ &= \langle \lambda \mathbf{x}_1, \mathbf{y}\rangle && \text{since } \mathbf{x}_1 \text{ is an eigenvector of } T\\ &= 0 && \text{since } \mathbf{y} \in W^\perp\end{aligned}$$

Hence $T(\mathbf{y}) \in W^\perp$, so $T(W^\perp) \subseteq W^\perp$.

To see that the restriction of T to $W^\perp$ is still symmetric, note that if $\mathbf{y}_1$, $\mathbf{y}_2 \in W^\perp$, then $\langle T(\mathbf{y}_1), \mathbf{y}_2\rangle = \langle \mathbf{y}_1, T(\mathbf{y}_2)\rangle$, since this holds more generally for all vectors in $\mathbf{R}^{k+1}$. Hence, by definition the restriction of T to $W^\perp$ is also symmetric.

Hence, by the induction hypothesis, applied to $T\,|\,_{W^\perp}$, there exists an orthonormal basis $\{\mathbf{x}_2, \ldots, \mathbf{x}_{k+1}\}$ of $W^\perp$, consisting of eigenvectors of the restricted mapping. But these vectors are also eigenvectors of T considered as a mapping on the whole space. Furthermore, since $\langle \mathbf{x}_1, \mathbf{x}_j\rangle = 0$ for each j, $2 \leq j \leq k + 1$, the set $\alpha = \{\mathbf{x}_1, \mathbf{x}_2, \ldots, \mathbf{x}_{k+1}\}$ is an orthonormal set, hence an orthonormal basis of $\mathbf{R}^{k+1}$. Each vector in α is an eigenvector of T, as claimed. ■

A practical way to produce such a basis for a given symmetric linear mapping, if that is required, is to use the result of Theorem (4.5.7). Eigenvectors corresponding to distinct eigenvalues of a symmetric mapping are automatically orthogonal, so to find an orthonormal basis of $\mathbf{R}^n$, all we need to do is to find orthonormal bases in each eigenspace of T and to take their *union*. Indeed, it may be shown that if E_λ is any one of the eigenspaces of T, then $E_\lambda^\perp = \sum_{\mu \neq \lambda} E_\mu$ (i.e., the sum of the other eigenspaces of T), so that $\mathbf{R}^n = \sum_\lambda E_\lambda$, the sum taken over all the distinct eigenvalues of T.

(4.6.2) Examples

a) Consider the mapping $T: \mathbf{R}^3 \to \mathbf{R}^3$ whose matrix with respect to the standard basis is $A = \begin{bmatrix} 1 & 1 & 0 \\ 1 & 1 & 0 \\ 0 & 0 & 2 \end{bmatrix}$. Since $A^t = A$ and the standard basis is orthonormal, T is a symmetric mapping. Computing, we find

$$\det(A - \lambda I) = -\lambda(\lambda - 2)^2$$

The eigenvalues of T are $\lambda = 0$ (with multiplicity 1) and $\lambda = 2$ (with multiplicity 2).

For $\lambda = 0$ we have $E_0 = \text{Span}\,\{(-1, 1, 0)\}$, whereas for $\lambda = 2$, $E_2 = \text{Span}\,\{(1, 1, 0), (0, 0, 1)\}$. These vectors are found, as usual, by solving the appropriate systems of equations. In E_0 $\|(-1, 1, 0)\| = \sqrt{2}$, so $\{(1/\sqrt{2})(-1, 1, 0)\}$ is an orthonormal basis of E_0. Similarly, since our two basis vectors for E_2 are already orthogonal, we only need to "normalize" and we see that $\{(1/\sqrt{2})(1, 1, 0), (0, 0, 1)\}$ is an orthonormal basis of E_2.

Hence, $\alpha = \{(1/\sqrt{2})(-1, 1, 0), (1/\sqrt{2})(1, 1, 0), (0, 0, 1)\}$ is an orthonormal basis of $\mathbf{R}^3$. With respect to this basis, $[T]_\alpha^\alpha = \begin{bmatrix} 0 & 0 & 0 \\ 0 & 2 & 0 \\ 0 & 0 & 2 \end{bmatrix}$, a diagonal matrix. As predicted by our theorem, T is diagonalizable.

b) Consider the mapping $T: \mathbf{R}^4 \to \mathbf{R}^4$ whose matrix with respect to the standard basis is

$$A = \begin{bmatrix} 1 & 1 & 1 & 1 \\ 1 & 1 & 1 & 1 \\ 1 & 1 & 1 & 1 \\ 1 & 1 & 1 & 1 \end{bmatrix}$$

The characteristic polynomial of A is $\det(A - \lambda I) = \lambda^3(\lambda - 4)$, so A has one eigenvalue $\lambda = 4$ of multiplicity one, and a second eigenvalue $\lambda = 0$ of multiplicity three. For $\lambda = 4$, we solve the system $(A - 4I)\mathbf{x} = \mathbf{0}$, and we find that $\mathbf{x} = (1,1,1,1)$ is a solution. For $\lambda = 0$, solving the system $(A - 0I)\mathbf{x} = \mathbf{0}$ by reducing to echelon form and substituting for the free variables, we have a basis $\beta = \{(-1,1,0,0), (-1,0,1,0), (-1,0,0,1)\}$ for E_0. The eigenvector $\mathbf{x}$ found previously is orthogonal to each vector in β, as we expect from Theorem (4.5.7). However, β itself is not orthogonal, since our standard method for solving homogeneous systems of equations does not guarantee an orthogonal basis for the space of solutions.

We may, however, apply the Gram-Schmidt procedure (4.4.8) to find an orthogonal basis for E_0, starting from β. We will call the three vectors in β $\mathbf{u}_1$, $\mathbf{u}_2$, and $\mathbf{u}_3$, respectively. Using (4.4.8), we compute:

$$\mathbf{v}_1 = \mathbf{u}_1 = (-1,1,0,0)$$

$$\begin{aligned}\mathbf{v}_2 &= \mathbf{u}_2 - (\langle \mathbf{v}_1,\mathbf{u}_2\rangle/\langle \mathbf{v}_1,\mathbf{v}_1\rangle)\mathbf{v}_1 = (-1,0,1,0) - (1/2)(-1,1,0,0) \\ &= (-1/2,-1/2,1,0)\end{aligned}$$

$$\begin{aligned}\mathbf{v}_3 &= \mathbf{u}_3 - (\langle \mathbf{v}_1,\mathbf{u}_3\rangle/\langle \mathbf{v}_1,\mathbf{v}_1\rangle)\mathbf{v}_1 - (\langle \mathbf{v}_2,\mathbf{u}_3\rangle/\langle \mathbf{v}_2,\mathbf{v}_2\rangle)\mathbf{v}_2 \\ &= (-1,0,0,1) - (1/2)(-1,1,0,0) - (1/3)(-1/2,-1/2,1,0) \\ &= (-1/3,-1/3,-1/3,1).\end{aligned}$$

Then

$$\begin{aligned}\alpha &= \{\mathbf{x}/\|\mathbf{x}\|, \mathbf{u}_1/\|\mathbf{u}_1\|, \mathbf{u}_2/\|\mathbf{u}_2\|, \mathbf{u}_3/\|\mathbf{u}_3\|\} \\ &= \{(1/2)(1,1,1,1), (1/\sqrt{2})(-1,1,0,0), (\sqrt{2/3})(-1/2,-1/2,1,0), \\ &\quad (\sqrt{3}/2)(-1/3,-1/3,-1/3,1)\}\end{aligned}$$

is an orthonormal basis of $\mathbf{R}^4$ consisting of eigenvectors of T. With respect to this basis, we have

$$[T]_\alpha^\alpha = \begin{bmatrix} 4 & 0 & 0 & 0 \\ 0 & 0 & 0 & 0 \\ 0 & 0 & 0 & 0 \\ 0 & 0 & 0 & 0 \end{bmatrix}$$

To introduce our geometric version of the spectral theorem, we carry our computations from the previous examples one step farther. In part a, note that the diagonal matrix $[T]_\alpha^\alpha$ may be "decomposed" in the following way:

$$[T]_\alpha^\alpha = 0 \cdot \begin{bmatrix} 1 & 0 & 0 \\ 0 & 0 & 0 \\ 0 & 0 & 0 \end{bmatrix} + 2 \cdot \begin{bmatrix} 0 & 0 & 0 \\ 0 & 1 & 0 \\ 0 & 0 & 1 \end{bmatrix}$$
$$= 0 \cdot A_1 + 2 \cdot A_2$$

where A_1 and A_2 are the two matrices appearing in the previous line. By our definitions, it follows that the matrix A_1 is nothing other than the matrix of the orthogonal projection onto E_0 with respect to the basis α—we are simply taking the orthogonal projection on the subspace spanned by the first vector in the basis. By the same token, A_2 is the matrix of the orthogonal projection onto E_2 with respect to the basis α.

Similarly, in Example (4.6.2b), we have that the diagonal matrix $[T]_\alpha^\alpha$ may be decomposed in the following way:

$$[T]_\alpha^\alpha = 4 \cdot \begin{bmatrix} 1 & 0 & 0 & 0 \\ 0 & 0 & 0 & 0 \\ 0 & 0 & 0 & 0 \\ 0 & 0 & 0 & 0 \end{bmatrix} + 0 \cdot \begin{bmatrix} 0 & 0 & 0 & 0 \\ 0 & 1 & 0 & 0 \\ 0 & 0 & 1 & 0 \\ 0 & 0 & 0 & 1 \end{bmatrix}$$
$$= 4 \cdot A_1 + 0 \cdot A_2,$$

where A_1 is the matrix of the orthogonal projection P_{E_4} with respect to α, and A_2 is the matrix of P_{E_0} with respect to α.

In general, if $T: \mathbf{R}^n \to \mathbf{R}^n$ is any symmetric linear mapping, and α is an orthonormal basis of $\mathbf{R}^n$ consisting of eigenvectors of T, constructed as in Example (4.6.2), then we can write

$$[T]_\alpha^\alpha = \left[\begin{array}{ccc|ccc|c} \lambda_1 & & 0 & & & & \\ & \ddots & & & 0 & & \\ 0 & & \lambda_1 & & & & \\ \hline & & & \lambda_2 & & 0 & \\ & & & & \ddots & & \\ & & & 0 & & \lambda_2 & \\ \hline & 0 & & & & & \ddots \end{array}\right]$$

where the λ_i are the distinct eigenvalues of T, and for each i the number of diagonal entries λ_i is equal to $\dim(E_{\lambda_i})$. From this we see that

$$[T]_\alpha^\alpha = \lambda_1 A_1 + \lambda_2 A_2 + \cdots + \lambda_k A_k$$

where the matrices A_i are diagonal matrices with 1's in the entries corresponding to the vectors in the basis α from E_{λ_i}, and 0's elsewhere. As we saw in Example (4.5.4), these diagonal matrices of 0's and 1's are the matrices of orthogonal projections on spaces spanned by subsets of the basis. In particular, for each i we have $A_i = [P_i]_\alpha^\alpha$, where P_i is the orthogonal projection on the subspace E_{λ_i}. Hence

$$[T]_\alpha^\alpha = \lambda_1 [P_1]_\alpha^\alpha + \cdots + \lambda_k [P_k]_\alpha^\alpha$$

As a consequence, the linear mapping T may be written as the linear combination

$$T = \lambda_1 P_1 + \cdots + \lambda_k P_k$$

This is called the *spectral decomposition* of T. We have proved the first part of Theorem (4.6.3).

(4.6.3) Theorem. Let T: $\mathbf{R}^n \to \mathbf{R}^n$ be a symmetric linear mapping, and let $\lambda_1, \ldots, \lambda_k$ be the distinct eigenvalues of T. Let P_i be the orthogonal projection of $\mathbf{R}^n$ onto the eigenspace E_{λ_i}. Then

a) $T = \lambda_1 P_1 + \cdots + \lambda_k P_k$, and

b) $I = P_1 + \cdots + P_k$.

The proof of part b of the theorem will be left as an exercise for the reader.

(4.6.4) Remark. Part b of the theorem says in particular that for every vector $\mathbf{x} \in \mathbf{R}^n$, $I(\mathbf{x}) = \mathbf{x} = P_1(\mathbf{x}) + \cdots + P_k(\mathbf{x})$. That is, $\mathbf{x}$ can be recovered or built up from its projections on the various eigenspaces of T. Although this result is almost trivial in $\mathbf{R}^n$, in the generalizations of the spectral theorem to mappings on function spaces, the analogous property of symmetric mappings (suitably defined) assumes a fundamental importance. It is the algebraic basis of the theory of Fourier series and the other orthogonal eigenfunction expansions of analysis.

We close this section with a geometric application of the spectral theorem. Recall that any conic section (other than a parabola), when centered at the origin, can be defined by an equation of the form

$$Ax_1^2 + 2Bx_1x_2 + Cx_2^2 = F$$

The quadratic terms can be interpreted as a matrix product

$$Ax_1^2 + 2Bx_1x_2 + Cx_2^2 = [x_1\ x_2]\begin{bmatrix} A & B \\ B & C \end{bmatrix}\begin{bmatrix} x_1 \\ x_2 \end{bmatrix}$$

involving the symmetric matrix $M = \begin{bmatrix} A & B \\ B & C \end{bmatrix}$. From high school geometry you should recall how to determine what type of conic section the equation defines, and how to graph the curve, in the case that $B = 0$ (i.e., when the matrix M is diagonal). (See Exercise 10 of Section 4.5.) We will now show that our results on diagonalizability may be applied to change coordinates (rotating the axes) so that the matrix M *becomes diagonal*. This is most easily described by giving an example.

(4.6.5) Example. Consider the conic section defined by $4x_1^2 + 6x_1x_2 - 4x_2^2 = 1$. Here the associated symmetric matrix is $M = \begin{bmatrix} 4 & 3 \\ 3 & -4 \end{bmatrix}$. By Theorem (4.6.1) we know that M is diagonalizable, and we can proceed to diagonalize it as usual. First, by computation, we find that the characteristic polynomial of M is $\lambda^2 - 25$, so that M has the eigenvalues $\lambda = \pm 5$. Since the eigenvalues are distinct, the corresponding eigenvectors will be orthogonal, and we find that $E_5 = \text{Span}\left\{\begin{bmatrix} 3 \\ 1 \end{bmatrix}\right\}$, whereas $E_{-5} = \text{Span}\left\{\begin{bmatrix} 1 \\ -3 \end{bmatrix}\right\}$. Hence,

$\alpha = \left\{(1/\sqrt{10})\begin{bmatrix}3\\1\end{bmatrix}, (1/\sqrt{10})\begin{bmatrix}1\\-3\end{bmatrix}\right\}$ is an orthonormal basis of $\mathbf{R}^2$. We let

$Q = \begin{bmatrix}3/\sqrt{10} & 1/\sqrt{10}\\ 1/\sqrt{10} & -3/\sqrt{10}\end{bmatrix}$, the change of basis matrix from α to the standard basis. Then, we find that

$$Q^{-1}MQ = \begin{bmatrix}5 & 0\\ 0 & -5\end{bmatrix}$$

Now, we introduce new coordinates in $\mathbf{R}^2$ by setting $\begin{bmatrix}x_1'\\x_2'\end{bmatrix} = Q^{-1}\cdot\begin{bmatrix}x_1\\x_2\end{bmatrix}$.

We note that $QQ^t = (1/\sqrt{10})\cdot\begin{bmatrix}3 & 1\\ 1 & -3\end{bmatrix}\cdot(1/\sqrt{10})\cdot\begin{bmatrix}3 & 1\\ 1 & -3\end{bmatrix} = \begin{bmatrix}1 & 0\\ 0 & 1\end{bmatrix}$. Hence, $Q^t = Q^{-1}$ for this change of basis matrix Q, reflecting the fact that α is an orthonormal basis. (See Exercise 4 for a fuller discussion of this phenomenon.) If we substitute into our previous expression for the equation of the conic as a matrix product and use the fact that $Q^t = Q^{-1}$, the equation of the conic in the new coordinates becomes

$$[x_1' \ x_2']\begin{bmatrix}5 & 0\\ 0 & -5\end{bmatrix}\begin{bmatrix}x_1'\\x_2'\end{bmatrix} = 5x_1'^{\,2} - 5x_2'^{\,2} = 1$$

From this we see that the conic section is a hyperbola. (See Figure 4.7.)

In (x_1, x_2)-coordinates, the vertices of the hyperbola are at the points $(3/\sqrt{50}, 1/\sqrt{50})$ and $(-3/\sqrt{50}, -1/\sqrt{50})$. The asymptotes are the lines $x_1' = \pm x_2'$ [or $x_1 = -2x_2$ and $x_1 = (1/2)x_2$]. Note that as we expect, the picture shows clearly that the new (x_1' and x_2') axes are simply rotated from the standard

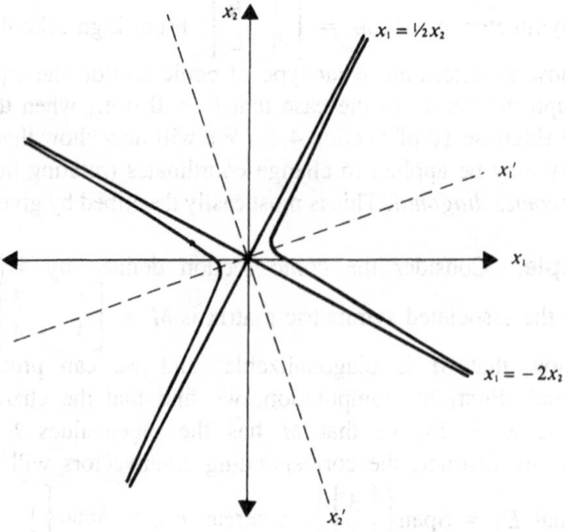

Figure 4.7

coordinate axes. This is the geometric meaning of the fact that the eigenspaces of the symmetric matrix M are orthogonal to each other.

The same technique may be applied to study quadric surfaces centered at the origin in $\mathbf{R}^3$.

(4.6.6) Example. For example, consider the quadric surface defined by

$$x_1^2 + 2x_1x_2 + x_3^2 = 1$$

in $\mathbf{R}^3$. Again, we can express the equation in matrix form:

$$x_1^2 + 2x_1x_2 + x_3^2 = [x_1\ x_2\ x_3]\begin{bmatrix}1 & 1 & 0\\ 1 & 0 & 0\\ 0 & 0 & 1\end{bmatrix}\begin{bmatrix}x_1\\ x_2\\ x_3\end{bmatrix} = 1$$

Here we have a symmetric 3×3 matrix $M = \begin{bmatrix}1 & 1 & 0\\ 1 & 0 & 0\\ 0 & 0 & 1\end{bmatrix}$, which will determine the "shape" of the surface. The eigenvalues of M are the roots of the characteristic polynomial equation, $(1 - \lambda)(\lambda^2 - \lambda - 1) = 0$. Using the quadratic formula, we find $\lambda = 1$ and $(1/2)(1 \pm \sqrt{5})$ as the roots. If we change coordinates to the coordinates given by an orthonormal basis of $\mathbf{R}^3$ consisting of corresponding eigenvectors of M, then the equation of the surface takes the form

$$x_1'^{\,2} + (1/2)(1 + \sqrt{5})x_2'^{\,2} + (1/2)(1 - \sqrt{5})x_3'^{\,2} = 1$$

Since two of the eigenvalues are positive, and one $[(1/2)(1 - \sqrt{5})]$ is negative, our surface is an elliptic hyperboloid. In addition, the positive term on the right-hand side of the equation shows that we have an elliptic hyperboloid of one sheet. (The axes in Figure 4.8 are again rotated with respect to the standard coordinate

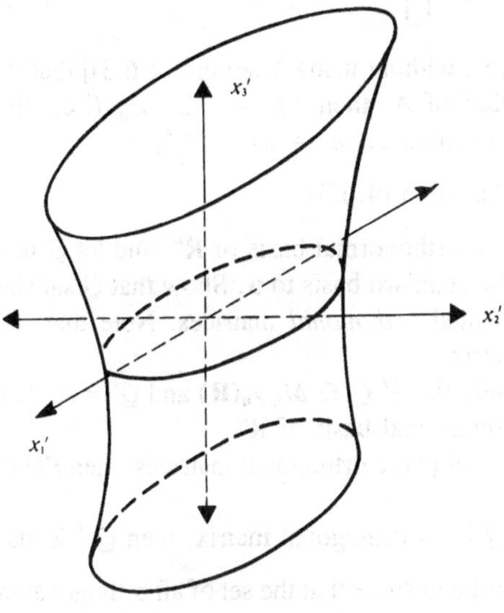

Figure 4.8

axes.) The sections of our surface by planes $x_3' =$ const. are all ellipses, whereas the sections by planes containing the x_3'-axis are hyperbolas.

Analogous analyses for conic sections or quadric surfaces centered at other points may be carried out by first rotating the coordinate axes as in these examples, then translating the center to the origin (completing the square in each variable).

EXERCISES

1. Verify the spectral theorem for each of the following symmetric matrices, by finding an orthonormal basis of the appropriate vector space, the change of basis matrix to this basis, and the spectral decomposition.

   a) $A = \begin{bmatrix} 2 & 3 \\ 3 & 2 \end{bmatrix}$

   b) $A = \begin{bmatrix} 1 & -1 \\ -1 & 4 \end{bmatrix}$

   c) $A = \begin{bmatrix} -1 & 2 & 2 \\ 2 & -1 & 2 \\ 2 & 2 & -1 \end{bmatrix}$

   d) $A = \begin{bmatrix} 1 & 0 & -1 \\ 0 & 1 & 0 \\ -1 & 0 & 1 \end{bmatrix}$

   e) $A = \begin{bmatrix} 2 & 2 & 2 & 2 \\ 2 & 2 & 2 & 2 \\ 2 & 2 & 2 & 2 \\ 2 & 2 & 2 & 1 \end{bmatrix}$

2. Show directly [i.e., without using Theorem (4.6.3)] that if A is symmetric and λ is an eigenvalue of A, then $E_\lambda^\perp = \sum_{\mu \neq \lambda} E_\mu$ (i.e., the sum of the other eigenspaces, for eigenvalues $\mu \neq \lambda$).

3. Prove part b of Theorem (4.6.3).

4. a) Suppose α is an orthonormal basis of $\mathbf{R}^n$, and let Q be the change of basis matrix from the standard basis to α. Show that Q satisfies $Q^t = Q^{-1}$. (Such matrices are called *orthogonal* matrices. Note that, in particular, I is an orthogonal matrix.)

   b) Show conversely that if $Q \in M_{n \times n}(\mathbf{R})$ and $Q^t = Q^{-1}$, then the columns of Q form an orthonormal basis of $\mathbf{R}^n$.

   c) Show that if P and Q are orthogonal matrices, then PQ is also an orthogonal matrix.

   d) Show that if Q is an orthogonal matrix, then Q^{-1} is also orthogonal.

Parts c and d of this problem show that the set of all orthogonal matrices in $M_{n \times n}(\mathbf{R})$ forms a *group*—a set with one operation (here matrix multiplication) that is associative, which possesses an identity element (here the identity matrix), and for which each element of the set has an inverse.

e) Show that if Q is an orthogonal matrix, then $\det(Q) = \pm 1$.
f) Find examples of 2×2 orthogonal matrices with det $= 1$ and det $= -1$. Can you see the geometric difference between these two cases?

5. (Continued from 4.)
   a) Show that Q is an orthogonal matrix if and only if for all vectors $\mathbf{x}, \mathbf{y} \in \mathbf{R}^n$, we have $\langle Q\mathbf{x}, Q\mathbf{y}\rangle = \langle \mathbf{x}, \mathbf{y}\rangle$. [This shows that the linear mappings defined by orthogonal matrices (with respect to an orthonormal basis) are precisely those mappings that *preserve inner products* of vectors.]
   b) From part a deduce that if Q is orthogonal, $\| Q\mathbf{x} \| = \| \mathbf{x} \|$ for all $\mathbf{x} \in \mathbf{R}^n$.
   c) Show that conversely, if A is any matrix such that $\| A\mathbf{x} \| = \| \mathbf{x} \|$ for all $\mathbf{x} \in \mathbf{R}^n$, then A is orthogonal. (*Hint:* polarization identity.)

6. We say that matrices A and B are *orthogonally similar* if there is an orthogonal matrix Q such that $B = Q^{-1}AQ = Q^tAQ$.
   a) Show that if A is symmetric and B is orthogonally similar to A, then B is also symmetric.
   b) Show that the same is true with symmetric replaced by skew-symmetric.

7. Show that every symmetric matrix A is orthogonally similar to a diagonal matrix D.

8. Show that if A is symmetric, then the following statements are equivalent:
   (i) A is (also) orthogonal: $A^t = A^{-1}$.
   (ii) $A^2 = I$.
   (iii) All eigenvalues of A are $+1$ or -1.
   [*Hint:* Show that (i) → (ii), (ii) → (iii), and (iii) → (i), so that any one of the statements implies any other.]

9. a) Show that if A is skew-symmetric then $I + A$ is invertible.
   b) Show that if A is skew-symmetric, then $(I - A)(I + A)^{-1}$ is orthogonal.

10. Let $A \in M_{n \times n}(\mathbf{R})$ be a symmetric matrix. Show that $\langle \mathbf{x}, \mathbf{y}\rangle = \mathbf{x}^tA\mathbf{y}$ defines an inner product on $\mathbf{R}^n$ (see the exercises accompanying Section 4.3) if and only if all the eigenvalues of A are positive.

If the condition of Exercise 10 is satisfied, we say that the symmetric matrix A is *positive-definite*.

11. a) Show that if Q is an invertible matrix, then Q^tQ is symmetric and positive-definite. (*Hint:* Use Exercise 10.)
    b) Show that if A is invertible, then A is symmetric and positive-definite if and only if A^{-1} is.
    c) Show that if A is symmetric and positive-definite, then A^k is as well, for each $k \geqslant 0$.

12. An invertible symmetric matrix $A \in M_{n \times n}(\mathbf{R})$ is said to be *negative-definite* if all the eigenvalues of A are negative, and *indefinite* if it has both positive and negative eigenvalues.

a) Which powers of a negative-definite matrix A are positive-definite? negative-definite? indefinite?

b) Can a power of an indefinite matrix A ever be positive definite? negative-definite?

13. A positive real number has square roots in $\mathbf{R}$. In this exercise you will show that an analogous fact is true for positive-definite symmetric matrices.

a) Show that if A is symmetric and positive definite, then there is a matrix R such that $R^2 = A$. (*Hint:* Use Exercise 7 to reduce the problem to finding a square root of a diagonal matrix.)

b) Show that in part a we may take R to be symmetric.

c) How many different square roots can a positive-definite symmetric matrix have?

14. Sketch the graphs of the following conic sections, using the method of the text.

a) $5x_1^2 + 2\,x_1x_2 + x_2^2 = 1$

b) $x_1^2 + 4x_1x_2 + x_2^2 = 9$

c) $x_1^2 - 3x_1x_2 - x_2^2 = 1$

d) $x_1^2 + x_1x_2 = 4$

15. a) What type of quadric surface in $\mathbf{R}^3$ is defined by the equation

$$-x_1^2 - x_2^2 - x_3^2 + 4x_1x_2 + 4x_1x_3 + 4x_2x_3 = 1$$

b) Same question for

$$x_1^2 - x_3^2 + x_2x_3 = 9$$

16. Let $A = \begin{bmatrix} a & b \\ b & c \end{bmatrix}$ be a general symmetric 2×2 matrix.

a) Show that the eigenvalues of A have the same sign if and only if $\det(A) = ac - b^2 > 0$ and opposite signs if and only if $\det(A) < 0$.

b) Let $a > 0$ and consider the conic section

$$[x_1\ x_2]\begin{bmatrix} a & b \\ b & c \end{bmatrix}\begin{bmatrix} x_1 \\ x_2 \end{bmatrix} = ax_1^2 + 2bx_1x_2 + cx_2^2 = 1$$

Deduce from part a that the conic is an ellipse (or a circle) if and only if $\det(A) > 0$, and a hyperbola if and only if $\det(A) < 0$. [In analytic geometry you may have seen this criterion stated in terms of the discriminant: $b^2 - ac = -\det(A)$. If the discriminant is used, the inequalities are reversed.]

17. a) Show that if A is a symmetric matrix and $\mathbf{x}$ is a unit eigenvector of A with eigenvalue λ, then $\langle A\mathbf{x}, \mathbf{x}\rangle = \lambda$.

b) With the same hypotheses as in part 1, show that $\| A\mathbf{x} \| = |\lambda|$.

c) Show that the eigenvalue of A with the largest absolute value is equal to $\max_{\|\mathbf{x}\|=1} \| A\mathbf{x} \|$. (*Hint:* Use the conclusion of the spectral theorem, and expand $\mathbf{x}$ in terms of an orthonormal basis consisting of eigenvectors of A.)

d) Similarly, show that the eigenvalue of A with the smallest absolute value is equal to $\min_{\|\mathbf{x}\|=1} \|A\mathbf{x}\|$

CHAPTER SUMMARY

In the first three chapters of the text, our major concerns were *algebraic*. In studying vector spaces in Chapter 1, we were exploring the properties of the algebraic sum and scalar multiplication operations in those spaces. Similarly, in our discussion of linear mappings in Chapter 2 we were studying those mappings between vector spaces that are compatible with the algebraic operations defined in the vector spaces—the linear mappings are precisely those mappings that preserve linear combinations. Although we motivated the definition of the determinant function in Chapter 3 by geometric considerations, the definition itself was again an algebraic one, based on the property of multilinearity for scalar-valued functions of a matrix.

In the present chapter, however, the focus of our discussion has been the *geometry* of linear mappings, and we have seen how the geometry of $\mathbf{R}^n$ and the properties of linear mappings interact. We have seen one powerful technique for studying linear mappings from this perspective. By finding the *eigenvectors* and *eigenvalues* of a linear mapping [Definitions (4.1.2)], we can find subspaces of the domain—the *eigenspaces* [Definition (4.1.6)]—on which the mapping acts in the simplest possible way, by multiplying each vector by a scalar. We showed how the eigenvalues appear as the roots of the *characteristic polynomial* of the matrix or mapping [Definition (4.1.11)], and how once those roots had been isolated, finding the corresponding eigenvectors meant solving appropriate systems of linear equations.

The case in which this method of finding eigenvectors and eigenvalues succeeds completely in describing the geometric behavior of a mapping on a finite-dimensional space is that in which the mapping is *diagonalizable* [Definition (4.2.1)]. The terminology came from the observation that in this case we can take the matrix of the mapping in question to a diagonal matrix by an appropriate change of basis [Proposition (4.2.2)].

In Theorem (4.2.7) we found necessary and sufficient conditions for a matrix or mapping to be diagonalizable. The two properties that must be present are:

(i) All the roots of the characteristic polynomial must be real—in other words, the number of eigenvalues of the mapping, counted with their multiplicities, must equal the dimension of the space, and

(ii) The dimension of the eigenspace for each eigenvalue must equal the multiplicity of that eigenvalue.

Although we did not use the following terminology in the text, to understand the meaning of condition (ii), we should think of the multiplicity of the eigenvalue as a root of the characteristic polynomial as the *algebraic multiplicity*, and the dimension of the eigenspace as the *geometric multiplicity* of the eigenvalue. In order for a mapping to be diagonalizable, it must be true that for each eigenvalue, the algebraic and geometric multiplicities are *equal*.

Starting in Section 4.3, we brought the geometry of the vector space $\mathbf{R}^n$ into the picture as well. The basic geometric operations of measuring lengths and angles both lead us to consider the *standard inner product* on $\mathbf{R}^n$ [Definition (4.3.1)]. In a sense, the relationship between vectors with the most far-reaching consequences is the geometric property of *orthogonality* [Definition (4.3.7)].

In Section 4.4 we saw that for each subspace $W \subset \mathbf{R}^n$, the set of all vectors orthogonal to every vector in W is itself a subspace, called the *orthogonal complement*, $W^\perp$, of W [Definition (4.4.1)]. This fact leads directly to the definition of the *orthogonal projection* mapping on W, a linear mapping whose effect is to decompose each vector into its "components" in the directions of W and $W^\perp$, and to discard the component in the direction of $W^\perp$.

We saw two important uses of these orthogonal projection mappings in Sections 4.4 and 4.6. The first of these was the *Gram-Schmidt process* [(4.4.8)], which uses orthogonal projections to "orthogonalize" a given set of vectors—that is, produce an orthogonal set spanning the same subspace of $\mathbf{R}^n$ as the original given set. A major consequence of Gram-Schmidt is the existence of convenient orthonormal bases for all subspaces of $\mathbf{R}^n$ [Theorem (4.4.9)].

The second use of orthogonal projections is in the *spectral theorem* [(4.6.1)], the culmination of our geometric study of linear mappings in this chapter. One geometric interpretation of the fact that all symmetric matrices are diagonalizable (the "algebraic version" of the spectral theorem) is that every symmetric mapping is a linear combination of orthogonal projections onto its eigenspaces. Thus, the orthogonal projections are the "building blocks" of symmetric mappings.

Finally, we also saw how the spectral theorem can be used to understand general conic sections and quadric surfaces, in which the coordinate axes must be rotated to put the equations in the standard forms.

SUPPLEMENTARY EXERCISES

1. **True-False**. For each true statement, give a brief proof. For each false statement, give a counterexample.
   a) If $\lambda \in \mathbf{R}$ is a root of the polynomial equation $\det(A - tI) = 0$, then A has an eigenvector with eigenvalue λ.
   b) If A is a diagonalizable matrix, then A is invertible.
   c) Every orthogonal matrix in $M_{2\times 2}(\mathbf{R})$ is the matrix of a rotation of $\mathbf{R}^2$ with respect to the standard basis.
   d) If $\lambda \neq \mu$ are eigenvalues of a linear mapping T, then $E_\lambda \cap E_\mu = \{\mathbf{0}\}$.
   e) The function $f(x) = e^{5x}$ is an eigenvector of the linear mapping T: $C^\infty(\mathbf{R}) \to C^\infty(\mathbf{R})$ defined by $T(f) = f' - 3f$.
   f) If an $n \times n$ matrix A is diagonalizable, the characteristic polynomial of A has n distinct roots.
   g) If $W \subset \mathbf{R}^n$ is a subspace, $W \cap W^\perp = \{\mathbf{0}\}$.

h) If $A \in M_{n \times n}(\mathbf{R})$ is skew-symmetric, then all the roots of the characteristic polynomial of A are real.

i) If $A \in M_{n \times n}(\mathbf{R})$ is symmetric, with two distinct eigenvalues λ and μ, then in $\mathbf{R}^n$, $E_\lambda \subset E_\mu^\perp$.

j) Every orthogonal set in $\mathbf{R}^n$ may be extended to an orthogonal basis of $\mathbf{R}^n$.

2. Let V be a vector space of dimension two with basis $\{\mathbf{x}, \mathbf{y}\}$. Let $T: V \to V$ be the linear mapping with $T(\mathbf{x}) = -\alpha\mathbf{y}$, and $T(\mathbf{y}) = \beta\mathbf{x}$, where α, β are *positive*. Is T diagonalizable?

3. Let V be a finite-dimensional vector space. Let $V = W_1 \oplus W_2$, and let $T: V \to V$ be a linear mapping such that $T(W_1) \subseteq W_1$ and $T(W_2) \subseteq W_2$.

a) Show that the set of distinct eigenvalues of T is the union of the sets of eigenvalues of $T|_{W_1}$ and $T|_{W_2}$.

b) Show that T is diagonalizable if and only if $T|_{W_1}$ and $T|_{W_2}$ are diagonalizable.

4. Are the following matrices diagonalizable? If so, give the diagonal matrix and the change of basis matrix.

a) $\begin{bmatrix} 6 & -2 \\ -2 & 1 \end{bmatrix}$ b) $\begin{bmatrix} 2 & 3 \\ -2 & -3 \end{bmatrix}$

c) $\begin{bmatrix} 0 & 0 & -1 \\ 1 & 0 & -3 \\ 0 & 1 & -3 \end{bmatrix}$ d) $\begin{bmatrix} 1 & 0 & 1 & -4 \\ 0 & 2 & 0 & 1 \\ 0 & 0 & 3 & 0 \\ 0 & 0 & 0 & 4 \end{bmatrix}$

5. Find an orthonormal basis of each eigenspace and the spectral decomposition for each of the following symmetric matrices.

a) $\begin{bmatrix} 3 & 5 \\ 5 & -1 \end{bmatrix}$ b) $\begin{bmatrix} 1 & 0 & 1 \\ 0 & 2 & 0 \\ 1 & 0 & -1 \end{bmatrix}$

c) $\begin{bmatrix} 2 & 1 & 1 \\ 1 & 2 & 1 \\ 1 & 1 & 2 \end{bmatrix}$

6. Let $V = M_{2 \times 2}(\mathbf{R})$ and consider the basis $\beta = \left\{ \begin{bmatrix} 1 & 0 \\ 0 & 0 \end{bmatrix}, \begin{bmatrix} 0 & 1 \\ 0 & 0 \end{bmatrix}, \begin{bmatrix} 0 & 0 \\ 1 & 0 \end{bmatrix}, \begin{bmatrix} 0 & 0 \\ 0 & 1 \end{bmatrix} \right\}$ for V. Let $A = \begin{bmatrix} a & b \\ c & d \end{bmatrix}$ be a fixed 2×2 matrix, and consider the mapping $T: V \to V$ defined by $T(X) = AX$ (matrix product).

a) Show that T is linear.

b) Compute $[T]_\beta^\beta$.

c) What are the eigenvalues and the corresponding eigenspaces of T?

d) Show that T is diagonalizable if and only if A is diagonalizable.
e) Find necessary and sufficient conditions on A for T to be symmetric with respect to the inner product $\langle X, Y\rangle = \text{Tr}(X^tY)$ from Exercise 10c) of Section 4.3.

7. a) Show that if $\text{Tr}(A)^2 \neq 4 \cdot \det(A)$, then $\begin{bmatrix} a & b \\ c & d \end{bmatrix}$ is similar to $\begin{bmatrix} 0 & -\det(A) \\ 1 & \text{Tr}(A) \end{bmatrix}$ in $M_{2\times 2}(\mathbf{R})$.
This gives a *standard form* for general matrices in $M_{2\times 2}(\mathbf{R})$, which is valid even if the characteristic polynomial has *no real roots*.
b) Let $A, B \in M_{2\times 2}(\mathbf{R})$. Assume that A and B have the same characteristic polynomial, and that the polynomial has distinct roots (not necessarily real). Show that A and B are similar.
c) What happens if A and B have the same characteristic polynomial, but that polynomial has a repeated root?

8. Let $SL_2(\mathbf{R})$ denote the set of matrices $A \in M_{2\times 2}(\mathbf{R})$ with $\det(A) = 1$.
a) Show that if $A \in SL_2(\mathbf{R})$ and $|\text{Tr}(A)| > 2$, then A is diagonalizable, and that, in fact, there is an invertible matrix Q such that $Q^{-1}AQ = \begin{bmatrix} \alpha & 0 \\ 0 & 1/\alpha \end{bmatrix}$ for some $\alpha \neq 0, \pm 1$ in $\mathbf{R}$.
b) Show that if $A \in SL_2(\mathbf{R})$, $|\text{Tr}(A)| = 2$, and $A \neq \pm I$, then there is an invertible matrix Q such that

$$Q^{-1}AQ = \begin{bmatrix} 1 & 1 \\ 0 & 1 \end{bmatrix} \quad \text{or} \quad \begin{bmatrix} -1 & 1 \\ 0 & -1 \end{bmatrix}$$

c) Show that if $A \in SL_2(\mathbf{R})$ and $|\text{Tr(A)}| < 2$, then A has no real eigenvalues, but there exists an invertible matrix Q such that $Q^{-1}AQ = \begin{bmatrix} \cos(\theta) & -\sin(\theta) \\ \sin(\theta) & \cos(\theta) \end{bmatrix}$ for some θ. [*Hint:* Since $|Tr(A)| < 2$, there is a unique θ, $0 < \theta < \pi$, with $\text{Tr}(A) = 2 \cdot \cos(\theta)$.]

9. Let $V = C[-1, 1]$ and define $\langle f, g\rangle = \int_{-1}^{1} f(t)g(t)\,dt$ for $f,g \in V$.
a) Show that this defines an inner product on V. (See Exercise 10 of Section 4.3.)
b) Show that the mapping $T: V \to V$ that takes the function $f(x)$ to the function $f(-x)$ is a linear mapping.

In the text we stated results about diagonalizability only for mappings on finite-dimensional spaces. In the next parts of this exercise we will see that there are *some* mappings on infinite-dimensional spaces that have properties that are very similar to those we know from the finite-dimensional case. (The general situation is much more complicated, however.)

c) Show that $T^2 = I$, so that the eigenvalues of T are $+1$ and -1.
d) Describe the eigenspaces E_1 and E_{-1}, and show that $C[-1, 1] = E_1 + E_{-1}$.
e) Show that T is symmetric with respect to the inner product defined before: $\langle T(f), g\rangle = \langle f, T(g)\rangle$ for all $f, g \in V$.

f) Show that every function in E_1 is orthogonal to every function in E_{-1} using this inner product.

10. Apply the Gram-Schmidt process to find a basis of Span$\{1, x, x^2\} \subset C[-1, 1]$, which is orthonormal with respect to the inner product of Exercise 9.

11. Let $\mathbf{x}_1, \ldots, \mathbf{x}_n \in \mathbf{R}^n$, and let $A \in M_{n \times n}(\mathbf{R})$ be the matrix whose rows are the vectors $\mathbf{x}_i$.
 a) Show that $|\det(A)| \leq \|\mathbf{x}_1\| \cdot \cdots \cdot \|\mathbf{x}_n\|$. (*Hint:* Think of the geometric interpretation of $|\det(A)|$.)
 b) When is the inequality in part (a) an equality? Prove your assertion.

12. In Section 4.5 of the text we mentioned that the Hessian matrix H_f of a function $f: \mathbf{R}^2 \to \mathbf{R}$ was useful in a second derivative test for maxima and minima of functions of two variables. The basis of this application is given in the following exercise. We consider the *graph* of

$$\begin{aligned} x_3 &= q(x_1, x_2) \\ &= (1/2)(ax_1^2 + 2bx_1x_2 + cx_2^2) \end{aligned}$$

in $\mathbf{R}^3$. Let A be the symmetric 2×2 matrix $\begin{bmatrix} a & b \\ b & c \end{bmatrix}$ for which $2 \cdot q(x_1\ x_2) = [x_1\ x_2] \cdot A \cdot \begin{bmatrix} x_1 \\ x_2 \end{bmatrix}$ (matrix product).
 a) Show that the graph of $x_3 = q(x_1, x_2)$ has
  (i) a minimum at (0, 0, 0) if and only if A is positive-definite,
  (ii) a maximum at (0, 0, 0) if and only if A is negative-definite,
  (iii) a saddle point (neither a maximum nor a minimum) at (0, 0, 0) if $\det(A) \neq 0$ and A is indefinite.
  [What happens if $\det(A) = 0$?]
 b) Show that we have case:
  (i) if and only if $\det(A) > 0$ and $a > 0$,
  (ii) if and only if $\det(A) > 0$ and $a < 0$,
  (iii) if and only if $\det(A) < 0$.
 c) Show that the Hessian matrix of q is the matrix A.

For a general function f, near a point where $\partial f/\partial x_1$ and $\partial f/\partial x_2$ are both zero, Taylor's theorem implies that the graph of $x_3 = f(x_1, x_2)$ is closely approximated by a graph of the form studied in this exercise, with

$$a = \partial^2 f/\partial x_1^2, \qquad b = \partial^2 f/\partial x_1 \partial x_2, \qquad c = \partial^2 f/\partial x_2^2$$

evaluated at the critical point. This may be used to give a proof of the second derivative test for functions of two variables.

13. Let $P: \mathbf{R}^n \to \mathbf{R}^n$ be a linear mapping. Show that $P = P_W$, the orthogonal projection on a subspace $W \subset \mathbf{R}^n$, if and only if $P^2 = P$ and P is a symmetric mapping. [*Hint:* We already know one direction. For the other, if the statement is to be true, then the only possibility for W is the subspace $\text{Im}(P) \subset \mathbf{R}^n$.]

14. Show that if T_1 and T_2 are linear mappings on $\mathbf{R}^n$ and $\langle T_1(\mathbf{x}), \mathbf{y}\rangle = \langle T_2(\mathbf{x}), \mathbf{y}\rangle$ for all $\mathbf{x}, \mathbf{y} \in \mathbf{R}^n$, then $T_1 = T_2$.

15. Let W_1 and W_2 be subspaces of $\mathbf{R}^n$, and let $P_1 = P_{W_1}$, $P_2 = P_{W_2}$.
   a) Show that P_1P_2 is an orthogonal projection if and only if the mappings P_1 and P_2 commute: $P_1P_2 = P_2P_1$. (You may need the result of Exercise 14.)
   b) If P_1P_2 is an orthogonal projection, what subspace is its image?
   c) Let W_1 and W_2 be distinct two-dimensional subspaces (planes) in $\mathbf{R}^3$. Show that $P_1P_2 = P_2P_1$ if and only if $W_1 \perp W_2$ (i.e., they meet at a right angle).
   d) Show, in general, that $P_1P_2 = P_2P_1$ if and only if, letting $V_1 = W_1 \cap (W_1 \cap W_2)^\perp$ and $V_2 = W_2 \cap (W_1 \cap W_2)^\perp$, we have that V_1 and V_2 are orthogonal to each other (i.e., $V_1 \subset V_2^\perp$ and vice versa).
   e) Show that if $P_1P_2 = P_2P_1$, then

$$P_{W_1+W_2} = P_1 + P_2 - P_1P_2$$

16. Let $A \in M_{m\times n}(\mathbf{R})$ be any matrix (not necessarily square). Assume that $n < m$ and $\operatorname{rank}(A) = n$ (the maximum possible in this situation).
   a) Show that $A^tA \in M_{n\times n}(\mathbf{R})$ is invertible.
   b) Show further that A^tA is symmetric and positive-definite.

17. Let $k \leqslant n$, let $A \in M_{n\times k}(\mathbf{R})$ be matrix of rank k, and let $W \subset \mathbf{R}^n$ be the subspace spanned by the columns of A. Show that $P = A \cdot (A^tA)^{-1} \cdot A^t$ is the matrix of the orthogonal projection mapping P_W with respect to the standard basis. (By Exercise 16, A^tA is an invertible $k \times k$ matrix, so this makes sense.)

The following exercises refer to our study of bilinear mappings in Exercises 14 through 17 of Section 4.3. Let $B: V \times V \to \mathbf{R}$ be a symmetric bilinear mapping on the finite-dimensional vector space V.

Definition. The *quadratic form associated to B* is the mapping $Q: V \to \mathbf{R}$ defined by $Q(\mathbf{x}) = B(\mathbf{x},\mathbf{x})$.

18. a) Show that if α is a basis for V, $[B]_\alpha = (a_{ij})$, and $[\mathbf{x}]_\alpha = (x_1, x_2, \ldots, x_n)$, then

$$Q(\mathbf{x}) = \sum_{i,j=1}^{n} a_{ij}x_ix_j, \tag{4.4}$$

   a homogeneous polynomial of degree two in the components of $\mathbf{x}$. (Such polynomials are called *quadratic forms*, which explains the terminology for the mapping Q.)
   b) Show that conversely, if a_{ij} is any collection of scalars with $a_{ij} = a_{ji}$ for all pairs i, j, and Q is the homogeneous polynomial $Q(\mathbf{x}) = \sum_{i,j=1}^{n} a_{ij}x_ix_j$, then

$$B(\mathbf{x},\mathbf{y}) = (1/4)(Q(\mathbf{x} + \mathbf{y}) - Q(\mathbf{x} - \mathbf{y}))$$

is a symmetric bilinear form. (This is known as the *polarization identity* for quadratic forms.)

c) Show that if Q is the quadratic form associated to a bilinear form B, then applying the polarization identity from part b, we recover B itself.

19. a) Using Exercise 17 of Section 4.3, show that by an orthogonal change of coordinates, every quadratic form as in Eq. (4.4) may be transformed to a "diagonal form":

$$Q = c_1y_1^2 + \cdots + c_ky_k^2,$$

where $k \leq n$, $c_j \in \mathbf{R}$, and the y_j are appropriate linear functions of the x_i.

b) Show that by a further change of coordinates, any quadratic form may be brought to the form

$$Q = z_1^2 + \cdots + z_p^2 - z_{p+1}^2 - \cdots - z_{p+q}^2 \qquad (4.5)$$

where $k = p + q \leq n$.

c) Show that the number of positive terms in Eq. (4.5) (that is, p) and the number of negative terms (q) are independent of the choice of coordinates, as long as Q takes the form given in Eq. (4.5). (This result is called *Sylvester's Law of Inertia*.)

The pair of integers (p,q)—the numbers of positive and negative terms in the form in Eq. (4.5)—is called the *signature* of the quadratic form Q.

20. Compute the signature of each of the following quadratic forms.

a) $x_1^2 - 2x_1x_2 + x_3^2$ on $\mathbf{R}^3$

b) $4x_1x_3 + 6x_2x_4$ on $\mathbf{R}^4$

c) $x_1^2 + 6x_1x_2 + 9x_2^2$ on $\mathbf{R}^3$

CHAPTER 5

Complex Numbers and Complex Vector Spaces

Introduction

In this chapter we extend the work of the first four chapters to vector spaces over the complex numbers. It is one of the strengths of the approach that we have taken that very little additional work is needed to apply our results to the more general situation. In fact, our results are not only applicable to vector spaces over the real numbers and complex numbers, but also to vector spaces over a field. A *field* is an algebraic object consisting of a set with two operations on pairs of elements, addition and multiplication, which are modeled on addition and multiplication of real numbers. This generalization from the real numbers to an abstract field is analogous to the generalization we made in the first chapter from $\mathbf{R}^n$ to an abstract real vector space. Of course, both the real and complex numbers are fields. The extension of our work to complex vector spaces is necessary because the theory of vector spaces over the real numbers suffers from one glaring deficiency, which is a consequence of the fact that not all polynomials of degree n with real coefficients have n real roots. Let us explain.

We saw in Chapter 4 that there are real matrices that have eigenvalues that are not real numbers, for example

$$A = \begin{bmatrix} 0 & -1 \\ 1 & 0 \end{bmatrix}$$

The characteristic polynomial of A is $\lambda^2 + 1$, whose roots are $\pm\sqrt{-1}$, that is, the roots are complex numbers. Although the eigenvalues are distinct, Corollary (4.2.8) does not apply because the eigenvalues are not real. However, this matrix is diagonalizable over the complex numbers. The complex numbers are of special importance since they are algebraically closed, that is, every polynomial with complex coefficients of degree n has n complex roots. So we see that this example does illustrate the general case.

We develop the algebra and geometry of the complex numbers from scratch. Much of our work so far relies only on the algebra that is common to real and complex numbers and depends on the fact that each is a field. Thus, virtually all our results hold if we replace vector spaces over **R** by vector spaces over a field. Rather than duplicate all these results, we simply run through several representative examples.

The only results that rely on the field being the real numbers are those that use the geometry of $\mathbf{R}^n$ (Sections 4.3, 4.4, and 4.5). In order to generalize this material, we develop the basic notions of geometry in a complex vector space. That is, we introduce Hermitian inner products that will allow us to define length and orthogonality as we did in $\mathbf{R}^n$. As a consequence, we will be able to extend the spectral theorem to complex vector spaces with Hermitian inner products.

---

§5.1. COMPLEX NUMBERS

In this section we give a careful development of the basic algebra and geometry of the complex numbers. The goal to keep in mind is that we want to construct a set F that contains the real numbers, $\mathbf{R} \subset F$, and, like the real numbers, possesses two operations on pairs of elements of F, which we also call addition and multiplication. These operations must satisfy two general "rules":

(a) When restricted to the subset $\mathbf{R} \subset F$, these are the usual operations of addition and multiplication of real numbers.

(b) The operations on F must satisfy the same properties as do addition and multiplication of real numbers.

As an example of (b), the operation of addition of elements of F must be commutative, $a + b = b + a$ for all $a, b \in F$, just as addition of real numbers is commutative. Shortly, we will be precise about which properties of **R** are of interest.

These two rules do not determine the complex numbers uniquely. For example, we could take $F = \mathbf{R}$ and satisfy (a) and (b). In order to obtain the complex numbers, we must add a third condition:

(c) Every polynomial $p(z) = a_n z^n + \cdots + a_1 z + a_0$ with coefficients in F, $a_i \in F$ for $i = 0, \ldots, n$, has n roots in F.

A set F satisfying the additional condition (c) is called *algebraically closed*. As usual, by a root of a polynomial we mean an element z of F that satisfies $p(z) = 0$.

It is easy to see that $\mathbf{R}$ itself does not satisfy (c) since $p(z) = z^2 + 1$ has no real roots. If such a set F exists, we at least know that it is strictly larger than $\mathbf{R}$. We begin with the definition.

(5.1.1) Definition. The set of *complex numbers*, denoted $\mathbf{C}$, is the set of ordered pairs of real numbers (a, b) with the operations of addition and multiplication defined by

For all (a, b) and $(c, d) \in \mathbf{C}$, the *sum* of (a, b) and (c, d) is the complex number defined by

$$(a, b) + (c, d) = (a + c, b + d) \tag{5.1}$$

and the *product* of (a, b) and (c, d) is the complex number defined by

$$(a, b)(c, d) = (ac - bd, ad + cb) \tag{5.2}$$

Note that on the right-hand sides of Eq. (5.1) and Eq. (5.2) we use the usual operations of addition and multiplication of real numbers in each coordinate of the ordered pair. Also, note that the operation of addition is the usual operation of addition of ordered pairs of real numbers (which was introduced in Section 1.1).

The subset of $\mathbf{C}$ consisting of those elements with second coordinate zero, $\{(a, 0) \mid a \in \mathbf{R}\}$, will be identified with the real numbers in the obvious way, $a \in \mathbf{R}$ is identified with $(a, 0) \in \mathbf{C}$. If we apply our rules of addition and multiplication to the subset $\mathbf{R} \subset \mathbf{C}$, we obtain

$$(a, 0) + (c, 0) = (a + c, 0)$$

and

$$(a, 0)(c, 0) = (ac - 0 \cdot 0, a \cdot 0 + c \cdot 0) = (ac, 0)$$

Therefore, we have at least satisfied our first general rule.

From our study of linear algebra in Chapter 1, we know that every ordered pair (a, b) is of the form $a(1, 0) + b(0, 1)$. From the previous paragraph we know that $(1, 0)$ is the real number 1. What is the complex number $(0, 1)$? To answer this question, let us square $(0, 1)$:

$$(0, 1)(0, 1) = (0 \cdot 0 - 1 \cdot 1, 0 \cdot 1 + 0 \cdot 1) = (-1, 0)$$

So $(0, 1)$ is a complex number whose square is -1. Immediately we see that we have found a root for the polynomial $p(z) = z^2 + 1$! What is the other root?

It is standard practice to denote the complex number $(1, 0)$ by 1 and the complex number (0, 1) by i or $\sqrt{-1}$. Using these conventions we write $a + bi$ for $a(1, 0) + b(0, 1)$ so that $a + bi$ is the ordered pair $(a, b) \in \mathbf{C}$.

(5.1.2) Definition. Let $z = a + bi \in \mathbf{C}$. The *real part of* z, denoted $\text{Re}(z)$, is the real number a. The *imaginary part of* z, denoted $\text{Im}(z)$, is the real number b. z is called a *real number* if $\text{Im}(z) = 0$ and *purely imaginary* if $\text{Re}(z) = 0$.

Finally, note that in this form addition and multiplication of complex numbers become

$$(a + bi) + (c + di) = (a + c) + (b + d)i$$

and

$$(a + bi)(c + di) = (ac - bd) + (ad + bc)i$$

(5.1.3) Example. The reader should verify the following calculations:

a) $(1 + i)(3 - 7i) = 10 - 4i$,

b) $i^4 = 1$,

c) $(-i)i = 1$.

Our informal rule (b) required that addition and multiplication of complex numbers satisfy the same properties as addition and multiplication of real numbers. As we did in the text with the definition of a vector space, we set down these properties in the form of a definition, the definition of an algebraic object called a *field*. There are numerous examples of fields besides the real and complex numbers, for instance, the rational numbers, and the integers modulo a prime number (see Exercises 9 through 14), to name a few of the simpler examples. The study of the properties of fields is a large part of the subject matter of a course in abstract algebra.

(5.1.4) Definition. A *field* is a set F with two operations, defined on ordered pairs of elements of F, called addition and multiplication. Addition assigns to the pair x and $y \in F$ their *sum*, which is denoted by $x + y$ and multiplication assigns to the pair x and $y \in F$ their *product*, which is denoted by $x \cdot y$ or xy. These two operations must satisfy the following properties for all x, y, and $z \in F$:

(i) Commutativity of addition: $x + y = y + x$.

(ii) Associativity of addition: $(x + y) + z = x + (y + z)$.

(iii) Existence of an additive identity: There is an element $0 \in F$, called zero, such that $x + 0 = x$.

(iv) Existence of additive inverses: For each x there is an element $-x \in F$ such that $x + (-x) = 0$.

(v) Commutativity of multiplication: $xy = yx$.

(vi) Associativity of multiplication: $(xy)z = x(yz)$.

(vii) Distributivity: $(x + y)z = xz + yz$ and $x(y + z) = xy + xz$.

(viii) Existence of a multiplicative identity: There is an element $1 \in F$, called 1, such that $x \cdot 1 = x$.

(ix) Existence of multiplicative inverses: If $x \neq 0$, then there is an element $x^{-1} \in F$ such that $xx^{-1} = 1$.

It should be clear to the reader that the real numbers are a field. It is somewhat less obvious that the complex numbers are a field.

(5.1.5) Proposition. The set of complex numbers is a field with the operations of addition and scalar multiplication as defined previously.

Proof: (i), (ii), (v), (vi), and (vii) follow immediately.

(iii) The additive identity is $0 = 0 + 0i$ since

$$(0 + 0i) + (a + bi) = (0 + a) + (0 + b)i = a + bi$$

(iv) The additive inverse of $a + bi$ is $(-a) + (-b)i$.

$$(a + bi) + ((-a) + (-b)i) = (a + (-a)) + (b + (-b))i = 0 + 0i = 0$$

(viii) The multiplicative identity is $1 = 1 + 0 \cdot i$ since

$$(1 + 0 \cdot i)(a + bi) = (1 \cdot a - 0 \cdot b) + (1 \cdot b + 0 \cdot a)i = a + bi$$

(ix) Note first that if $a + bi \neq 0$, then either $a \neq 0$ or $b \neq 0$ and $a^2 + b^2 \neq 0$. Further, note that $(a + bi)(a + (-b)i) = a^2 + b^2$. Therefore

$$(a + bi)\frac{a - ib}{a^2 + b^2} = 1$$

Thus

$$(a + ib)^{-1} = (a - ib)/(a^2 + b^2). \quad \blacksquare$$

As with real numbers, we write $a - bi$ for $a + (-b)i$, a and $b \in \mathbf{R}$; $x - y$ for $x + (-y)$, with x and $y \in \mathbf{C}$; and $1/z$ for z^{-1}, $z \in \mathbf{C}\setminus\{0\}$.

The complex number $a - bi$ is called the *complex conjugate* of $z = a + bi$ and is denoted by $\bar{z}$. With this notation $z^{-1} = \bar{z}/(z\bar{z})$ since $z\bar{z} = a^2 + b^2$.

(5.1.6) Example. Verify the following calculations:

a) $(i)^{-1} = -i$,

b) $(3 - i)/(4 + i) = (11 - 7i)/17$.

The following properties of a field follow directly from the definition and will be left as an exercise.

(5.1.7) Proposition

(i) The additive identity in a field is unique.

(ii) The additive inverse of an element of a field is unique.

(iii) The multiplicative identity of a field is unique.

(iv) The multiplicative inverse of a nonzero element of a field is unique.

We have already seen that addition of complex numbers corresponds to coordinate-wise addition of ordered pairs of real numbers, which we know is simply addition of vectors in the plane. We now want to obtain a similar geometric interpretation of multiplication of complex numbers.

Recall that if $(a, b) \in \mathbf{R}^2$, the length r of the line segment $\overline{(0,0)(a,b)}$ is $r = \sqrt{a^2 + b^2}$. This is also the length of the vector $(a,b) \in \mathbf{R}^2$. If θ is the angle that this segment makes with the positive first coordinate axis, then $a = r{\cdot}\cos(\theta)$ and $b = r{\cdot}\sin(\theta)$. (As in Section 2.1 we make the convention that an angle is positive if it is measured counterclockwise from the positive first coordinate axis.) Or, we could say that the polar coordinates of (a, b) are (r, θ). We now have Definition (5.1.8).

(5.1.8) Definition. The *absolute value* of the complex number $z = a + bi$ is the nonnegative real number $\sqrt{a^2 + b^2}$ and is denoted by $|z|$ or $r = |z|$. The *argument* of the complex number z is the angle θ of the polar coordinate representation of z.

It should be clear that we can now write

$$z = |z|\,(\cos(\theta) + i\,\sin(\theta))$$

for $z = a + bi$. Notice that if θ is replaced by $\theta \pm 2\pi k$, k an integer, we have defined the same complex number z. For example, $i = i{\cdot}\sin(\pi/2 + 2\pi)$.

(5.1.9) Example. In $\mathbf{R}$ the equation $|x| = c$ for $c \geq 0$ has exactly two solutions if $c > 0$, $x = c$, and $x = -c$, and exactly one solution if $c = 0$, $x = 0$. In $\mathbf{C}$, however, the situation is much more interesting. Let c be a real number greater than or equal to zero. Then if $z = a + ib$ satisfies $|z| = c > 0$, we know that $\sqrt{a^2 + b^2} = c$, so that $a^2 + b^2 = c^2$. That is, z lies on the circle of radius c centered at the origin and every point on this circle satisfies $|z| = c$. If $c = 0$, we still have the unique solution $z = 0$. Similarly, if z_0 is a fixed nonzero complex number, the equation $|z - z_0| = c > 0$ defines a circle of radius c centered at $z_0 \in \mathbf{C}$ and if $c = 0$, $|z - z_0| = 0$ is solved uniquely by $z = z_0$. (See Figure 5.1.)

If $z_j = r_j(\cos(\theta_j) + i{\cdot}\sin(\theta_j))$, $j = 1, 2$, are two complex numbers, then using the trigonometric identities for the cosine and sine of the sum of two angles (see Figure 5.2), we obtain

$$\begin{aligned} z_1 z_2 &= (r_1 r_2)[\cos(\theta_1)\cos(\theta_2) - \sin(\theta_1)\sin(\theta_2) \\ &\quad + i(\cos(\theta_1)\sin(\theta_2) + \cos(\theta_2)\sin(\theta_1))] \qquad (5.3) \\ &= r_1 r_2(\cos(\theta_1 + \theta_2) + i{\cdot}\sin(\theta_1 + \theta_2)) \end{aligned}$$

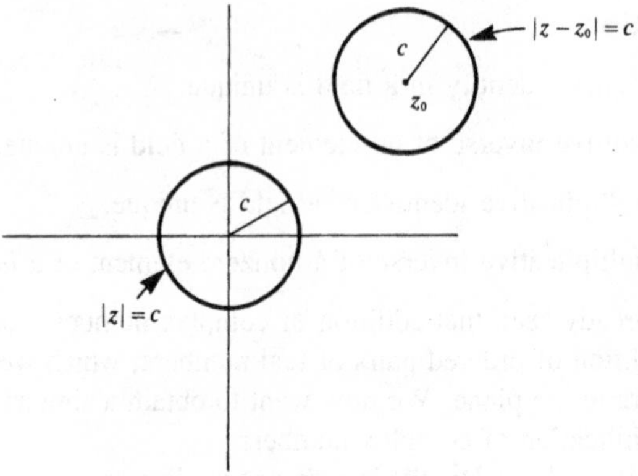

Figure 5.1

This formula gives a clear geometric interpretation of multiplication of complex numbers: To multiply two complex numbers, we add their arguments and take the product of their absolute values.

Formula (5.3) for the product of two complex numbers gives a convenient form for the powers of complex numbers. Thus, if $z = r(\cos(\theta) + i \sin(\theta))$

$$z^2 = r^2(\cos(2\theta) + i \sin(2\theta))$$

and if $z \neq 0$

$$z^{-1} = r^{-1} (\cos(-\theta) + i \cdot \sin(-\theta))$$

In general, if n is an integer

$$z^n = r^n (\cos(n\theta) + i \sin(n\theta)) \tag{5.4}$$

This is known as *de Moivre's formula*.

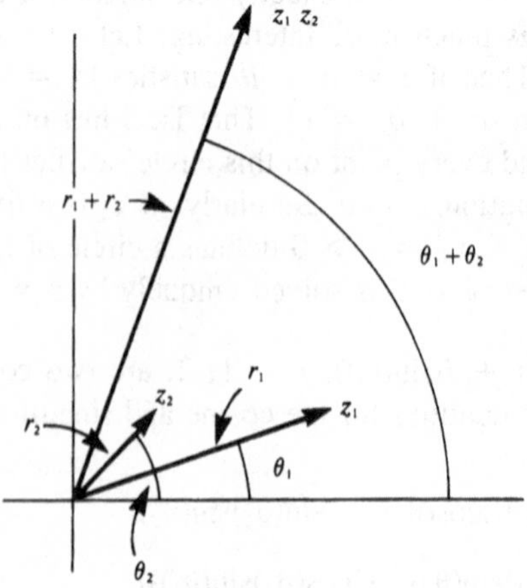

Figure 5.2

Formula (5.4) can be used to find roots of complex numbers. Let $z_0 = r_0(\cos(\theta_0) + i\sin(\theta_0))$ be a fixed complex number. Let n be a positive integer. An nth root of z_0 is a complex number z such that $z^n = z_0$, or equivalently, $z^n - z_0 = 0$. From de Moivre's formula we see that $z = r(\cos(\theta) + i\cdot\sin(\theta))$ must satisfy

$$r^n = r_0 \text{ and } n\theta = \theta_0 + 2\pi k \text{ for } k \text{ an integer} \tag{5.5}$$

[Recall that the argument of a complex number is not unique, so that, in fact, $n\theta$ need only satisfy $n\theta = \theta_0 + (2\pi k)$ for k an integer.]

Since we require both r and r_0 to be nonnegative, it is clear that $r = r_0^{1/n}$ defines r uniquely. On the other hand, θ is of the form

$$\theta = \theta_0/n + 2\pi k/n$$

for k an integer. The values $k = 0, 1, 2, \ldots, n-1$ determine n distinct values for θ. Any other value of k, for example, $k = n$, would yield one of the n values of θ we have already determined. Therefore, the n nth roots of the complex number $z_0 \neq 0$ are

$$r_0^{1/n}\,(\cos(\theta_0/n + 2\pi k/n) + i\sin(\theta/n + 2\pi k/n)) \tag{5.6}$$

$k = 0, 1, \ldots, n-1$.

(5.1.10) Example. We compute the square roots of i using formula (5.6). The complex number i can be expressed in polar coordinates with $r = 1$ and $\theta = \pi/2$, that is, $i = \cos(\pi/2) + i\sin(\pi/2)$. Then in polar coordinates the square roots of i are given by $r = 1$ and $\theta = \pi/4$ and $r = 1$ and $\theta = 5\pi/4$. Thus, the square roots of i are $\pm(\sqrt{2}/2)(1 + i)$. We leave it as an exercise for the reader to verify that the cube roots of i are $(\sqrt{3} + i)/2$, $(-\sqrt{3} + i)/2$, and $-i$.

We have shown that every complex number z_0 has n distinct nth roots. This is equivalent to saying that every polynomial of the form

$$p(z) = z^n - z_0,\ z_0 \in \mathbf{C}$$

has n distinct roots. It is clear that this is not true for the real numbers. For example, 1 is the only real cube root of 1, whereas 1, $\cos(2\pi/3) + i\cdot\sin(2\pi/3)$, and $\cos(4\pi/3) + i\cdot\sin(4\pi/3)$ are complex cube roots of 1.

These facts might lead us to expect that every polynomial of degree n with complex coefficients has n roots. This is true although our previous arguments are not in any sense a proof of this fact. We will state this formally.

(5.1.11) Definition. A field F is called *algebraically closed* if every polynomial $p(z) = a_nz^n + \cdots + a_1z + a_0$ with $a_i \in F$ and $a_n \neq 0$ has n roots counted with multiplicity in F.

(5.1.12) Theorem. $\mathbf{C}$ is algebraically closed and $\mathbf{C}$ is the smallest algebraically closed field containing $\mathbf{R}$.

There are many proofs of this theorem, but all of them are beyond the scope of this course. We will have to be content with the statement of this beautiful result, which is known as the *fundamental theorem of algebra*. It is worth noting that Definition (5.1.11) does not indicate how to find the roots. In general, it is a difficult if not impossible matter to find explicitly the roots of a polynomial whose coefficients are arbitrarily chosen.

EXERCISES

1. Perform the following algebraic operations involving complex numbers:
   a) $(6 + \sqrt{7}i) + (3 + 2i) = ?$
   b) $(5 + 2i)(-1 + i) = ?$
   c) $|17 - i| = ?$
   d) $(3 - 5i)^{-1} = ?$
   e) $(3 + 5i)^{-1} = ?$
   f) $(4 - 2i)/(3 + 5i) = ?$

2. Verify that addition and multiplication of complex numbers satisfy properties (i), (ii), (v), (vi), and (vii) of a field.

3. Prove that if z and w are complex numbers.
   a) $\bar{z} + \bar{w} = \overline{z + w}$
   b) $\bar{z} \cdot \bar{w} = \overline{zw}$
   c) $\bar{z}/\bar{w} = \overline{z/w}$ for $w \neq 0$.
   d) $\bar{z}^{-1} = \overline{(z^{-1})}$ for $z \neq 0$.

4. a) Prove $z \in \mathbf{C}$ is real if and only if $z = \bar{z}$.
   b) Prove $z \in \mathbf{C}$ is purely imaginary if and only if $z = -\bar{z}$.

5. a) Let $p(z)$ be a polynomial with real coefficients; prove that if z is a complex number satisfying $p(z) = 0$, then $p(\bar{z}) = 0$.
   b) Prove that every polynomial with real coefficients has an even number of roots in $\mathbf{C} \setminus \mathbf{R}$.
   c) Deduce from part b that every polynomial of odd degree with real coefficients has a real root.

6. Find all the roots of the following polynomials:
   a) $p(z) = z^2 + 2z + 2$
   b) $p(z) = z^3 - z^2 + 2z - 2$
   c) $p(z) = z^2 + i$
   d) $p(z) = z^3 + (\sqrt{3}/2) + i(1/2)$

7. Prove Proposition (5.1.7).

8. (This problem deals with fields other than $\mathbf{R}$ or $\mathbf{C}$.)
   a) Prove that the set of rational numbers with the usual operations of addition and multiplication is a field.

b) Prove that the set of complex numbers of the form $a + ib$, a and b rational numbers, with the usual operations of complex addition and multiplication is a field.

c) Prove that the set of real numbers of the form $a + (\sqrt{2})b$ where a and b are rational numbers is a field with the usual operations of addition and multiplication of real numbers.

The Finite Fields F_p

In this extended problem we construct a collection of fields, one for each prime number p, which we denote by F_p. These fields prove to be very important in algebra and in the theory of numbers. It is worth noting that unlike the previous examples, the fields F_p are *not* contained in the real numbers or the complex numbers.

First, recall that an integer $p > 1$ is called a prime number if whenever p is factored into a product of integers $p = rs$, then either r or s is equal to 1 or -1. If m is an integer, m modulo p, which we denote by $[m]_p$, is defined to be the remainder r of the division of m by p. That is, we write $m = qp + r$, where r is a nonnegative integer less than p, $r = 0, 1, 2, \ldots$, or $p - 1$. $[m]_p$ is also called the residue class of m modulo p. We call m a representative of $[m]_p$.

9. Prove that $[m]_p = [m']_p$ if and only if $m - m' = qp$ for some integer q.

10. Let m, n, m', n' be integers. Prove that if $[m]_p = [m']_p$ and $[n]_p = [n']_p$, then
   a) $[m + n]_p = [m' + n']_p$. (This proves that addition of residue classes modulo p is well defined, that is, it does not depend on which integer we use to represent $[m]_p$.)
   b) $[mn]_p = [m'n']_p$. (This proves that multiplication of residue classes modulo p is well defined, that is, it does not depend on which integer we use to represent $[m]_p$.)

Definition. F_p is the set of residue classes modulo p, $F_p = \{[0]_p, [1]_p, [2]_p, \ldots, [p-1]_p\}$ with the operations of addition and multiplication defined by $[m]_p + [n]_p = [m + n]_p$ and $[m]_p[n]_p = [mn]_p$.

The operations of addition (multiplication) in F_p can be recorded in a table. Each entry of the table is the sum (product) of the first entry of its row and the first entry of its column. The addition and multiplication tables for F_2 are

$+$	$[0]_2$	$[1]_2$
$[0]_2$	$[0]_2$	$[1]_2$
$[1]_2$	$[1]_2$	$[0]_2$

$\cdot$	$[0]_2$	$[1]_2$
$[0]_2$	$[0]_2$	$[0]_2$
$[1]_2$	$[0]_2$	$[1]_2$

11. Write out the addition and multiplication tables for the fields F_3 and F_5.

12. Verify that F_p satisfies properties (i) through (viii) of a field with $[0]_p$ and $[1]_p$, respectively as the additive and multiplicative identities.

13. In this problem we verify that F_p satisfies property (ix) of a field so that along with Exercise 12 we have shown that F_p is a field. It must be shown that for every $[m]_p \neq [0]_p$, there is an $[n]_p$ so that $[m]_p\,[n]_p = [1]_p$. To simplify the proof, let us assume that $1 \leqslant m \leqslant p - 1$.
    a) For fixed m, prove that if $1 \leqslant n \leqslant p - 1$, then $[m]_p[n]_p \neq [0]_p$.
    b) Prove that if n and n' satisfy $1 \leqslant n,n' \leqslant p - 1$, then $[m]_p[n]_p = [m]_p[n']_p$ implies that $n = n'$.
    c) Conclude that (ix) holds.

14. What are the roots of the following polynomials in F_3 (we use the shorthand notation of writing a for the element $[a]_p$ of F_p.)
    a) $x^2 - x - 2$.
    b) $x^2 - 2x - 1$.
    c) $x^2 - 1$.

15. Prove that F_3 is not algebraically closed.
    (*Hint:* Construct a polynomial with coefficients in F_3 that has fewer roots in F_3 counted with multiplicity than its degree.)

§5.2. VECTOR SPACES OVER A FIELD

In this section we extend the results of the first four chapters to vector spaces over an abstract field F. Although we have in mind that F will be either **R** or **C**, it is important to realize that the definition of a vector space and much of the material developed so far rely on the abstract properties of the operations of addition and multiplication embodied in the definition of a field, rather than on the particulars of real or complex numbers. This program of abstracting the essential properties of familiar objects—like the real or complex numbers—to obtain powerful general results is a cornerstone of modern mathematical thought. In this spirit we begin our discussion with vector spaces over a field and then specialize to vector spaces over **C** in the next section.

Let F be a field [see Definition (5.1.4)]. The definition of a real vector space, Definition (1.1.1) of Chapter 1, used only the fact that **R** is a field. Thus, we may repeat this definition essentially verbatim to define a vector space over a field F.

(5.2.1) Definition. A *vector space* over a field F is a set V (whose elements are called *vectors*) together with

a) an operation called *vector addition*, which for each pair of vectors $\mathbf{x}$, $\mathbf{y} \in V$ produces a vector denoted $\mathbf{x} + \mathbf{y} \in V$, and

b) an operation called *multiplication by a scalar* (a field element), which for each vector $\mathbf{x} \in V$, and each scalar $c \in F$ produces a vector denoted $c\mathbf{x} \in V$.

Furthermore, the two operations must satisfy the following *axioms:*

1) For all vectors $\mathbf{x}$, $\mathbf{y}$, and $\mathbf{z} \in V$, $(\mathbf{x} + \mathbf{y}) + \mathbf{z} = \mathbf{x} + (\mathbf{y} + \mathbf{z})$.

2) For all vectors $\mathbf{x}$ and $\mathbf{y} \in V$, $\mathbf{x} + \mathbf{y} = \mathbf{y} + \mathbf{x}$.

3) There exists a vector $\mathbf{0} \in V$ with the property that $\mathbf{x} + \mathbf{0} = \mathbf{x}$ for all vectors $\mathbf{x} \in V$.

4) For each vector $\mathbf{x} \in V$, there exists a vector denoted $-\mathbf{x}$ with the property that $\mathbf{x} + -\mathbf{x} = \mathbf{0}$.

5) For all vectors $\mathbf{x}$ and $\mathbf{y} \in V$ and all scalars $c \in F$, $c(\mathbf{x} + \mathbf{y}) = c\mathbf{x} + c\mathbf{y}$.

6) For all vectors $\mathbf{x} \in V$, and all scalars c and $d \in F$, $(c + d)\mathbf{x} = c\mathbf{x} + d\mathbf{x}$

7) For all vectors $\mathbf{x} \in V$, and all scalars c and $d \in F$, $(cd)\mathbf{x} = c(d\mathbf{x})$.

8) For all vectors $\mathbf{x} \in V$, $1\mathbf{x} = \mathbf{x}$.

As we noted in the introduction to this chapter, all of our results, except those in Sections 4.3, 4.4, and 4.5, which rely on the inner product in $\mathbf{R}^n$, depend only on this definition and not on the choice of F. Thus, any of these results can be restated, as we did Definition (5.2.1), replacing any reference to real numbers and real vector spaces by a field F and a vector space over that field. Consequently, we give several examples to illustrate the more general setup. The verifications that these examples require will be left for the reader in the exercises.

(5.2.2) Example. Our first example will again be the collection of n-tuples of elements of F, which we denote by F^n,

$$F^n = \{\mathbf{x} = (x_1, \ldots, x_n) | x_i \in F, \quad \text{for } i = 1, \ldots, n\}$$

As before, we define addition and scalar multiplication coordinate by coordinate.

If $\mathbf{x} = (x_1, \ldots, x_n)$ and $\mathbf{y} = (y_1, \ldots, y_n) \in F^n$, we define $\mathbf{x} + \mathbf{y}$ by $\mathbf{x} + \mathbf{y} = (x_1 + y_1, \ldots, x_n + y_n)$. Notice we are defining addition in V, $\mathbf{x} + \mathbf{y}$, using addition in F, $x_i + y_i$. And if $c \in F$, we define $c\mathbf{x}$ by $c\mathbf{x} = (cx_1, \ldots, cx_n)$. Again, $c\mathbf{x}$ is multiplication in V and cx_i is multiplication in F.

It is an easy exercise to verify that F^n is a vector space over F.

(5.2.3) Example. A polynomial with coefficients in F is an expression of the form $p(x) = a_nx^n + a_{n-1}x^{n-1} + \cdots + a_1x + a_0$, where each coefficient $a_i \in F$ and x is an indeterminate. Let $P_n(F)$ be the set of all polynomials of degree less than or equal to n with coefficients in F. Define addition and scalar multiplication in the expected manner.

If $p(x) = a_nx^n + \cdots + a_0$ and $q(x) = b_nx^n + \cdots + b_0 \in P_n(F)$, then we define $(p + q)(x)$ by

$$(p + q)(x) = (a_n + b_n)x^n + \cdots + (a_0 + b_0)$$

If $c \in F$, we define $(cp)(x)$ by $(cp)(x) = (ca_n)x^n + \cdots + (ca_0)$. Again, it is an easy exercise to verify that $P_n(F)$ is a vector space.

(5.2.4) Example. In this example we specialize to the case $F = \mathbf{C}$. Let V be the set of all complex-valued functions of a real variable, $f: \mathbf{R} \to \mathbf{C}$. That is, f assigns to each real number t a complex number $f(t)$. Since $f(t) \in \mathbf{C}$ it can be expressed in the form $a + ib$, and we write $f(t) = u(t) + iv(t)$, where u and v are real-valued functions of t. If $f_j(t) = u_j(t) + iv_j(t)$ for $j = 1, 2$, we define $(f_1 + f_2)(t)$ by $(f_1 + f_2)(t) = (u_1(t) + u_2(t)) + i(v_1(t) + v_2(t))$. If $c = a + ib \in \mathbf{C}$, we define $(cf)(t) = c(u(t) + iv(t)) = (au(t) - bv(t)) + i(bu(t) + av(t))$. Again, it is easy to prove that V is a vector space over $\mathbf{C}$.

If V is a vector space over a field F, then the definition of a vector subspace still holds: $W \subset V$ is a vector subspace if W is a vector space in its own right using the operations inherited from V.

(5.2.5) Examples

a) Let $V = F^n$, and let $a_1, \ldots, a_n \in F$. Define W by $W = \{\mathbf{v} \in V | \; a_1v_1 + \cdots + a_nv_n = 0\}$. We leave it as an exercise for the reader to show that W is a vector subspace of V. Of course, this can be extended to the set of solutions of a system of homogeneous linear equations over the field F.

b) Let $V = P_n(F)$, and let $c \in F$. Define W by $W = \{p(x) \in P_n(F) | p(c) = 0\}$. We leave it to the reader to show that W is a vector subspace of V.

c) Let V be the vector space of functions $f: \mathbf{R} \to \mathbf{C}$ of Example (5.2.4). Let W be the subset of V consisting of differentiable functions. We say that $f: \mathbf{R} \to \mathbf{C}$ is differentiable if its real and imaginary parts are ordinary differentiable functions of t. So if $f(t) = u(t) + iv(t)$, $u(t)$ and $v(t)$ must be differentiable. We leave it as an exercise to verify that W is a vector subspace of V.

Linear dependence and independence are defined as before and therefore basis, dimension, and related concepts still make sense.

(5.2.6) Example. Let $V = F^n$. Let $\mathbf{e}_i$ denote the vector $(0, \ldots, 1, \ldots, 0)$, which has 1 in the ith position and zero in all the other positions. $\alpha = \{\mathbf{e}_1, \ldots, \mathbf{e}_n\}$ is a linearly independent set that spans F^n, that is, it is a basis for F^n. Therefore, $\dim(F^n) = n$ as a vector space over F. Before leaving this example it is important to note two things. First, every field F has both an additive identity and a multiplicative identity, and they are unique so that $\mathbf{e}_i$ makes sense for F any field. Second, the dimension always refers to the dimension of the vector space with respect to the field F. (See Exercise 6 for a related example.)

In Chapter 1 we developed the elimination algorithm for solving systems of equations with real coefficients. This was the primary method in numerical examples for showing that sets of vectors were independent or dependent and for determining the dimensions of subspaces. Since elimination only uses the properties of $\mathbf{R}$ as a field, it can be applied to solve systems of equations over any field:

(5.2.7) Example. We will find a solution to the following system of equations over **C**.

$$\begin{aligned} (1 + i)x_1 + \qquad\quad (3 + i)x_3 &= 0 \\ x_1 - ix_2 + (2 + i)x_3 &= 0 \end{aligned}$$

We add $-(1 + i)^{-1}$ times the first equations to the second. $-(1 + i)^{-1} = (-1 + i)/2$, so we obtain

$$\begin{aligned} (1 + i)x_1 + \qquad\quad (3 + i)x_3 &= 0 \\ - ix_2 + \quad (2i)x_3 &= 0 \end{aligned}$$

Multiply through to make the leading coefficients 1:

$$\begin{aligned} x_1 + \quad + (2 - i)x_3 &= 0 \\ x_2 - \qquad 2x_3 &= 0 \end{aligned}$$

x_3 is the only free variable, so that setting $x_3 = 1$, we obtain the solution $(-2 + i, 2, 1)$.

Now let V and W be vector spaces over a field F. A linear transformation T: $V \to W$ is a function from V to W, which satisfies

$$T(a\mathbf{u} + b\mathbf{v}) = aT(\mathbf{u}) + bT(\mathbf{v})$$

for $\mathbf{u}$ and $\mathbf{v} \in V$ and a and $b \in F$.

(5.2.8) Examples

a) Let $V = F^n$, let $W = F^m$, and let A be an $m \times n$ matrix whose entries are elements of F. If we write elements of V and W as column vectors, then matrix multiplication by A defines a linear transformation from V to W. Given any linear transformation between finite-dimensional vector spaces V and W, we may choose bases for V and W and construct the matrix of the transformation. Of course, we can use this matrix to determine the kernel and image of T.

b) If T: $\mathbf{C}^3 \to \mathbf{C}^2$ is defined by matrix multiplication by the matrix

$$\begin{bmatrix} 1 + i & 0 & 3 + i \\ 1 & -i & 2 + i \end{bmatrix}$$

the calculations of Example (5.2.7) show that $\text{Ker}(T) = \text{Span}\{(-2 + i, 2, 1)\}$ and $\dim(\text{Ker}(T)) = 1$. The dimension theorem then implies that $\dim(\text{Im}(T)) = 2$. We conclude that $\text{Im}(T) = \mathbf{C}^2$ and T is surjective.

c) The determinant of an $n \times n$ matrix with entries in a field F is computed using the formula of Equation (3.1). That is, if A is an $n \times n$ matrix

$$\det(A) = \sum_{j=1}^{n} (-1)^{i+j} a_{ij} \det(A_{ij})$$

We also apply Theorem (3.2.14) to determine whether or not a matrix is invertible. For example, let $T: \mathbf{C}^2 \to \mathbf{C}^2$ be defined by matrix multiplication by the matrix

$$\begin{bmatrix} -i & 2+3i \\ 1 & 1-i \end{bmatrix}$$

In order to determine if T is invertible, we calculate the determinant of the matrix. Since $\det(T) = (-i) \cdot (1 - i) - (2 + 3i) \cdot 1 = -3 - 4i \neq 0$, we conclude that T is invertible. Further, we can apply Corollary (3.3.2) to calculate the inverse of T. In this example, the matrix of T^{-1} is

$$\frac{-3+4i}{25}\begin{bmatrix} 1-i & -2-3i \\ -1 & -i \end{bmatrix}$$

As we emphasized in Section 3.2, if the matrix is larger than 3×3, this is not an effective method for calculating the inverse of the matrix. It is simpler to calculate the inverse using the Gauss-Jordan procedure for larger matrices.

(5.2.9) Example. Let V be the vector space of differentiable functions $f: \mathbf{R} \to \mathbf{C}$. Define $T: V \to V$ by $T(f) = df/dt$ $(= du/dt + i\, dv/dt)$. We leave it to the reader to verify that T is a complex linear transformation.

Now let us consider two eigenvalue-eigenvector calculations. The results of Sections 4.1 and 4.2 still hold, in particular, Theorem (4.2.7) and its corollaries. To apply this result to a transformation over another field F, we proceed with the algebra as we did in Chapter 4. Let us consider an example over $F = \mathbf{C}$.

(5.2.10) Example. Let $T: \mathbf{C}^2 \to \mathbf{C}^2$ be defined by the 2×2 matrix

$$A = \begin{bmatrix} i & 1-i \\ 1 & -i \end{bmatrix}$$

To find the eigenvalues of A, evaluate $\det(A - \lambda I)$ and solve for λ. The characteristic polynomial is $\lambda^2 + i$ so that the roots are the complex square roots of $-i$, that is, $\lambda = \pm(\sqrt{2}/2)(1 - i)$. Since the eigenvalues are distinct, we conclude that A is diagonalizable by the complex version of Corollary (4.2.8). The eigenvectors of A are nonzero elements of $\text{Ker}(A - \lambda I)$. We leave it to the reader to find the eigenvectors of A.

We have not yet seen how to treat the matrix $A = \begin{bmatrix} 0 & -1 \\ 1 & 0 \end{bmatrix}$, which was the initial example in our introduction. This matrix has real entries and distinct complex eigenvalues, $\pm i$. As we noted, this matrix cannot be diagonalized as a real matrix.

In general, if A is any $n \times n$ matrix with real entries, we may view A as a matrix with complex entries, since $\mathbf{R} \subset \mathbf{C}$. Viewed as a linear transformation, this matrix will act on $\mathbf{C}^n$ rather than $\mathbf{R}^n$. This allows us to utilize the algebra of the complex numbers. The fundamental theorem of algebra [Theorem (5.1.12)] can now be applied to the characteristic polynomial of the matrix A, $\det(A - \lambda I)$. As a consequence, $\det(A - \lambda I)$ has n complex roots counted with multiplicity. Thus, in order to apply Theorem (4.2.7) to a particular matrix or transformation, we need only determine if the dimension of the eigenspace of the eigenvalue λ is equal to the multiplicity of λ for each eigenvalue λ of the matrix or transformation. If A is a real matrix with complex eigenvalues that satisfies the hypothesis of Theorem (4.2.7) over **C**, the resulting diagonal matrix will necessarily have complex entries on the diagonal (the eigenvalues) and the eigenvectors will have complex entries. Let us return to our 2×2 example.

The eigenvalues of $A = \begin{bmatrix} 0 & -1 \\ 1 & 0 \end{bmatrix}$ are distinct complex numbers, $\pm i$, so by Corollary (4.2.8) applied to complex vector spaces and transformations, the matrix A is diagonalizable. The matrix A is similar to the diagonal matrix $\Lambda = \begin{bmatrix} i & 0 \\ 0 & -i \end{bmatrix}$. The eigenvectors of A are found by finding vectors in the kernel of $A - (\pm i)I$. The eigenvectors for i and $-i$ are $(i, 1)$ and $(-i, 1)$, respectively. If we let $Q = \begin{bmatrix} i & -i \\ 1 & 1 \end{bmatrix}$, the matrix of eigenvectors, it is easy to verify that $Q^{-1}AQ = \Lambda$.

These arguments concerning the diagonalization of real matrices over the complex numbers can also be applied to linear transformations. If $T: \mathbf{R}^n \to \mathbf{R}^n$ is a linear transformation, we extend T to a complex linear transformation $T: \mathbf{C}^n \to \mathbf{C}^n$. It suffices to give an $n \times n$ matrix with complex entries that represents T with respect to the standard basis α' for $\mathbf{C}^n$. As in the previous discussion, we take $[T]_{\alpha'}^{\alpha'}$ to be the matrix $[T]_{\alpha}^{\alpha}$ of T with respect to the standard basis of $\mathbf{R}^n$. Since the entries of $[T]_{\alpha}^{\alpha}$ are real numbers, hence complex numbers, $[T]_{\alpha'}^{\alpha'} = [T]_{\alpha}^{\alpha}$ is an $n \times n$ complex matrix. We can proceed to find the eigenvalues and eigenvectors of $T: \mathbf{C}^n \to \mathbf{C}^n$ by working with $[T]_{\alpha'}^{\alpha'}$ over **C**.

Although this approach to real matrices involves a trade-off—in order to diagonalize the matrix or transformation (when possible), we must replace **R** by **C**—this trade-off does have two benefits. First, as we will see in Chapter 6, we will be able to find a "canonical form" for all matrices and transformations over **C** that is close to diagonal form. Thus, from the previous comments these results will also apply to real matrices and transformations. Second, there are explicit calculations with real matrices, for example, the calculation of the exponential of a matrix in the theory of differential equations (which we will see in Chapter 7), which are best accomplished by using the diagonal form (or the canonical form) of a real matrix whether or not this diagonal matrix has complex entries.

In Exercise 12 we will consider other examples of diagonalizing real matrices over **C**.

EXERCISES

1. a) Verify that F^n is a vector space over F [See Example (5.2.2)].
   b) Verify that $P_n(F)$ is a vector space over F [See Example (5.2.3)].
   c) Verify that V of Example (5.2.4) is a complex vector space.

2. Verify that each set W in Example (5.2.5) is a subspace of the given V.

3. a) Prove that $\alpha = \{\mathbf{e}_1, \ldots, \mathbf{e}_n\}$ is a basis for F^n.
   b) Prove that $\alpha = \{1, x, \ldots, x^n\}$ is a basis for $P_n(F)$.

4. Are the following sets of vectors linearly independent or dependent in $\mathbf{C}^3$? What is the dimension of their span?
   a) $\{(1, i, 0), (i, -1, 0), (i, i, i)\}$
   b) $\{(1 - i, 0, i), (1 + i, 0, -i), (0, i, 1)\}$
   c) $\{(3 - i, 2, -i), (1, 0, 1), (i, -i, 0), (0, 1 + i, 1 - i)\}$.

5. Find a basis for the kernel and the image of the transformation $T: \mathbf{C}^k \to \mathbf{C}^l$ whose matrix is

   a) $\begin{bmatrix} 1 - i & i \\ 2 & -1 + i \end{bmatrix}$ b) $\begin{bmatrix} 2 & i & 1 + i \\ 1 - i & 1 & 2 \\ 0 & 1 - i & 2 \end{bmatrix}$

   c) $\begin{bmatrix} i & 0 & 1 & 0 \\ 0 & -i & 0 & -2 \\ i & 0 & 1 & 0 \\ 0 & i & 0 & 2 \end{bmatrix}$

6. This exercise explores the relationship between $\mathbf{R}^n$ and $\mathbf{C}^n$.
   a) Let $\hat{\mathbf{C}}^n$ be the set of vectors in $\mathbf{C}^n$ with the operations of coordinatewise addition and scalar multiplication by real scalars. Prove that $\hat{\mathbf{C}}^n$ is a real vector space.
   b) Prove that $\beta = \{\mathbf{e}_1, i\mathbf{e}_1, \ldots, \mathbf{e}_n, i\mathbf{e}_n\}$ as a basis for $\hat{\mathbf{C}}^n$. Thus $\dim(\hat{\mathbf{C}}^n) = 2n$ is a real vector space.
   c) Define the inclusion mapping Inc: $\mathbf{R}^n \to \hat{\mathbf{C}}^n$ by $\text{Inc}(\mathbf{x}) = (x_1, 0, x_2, 0, \ldots, x_n, 0) \in \hat{\mathbf{C}}^n$, where $\mathbf{x}$ is expressed in terms of the standard coordinates in $\mathbf{R}^n$ and Inc($\mathbf{x}$) is expressed in terms of β. Prove that Inc is an injective linear transformation.
   d) Since $\hat{\mathbf{C}}^n$ and $\mathbf{C}^n$ are equal as sets and Inc is injective, we can view $\mathbf{R}^n$ as a subset of $\hat{\mathbf{C}}^n$ and of $\mathbf{C}^n$ by identifying $\mathbf{R}^n$ with the image of Inc. Prove that $\mathbf{R}^n \subset \hat{\mathbf{C}}^n$ is a real vector subspace of $\hat{\mathbf{C}}^n$ but that $\mathbf{R}^n$ is not a complex vector subspace of $\mathbf{C}^n$.

7. Verify that the transformation $T = d/dt$ of Example (5.2.9) is a complex linear transformation.

8. Let $T: \mathbf{C}^n \to \mathbf{C}^n$ be defined by $T(\mathbf{x}) = \bar{\mathbf{x}}$, that is, if $\mathbf{x} = (x_1, \ldots, x_n)$, then $T(\mathbf{x}) = (\bar{x}_1, \ldots, \bar{x}_n)$.
   a) Prove that T is *not* a linear transformation over $\mathbf{C}$.
   b) Prove that T is a linear transformation over $\mathbf{R}$ if we consider $\mathbf{C}^n$ to be a real vector space (see Exercise 6). What is the matrix of T with respect to the basis of Exercise 6?

9. Find the eigenvectors of the matrix of Example (5.2.10).

10. Find the eigenvalues and eigenvectors of the following matrices over **C**.

a) $\begin{bmatrix} 1 & i \\ -i & 1 \end{bmatrix}$ b) $\begin{bmatrix} 0 & 2 \\ -2 & 0 \end{bmatrix}$

c) $\begin{bmatrix} i & 2 & 1 \\ & -i & 1 \\ & & i \end{bmatrix}$ d) $\begin{bmatrix} 0 & 1 & 0 & 1 \\ -1 & 0 & 0 & 0 \\ 0 & 0 & i & 1 \\ 0 & 0 & 0 & -i \end{bmatrix}$

11. Let $V = F^2$, where $F = F_3$ the field with three elements. For each of the following matrices A with entries in F if possible (!) find the eigenvalues and eigenvectors in F^2. Which of these matrices are diagonalizable over F_3?

a) $\begin{bmatrix} 1 & 2 \\ 0 & 1 \end{bmatrix}$ b) $\begin{bmatrix} 2 & 1 \\ 1 & 1 \end{bmatrix}$ c) $\begin{bmatrix} 2 & 1 \\ 1 & 2 \end{bmatrix}$

12. For the following types of 2×2 real matrices, find their eigenvalues and eigenvectors over **C**.

a) Rotation matrices, $R_\theta = \begin{bmatrix} \cos(\theta) & -\sin(\theta) \\ \sin(\theta) & \cos(\theta) \end{bmatrix}$

b) Skew symmetric matrices, $A = \begin{bmatrix} 0 & -a \\ a & 0 \end{bmatrix}$, $a \neq 0$.

13. Let A be an $n \times n$ matrix over the field F_p. Is it still true that A invertible if and only if $\det(A) \neq 0$?

14. Find the inverses of the following matrices with entries in F_3.

a) $\begin{bmatrix} 1 & 2 \\ 0 & 1 \end{bmatrix}$ b) $\begin{bmatrix} 2 & 1 \\ 1 & 1 \end{bmatrix}$ c) $\begin{bmatrix} 1 & 0 & 1 \\ 1 & 2 & 1 \\ 1 & 1 & 2 \end{bmatrix}$

15. Let $F = \mathbf{Q}$, the field of rational numbers, and let $V = \mathbf{R}$.

a) Prove that **R** with the usual operation of addition of real numbers and scalar multiplication of a real number by a rational number (the rational numbers are the scalars!) is a rational vector space.

b) Prove that the set $\{1, \sqrt{2}\}$ is a linearly independent set in V.

c) Prove that the set $\{\sqrt{2}, \sqrt{3}\}$ is a linearly independent set in V.

d) Prove that the set $\{\sqrt{p}, \sqrt{q}\}$ is a linearly independent set in V for any pair of prime numbers p and q, $p \neq q$. Does this suggest anything about the dimension of V as a vector space over **Q**?

§5.3. GEOMETRY IN A COMPLEX VECTOR SPACE

The development of complex vector spaces, like that of real vector spaces, diverges from that of a vector space over an arbitrary field F when we introduce an inner product and consider questions of a geometric nature. Before reading this section the reader may wish to review Section 4.3. The inner products that we will discuss here are called Hermitian inner products to distinguish them from inner products in real vector spaces.

(5.3.1) Definition. Let V be a complex vector space. A *Hermitian inner product* on V is a complex valued function on pairs of vectors in V, denoted by $\langle \mathbf{u}, \mathbf{v} \rangle \in \mathbf{C}$ for $\mathbf{u}, \mathbf{v} \in V$, which satisfies the following properties:

a) For all $\mathbf{u}$, $\mathbf{v}$, and $\mathbf{w} \in V$ and $a, b, \in \mathbf{C}$, $\langle a\mathbf{u} + b\mathbf{v}, \mathbf{w} \rangle = a\langle \mathbf{u}, \mathbf{w} \rangle + b\langle \mathbf{v}, \mathbf{w} \rangle$,

b) For all $\mathbf{u}, \mathbf{v}, \in V$, $\langle \mathbf{u}, \mathbf{v} \rangle = \overline{\langle \mathbf{v}, \mathbf{u} \rangle}$, and

c) For all $\mathbf{v} \in V$, $\langle \mathbf{v}, \mathbf{v} \rangle \geqslant 0$ and $\langle \mathbf{v}, \mathbf{v} \rangle = 0$ implies $\mathbf{v} = \mathbf{0}$.

We call a complex vector space with a Hermitian inner product a Hermitian inner product space.

As a consequence of part b, $\langle \mathbf{v}, \mathbf{v} \rangle = \overline{\langle \mathbf{v},\mathbf{v} \rangle}$, which implies that $\langle \mathbf{v}, \mathbf{v} \rangle \in \mathbf{R}$ (see Exercise 4 of Section 5.1). Thus, it makes sense in part c to say that $\langle \mathbf{v}, \mathbf{v} \rangle \geqslant 0$. This is an important point because it ensures that the length of a vector [Definition (5.3.3) below] is a non-negative real number. A further consequence of part b is that the inner product is conjugate linear in the second variable rather than linear. For if $\mathbf{u}, \mathbf{v} \in V$ and $a \in \mathbf{C}$, $\langle \mathbf{u}, a\mathbf{v} \rangle = \overline{\langle a\mathbf{v}, \mathbf{u} \rangle} = \overline{a\langle \mathbf{v}, \mathbf{u} \rangle} = \overline{a}\,\overline{\langle \mathbf{v}, \mathbf{u} \rangle} = \overline{a}\,\langle \mathbf{u}, \mathbf{v} \rangle$. (This property is sometimes called *sesquilinearity*.) So we see that part b makes both a Hermitian inner product similar to a real inner product—$\langle \mathbf{v}, \mathbf{v} \rangle \in \mathbf{R}$—and different from a real inner product—$\langle\ ,\ \rangle$ is not linear in both variables.

(5.3.2) Example. Our first example is the standard Hermitian inner product on $\mathbf{C}^n$. For $\mathbf{x} = (x_1, \ldots, x_n)$ and $\mathbf{y} = (y_1, \ldots, y_n) \in \mathbf{C}^n$ we define their inner product by $\langle \mathbf{x}, \mathbf{y} \rangle = x_1\overline{y}_1 + \cdots + x_n\overline{y}_n$. We leave the proof that this a Hermitian inner product to the reader. We will see other examples in Exercises 12, 13, and 14.

Our development of Hermitian inner products will parallel the development of real inner products in Chapter 4. We begin with the basic definitions concerning vectors and sets of vectors.

(5.3.3) Definition. Let V be a complex vector space with a Hermitian inner product. The *norm* or *length* of a vector $\mathbf{v} \in V$ is $\|\mathbf{v}\| = \langle \mathbf{v}, \mathbf{v} \rangle^{1/2}$. A set of nonzero vectors $\mathbf{v}_1, \ldots, \mathbf{v}_k \in V$ is called *orthogonal* if $\langle \mathbf{v}_i, \mathbf{v}_j \rangle = 0$ for $i \neq j$. If in addition $\langle \mathbf{v}_i, \mathbf{v}_i \rangle = 1$, for all i, the vectors are called *orthonormal*.

It is an easy exercise to verify that the standard basis in $\mathbf{C}^n$ is an orthonormal basis. The advantage of an orthonormal basis is that it simplifies most calculations. For example, as in the real case, if $\alpha = \{\mathbf{x}_1, \ldots, \mathbf{x}_n\}$ is an orthonormal basis for a complex vector space V with Hermitian inner product $\langle\ ,\ \rangle$, and if

$\mathbf{v} = v_1\mathbf{x}_1 + \cdots + v_n\mathbf{x}_n$, then $v_i = \langle \mathbf{v}, \mathbf{x}_i \rangle$. Further, if $\mathbf{u} = u_1\mathbf{x}_1 + \cdots + u_n\mathbf{x}_n$, then $\langle \mathbf{u}, \mathbf{v} \rangle = u_1\bar{v}_1 + \cdots + u_n\bar{v}_n$. Thus, if we choose an orthonormal basis, the calculation of inner products is the same as that with the standard basis in $\mathbf{C}^n$.

(5.3.4) Example. Since the Gram-Schmidt process depends only on the properties of an inner product, it applies in the Hermitian case. Let us apply Gram-Schmidt to the pair of vectors $\mathbf{v}_1 = (1 + i, i)$ and $\mathbf{v}_2 = (-i, 2 - i)$ in $\mathbf{C}^2$. The corresponding orthonormal pair of vectors is

$$\mathbf{w}_1 = \mathbf{v}_1/\|\mathbf{v}_1\| = (1/\sqrt{3})(1 + i, i)$$

$$\mathbf{w}_2 = (\mathbf{v}_2 - \langle \mathbf{v}_2, \mathbf{w}_1 \rangle \mathbf{w}_1)/\|\mathbf{v}_2 - \langle \mathbf{v}_2, \mathbf{w}_1 \rangle \mathbf{w}_1\| = (1/\sqrt{15})(-1 + 2i, 3 - i)$$

Since the Gram-Schmidt process works in this new context, orthogonal projections do as well.

Notice that if $\mathbf{v}$ is a vector in a Hermitian inner product space and $a \in \mathbf{C}$, then $\|a\mathbf{v}\| = \langle a\mathbf{v}, a\mathbf{v} \rangle^{1/2} = (a\bar{a})^{1/2} \cdot \langle \mathbf{v}, \mathbf{v} \rangle^{1/2} = |a| \cdot \|\mathbf{v}\|$. If $a \in \mathbf{C}$ and $|a| = 1$, $\|a\mathbf{v}\| = \|\mathbf{v}\|$; further, if $\mathbf{v}$ is a unit vector, then $a\mathbf{v}$ is also a unit vector. However, there is a difference between the Hermitian case and the real case. There is an infinite number of choices of unit vector $\mathbf{w} = a\mathbf{v}$, $|a| = 1$, having the same span as the unit vector $\mathbf{v}$, because the set of complex numbers of length one is a circle in $\mathbf{C}$ [Example (5.1.9)], not just $\{+1, -1\}$ as in $\mathbf{R}$. Thus, if $\{\mathbf{w}_1, \ldots, \mathbf{w}_k\}$ is an orthonormal set in a Hermitian inner product space, there is an infinite number of orthonormal sets $\{\mathbf{w}_1', \ldots, \mathbf{w}_k'\}$ with $\mathbf{w}_i'$ equal to a scalar multiple of $\mathbf{w}_i$, for $i = 1, \ldots, k$.

Just as in the case of real matrices and real linear transformations, introducing an inner product in the complex case will allow us to obtain more detailed information about the structure of an individual transformation. Let us review a few of the basics of the real case.

If A is a real $n \times n$ matrix with entries a_{ij}, its transpose A^t is the $n \times n$ matrix with entries a_{ji}. Proposition (4.5.2a), part 1 reinterprets this definition in terms of the standard inner product on $\mathbf{R}^n$; that is, A^t satisfies $\langle A\mathbf{x}, \mathbf{y} \rangle = \langle \mathbf{x}, A^t\mathbf{y} \rangle$ for all $\mathbf{x}$ and $\mathbf{y} \in \mathbf{R}^n$. The matrix A is called symmetric if $A = A^t$, or equivalently, if $\langle A\mathbf{x}, \mathbf{y} \rangle = \langle \mathbf{x}, A\mathbf{y} \rangle$ for all $\mathbf{x}$ and $\mathbf{y} \in \mathbf{R}^n$ [Proposition (4.5.2a) part 2]. Similarly, if $T: \mathbf{R}^n \to \mathbf{R}^n$ is a linear transformation, we say T is symmetric if $\langle T(\mathbf{x}), \mathbf{y} \rangle = \langle \mathbf{x}, T(\mathbf{y}) \rangle$ for all $\mathbf{x}$ and $\mathbf{y} \in \mathbf{R}^n$. If α is an orthonormal basis for $\mathbf{R}^n$, a symmetric transformation T has a symmetric matrix with respect to α, $[T]_\alpha^\alpha = ([T]_\alpha^\alpha)^t$.

Before we proceed with the Hermitian version of the preceding definitions, we introduce the definition of the transpose of a linear transformation. Let $T: \mathbf{R}^n \to \mathbf{R}^n$ be a linear transformation. We define the transpose of T to be the unique linear transformation T^t that satisfies $\langle T(\mathbf{x}), \mathbf{y} \rangle = \langle \mathbf{x}, T^t(\mathbf{y}) \rangle$ for all $\mathbf{x}$ and $\mathbf{y} \in \mathbf{R}^n$. (This definition was not needed for the work in Chapter 4.) Let α be an orthonormal basis for $\mathbf{R}^n$, then the proof of Proposition (4.5.2) shows that the matrix of T^t with respect to α must be the matrix $([T]_\alpha^\alpha)^t$. Since we have specified a matrix for the transformation T^t we have in fact constructed the transformation.

We require the Hermitian version of the calculation which occurs in Proposition (4.5.2). Let V be a Hermitian inner product space and let $T: V \to V$ be a linear transformation. If $\alpha = \{\mathbf{v}_1, \ldots, \mathbf{v}_n\}$ is an orthonormal basis for V, we can compute $[T]_\alpha^\alpha$ in a particularly simple manner. First, the columns of $[T]_\alpha^\alpha$ are the coordinate vectors $[T(\mathbf{v}_j)]_\alpha$, which express $T(\mathbf{v}_j)$ in terms of the basis

$$[T(\mathbf{v}_j)]_\alpha = \begin{bmatrix} a_{1j} \\ \vdots \\ a_{nj} \end{bmatrix}$$

where $T(\mathbf{v}_j) = a_{1j}\mathbf{v}_1 + \cdots + a_{nj}\mathbf{v}_n$. Since α is an orthonormal basis $\langle T(\mathbf{v}_j), \mathbf{v}_i\rangle = a_{ij}$. Thus, we have proven the following lemma.

(5.3.5) Lemma. If $\alpha = \{\mathbf{v}_1, \ldots, \mathbf{v}_n\}$ is an orthonormal basis for the Hermitian inner product space V and $T: V \to V$ is a linear transformation, then the ijth entry of $[T]_\alpha^\alpha$ is $\langle T(\mathbf{v}_j), \mathbf{v}_i\rangle$.

(5.3.6) Example. Let $T: \mathbf{C}^2 \to \mathbf{C}^2$ be defined by multiplication in standard coordinates by the matrix $A = \begin{bmatrix} i & 2i \\ 0 & 1-i \end{bmatrix}$. Consider the orthonormal basis $\alpha = \{\mathbf{v}_1, \mathbf{v}_2\}$ with $\mathbf{v}_1 = 1/\sqrt{2}\,(1, i)$, $\mathbf{v}_2 = 1/\sqrt{2}\,(i, 1)$. Let us compute $[T]_\alpha^\alpha$ by computing $\langle T(\mathbf{v}_j), \mathbf{v}_i\rangle$:

$$\langle T(\mathbf{v}_1), \mathbf{v}_1\rangle = \langle (1/\sqrt{2})(-2+i, 1+i), (1/\sqrt{2})(1, i)\rangle = -1/2$$

$$\langle T(\mathbf{v}_1), \mathbf{v}_2\rangle = \langle (1/\sqrt{2})(-2+i, 1+i), (1/\sqrt{2})(i, 1)\rangle = (2+3i)/2$$

$$\langle T(\mathbf{v}_2), \mathbf{v}_1\rangle = \langle (1/\sqrt{2})(-1+2i, 1-i), (1/\sqrt{2})(1, i)\rangle = (i-2)/2$$

$$\langle T(\mathbf{v}_2), \mathbf{v}_2\rangle = \langle (1/\sqrt{2})(-1+2i, 1-i), (1/\sqrt{2})(i, 1)\rangle = 3/2$$

Therefore

$$[T]_\alpha^\alpha = \begin{bmatrix} -1/2 & (-2+i)/2 \\ (2+3i)/2 & 3/2 \end{bmatrix}$$

Let V be a finite dimensional Hermitian inner product space. We want to apply the lemma to compute the "complex transpose" of $T: V \to V$. Let us denote the "complex transpose" of T by T^*. We want to construct T^* so that it satisfies the equation $\langle T(\mathbf{v}), \mathbf{w}\rangle = \langle \mathbf{v}, T^*(\mathbf{w})\rangle$ for all $\mathbf{v}$ and $\mathbf{w} \in V$. It is sufficient that we construct a matrix for T^* with respect to a basis for V. If $\alpha = \{\mathbf{v}_1, \ldots, \mathbf{v}_n\}$ is an orthonormal basis for V, the ijth entry of $[T^*]_\alpha^\alpha$ is the complex number $\langle T^*(\mathbf{v}_j), \mathbf{v}_i\rangle$. Let us use the properties of the Hermitian inner product and the equation $\langle T(\mathbf{v}), \mathbf{w}\rangle = \langle \mathbf{v}, T^*(\mathbf{w})\rangle$ to simplify this expression.

$$\langle T^*(\mathbf{v}_j), \mathbf{v}_i\rangle = \overline{\langle \mathbf{v}_i, T^*(\mathbf{v}_j)\rangle} = \overline{\langle T(\mathbf{v}_i), \mathbf{v}_j\rangle}$$

Now let us rewrite this in terms of matrices. The transpose of a matrix A with complex entries is defined as in the real case. If A has entries a_{ij}, A^t has entries a_{ji}. Further, if A is a complex matrix, we denote by $\overline{A}$ the matrix whose entries are $\overline{a}_{ij}$. Thus we could say that the ijth entry of $[T^*]_\alpha^\alpha$ is the ijth entry of the transpose of the conjugate of $[T]_\alpha^\alpha$, or $[T^*]_\alpha^\alpha = ([\overline{T}]_\alpha^\alpha)^t$. Thus we define T^* in Definition (5.3.7).

(5.3.7) Definition. Let V be a finite dimensional Hermitian inner product space and let α be an orthonormal basis for V. The *adjoint* of the linear transformation $T: V \to V$ is the linear transformation T^* whose matrix with respect to the orthonormal basis α is the matrix $([\overline{T}]_\alpha^\alpha)^t$; that is, $[T^*]_\alpha^\alpha = ([\overline{T}]_\alpha^\alpha)^t$.

The following proposition is a consequence of our previous calculations.

(5.3.8) Proposition. Let V be a finite dimensional Hermitian inner product space. The adjoint of $T: V \to V$ satisfies $\langle T(\mathbf{v}), \mathbf{w}\rangle = \langle \mathbf{v}, T^*(\mathbf{w})\rangle$ for all $\mathbf{v}$ and $\mathbf{w} \in V$.

Proof: From the sesquilinearity of $\langle\ ,\ \rangle$, it suffices to prove the proposition for a basis of V. However, this is exactly the calculation which preceded the definition. ■

As a consequence of the proposition, we define the *adjoint* of an $n \times n$ complex matrix A to be the complex conjugate of the transpose of A, $A^* = \overline{A}^t$. Thus, in Example (5.3.6),

$$A^* = \begin{bmatrix} -i & 0 \\ -2i & 1+i \end{bmatrix} \quad \text{and} \quad [T^*]_\alpha^\alpha = \begin{bmatrix} -1/2 & (2-3i)/2 \\ (-2-i)/2 & 3/2 \end{bmatrix}$$

Notice that if A is an $n \times n$ matrix with real entries—remember that these are also complex numbers since $\mathbf{R} \subset \mathbf{C}$—then $A^* = \overline{A}^t = A^t$. Thus the adjoint of a real matrix is just its transpose. Conversely, if A is an $n \times n$ matrix with complex entries and $A^* = A^t$, then $\overline{a}_{ij} = a_{ij}$. By Exercise 4a of Section 5.1, we conclude that $a_{ij} \in \mathbf{R}$ and A is a matrix of real numbers.

Of course, we now wish to extend the concept of a symmetric matrix to transformations on Hermitian vector spaces.

(5.3.9) Definition. $T: V \to V$ is called *Hermitian* or *self-adjoint* if $\langle T(\mathbf{u}), \mathbf{v}\rangle = \langle \mathbf{u}, T(\mathbf{v})\rangle$ for all $\mathbf{u}$ and $\mathbf{v} \in V$. Equivalently, T is Hermitian or self-adjoint if $T = T^*$ or $[\overline{T}]_\alpha^{\alpha t} = [T]_\alpha^\alpha$ for an orthonormal basis α. An $n \times n$ complex matrix is called *Hermitian* or *self-adjoint* if $A = A^*$.

In particular, if T (or A) is self-adjoint, then the diagonal entries of $[T]_\alpha^\alpha$ (or A) are necessarily real numbers since they are equal to their own conjugates. Let us consider an example: The matrix $A = \begin{bmatrix} 2 & 1+i \\ 1-i & -2 \end{bmatrix}$ is self-adjoint, $A^* = A$. Further, notice that its eigenvalues are $\lambda = \pm\sqrt{6}$. They are real numbers! Even better, we can verify that the eigenvectors are orthogonal. Naturally, this suggests that we consider extending the results of Chapter 4 on symmetric matrices and transformations to self-adjoint matrices and transformations.

First, we must show that the eigenvalues of a self-adjoint transformation are real.

(5.3.10) Proposition. If λ is an eigenvalue of the self-adjoint linear transformation T, then $\lambda \in \mathbf{R}$.

Proof: Suppose that λ is an eigenvalue for T with corresponding nonzero eigenvector $\mathbf{v}$ of unit length $T(\mathbf{v}) = \lambda\mathbf{v}$ and $\langle \mathbf{v}, \mathbf{v} \rangle = 1$. Using the fact that T is self-adjoint, we see that $\lambda = \lambda\langle \mathbf{v}, \mathbf{v} \rangle = \langle \lambda\mathbf{v}, \mathbf{v} \rangle = \langle T(\mathbf{v}), \mathbf{v} \rangle = \langle \mathbf{v}, T(\mathbf{v}) \rangle = \langle \mathbf{v}, \lambda\mathbf{v} \rangle = \bar{\lambda}\langle \mathbf{v}, \mathbf{v} \rangle = \bar{\lambda}$. Therefore, $\lambda = \bar{\lambda}$ and λ must be a real number. ■

Notice how much simpler this is than the proof we used in the case of a symmetric transformation in Chapter 4.

To show that eigenvectors of distinct eigenvalues are orthogonal, we proceed exactly as we did in the symmetric case. If $\lambda \neq \mu$ are eigenvalues for the self-adjoint transformation T with eigenvectors $\mathbf{u}$ and $\mathbf{v}$, $T(\mathbf{u}) = \lambda\mathbf{u}$ and $T(\mathbf{v}) = \mu\mathbf{v}$, then repeating the proof of the proposition, we have $\lambda\langle \mathbf{u}, \mathbf{v} \rangle = \langle \lambda\mathbf{u}, \mathbf{v} \rangle = \langle T(\mathbf{u}), \mathbf{v} \rangle = \langle \mathbf{u}, T(\mathbf{v}) \rangle = \langle \mathbf{u}, \mu\mathbf{v} \rangle = \mu\langle \mathbf{u}, \mathbf{v} \rangle$. Since $\lambda \neq \mu$, it must be the case that $\langle \mathbf{u}, \mathbf{v} \rangle = 0$; that is, $\mathbf{u}$ and $\mathbf{v}$ are orthogonal. Thus, we have proven Proposition (5.3.11) for self-adjoint mappings.

(5.3.11) Proposition. If $\mathbf{u}$ and $\mathbf{v}$ are eigenvectors, respectively, for the distinct eigenvalues λ and μ of $T: V \to V$, then $\mathbf{u}$ and $\mathbf{v}$ are orthogonal, $\langle \mathbf{u}, \mathbf{v} \rangle = 0$.

Notice that Proposition (5.3.11) also applies to real symmetric matrices and shows that their eigenvectors must be orthogonal. This is considerably simpler than the proof of this fact which we gave in Theorem (4.5.7).

Finally, the major result of Chapter 4, the spectral theorem for a symmetric linear transformation of a finite-dimensional real vector space with a real inner product extends to a self-adjoint linear transformation of a finite-dimensional complex vector space with Hermitian inner product.

(5.3.12) Theorem. Let $T: V \to V$ be a self-adjoint transformation of a complex vector space V with Hermitian inner product $\langle\ ,\ \rangle$. Then there is an orthonormal basis of V consisting of eigenvectors for T and, in particular, T is diagonalizable.

The proof of the spectral theorem in the self-adjoint case is identical to the proof in the symmetric case when we replace real vector spaces and inner products with complex vector spaces and Hermitian inner products so we will not repeat the proof here. It is interesting to note that the spectral theorem for a symmetric matrix or transformation on $\mathbf{R}^n$ follows from the complex case simply by viewing the matrix or transformation as acting on $\mathbf{C}^n$ rather than $\mathbf{R}^n$, because a real symmetric matrix is also a self-adjoint matrix.

The spectral decomposition of a self-adjoint transformation follows from the spectral theorem.

(5.3.13) Theorem. Let $T: V \to V$ be a self-adjoint transformation of a complex vector space V with Hermitian inner product $\langle\ ,\ \rangle$. Let $\lambda_1, \ldots, \lambda_k \in \mathbf{R}$ be the distinct eigenvalues of T, and let P_i be the orthogonal projections of V onto the eigenspaces E_{λ_i}, then

a) $T = \lambda_1 P_1 + \cdots + \lambda_k P_k$, and

b) $I = P_1 + \cdots + P_k$

The proof of the spectral decomposition in the self-adjoint case also mimics the proof in the symmetric case. Since the algebra of self-adjoint transformations is the same as that of symmetric transformations, diagonalizing a self-adjoint transformation or constructing the spectral decomposition of a self-adjoint transformation is the same as for a symmetric transformation. This includes the construction of the projections P_{λ_i}.

(5.3.14) Remark. It is important to note that the spectral theorem as we have stated it depends strongly on the assumption that the dimension of V is finite. If V is allowed to be infinite-dimensional, we must place additional conditions on the Hermitian inner product, called completeness and separability, which, loosely speaking, allow us to consider infinite linear combinations in some reasonable way. However, even with these added assumptions, the behavior of self-adjoint transformations is far more complicated. This is so partially because there can now be an infinite number of distinct eigenvalues, and infinite subsets of the real numbers can be far more complicated than finite subsets. These difficulties make it impossible with our current knowledge to discuss adequately the spectral theorem for infinite-dimensional vector spaces. On the other hand, there are self-adjoint transformations on infinite-dimensional vector complex vector spaces that behave much like those in the finite-dimensional case. One such transformation is studied in some detail in Section 7.5.

EXERCISES

1. Which of the following pairs of vectors form orthogonal bases for $\mathbf{C}^2$?
   a) $(1 - i, 1)\ (-1, 1 + i)$.
   b) $(1/\sqrt{2}, 1/\sqrt{2})\ (-1/\sqrt{2}, 1/\sqrt{2})$.
   c) $(1/\sqrt{5})(1, 2i)\ (1/\sqrt{65})(-4 + 6i, 3 + 2i)$.

2. Let $\mathbf{v}_1 = (1, i, 2 + i)$. Find vectors $\mathbf{v}_2$ and $\mathbf{v}_3$ so that $\alpha = \{\mathbf{v}_1, \mathbf{v}_2, \mathbf{v}_3\}$ is an orthogonal basis for $\mathbf{C}^3$.

3. a) Apply the Gram-Schmidt process to the vectors $\mathbf{v}_1 = (i, -1, i)$ and $\mathbf{v}_2 = (1, 1, 0)$ to produce an orthonormal pair of vectors in $\mathbf{C}^3$.
   b) Find the matrix of the orthogonal projection to the subspace spanned by the vectors $\mathbf{v}_1$ and $\mathbf{v}_2$ of part a.

4. Prove that the standard Hermitian inner product on $\mathbf{C}^n$ [see Example (5.3.2)] satisfies Definition (5.3.1).

5. Prove that if $\alpha = \{\mathbf{x}_1, \ldots, \mathbf{x}_n\}$ is an orthonormal basis for the complex vector space V with Hermitian inner product $\langle\ ,\ \rangle$ and if $\mathbf{v} = v_1\mathbf{x}_1 + \cdots + v_n\mathbf{x}_n$, then $v_i = \langle \mathbf{v}, \mathbf{x}_i \rangle$.

6. Find the spectral decomposition of the matrix A:

a) $A = \begin{bmatrix} 1 & i \\ -i & 0 \end{bmatrix}$ b) $A = \begin{bmatrix} 1 & 0 & -i \\ 0 & 2 & 0 \\ i & 0 & 2 \end{bmatrix}$

For the following problems let V be an n-dimensional complex vector space with Hermitian inner product $\langle \, , \rangle$.

7. A linear transformation $T: V \to V$ is called *skew-adjoint* or *skew-Hermitian* if $T = -T^*$. Prove that
   a) the eigenvalues of a skew-Hermitian transformation are purely imaginary, and
   b) the eigenvectors corresponding to distinct eigenvalues of a skew-Hermitian transformation are orthogonal.

8. A linear transformation $T: V \to V$ is called *unitary* if $\langle T(\mathbf{v}), T(\mathbf{w}) \rangle = \langle \mathbf{v}, \mathbf{w} \rangle$ for all $\mathbf{v}, \mathbf{w} \in V$.
   a) Prove that if T is unitary then T is invertible and $T^{-1} = T^*$.
   b) Prove that a matrix satisfies $U^{-1} = U^*$, that is, it is unitary, if and only if the columns form an orthonormal basis for $\mathbf{C}^n$.
   c) Prove that if A is a self-adjoint matrix, then A is similar to a diagonalizable matrix $D = Q^{-1}AQ$, where Q is a unitary matrix.

9. a) Prove that the eigenvalues of a unitary transformation are complex numbers of length 1.
   b) Prove that the eigenvectors corresponding to distinct eigenvalues of a unitary transformation are orthogonal.

10. A linear transformation $T: V \to V$ is called *normal* if $TT^* = T^*T$.
    a) Prove that self-adjoint, skew-adjoint, and unitary transformations are normal.
    b) Give an example of a normal transformation that is neither self-adjoint, skew-adjoint, nor unitary.
    c) Prove that the eigenvectors corresponding to distinct eigenvalues of a normal transformation are orthogonal.

11. Let $T: V \to V$ be a normal transformation of a Hermitian inner product space V.
    a) Prove that if $\mathbf{v} \in \text{Ker}(T)$, then $\mathbf{v} \in \text{Ker}(T^*)$. [*Hint:* Prove $\langle T^*(\mathbf{v}), T^*(\mathbf{v}) \rangle = 0$.]
    b) Prove that $T - \lambda I$ is a normal transformation.
    c) Prove that if $\mathbf{v}$ is an eigenvector of T for the eigenvalue λ, $T(\mathbf{v}) = \lambda\mathbf{v}$, then $\mathbf{v}$ is an eigenvector of T^* for the eigenvalue $\bar{\lambda}$. (*Hint:* Apply part a of this exercise to the normal transformation $T - \lambda I$.)
    d) Prove that if $\mathbf{v}$ and $\mathbf{w}$ are respectively eigenvectors for the distinct eigenvalues λ and μ of T, then $\langle \mathbf{v}, \mathbf{w} \rangle = 0$.

12. The spectral theorem for self-adjoint (symmetric) transformation uses the fact that eigenvectors for distinct eigenvalues of a self-adjoint (symmetric) trans-

formation are orthogonal. By virtue of Exercise 11d, we are able to prove a spectral theorem for normal transformations. This problem outlines the proof. Let $T: V \to V$ be a normal transformation of the Hermitian inner product space V.

a) Let $\mathbf{v} \neq \mathbf{0}$ be an eigenvector of T for λ. Prove that if $\mathbf{w}$ is orthogonal to $\mathbf{v}$, then $T(\mathbf{w})$ is also orthogonal to $\mathbf{v}$. Thus, if $\mathbf{v}$ is an eigenvector of T, $W = \{\mathbf{w} \in V \mid \langle \mathbf{w}, \mathbf{v}\rangle = 0\}$ satisfies $\text{Im}(T|_W) \subset W$.

b) With $\mathbf{v}$ and W defined as in part a, prove that $(T|_W)^* = (T^*)|_W$ and that $T|_W$ is normal.

c) Prove the spectral theorem for normal transformations of a finite dimensional inner product space V. That is, prove that V has an orthonormal basis consisting of eigenvectors of T. [*Hint:* Proceed by induction on the dimension of V as we did in the proof of Theorem (4.6.1).]

13. In this problem we indicate how to construct other Hermitian inner products on $V = \mathbf{C}^n$. Let A be an $n \times n$ self-adjoint matrix. For $\mathbf{x}$ and $\mathbf{y} \in \mathbf{C}^n$, define $\langle\ ,\ \rangle_A$ by $\langle \mathbf{x}, \mathbf{y}\rangle_A = \langle \mathbf{x}, A\mathbf{y}\rangle$, where $\langle\ ,\ \rangle$ is the usual Hermitian inner product on $\mathbf{C}^n$.

a) Prove that $\langle\ ,\ \rangle_A$ is complex linear in the first entry and conjugate linear in the second entry.

b) Prove that $\langle \mathbf{x}, \mathbf{y}\rangle_A = \overline{\langle \mathbf{y}, \mathbf{x}\rangle_A}$ for all $\mathbf{x}$ and $\mathbf{y}$ in $\mathbf{C}^n$.

c) Prove that $\langle\ ,\ \rangle_A$ is positive-definite if and only if all the eigenvalues of A are positive. (*Hint:* Expand $\mathbf{x}$ in terms of a basis for $\mathbf{C}^n$ consisting of eigenvectors of A.)

d) Conclude that $\langle\ ,\ \rangle_A$ is a Hermitian inner product on $\mathbf{C}^n$ if and only if the eigenvalues of A are positive.

14. Let $A = \begin{bmatrix} 1 & i \\ -i & 2 \end{bmatrix}$. Are the following pairs of vectors orthogonal with respect to $\langle\ ,\ \rangle_A$?

a) $(1, 0)$ $(-i, 1)$.

b) $(1, 0)$ $(0, 1)$.

c) $(1, 1)$ $(-1, i)$.

15. Find an orthonormal basis for $\mathbf{C}^2$ with respect to the inner product defined by the matrix A of Exercise 14.

CHAPTER SUMMARY

Despite the brevity of this chapter, we have introduced several crucial ideas that make our entire theory much stronger. We extended the general framework of linear algebra to encompass vector spaces over a field F, which is not necessarily the field of real numbers. This construction is of particular importance to us when $F = \mathbf{C}$, the field of complex numbers. In this case, since $\mathbf{R} \subset \mathbf{C}$, we applied results from linear algebra over $\mathbf{C}$ to obtain information about linear algebra over $\mathbf{R}$. For example, Theorem (4.2.7) can be applied to real linear transformations with non-real eigenvalues. In addition, we developed the geometry of Hermitian inner product spaces in complete analogy with the geometry of real inner product spaces. Let us review the details of this chapter.

Our initial goal was to introduce the *complex numbers*. The complex numbers **C** can be identified as a set with $\mathbf{R}^2$ with the usual notion of vector addition and a new operation of multiplication [Definition (5.1.1)]. These operations make **C** into a *field* [Definition (5.1.4) and Proposition (5.1.5)]. We saw that a field is a direct generalization of the real numbers and the operations of addition and multiplication. Since we identified **C** with $\mathbf{R}^2$, we were able to transfer the geometry of $\mathbf{R}^2$ to **C** [Definition (5.1.8)]. Using polar coordinates in **C**, we obtained a clear geometric interpretation of multiplication of complex numbers. Most importantly, we have the *fundamental theorem of algebra*, which showed that **C** is indeed the correct field to be using in order to extend our results over **R**. No larger field is necessary and no smaller field will suffice.

In the second section we defined a *vector space over a field* [Definition (5.2.1)]. Through a series of examples we indicated that excluding our work involving the geometry of $\mathbf{R}^n$, all the results of the first four chapters extend to vector spaces over a field. Thus, for example, if V and W are finite dimensional vector spaces over a field F and $T: V \to W$ is a linear transformation with respect to the field F, we choose bases for V and W in order to construct the matrix for T. We apply the elimination algorithm to the matrix of T in order to determine the dimensions of the kernel and image of T or to find the solutions of $T(\mathbf{v}) = \mathbf{w}$ for $\mathbf{w} \in \text{Im}(T)$. If $V = W$ we determine whether or not T is invertible by calculating whether or not $\det(T) \neq 0$. Further, if all the eigenvalues of T lie in the field F, if the multiplicity of each eigenvalue equals the dimension of the corresponding eigenspace, and if the sum of the dimensions of the eigenspaces equals the dimension of V, then T is diagonalizable [Theorem (4.2.7).]

In the third section we returned to a geometric point of view with the introduction of *Hermitian inner products* on a complex vector space [Definition (5.3.1)]. Using the standard inner product on $\mathbf{R}^n$ as our model, we extended our basic notions of *length* and *orthogonality* to complex vector spaces. We also made an addition to the theory by defining the *adjoint of a linear transformation* [Definition (5.3.7)] and using this as the starting point for our discussion of transformations rather than the adjoint of a matrix. We used this to define *Hermitian* or *self-adjoint* transformations—the analog of symmetric matrices. Finally, we stated the *Hermitian version of the spectral theorem*.

SUPPLEMENTARY EXERCISES

1. **True-False.** For each true statement, give a brief proof or reason. For each false statement give a counterexample.
   a) If a complex number is not a real number, then it is a purely imaginary number.
   b) If the sum of two complex numbers is a real number, then each of the numbers is actually a real number.

c) If a complex polynomial has all real roots, then all the coefficients of the polynomial are real numbers.
d) Let $T: \mathbf{C}^n \to \mathbf{C}^n$ be a linear transformation. Then $\overline{T}: \mathbf{C}^n \to \mathbf{C}^n$ defined by $\overline{T}(\mathbf{x}) = \overline{T(\mathbf{x})}$ is a linear transformation.
e) If F is a field and $a, b \in F$ with $ab = 0$, then either $a = 0$ or $b = 0$.
f) If $\mathbf{v} \in \mathbf{C}^n$, $\mathbf{v} = \overline{\mathbf{v}}$ implies that every entry of $\mathbf{v}$ is real.
g) If V is a complex vector space with a Hermitian inner product, $\langle\ ,\ \rangle$, then $\langle \mathbf{v}, \mathbf{w}\rangle = \langle \mathbf{w}, \mathbf{v}\rangle$ for all $\mathbf{v}$ and $\mathbf{w} \in V$.
h) Every 2×2 rotation matrix is diagonalizable over $\mathbf{C}$.
i) If A is an $n \times n$ complex matrix such that $\langle A\mathbf{v}, \mathbf{w}\rangle = \langle \mathbf{v}, A\mathbf{w}\rangle$ for all $\mathbf{v}$ and $\mathbf{w} \in \mathbf{C}^n$, then the entries a_{ij} of A satisfy $a_{ij} = a_{ji}$.

2. Express the following complex numbers in the form $a + bi$.
a) $(1 + i)^5$.
b) $(1 - 7i)(1 + 5i)/(2 - i)$.
c) $\sqrt{2 + 2i}$ (find all roots).
d) $(-i)^{1/3}$ (find all roots).

3. Find all the roots of the following polynomials:
a) $p(z) = z^3 - i$.
b) $p(z) = z^2 - (3 - 4i)$.
c) $p(z) = z^2 + iz + 1$.

4. Let $F = F_p$, the finite field with p elements, p a prime number. How many vectors are in F^n? (See Exercises 11 through 15 of Section 5.1.)

5. Let F be any field. A rational function over F in one variable is a quotient of polynomials. For example, $r(z) = (z^2 + i)/(z - 3)$ is a complex rational function. Let $F(z)$ denote the set of all rational functions over F with the usual operations of addition and multiplication of quotients of polynomials. Prove that $F(z)$ is a field.

In Exercise 6 of Section 5.2 it was shown that $\mathbf{C}^n$ can be given the structure of a real vector space of dimension $2n$ and that a basis is given by $\alpha = \{\mathbf{e}_1, i\mathbf{e}_1, \ldots, \mathbf{e}_n, i\mathbf{e}_n\}$. Let us denote this real vector space by $\hat{\mathbf{C}}^n$.

6. Let $T: \mathbf{C}^n \to \mathbf{C}^n$ be a complex linear transformation.
a) Prove that $T: \hat{\mathbf{C}}^n \to \hat{\mathbf{C}}^n$ is a real linear transformation. Denote this real linear transformation by T.
b) Let $J: \mathbf{C}^n \to \mathbf{C}^n$ be multiplication by i, $J(\mathbf{x}) = i\mathbf{x}$. Let $\hat{S}: \hat{\mathbf{C}}^n \to \hat{\mathbf{C}}^n$ be any real linear transformation. $\hat{S}$ also defines a function on $\mathbf{C}^n$. Call this function S. Prove that S is a complex linear transformation if and only if $SJ = JS$.
c) If A is the matrix of T with respect to the standard basis of $\mathbf{C}^n$, with entries $a_{ij} \in \mathbf{C}$, what is the matrix $[\hat{T}]_\alpha^\alpha$? (Note that $[\hat{T}]_\alpha^\alpha$ must have real entries.)

7. a) Let $T_\theta: \mathbf{C} \to \mathbf{C}$ be the complex linear transformation defined by multiplication by $z = \cos(\theta) + i\sin(\theta)$, $T_\theta(\mathbf{w}) = z\mathbf{w}$. Show that $[\hat{T}_\theta]_\alpha^\alpha$ is a rotation matrix.

b) Let $\hat{S}: \hat{C} \to \hat{C}$ be a projection to the line spanned by a nonzero vector **a** (remember $\hat{\mathbf{C}}$ is identified with $\mathbf{R}^2$). Is $S: \mathbf{C} \to \mathbf{C}$ a complex linear transformation?

8. Let $\langle\ ,\ \rangle$ denote the standard Hermitian inner product on $\mathbf{C}^n$.
   a) Prove that $\text{Re}\langle\ ,\ \rangle$ defines a real inner product on $\hat{\mathbf{C}}^n$.
   b) Prove that if **x** and **y** are orthogonal in $\mathbf{C}^n$, then they are orthogonal in $\hat{\mathbf{C}}^n$ with respect to the inner product $\text{Re}\langle\ ,\ \rangle$.
   c) If **x** and **y** are orthogonal in $\hat{\mathbf{C}}^n$, are they orthogonal in $\mathbf{C}^n$?

9. Let $T: \mathbf{C}^n \to \mathbf{C}^n$ be a linear transformation. Let $\hat{T}$ be defined as before.
   a) Prove that $(T^*)\hat{} = (\hat{T})^t$; that is, $\hat{}$ of T^* is equal to the transpose of $\hat{T}$.
   b) Prove that if T is self-adjoint, then $\hat{T}$ is symmetric.
   c) Prove that if T is unitary, then $\hat{T}$ is orthogonal. (See Exercise 8 of Section 5.3.)

10. a) Let $T: \mathbf{C}^n \to \mathbf{C}^n$ be given by a diagonal matrix with respect to the standard basis with diagonal entries $\lambda_1, \ldots, \lambda_n$. Prove that the eigenvalues of $\hat{T}$ are $\lambda_1, \bar{\lambda}_1, \ldots, \lambda_n, \bar{\lambda}_n$.
    b) Can you use part a to show that if $T: \mathbf{C}^n \to \mathbf{C}^n$ is diagonalizable with eigenvalues $\lambda_1, \ldots, \lambda_n$, then $\hat{T}$ has eigenvalues $\lambda_1, \bar{\lambda}_1, \ldots, \lambda_n, \bar{\lambda}_n$?

CHAPTER 6

Jordan Canonical Form

Introduction

In many ways the simplest form the matrix of a linear mapping can take is a diagonal form, and diagonalizable mappings are correspondingly easy to understand and work with. However, as we know, not every linear mapping is diagonalizable. Our main goal in this chapter is to develop a canonical form for all linear mappings—that is, so to speak, a *next best* form after a diagonal form for the matrices of linear mappings that are not necessarily diagonalizable.

We begin by showing that if all the eigenvalues of a mapping $T: V \to V$ lie in the field F over which V is defined, then T can be triangularized; that is, there exists basis β for V such that $[T]_\beta^\beta$ is an upper-triangular matrix. This reduction to upper-triangular form may be viewed as a first step in our program.

It is possible to force many *more* entries from the matrix to be zero, however. In fact, we will ultimately produce a canonical form in which the only possible nonzero entries are on the diagonal and immediately above the diagonal. Our final result here is known as the Jordan canonical form, and it gives an extremely powerful (though, unavoidably, rather intricate) tool for studying linear mappings. Indeed, we will see that this Jordan form gives what amounts to a complete classification of linear mappings when all the eigenvalues lie in the field of scalars.

In this chapter we will again be using the field of complex numbers and complex vector spaces, which were introduced in Chapter 5. The results of this chapter will be stated for a field F and a vector space over that field. Thus, we are allowing the option that $F = \mathbf{C}$ and V is a complex vector space in addition to the familiar case $F = \mathbf{R}$ and V a real vector space. It should be emphasized, however, that most of these results hold for a field F without assuming that $F = \mathbf{R}$ or $\mathbf{C}$.

In this chapter V will always denote a *finite-dimensional* vector space, and this hypothesis will be understood in all statements. We will include it in the statements of definitions and theorems, though, so that this hypothesis will be clearly stated in results that are used later in the text.

---

§6.1. TRIANGULAR FORM

In this section we show that if V is a vector space over a field F, and $T: V \to V$ is a linear mapping whose characteristic polynomial has $\dim(V)$ roots (counting multiplicities) in the field F (we will also sometimes say, for brevity, that all the eigenvalues of T are contained in F), then there is a basis β for V with the property that the matrix of T with respect to β is an *upper-triangular* matrix:

$$[T]_\beta^\beta = \begin{bmatrix} a_{11} & \cdots & a_{1n} \\ 0 & \ddots & \vdots \\ \vdots & \ddots & \vdots \\ 0 & \cdots \; 0 & a_{nn} \end{bmatrix}$$

To prove this, we start by interpreting this property of the matrix of T in a more geometric fashion, beginning with an example.

(6.1.1) Example. Consider $F = \mathbf{C}$, $V = \mathbf{C}^4$, let β be the standard basis, and let

$$[T]_\beta^\beta = \begin{bmatrix} 1 & 1-i & 2 & 1 \\ 0 & 1 & i & 1 \\ 0 & 0 & 1-i & 3+i \\ 0 & 0 & 0 & 1-i \end{bmatrix}$$

Since the matrix is upper-triangular already, we see that the basis β itself has the desired property. Notice first that $T(\mathbf{e}_1) = \mathbf{e}_1$, so that $\mathbf{e}_1$ is an eigenvector of T. Next, $T(\mathbf{e}_2) = (1 - i)\mathbf{e}_1 + \mathbf{e}_2 \in \text{Span}(\{\mathbf{e}_1, \mathbf{e}_2\})$. Similarly, $T(\mathbf{e}_3) \in \text{Span}(\{\mathbf{e}_1, \mathbf{e}_2, \mathbf{e}_3\})$ and $T(\mathbf{e}_4) \in \text{Span}(\{\mathbf{e}_1, \mathbf{e}_2, \mathbf{e}_3, \mathbf{e}_4\})$.

In particular, if we let W_i be the subspace of $\mathbf{C}^4$ spanned by the first i vectors in the standard basis, then $T(\mathbf{e}_i) \in W_i$ for all i. As a result, we have that $T(W_i) \subset W_i$ for all i. (Recall that $T(W) = \{ T(\mathbf{w}) | \mathbf{w} \in W\} = \text{Im}(T \mid_W)$ for any subspace W of V.)

This example illustrates a general pattern that will hold whenever we have $T: V \to V$ and a basis β for V such that $[T]_\beta^\beta$ is upper-triangular. If we write $\beta = \{\mathbf{x}_1, \ldots, \mathbf{x}_n\}$, and let $W_i = \text{Span}(\{\mathbf{x}_1, \ldots, \mathbf{x}_i\})$ for each $i \leq n$, then by the definition of the matrix of a linear mapping, it may be seen that $[T]_\beta^\beta$ is upper-triangular if and only if $T(W_i) \subset W_i$ for each of the subspaces W_i.

To help describe this situation, we introduce some new terminology.

(6.1.2) Definition. Let $T: V \to V$ be a linear mapping. A subspace $W \subset V$ is said to be *invariant* (or *stable*) under T if $T(W) \subset W$.

(6.1.3) Examples

a) $\{\mathbf{0}\}$ and V itself are invariant under all linear mappings $T: V \to V$.

b) Ker(T) and Im(T) are invariant subspaces as well (see Exercise 2).

c) If λ is an eigenvalue of T, then the eigenspace E_λ is invariant under T as well. This is true since for $\mathbf{v} \in E_\lambda$, $T(\mathbf{v}) = \lambda\mathbf{v} \in E_\lambda$.

d) Let $T: \mathbf{C}^4 \to \mathbf{C}^4$ be the linear mapping defined by the matrix

$$A = \begin{bmatrix} 1 & 1 & 1 & 1 \\ -1 & 1 & -1 & 1 \\ 1 & -1 & 1 & -1 \\ -1 & -1 & -1 & -1 \end{bmatrix}$$

and let $W \subset \mathbf{C}^4$ be the subspace $W = \text{Span}\{(1, 1, 1, 1), (1, 0, 0, -1), (0, 1, -1, 0)\}$. To determine if W is invariant under T, we must see if the image under T of each of the vectors in the spanning set for W is again a vector in W. Here we see that $T(1, 1, 1, 1) = (4, 0, 0, -4)$ by computing the product of A and the vector $(1, 1, 1, 1)$. Since this is just $4 \cdot (1, 0, 0, -1)$, we see that $T(1, 1, 1, 1) \in W$. Next, $T(1, 0, 0, -1) = (0, -2, 2, 0) = -2 \cdot (0, 1, -1, 0) \in W$. Finally, $T(0, 1, -1, 0) = (0, 2, -2, 0) \in W$ as well, since W is a subspace of $\mathbf{C}^4$. As a consequence, W is invariant under T.

Summarizing our previous discussion of the linear mappings defined by upper-triangular matrices, we have Proposition (6.1.4).

(6.1.4) Proposition. Let V be a vector space, let $T: V \to V$ be a linear mapping, and let $\beta = \{\mathbf{x}_1, \ldots, \mathbf{x}_n\}$ be a basis for V. Then $[T]_\beta^\beta$ is upper-triangular if and only if each of the subspaces $W_i = \text{Span}(\{\mathbf{x}_1, \ldots, \mathbf{x}_i\})$ is invariant under T.

Note that the subspaces W_i in the proposition are related as follows:

$$\{\mathbf{0}\} \subset W_1 \subset W_2 \subset \cdots \subset W_{n-1} \subset W_n = V$$

The W_i form an *increasing sequence* of subspaces.

(6.1.5) Definition. We say that a linear mapping $T: V \to V$ on a finite-dimensional vector space V is *triangularizable* if there exists a basis β such that $[T]_\beta^\beta$ is upper-triangular.

Just as a diagonalizable mapping may be diagonalized by changing basis to a basis consisting of eigenvectors, a triangularizable mapping may be *triangularized* by a change of basis. To show that a given mapping is triangularizable, we must produce the increasing sequence of invariant subspaces

$$\{\mathbf{0}\} \subset W_1 \subset W_2 \subset \cdots \subset W_{n-1} \subset W_n = V$$

spanned by the subsets of the basis β and show that each W_i is invariant under T.

To be able to determine when it is possible to do this, we need to study in greater detail the restriction of T to an invariant subspace $W \subset V$. Note that in this case $T|_W : W \to W$ defines a new linear mapping, and we can ask how T and $T|_W$ are related. In keeping with the general philosophy, that we learned in Chapter 4, of studying linear mappings by considering their eigenvalues and eigenvectors, we should begin by considering how the eigenvalues of $T|_W$ are related to those of T. The key facts here are given in the following proposition.

(6.1.6) Proposition. Let $T: V \to V$, and let $W \subset V$ be an invariant subspace. Then the characteristic polynomial of $T|_W$ divides the characteristic polynomial of T.

Proof: Let $\alpha = \{\mathbf{x}_1, \ldots, \mathbf{x}_k\}$ be a basis for W, and extend α to a basis $\beta = \{\mathbf{x}_1, \ldots, \mathbf{x}_k, \mathbf{x}_{k+1}, \ldots, \mathbf{x}_n\}$ for V. Since W is invariant under T, for each i $(i \leqslant k)$, we have

$$T(\mathbf{x}_i) = a_{1i}\mathbf{x}_1 + \cdots + a_{ki}\mathbf{x}_k + 0\mathbf{x}_{k+1} + \cdots + 0\mathbf{x}_n$$

Hence the matrix $[T]_\beta^\beta$ has a block form composition:

$$[T]_\beta^\beta = \begin{bmatrix} A & B \\ 0 & C \end{bmatrix}$$

where $A = [T|_W]_\alpha^\alpha$ is a $k \times k$ matrix and B and C include the coefficients in the expansions of $T(\mathbf{x}_j)$, where $k + 1 \leqslant j \leqslant n$. When we compute the characteristic polynomial of T using this matrix, we have (by Section 3.2, Exercise 8) $\det(T - \lambda I) = \det(A - \lambda I)\det(C - \lambda I)$. But $\det(A - \lambda I)$ is just the characteristic polynomial of $T|_W$, so this polynomial divides the characteristic polynomial of T, as claimed. ■

An important corollary of this proposition is that every eigenvalue of $T|_W$ is also an eigenvalue of T, or phrased in a slightly different way, the set of eigenvalues of $T|_W$ is some *subset* of the eigenvalues of T on the whole space.

(6.1.7) Example. Let $V = \mathbf{R}^3$, and let $T: V \to V$ be the linear mapping defined by $A = \begin{bmatrix} 1 & 2 & 0 \\ 1 & 0 & 1 \\ 0 & 2 & -1 \end{bmatrix}$. The reader should check that the two-dimensional subspace

$W = \text{Span}(\{(1, 1, 0), (1, 0, 1)\})$ is invariant under T. When we compute the characteristic polynomial of T, we find $\det(T - \lambda I) = -\lambda^3 + 5\lambda$. On the other hand, if we use the basis $\alpha = \{(1, 1, 0), (1, 0, 1)\}$ for W, then we have $[T|_W]_\alpha^\alpha = \begin{bmatrix} 1 & 2 \\ 2 & -1 \end{bmatrix}$, so $\det(T|_W - \lambda I) = \lambda^2 - 5$. Note that $-\lambda^3 + 5\lambda = -\lambda(\lambda^2 - 5)$, so that as we expect from the proposition, $\det(T|_W - \lambda I)$ does divide $\det(T - \lambda I)$. As a result, the eigenvalues of T are $\lambda = \sqrt{5}, -\sqrt{5}, 0$, whereas those of $T|_W$ are $\lambda = \sqrt{5}, -\sqrt{5}$.

We are now ready to prove a criterion that can be used to determine if a linear mapping is triangularizable. Since triangularizability is a weaker property than diagonalizability, it is perhaps not surprising that the hypotheses needed here are much less stringent than those of our criterion for diagonalizability, Theorem (4.2.7).

(6.1.8) Theorem. Let V be a finite-dimensional vector space over a field F, and let $T: V \to V$ be a linear mapping. Then T is triangularizable if and only if the characteristic polynomial equation of T has $\dim(V)$ roots (counted with multiplicities) in the field F.

(6.1.9) Remarks. This hypothesis is satisfied automatically if $F = \mathbf{C}$, for example. By the fundamental theorem of algebra [Theorem (5.1.12)] a polynomial equation of degree n with complex coefficients has n roots in $\mathbf{C}$. As a result, our theorem will imply that *every* matrix $A \in M_{n\times n}(\mathbf{C})$ may be triangularized. The same is not true for every matrix with real entries, however (as long as we take $F = \mathbf{R}$). For example, the matrix of a rotation R_θ (θ not an integer multiple of π) in $\mathbf{R}^2$ cannot be triangularized over $F = \mathbf{R}$, since the roots of the characteristic polynomial equation are not real.

Our proof of Theorem (6.1.8) will be based on the following lemma.

(6.1.10) Lemma. Let $T: V \to V$ be as in the theorem, and assume that the characteristic polynomial of T has $n = \dim(V)$ roots in F. If $W \subsetneqq V$ is an invariant subspace under T, then there exists a vector $\mathbf{x} \neq \mathbf{0}$ in V such that $\mathbf{x} \notin W$ and $W + \text{Span}(\{\mathbf{x}\})$ is also invariant under T.

Proof: Let $\alpha = \{\mathbf{x}_1, \ldots, \mathbf{x}_k\}$ be a basis for W and extend α by adjoining the vectors in $\alpha' = \{\mathbf{x}_{k+1}, \ldots, \mathbf{x}_n\}$ to form a basis $\beta = \alpha \cup \alpha'$ for V. Let $W' = \text{Span}(\alpha')$. Now, consider the linear mapping $P: V \to V$ defined by

$$P(a_1\mathbf{x}_1 + \cdots + a_n\mathbf{x}_n) = a_1\mathbf{x}_1 + \cdots + a_k\mathbf{x}_k$$

The attentive reader will note the close analogy between this mapping P and the orthogonal projection mappings we studied in Chapter 4. (See also Exercise 6 of Section 2.2.) Indeed, it is easily checked (Exercise 12 of this section) that $\text{Ker}(P) = W'$, that $\text{Im}(P) = W$, and that $P^2 = P$. P is called the *projection on W with kernel W'*. It follows from the definition of P that $I - P$ is the projection on W' with kernel W. Note that P will be the orthogonal projection on W in the case that $W' = W^\perp$.

Let $S = (I - P)T$. Since $\text{Im}(I - P) = W'$, we see by Proposition (2.5.6) that $\text{Im}(S) \subset \text{Im}(I - P) = W'$. Thus, W' is an invariant subspace of S.

We claim that the set of eigenvalues of $S|_{W'}$, is a subset of the set of eigenvalues of T. From the proof of Proposition (6.1.6), we see that since W is an invariant subspace of T, $[T]_\beta^\beta$ has a block decomposition:

$$[T]_\beta^\beta = \begin{bmatrix} A & B \\ 0 & C \end{bmatrix}$$

It is an easy consequence of the definitions of $[T]_\beta^\beta$, $I - P$, and S that the $k \times k$ block $A = [T|_W]_\alpha^\alpha$ and the $(n - k) \times (n - k)$ block $C = [S|_{W'}]_{\alpha'}^{\alpha'}$. Hence, $\det(T - \lambda I) = \det(T|_W - \lambda I)\det(S|_{W'} - \lambda I)$, and the characteristic polynomial of $S|_{W'}$ divides the characteristic polynomial of T.

Since all the eigenvalues of T lie in the field F, the same is true of all the eigenvalues of $S|_{W'}$. Hence there is some nonzero vector $\mathbf{x} \in W'$ and some $\lambda \in F$ such that $S(\mathbf{x}) = \lambda\mathbf{x}$. However, this says

$$(I - P)T(\mathbf{x}) = \lambda\mathbf{x}$$

so

$$T(\mathbf{x}) - PT(\mathbf{x}) = \lambda\mathbf{x}$$

or

$$T(\mathbf{x}) = \lambda\mathbf{x} + PT(\mathbf{x})$$

Since $PT(\mathbf{x}) \in W$, we see that $W + \text{Span}(\{\mathbf{x}\})$ is also invariant under T and the lemma is proved. ■

We are now ready to complete the proof of Theorem (6.1.8).

Proof: $\rightarrow$: If T is triangularizable, then there exists a basis β for V such that $[T]_\beta^\beta$ is upper-triangular. The eigenvalues of T are the diagonal entries of this matrix, so they are elements of the field F.

$\leftarrow$: Conversely, if all the eigenvalues of T are in F, then we may proceed as follows. For the first step, let λ be any eigenvalue of T, and let $\mathbf{x}_1$ be an eigenvector with eigenvalue λ. Let $W_1 = \text{Span}(\{\mathbf{x}_1\})$. By definition W_1 is invariant under T. Now, assume by induction that we have constructed invariant subspaces $W_1 \subset W_2 \subset \cdots \subset W_k$ with $W_i = \text{Span}(\{\mathbf{x}_1, \ldots, \mathbf{x}_i\})$ for each i. By Lemma (6.1.10) there exists a vector $\mathbf{x}_{k+1} \notin W_k$ such that the subspace $W_{k+1} = W_k + \text{Span}(\{\mathbf{x}_{k+1}\})$ is also invariant under T. We continue this process until we have produced a basis for V. Hence, T is triangularizable. ■

As we know, in an upper-triangular matrix of T, the eigenvalues are the diagonal entries. When we apply Theorem (6.1.8) in the future, it will be most convenient to assume that the diagonal entries are arranged so that all the occurrences of each of the distinct eigenvalues are consecutive—in other words, if the eigenvalues of T are λ_1 with multiplicity m_1, λ_2 with multiplicity m_2, and so on, then the diagonal entries are m_1 λ_1's, followed by m_2 λ_2's, and so on. It is possible to achieve this form, since each time we apply Lemma (6.1.10) to extend our triangular basis, the eigenvalues of the mapping we called S in that proof are simply the remaining eigenvalues of the mapping T. For future reference we record this observation as Corollary (6.1.11).

(6.1.11) Corollary. If $T: V \to V$ is triangularizable, with eigenvalues λ_i with respective multiplicities m_i, then there exists a basis β for V such that $[T]_\beta^\beta$ is upper-triangular, and the diagonal entries of $[T]_\beta^\beta$ are m_1 λ_1's, followed by m_2 λ_2's, and so on.

As a by-product of our result on triangularizability, we are now able to give a proof of the general Cayley-Hamilton theorem, under the assumption that all the eigenvalues of $T: V \to V$ lie in the field F over which V is defined. Recall that from Chapter 4, Section 4.1, if T is a linear mapping (or a matrix) and $p(t) = a_n t^n + a_{n-1}t^{n-1} + \cdots + a_0$ is a polynomial, we can define a new linear mapping

$$p(T) = a_n T^n + a_{n-1}T^{n-1} + \cdots + a_1 I$$

In particular, we can do this for the characteristic polynomial of T.

(6.1.12) Theorem. (Cayley-Hamilton) Let $T: V \to V$ be a linear mapping on a finite-dimensional vector space V, and let $p(t) = \det(T - tI)$ be its characteristic polynomial. Assume that $p(t)$ has $\dim(V)$ roots in the field F over which V is defined. Then $p(T) = 0$ (the zero mapping on V).

Proof: It suffices to show that $p(T)(\mathbf{x}) = \mathbf{0}$ for all the vectors in some basis of V. By Theorem (6.1.8), though, we know there exists some basis $\beta = \{\mathbf{x}_1, \ldots, \mathbf{x}_n\}$ for V such that each $W_i = \text{Span}(\{\mathbf{x}_1, \ldots, \mathbf{x}_i\})$ is invariant under T. Since all the eigenvalues of T lie in F, we know that $p(t)$ factors as $p(t) = \pm (t - \lambda_1) \cdots (t - \lambda_n)$ for some $\lambda_i \in F$ (not necessarily distinct). If the factors here are ordered in the same fashion as the diagonal entries of $[T]_\beta^\beta$, then we have $T(\mathbf{x}_i) = \lambda_i \mathbf{x}_i + \mathbf{y}_{i-1}$, where $\mathbf{y}_{i-1} \in W_{i-1}$ for $i \geq 2$, and $T(\mathbf{x}_1) = \lambda_1 \mathbf{x}_1$.

We proceed by induction on i. For $i = 1$ we have

$$p(T)(\mathbf{x}_1) = \pm(T - \lambda_1 I)(T - \lambda_2 I) \cdots (T - \lambda_n I)(\mathbf{x}_1)$$

Since the powers of T commute with each other and with I, this can be rewritten as

$$\begin{aligned} p(T)(\mathbf{x}_1) &= \pm(T - \lambda_2 I) \cdots (T - \lambda_n I)(T - \lambda_1 I)(\mathbf{x}_1) \\ &= \pm(T - \lambda_2 I) \cdots (T - \lambda_n I)(\mathbf{0}) \\ &= \mathbf{0} \end{aligned}$$

Now assume that we have shown $p(T)(\mathbf{x}_i) = \mathbf{0}$ for all $i \leq k$, and consider $p(T)(\mathbf{x}_{k+1})$. It is clear that only the factors $T - \lambda_1 I, \ldots, T - \lambda_k I$ are needed to send $\mathbf{x}_i$ to $\mathbf{0}$ for $i \leq k$. As before, we can rearrange the factors in $p(T)$ to obtain

$$p(T)(\mathbf{x}_{k+1}) = \pm(T - \lambda_1 I) \cdots (T - \lambda_n I)(T - \lambda_{k+1} I)(\mathbf{x}_{k+1})$$

However, we know that $T(\mathbf{x}_{k+1}) = \lambda_{k+1}\mathbf{x}_{k+1} + \mathbf{y}_k$, where $\mathbf{y}_k \in W_k$. Hence, $(T - \lambda_{k+1}I)(\mathbf{x}_{k+1}) \in W_k$, and by induction we know that the other factors in $p(T)$ send all the vectors in this subspace to $\mathbf{0}$. Hence, $p(T)(\mathbf{x}_{k+1}) = \mathbf{0}$ as well. ■

(6.1.13) Remark. The conclusion of the Cayley-Hamilton theorem is valid even if some of the eigenvalues of T are contained in an extension of the field F (i.e., a field that contains F and whose operations restrict on F to the operations F.) See Exercise 10 for the case of linear mappings defined over $\mathbf{R}$.

We close this section with an application of the Cayley-Hamilton theorem, giving an alternate way to compute inverse matrices. Suppose $A \in M_{n\times n}(F)$. If A is invertible, then by Theorem (3.2.14) we know $\det(A) \neq 0$. If the characteristic polynomial of A is $\det(A - tI) = (-1)^n t^n + \cdots + a_1 t + \det(A)$, then by Cayley-Hamilton, we have

$$p(A) = (-1)^n A^n + \cdots + a_1 A + \det(A)I = 0$$

Multiplying both sides of this equation by A^{-1} and rearranging, we obtain

$$(-1/\det(A))((-1)^n A^{n-1} + \cdots + a_1 I) = A^{-1}$$

Thus, the inverse of A may be computed by performing matrix products and sums.

(6.1.14) Example. Let $A = \begin{bmatrix} 1 & 2 & 0 \\ 2 & -1 & 1 \\ 0 & 1 & 3 \end{bmatrix}$. We have $p(t) = \det(A - tI) = -t^3 + 3t^2 + 6t - 16$. Using the preceding formula, we have

$$\begin{aligned} A^{-1} &= (1/16)(-A^2 + 3A + 6I) \\ &= (1/16)\begin{bmatrix} 4 & 6 & -2 \\ 6 & -3 & 1 \\ -2 & 1 & 5 \end{bmatrix} \end{aligned}$$

(6.1.15) Remark. For computing the inverses of large matrices, this method is far less efficient than the Gauss-Jordan procedure we learned in Section 2.6. For this reason, it is not often used for numerical calculations. Nevertheless, it is rather interesting to note that the inverse of an invertible matrix may be computed using only matrix powers and matrix sums, and this observation is often useful in other, more theoretical applications of matrix algebra.

EXERCISES

1. Determine whether the given subspace $W \subset V$ is invariant under the given linear mapping $T\colon V \to V$.

a) $V = \mathbf{R}^3$, $T: V \to V$ defined by $\begin{bmatrix} 1 & 3 & 1 \\ 0 & 2 & -1 \\ 1 & 0 & 0 \end{bmatrix}$, $W = \text{Span}\{(1, -1, 1), (1, 2, 1)\}$

b) $V = C^\infty(\mathbf{R})$, $T: V \to V$ defined by $T(f) = f'$, $W = \text{Span}\{\sin(x), \cos(x)\}$

c) $V = \mathbf{C}^5$, $T: V \to V$ defined by $\begin{bmatrix} i & 1 & 0 & 0 & 0 \\ 0 & i & 0 & 0 & 0 \\ 0 & 0 & 2 & 1 & 0 \\ 0 & 0 & 0 & 2 & 1 \\ 0 & 0 & 0 & 0 & 2 \end{bmatrix}$, $W = \text{Span}\ \{\mathbf{e}_1, \mathbf{e}_3, \mathbf{e}_4\}$

d) $V = P_5(\mathbf{R})$, $T: V \to V$ defined by $T(p) = p'' - xp' + 3p$, $W = \text{Span}\{1 + x, x + x^4, x^2 - x^5\}$.

e) $V = M_{2\times 2}(\mathbf{R})$, $T: V \to V$ defined by $T(X) = AX - XA$, where $A = \begin{bmatrix} 2 & 1 \\ 1 & 3 \end{bmatrix}$, $W = \{X \in M_{2\times 2}(\mathbf{R}) \mid \text{Tr}(X) = 0\}$

2. Show that for any linear mapping $T: V \to V$, the subspaces $\text{Ker}(T)$ and $\text{Im}(T)$ are invariant under T.

3. Show that if $W \subset V$ is invariant under $T: V \to V$ and $\dim(W) = 1$, then W is spanned by an eigenvector for T.

4. a) Let $A = \begin{bmatrix} \sqrt{2}/2 & 0 & \sqrt{2}/2 \\ 0 & 4 & 0 \\ -\sqrt{2}/2 & 0 & \sqrt{2}/2 \end{bmatrix}$. Find all the subspaces of $\mathbf{R}^3$ that are invariant under the linear mapping defined by A.

b) Show that if $T: \mathbf{R}^3 \to \mathbf{R}^3$ is a linear mapping, then there always exists a subspace $W \subset \mathbf{R}^3$, $W \neq \{\mathbf{0}\}, \mathbf{R}^3$, which is invariant under T.

5. Show that if W_1 and W_2 are subspaces of V invariant under a linear mapping $T: V \to V$, then $W_1 \cap W_2$ and $W_1 + W_2$ are also invariant under T.

6. a) Let $V = W_1 \oplus W_2$, let $T: V \to V$ and assume that W_1 and W_2 are invariant under T. Let β be a basis for V of the form $\beta = \beta_1 \cup \beta_2$, where β_i is a basis for W_i $(i = 1, 2)$. Show that $[T]_\beta^\beta = \begin{bmatrix} A_1 & 0 \\ 0 & A_2 \end{bmatrix}$, where $A_i = [T|_{W_i}]_{\beta_i}^{\beta_i}$. (This decomposition of $[T]$ into blocks along the diagonal is called a *direct sum decomposition* corresponding to the decomposition of V: $V = W_1 \oplus W_2$.)

b) More generally, if W_i $(1 \leq i \leq k)$ are subspaces of a vector space V, we say that V is the direct sum of the W_i, written as $V = W_1 \oplus \cdots \oplus W_k$ if $V = W_1 + \cdots + W_k$ and for each i, $W_i \cap (\sum_{j\neq i} W_j) = \{\mathbf{0}\}$. Generalize the results of part a to the case where V is the direct sum of an arbitrary collection of subspaces.

7. Let $A \in M_{n \times n}(F)$ be an arbitrary upper-triangular matrix.
   a) Show that for all polynomials $p \in P(F)$, $p(A)$ is upper-triangular.
   b) What are the eigenvalues of A^k $(k \geq 1)$?
   c) What are the eigenvalues of $p(A)$, where $p(t) = a_m t^m + \cdots + a_0$ is any polynomial?

8. Let $A, Q \in M_{n \times n}(F)$, and let Q be invertible.
   a) Show that if $p(t)$ is any polynomial with coefficients in F, then $p(Q^{-1}AQ) = Q^{-1}p(A)Q$.
   b) Conclude that if A is diagonalizable (respectively triangularizable), then $p(A)$ is also diagonalizable (respectively triangularizable) for all polynomials $p(t)$.

9. Let $T, S: V \to V$ be linear mappings, and assume that $TS = ST$. Let W be the λ-eigenspace for T. Show that W is invariant under S.

10. Show that the conclusion of the Cayley-Hamilton theorem is valid for matrices $A \in M_{n \times n}(\mathbf{R})$ even if some of the eigenvalues of A are not real. (*Hint:* View A as the matrix of a linear mapping: $\mathbf{C}^n \to \mathbf{C}^n$. This proof will show, in fact, that whenever our original field F may be embedded in an *algebraically closed* field F', then Cayley-Hamilton is true for linear mappings on vector spaces defined over F. It may be shown that such a field F' *always* exists, but the proof of that fact is beyond the scope of this course.)

11. Compute the inverses of each of the following matrices by the method of Example (6.1.14).

    a) $\begin{bmatrix} 1 & 4 \\ 3 & 8 \end{bmatrix}$ b) $\begin{bmatrix} 1 & -1 & 1 \\ 2 & 0 & 1 \\ 3 & 1 & 0 \end{bmatrix}$

    c) $\begin{bmatrix} 1 & 4 & 2 \\ 2 & 0 & 1 \\ 7 & 1 & 1 \end{bmatrix}$ d) $\begin{bmatrix} 1 & -1 & 1 & 1 \\ -1 & 0 & 1 & 1 \\ 0 & 1 & -1 & 1 \\ 0 & 0 & 1 & 1 \end{bmatrix}$

12. Let $\beta = \{\mathbf{x}_1, \ldots, \mathbf{x}_n\}$ be a basis for a vector space V, and let P be the mapping $P(a_1\mathbf{x}_1 + \cdots + a_n\mathbf{x}_n) = a_1\mathbf{x}_1 + \cdots + a_k\mathbf{x}_k$.
    a) Show that $\text{Ker}(P) = \text{Span}(\{\mathbf{x}_{k+1}, \ldots, \mathbf{x}_n\})$ and $\text{Im}(P) = \text{Span}(\{\mathbf{x}_1, \ldots, \mathbf{x}_k\})$.
    b) Show that $P^2 = P$.
    c) Show conversely that if $P: V \to V$ is any linear mapping such that $P^2 = P$, then there exists a basis β for V such that P takes the form given in part a. (*Hint:* Show that $P^2 = P$ implies that $V = \text{Ker}(P) \oplus \text{Im}(P)$. These mappings are called *projections*. The orthogonal projections we studied in Chapter 4 are special cases.)

13. Show, without using the Cayley-Hamilton theorem, that given any matrix $A \in M_{n \times n}(F)$, there exist nonzero polynomials $p \in P(F)$ such that $p(A) = 0$. (*Hint:* Can the set $\{A^k \mid k \geq 0\}$ be linearly independent?)

14. Formulate and prove a result parallel to Proposition (6.1.4) giving necessary and sufficient conditions for $[T]_\beta^\beta$ to be *lower-triangular*; that is, if the entries of $[T]_\beta^\beta$ are denoted a_{ij}, then $a_{ij} = 0$ for $i < j$.

15. Show that if V is a finite-dimensional vector space over a field F, and $T: V \to V$ is a linear mapping, then there exists a basis β such that $[T]_\beta^\beta$ is lower-triangular if and only if the characteristic polynomial of T has $\dim(V)$ roots (counted with multiplicities) in F.

§6.2. A CANONICAL FORM FOR NILPOTENT MAPPINGS

As a second step in and test case for our program to find a canonical form for a general linear mapping, we look at linear mappings $N: V \to V$, which have only one distinct eigenvalue $\lambda = 0$, with multiplicity $n = \dim(V)$. If N is such a mapping, then by the Cayley-Hamilton theorem, $N^n = 0$. Hence, by the terminology introduced in Section 4.1, Exercise 14, N is *nilpotent*. Conversely, if N is nilpotent, with $N^k = 0$ for some $k \geq 1$, then it is easy to see that every eigenvalue of N is equal to 0. Hence, a mapping has one eigenvalue $\lambda = 0$ with multiplicity n if and only if N is nilpotent.

There are two reasons for considering nilpotent mappings. First, since there is only one distinct eigenvalue to deal with, we will be able to see much more clearly exactly what our canonical form should look like in this case. Second, we will see in the next section that the general case may be reduced to this one, so all of the ideas introduced here will be useful later as well. Now to work!

Let $\dim(V) = n$, and let $N: V \to V$ be a nilpotent mapping. By the preceding discussion, we know that $N^n = 0$. Hence, given any vector $\mathbf{x} \in V$, we will have $N^n(\mathbf{x}) = \mathbf{0}$. For a given $\mathbf{x}$, however, it may very well be true that $N^k(\mathbf{x}) = \mathbf{0}$ for some $k < n$, Now, for each $\mathbf{x} \in V$, either $\mathbf{x} = \mathbf{0}$ or there is a unique integer k, $1 \leq k \leq n$, such that $N^k(\mathbf{x}) = \mathbf{0}$, but $N^{k-1}(\mathbf{x}) \neq \mathbf{0}$. It follows then, that if $\mathbf{x} \neq \mathbf{0}$, the set $\{N^{k-1}(\mathbf{x}), N^{k-2}(\mathbf{x}), \ldots, N(\mathbf{x}), \mathbf{x}\}$ consists of distinct (why?) nonzero vectors. In fact, more is true—see Proposition (6.2.3).

(6.2.1) Definitions. Let N, $\mathbf{x} \neq \mathbf{0}$ and k be as before.

a) The set $\{N^{k-1}(\mathbf{x}), N^{k-2}(\mathbf{x}), \ldots, \mathbf{x}\}$ is called the *cycle* generated by $\mathbf{x}$. $\mathbf{x}$ is called the *initial vector* of the cycle.

b) The subspace Span($\{N^{k-1}(\mathbf{x}), N^{k-2}(\mathbf{x}), \ldots, \mathbf{x}\}$ is called the *cyclic subspace* generated by $\mathbf{x}$, and denoted $C(\mathbf{x})$.

c) The integer k is called the *length* of the cycle.

(6.2.2) Example. Consider the linear mapping $N\colon \mathbf{R}^4 \to \mathbf{R}^4$ defined by the upper-triangular matrix

$$A = \begin{bmatrix} 0 & 2 & 1 & -1 \\ 0 & 0 & 1 & 0 \\ 0 & 0 & 0 & 1 \\ 0 & 0 & 0 & 0 \end{bmatrix}$$

Since all the diagonal entries are zero, by the previous discussion A is nilpotent. Indeed, it is easily checked that $A^4 = 0$, but $A^3 \neq 0$. If we let $\mathbf{x} = (1, 1, -1, 0)$, then $A\mathbf{x} = (1, -1, 0, 0)$, $A^2\mathbf{x} = A(A\mathbf{x}) = (-2, 0, 0, 0)$, and $A^3\mathbf{x} = A(A^2\mathbf{x}) = (0, 0, 0, 0)$. Hence, $\{A^2\mathbf{x}, A\mathbf{x}, \mathbf{x}\}$ is a cycle of length 3.

Different vectors may generate cycles of different lengths. For instance, the reader should check that the vector $\mathbf{x} = (0, 1, 0, 0)$ generates a cycle of length 2 in this example.

The following general statements are consequences of the definitions.

(6.2.3) Proposition. With all notation as before:

a) $N^{k-1}(\mathbf{x})$ is an eigenvector of N with eigenvalue $\lambda = 0$.

b) $C(\mathbf{x})$ is an invariant subspace of V under N.

c) The cycle generated by $\mathbf{x} \neq \mathbf{0}$ is a linearly independent set. Hence $\dim(C(\mathbf{x})) = k$, the length of the cycle.

Proof:

a) and **b)** are left as exercises for the reader.

c) We will prove this by induction on the length of the cycle. If $k = 1$, then the cycle is $\{\mathbf{x}\}$. Since $\mathbf{x} \neq \mathbf{0}$, this set is linearly independent. Now, assume that the result has been proved for cycles of length m, and consider a cycle of length $m + 1$, say, $\{N^m(\mathbf{x}), \ldots, \mathbf{x}\}$. If we have a possible linear dependence

$$a_m N^m(\mathbf{x}) + \cdots + a_1 N(\mathbf{x}) + a_0 \mathbf{x} = \mathbf{0} \tag{6.1}$$

then applying N to both sides yields

$$a_{m-1} N^m(\mathbf{x}) + \cdots + a_1 N^2(\mathbf{x}) + a_0 N(\mathbf{x}) = \mathbf{0}$$

(Note that $N^{m+1}(\mathbf{x}) = \mathbf{0}$, so that the term drops out.) Now, the vectors $\{N^m(\mathbf{x}), \ldots, N(\mathbf{x})\}$ form a cycle of length m, so by induction, we must have $a_{m-1} = \cdots = a_0 = 0$. But then in the original linear combination (6.1), $a_m = 0$ as well, since by definition, $N^m(\mathbf{x}) \neq \mathbf{0}$. ■

The real importance of these cyclic subspaces is revealed by the following observation. Let N be a nilpotent mapping, let $C(\mathbf{x})$ be the cyclic subspace generated by some $\mathbf{x} \in V$, and let α be the cycle generated by $\mathbf{x}$, viewed as a basis for $C(\mathbf{x})$. Since $N(N^i(\mathbf{x})) = N^{i+1}(\mathbf{x})$ for all i, we see that in $[N|_{C(\mathbf{x})}]_\alpha^\alpha$, all the entries except those immediately above the diagonal are 0, and the entries above the diagonal are all 1's:

$$[N|_{C(\mathbf{x})}]_\alpha^\alpha = \begin{bmatrix} 0 & 1 & 0 & \cdots & & 0 \\ 0 & 0 & 1 & 0 & \cdot & 0 \\ \vdots & & & \ddots & & \\ 0 & \cdots & & 0 & & \\ 0 & & \cdots & & 0 & 1 \\ & & & & & 0 \end{bmatrix} \tag{6.2}$$

Thus, the matrix of N restricted to a cyclic subspace is not only upper-triangular, but an upper-triangular matrix of a *very special* form. The simplicity of this matrix suggests that it is this form that we should take as the model for our canonical form for nilpotent mappings.

Of course, V itself may not coincide with any of the cyclic subspaces $C(\mathbf{x})$, so our next task is to investigate how the different cyclic subspaces fit together in V. To do this, we must shift our attention from the initial vector of the cycle to the eigenvector at the other end.

(6.2.4) Proposition. Let $\alpha_i = \{N^{k_i - 1}(\mathbf{x}_i), \ldots, \mathbf{x}_i\}$ $(1 \leq i \leq r)$ be cycles of lengths k_i, respectively. If the set of eigenvectors $\{N^{k_1-1}(\mathbf{x}_1), \ldots, N^{k_r-1}(\mathbf{x}_r)\}$ is linearly independent, then $\alpha_1 \cup \cdots \cup \alpha_r$ is linearly independent.

Proof: To make this proof somewhat more visual, we introduce a convenient pictorial representation of the vectors in the cycles, called a *cycle tableau*. Let us assume that the cycles have been arranged so that $k_1 \geq k_2 \geq \cdots \geq k_r$. Then the cycle tableau consists of r rows of boxes, where the boxes in the ith row represent the vectors in the ith cycle. (Thus, there are k_i boxes in the ith row of the tableau.) We will always arrange the tableau so that the leftmost boxes in each row are in the same column. In other words, cycle tableaux are always left-justified. For example, if we have four cycles of lengths 3, 2, 2, 1, respectively, then the corresponding cycle tableau is

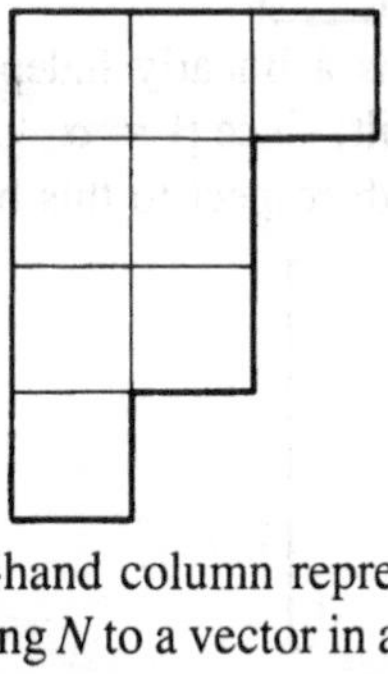

Note that the boxes in the left-hand column represent the eigenvectors.

Now, applying the mapping N to a vector in any one of the cycles (respectively a linear combination of those vectors) corresponds to shifting one box to the left on the corresponding row (respectively rows). Since the eigenvectors $N^{k_i-1}(\mathbf{x}_i)$ map to $\mathbf{0}$ under N, the vectors in the left-hand column get "pushed over the edge" and disappear.

Suppose we have a linear combination of the vectors in $\alpha_1 \cup \cdots \cup \alpha_r$ that sums to zero. (Imagine the coefficients as "filled in" in the corresponding boxes in the tableau.) There is some power of N, say, N^l that will shift the rightmost box or boxes in the tableau into the leftmost column. (In general $l = k_1 - 1$.) Applying N^l to the linear combination gives a linear combination of the eigenvectors $N^{k_i - 1}(\mathbf{x}_i)$ that sums to zero. Since those eigenvectors are linearly independent, those coefficients must be zero. In the same way, now, applying N^{l-j} to the original combination for each j $(1 \leq j \leq l)$ in turn will shift each column in the tableau into the leftmost column, so all the coefficients must be zero. As a consequence, $\alpha_1 \cup \cdots \cup \alpha_r$ is linearly independent. ■

(6.2.5) Definition. We say that the cycles $\alpha_i = \{N^{k_i - 1}(\mathbf{x}_i), \ldots, \mathbf{x}_i\}$ are *non-overlapping* cycles if $\alpha_1 \cup \cdots \cup \alpha_r$ is linearly independent.

(6.2.6) Example. Let $V = \mathbf{R}^6$, and let $N: V \to V$ be the mapping defined by the matrix

$$A = \begin{bmatrix} 0 & -1 & 1 & 0 & 0 & 0 \\ 0 & 0 & 2 & 0 & 0 & 0 \\ 0 & 0 & 0 & 0 & 0 & 0 \\ 0 & 0 & 0 & 0 & 4 & 0 \\ 0 & 0 & 0 & 0 & 0 & 0 \\ 0 & 0 & 0 & 0 & 0 & 0 \end{bmatrix}$$

It is easy to check that the standard basis vectors $\mathbf{e}_1$, $\mathbf{e}_4$, and $\mathbf{e}_6$ are eigenvectors of N, and that $\alpha_1 = \{\mathbf{e}_1, -\mathbf{e}_2, (-1/2)\mathbf{e}_2 + (-1/2)\mathbf{e}_3\}$, $\alpha_2 = \{\mathbf{e}_4, (1/4)\mathbf{e}_5\}$ and $\alpha_3 = \{\mathbf{e}_6\}$ are cycles of lengths 3, 2, 1, respectively. The corresponding cycle tableau is

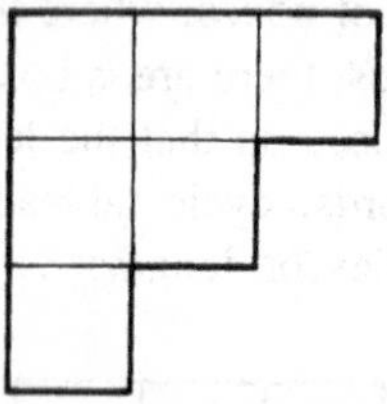

Furthermore, since $\{\mathbf{e}_1, \mathbf{e}_4, \mathbf{e}_6\}$ is a linearly independent set, α_i $(1 \leq i \leq 3)$ are nonoverlapping cycles. As a result, since $\beta = \alpha_1 \cup \alpha_2 \cup \alpha_3$ contains $6 = \dim(\mathbf{R}^6)$ vectors, β is a basis for $\mathbf{R}^6$. With respect to this basis

$$[N]_\beta^\beta = \begin{bmatrix} 0 & 1 & 0 & & & \\ 0 & 0 & 1 & & & \\ 0 & 0 & 0 & & & \\ & & & 0 & 1 & \\ & & & 0 & 0 & \\ & & & & & 0 \end{bmatrix}$$

(All the other entries in the matrix are 0.) This matrix is a direct sum of blocks of the form given in Eq. (6.2). (See Exercise 6 of Section 6.1.)

Note that once again we have found a form for the matrix of a nilpotent mapping in which only the entries immediately above the diagonal can be nonzero.

In general, if we can find a basis β for V that is the union of a collection of nonoverlapping cycles, then $[N]_\beta^\beta$ will be a matrix of this same form—with a direct sum decomposition into blocks of the form given in Eq. (6.2) on the diagonal. We will now show that such a basis always exists, and this fact leads to our canonical form for nilpotent mappings.

(6.2.7) Definition. Let $N: V \to V$ be a nilpotent mapping on a finite-dimensional vector space V. We call a basis β for V a *canonical basis* (with respect to N) if β is the union of a collection of nonoverlapping cycles for N.

(6.2.8) Theorem. (Canonical form for nilpotent mappings) Let $N: V \to V$ be a nilpotent mapping on a finite-dimensional vector space. There exists a canonical basis β of V with respect to N.

Proof: Our proof proceeds by induction on $\dim(V)$. If $\dim(V) = 1$, then $N = 0$, and any basis for V is canonical. Now assume the theorem has been proved for all spaces of dimension $< k$, and consider a vector space V of dimension k and a nilpotent mapping $N: V \to V$. Since N is nilpotent, $\dim(\text{Im}(N)) < k$. Furthermore, $\text{Im}(N)$ is an invariant subspace and $N|_{\text{Im}(N)}$ is also a nilpotent mapping, so by induction there exists a canonical basis γ for $N|_{\text{Im}(N)}$. Write $\gamma = \gamma_1 \cup \cdots \cup \gamma_r$, where the γ_i are nonoverlapping cycles for $N|_{\text{Im}(N)}$.

We need to extend γ in two different ways to find a basis for V. First, note that if $\mathbf{v}_i$ is the initial vector of the cycle γ_i, then since $\mathbf{v}_i \in \text{Im}(N)$, $\mathbf{v}_i = N(\mathbf{x}_i)$ for some vector $\mathbf{x}_i \in V$. Let $\sigma = \{\mathbf{x}_1, \ldots, \mathbf{x}_r\}$. Note that $\gamma \cup \sigma$ is a collection of r nonoverlapping cycles for N, with the ith cycle given by $\gamma_i \cup \{\mathbf{x}_i\}$.

Second, the final vectors of the cycles γ_i form a linearly independent subset of $\text{Ker}(N)$. However, there may be further eigenvectors in $\text{Ker}(N)$ that are not in $\text{Im}(N)$, so let α be a linearly independent subset of $\text{Ker}(N)$ such that α, together with the final vectors of the cycles γ_i forms a basis of $\text{Ker}(N)$.

We claim that $\beta = \alpha \cup \gamma \cup \sigma$ is the desired canonical basis. Since each of the vectors in α is a cycle of length 1, it follows that β $(= \alpha \cup (\gamma \cup \sigma))$ is the union of a collection of nonoverlapping cycles for N. [Note that we are using Proposition (6.2.4) here.] Now, β contains the $\dim(\text{Im}(N)) = \dim(V) - \dim(\text{Ker}(N))$ vectors in γ, together with the r vectors in σ and the $\dim(\text{Ker}(N)) - r$ vectors in α. As a result, β contains $(\dim(V) - \dim(\text{Ker}(N))) + r + (\dim(\text{Ker}(N)) - r) = \dim(V)$ vectors in all. It follows that β is a canonical basis for V, and the theorem is proved. ■

The proof of the theorem, while very clear, does not give a practical method for computing the canonical form of a nilpotent mapping. To develop such a method, we must make some observations about the cycle tableau corresponding to a canonical basis.

(6.2.9) Lemma. Consider the cycle tableau corresponding to a canonical basis for a nilpotent mapping $N: V \to V$. As before, let r be the number of rows, and let k_i be the number of boxes in the ith row $(k_1 \geq k_2 \geq \cdots \geq k_r)$. For each j ($1 \leqslant j \leqslant k_1$), the number of boxes in the jth column of the tableau is $\dim(\text{Ker}(N^j)) - \dim(\text{Ker}(N^{j-1}))$.

Proof: Suppose $N^l = 0$ but $N^{l-1} \neq 0$. Then we have (see Exercise 5) $\{\mathbf{0}\} \subset \text{Ker}(N) \subset \text{Ker}(N^2) \subset \cdots \subset \text{Ker}(N^{l-1}) \subset \text{Ker}(N^l) = V$. Thus, for each j, $\dim(\text{Ker}(N^j)) - \dim(\text{Ker}(N^{j-1}))$ measures the "new" vectors in $\text{Ker}(N^j)$ that are not in $\text{Ker}(N^{j-1})$. Now it is easy to see that if we represent the vectors in a canonical basis for V by the corresponding cycle tableau, then the vectors represented by the boxes in the first j columns (from the left) of the tableau are all in $\text{Ker}(N^j)$. As in the proof of Proposition (6.2.4), we see that N^j will "push all of these vectors off the left-hand edge of the tableau"—that is, N^j sends these vectors to $\mathbf{0}$. By the same reasoning, the vectors represented by the boxes in the jth column of the tableau represent a set of linearly independent vectors in $\text{Ker}(N^j) \setminus \text{Ker}(N^{j-1})$, which together with the basis of $\text{Ker}(N^{j-1})$ form a linearly independent set. As a result, we must have that

$$\text{The number of boxes in the } j\text{th column is} \leqslant \dim(\text{Ker}(N^j)) - \dim(\text{Ker}(N^{j-1})) \qquad (6.3)$$

On the other hand, if we add these differences of dimensions, we find that

$$\begin{aligned}\sum_{j=1}^{l} (\dim(\text{Ker}(N^j)) - \dim(\text{Ker}(N^{j-1}))) &= (\dim(\text{Ker}(N)) - 0) \\ &\quad + (\dim(\text{Ker}(N^2)) - \dim(\text{Ker}(N)) \\ &\quad + \cdots \\ &\quad + (\dim(\text{Ker}(N^l)) - \dim(\text{Ker}(N^{l-1}))) \\ &= \dim(\text{Ker}(N^l)) \\ &= \dim(V)\end{aligned}$$

since the other terms all cancel in pairs. By observation (6.3), and the fact that the total number of boxes in the tableau is equal to $\dim(V)$, it follows that the number of boxes in the jth column must be precisely equal to $\dim(\text{Ker}(N^j)) - \dim(\text{Ker}(N^{j-1}))$ for each j. ■

(6.2.10) Example. If the cycle tableau corresponding to a canonical basis for N is

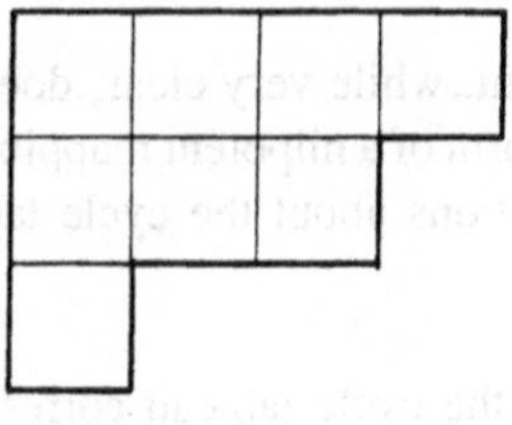

then $\dim(\text{Ker}(N)) = 3$, $\dim(\text{Ker}(N^2)) = 5$, $\dim(\text{Ker}(N^3)) = 7$ and $\dim(\text{Ker}(N^4)) = 8$.

Conversely, if we had computed these dimensions for a nilpotent mapping N on a vector space of dimension 8, then the cycle tableau would be uniquely determined by this information. Indeed, suppose we know $\dim(\mathrm{Ker}(N)) = 3$, $\dim(\mathrm{Ker}(N^2)) = 5$, $\dim(\mathrm{Ker}(N^3)) = 7$, and $\dim(\mathrm{Ker}(N^4)) = 8$. Then from Lemma (6.2.9) we know that there will be $\dim(\mathrm{Ker}(N)) = 3$ boxes in the first column of the tableau, $\dim(\mathrm{Ker}(N^2)) - \dim(\mathrm{Ker}(N)) = 5 - 3 = 2$ boxes in the second column, $\dim(\mathrm{Ker}(N^3)) - \dim(\mathrm{Ker}(N^2)) = 7 - 5 = 2$ boxes in the third column, and $\dim(\mathrm{Ker}(N^4)) - \dim(\mathrm{Ker}(N^3)) = 8 - 7 = 1$ boxes in the fourth column. When the columns are assembled to form the tableau, the result is exactly as before.

(6.2.11) Corollary. The canonical form of a nilpotent mapping is unique (provided the cycles in the canonical basis are arranged so the lengths satisfy $k_1 \geqslant k_2 \geqslant \cdots \geqslant k_r$).

Proof: The number of boxes in each column of the cycle tableau (and hence the entire tableau and the canonical form) is determined by the integers $\dim(\mathrm{Ker}(N^j))$, $j \geqslant 1$. ■

Once we have the cycle tableau corresponding to a canonical basis, to find an explicit canonical basis, we may proceed as follows. Since the final vector of a cycle of length k is in $\mathrm{Ker}(N) \cap \mathrm{Im}(N^{k-1})$, corresponding to a row of length k in the tableau, we find an eigenvector of N that is an element of $\mathrm{Im}(N^{k-1})$ [but not of $\mathrm{Im}(N^k)$]. Then the other vectors in the cycle may be found by solving systems of linear equations. For example, if $\mathbf{y}$ is to be the final vector of the cycle, to find the initial vector in the cycle solve $N^{k-1}(\mathbf{x}) = \mathbf{y}$ (there will always be more than one solution, but any particular solution will do), then the remaining (middle) vectors in the cycle will be $N(\mathbf{x}), N^2(\mathbf{x}), \ldots,$ and $N^{k-2}(\mathbf{x})$. These vectors may be found simply by computing the appropriate matrix products or applying the mapping N repeatedly. If there are several cycles in the basis, care should be taken to ensure that the final vectors of the cycles are linearly independent. The union of these cycles corresponding to the rows in the tableau will be the canonical basis.

(6.2.12) Examples

a) Let $V = \mathbf{R}^3$, and consider the mapping $N: V \to V$ defined by

$$A = \begin{bmatrix} 1 & 1 & 1 \\ -2 & -2 & -2 \\ 1 & 1 & 1 \end{bmatrix}$$

Note that $A^2 = 0$, so that N is indeed a nilpotent mapping. We have $\dim(\mathrm{Ker}(N)) = 2$ and $\dim(\mathrm{Ker}(N^2)) = 3$ in this case, so there will be $2 - 0 = 2$ boxes in the first column and $3 - 2 = 1$ box in the second column. The cycle tableau for a canonical basis will be

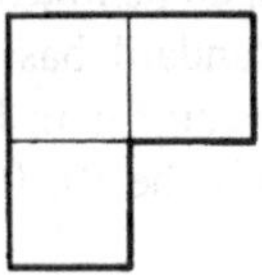

For this tableau we see immediately that the canonical form of N is

$$\begin{bmatrix} 0 & 1 & 0 \\ 0 & 0 & 0 \\ 0 & 0 & 0 \end{bmatrix} \tag{6.4}$$

A canonical basis may be constructed as follows. First, note that $\text{Ker}(N) = \text{Span}\{(-1, 1, 0), (-1, 0, 1)\}$ and $\text{Im}(N) = \text{Span}\{(1, -2, 1)\}$. (Look at the columns of A) This means that as the final vector of the cycle corresponding to the first row of the tableau, we should take any nonzero vector in $\text{Ker}(N) \cap \text{Im}(N)$, and $\mathbf{y} = (1, -2, 1)$ is the handiest one. Now the initial vector of the cycle of length 2 can be any vector that solves the system $A\mathbf{x} = \mathbf{y}$. Note there are many choices here that will lead to different canonical bases, but $\mathbf{x} = (1, 0, 0)$ will do as well as any other. For the second cycle we must find an eigenvector of N that together with the vector $\mathbf{y} = (1, -2, 1)$ gives a linearly independent set. Here again, there are many different choices that will work. One possible choice is the vector $(-1, 1, 0)$ from the basis of $\text{Ker}(N)$ found earlier. Hence, one canonical basis for N is $\beta = \{(1, -2, 1), (1, 0, 0),\} \cup \{(-1, 1, 0)\}$. It is easy to check that, in fact, $[N]_\beta^\beta$ has the form given in Eq. (6.4).

b) In some cases the final vector of a cycle will not be quite so easy to find. However, our methods for solving systems of linear equations may always be applied to win through to the desired result. For example, consider the matrix

$$N = \begin{bmatrix} 0 & 1 & 0 & 0 & 0 \\ 0 & 0 & 0 & 0 & 0 \\ 0 & 0 & 0 & 0 & 0 \\ 0 & 6 & 1 & 0 & 0 \\ 0 & 0 & 2 & 3 & 0 \end{bmatrix}$$

Computing, we find

$$N^2 = \begin{bmatrix} 0 & 0 & 0 & 0 & 0 \\ 0 & 0 & 0 & 0 & 0 \\ 0 & 0 & 0 & 0 & 0 \\ 0 & 0 & 0 & 0 & 0 \\ 0 & 18 & 3 & 0 & 0 \end{bmatrix},$$

and $N^3 = 0$. Hence, N is, in fact, nilpotent. Furthermore, by reducing the matrices N and N^2 to echelon form, we can see that $\dim(\text{Ker}(N)) = 2$ and $\dim(\text{Ker}(N^2)) = 4$. Hence, our cycle tableau will have one row with three boxes and one row with two boxes. This indicates that the canonical basis will consist of one cycle of length 3 and another cycle of length 2. To find the cycle of length 3, note that if $\mathbf{y}$ is the final vector in this cycle, then $\mathbf{y} \in \text{Ker}(N) \cap \text{Im}(N^2)$. Looking at the columns of N^2, which as we know span its image, we see that both nonzero columns are scalar multiples of the fifth standard basis vector $\mathbf{e}_5 = (0, 0, 0, 0, 1)$. In addition, it is easily seen that this vector is in $\text{Ker}(N)$. Hence, our final vector for the cycle of length 3 can be taken to be $(0, 0, 0, 0, 1)$. Now, to find the initial vector of that cycle, we must solve the system $N^2\mathbf{x} = \mathbf{e}_5$. As always, there are many solutions of this equation, but $\mathbf{x} = (0, 1/18, 0, 0, 0)$ is one. The middle vector of the cycle is then $N(\mathbf{x}) = (1/18, 0, 0, 1/3, 0)$ (just multiply N times $\mathbf{x}$).

Now, to complete the canonical basis, we must find a cycle of length 2. Let us call the final vector of this cycle $\mathbf{z}$. The vector $\mathbf{z}$ is in $\text{Ker}(N) \cap \text{Im}(N)$. Here, we find a slightly more complicated situation in that none of the columns of N (other than the fourth)) is an eigenvector of N. (We cannot use the fourth column, since we want $\mathbf{z}$ and the vector $\mathbf{y} = \mathbf{e}_5$ from the cycle of length 3 to be linearly independent.) Hence, we must look at other vectors in $\text{Im}(N) = \text{Span}\{\text{columns of } N\}$ in order to find $\mathbf{z}$. By the previous remark, we might guess that only the second and third columns are necessary here, and this is indeed the case. Our candidate for $\mathbf{z}$ is thus a vector of the form $\mathbf{z} = a(1, 0, 0, 6, 0) + b(0, 0, 0, 1, 2) = (a, 0, 0, 6a + b, 2b)$ such that $N\mathbf{z} = \mathbf{0}$. To determine the coefficients a and b, we just set up the corresponding system of equations, and we find that $18a + 3b = 0$ is the necessary condition. We may take any solution of this equation, so $a = 1$ and $b = -6$ is as good as any. Hence, our vector $\mathbf{z} = (1, 0, 0, 0, -12)$. To finish, we need to find the initial vector in the cycle, that is, to solve the system $N\mathbf{w} = \mathbf{z}$. It is easily seen that $\mathbf{w} = (0, 1, -6, 0, 0)$ is one solution.

The final result of this extended computation is that one canonical basis for our nilpotent matrix N is the basis given by $\beta = \{(0, 0, 0, 0, 1), (1/18, 0, 0, 1/3, 0), (0, 1/18, 0, 0, 0)\} \cup \{(1, 0, 0, 0, -12), (0, 1, -6, 0, 0)\}$. With respect to this basis N takes the canonical form

$$\begin{bmatrix} 0 & 1 & 0 & & \\ 0 & 0 & 1 & & \\ 0 & 0 & 0 & & \\ & & & 0 & 1 \\ & & & 0 & 0 \end{bmatrix}$$

EXERCISES

1. Verify that each of the following mappings is nilpotent, and find the smallest k such that $N^k = 0$.

a) $N: \mathbf{C}^4 \to \mathbf{C}^4$ defined by

$$\begin{bmatrix} 1 & -3 & 3 & -1 \\ 1 & -3 & 3 & -1 \\ 1 & -3 & 3 & -1 \\ 1 & -3 & 3 & -1 \end{bmatrix}$$

b) $N: \mathbf{C}^4 \to \mathbf{C}^4$ defined by

$$\begin{bmatrix} 0 & 0 & 0 & 0 \\ -2 & 0 & 0 & 0 \\ 0 & 0 & 0 & 0 \\ 1 & 4 & -2 & 0 \end{bmatrix}$$

c) $N: P_4(\mathbf{C}) \to P_4(\mathbf{C})$ defined by $N(p) = p'' - 3p'$.
d) $V = \text{Span}\{1, x, y, x^2, xy, y^2\}$, $N: V \to V$ defined by $N(p) = \partial p/\partial x + \partial p/\partial y$.

2. Find the length of the cycle generated by the given vector for the given nilpotent mapping.
a) $\mathbf{x} = (1, 2, 3, 1)$, N from Exercise 1a.
b) $\mathbf{x} = (3, 1, 0, 0)$, N from Exercise 1a.
c) $\mathbf{x} = (-i, 1, 0, 2 - i)$, N from Exercise 1b.
d) $p = x^3 - 3x$ N from Exercise 1c.
e) $p = x^2 + xy$, N from Exercise 1d.

3. Prove parts a and b of Proposition (6.2.3).

4. Show that if $N: V \to V$ is nilpotent, $\mathbf{x}, \mathbf{y} \in V$, and $\mathbf{y} \in C(\mathbf{x})$, then $C(\mathbf{y}) \subset C(\mathbf{x})$.

5. a) Let $N: V \to V$ be a nilpotent mapping such that $N^l = 0$ but $N^{l-1} \neq 0$. Show that

$$\{\mathbf{0}\} \subset \text{Ker}(N) \subset \text{Ker}(N^2) \subset \cdots \subset \text{Ker}(N^{l-1}) \subset \text{Ker}(N^l) = V.$$

b) Show that $V \supset \text{Im}(N) \supset \text{Im}(N^2) \supset \cdots \supset \text{Im}(N^{l-1}) \supset \text{Im}(N^l) = \{\mathbf{0}\}$.

6. Let $N: V \to V$ be any linear mapping. Show that $\dim(\text{Ker}(N^j)) - \dim(\text{Ker}(N^{j-1})) = \dim(\text{Im}(N^{j-1})) - \dim(\text{Im}(N^j))$. If N is nilpotent, interpret this statement in terms of the cycle tableau of a canonical basis for N. [*Hint:* Try to identify which boxes in the cycle tableau represent vectors in $\text{Im}(N^j)$.]

7. Find the canonical form and a canonical basis for each of the following nilpotent linear mappings:
a) the mapping N from Exercise 1a.
b) the mapping N from Exercise 1b.
c) the mapping N from Exercise 1c.
d) the mapping N from Exercise 1d.
e) $T: P_5(\mathbf{R}) \to P_5(\mathbf{R})$ defined by $T(p) = p''$ (second derivative).
f) $T: \mathbf{R}^4 \to \mathbf{R}^4$ defined by

$$\begin{bmatrix} 4 & 1 & -1 & 2 \\ -4 & -1 & 2 & -1 \\ 4 & 1 & -1 & 2 \\ -4 & -1 & 1 & -2 \end{bmatrix}$$

g) $T: \mathbf{C}^4 \to \mathbf{C}^4$ defined by

$$\begin{bmatrix} 0 & -2i & i & 4 \\ 0 & 0 & 0 & 2+i \\ 0 & 0 & 0 & 1 \\ 0 & 0 & 0 & 0 \end{bmatrix}$$

8. A linear mapping $U: V \to V$ is said to be *unipotent* if $I - U$ is nilpotent.
   a) Show that U is unipotent if and only if U has one eigenvalue $\lambda = 1$ with multiplicity $n = \dim(V)$.
   b) More generally, a mapping $U: V \to V$ is said to be *quasiunipotent* if U^k is unipotent for some $k \geqslant 1$. What can be said about the eigenvalues of a quasiunipotent mapping?

9. How many different possible cycle tableaux are there for nilpotent mappings $N: F^n \to F^n$ for $n = 3, 4, 5$, general n? (*Hint:* For general n you will not be able to find a formula for the number of different cycle tableaux. Instead, try to find another way to express this number.)

10. Show that if $A \in M_{n\times n}(F)$ has rank 1 and $\text{Tr}(A) = 0$, then A is nilpotent.

11. Show that $N \in M_{n\times n}(\mathbf{C})$ is nilpotent if and only if $\text{Tr}(N) = \text{Tr}(N^2) = \cdots = \text{Tr}(N^n) = 0$.

12. Show that if $N \in M_{n\times n}(F)$ is nilpotent, then there exists an invertible matrix $Q \in M_{n\times n}(F)$ such that $Q^{-1}NQ$ is in canonical form.

13. Let N_1 and N_2 be nilpotent matrices in $M_{n\times n}(F)$.
   a) Show that if N_1 and N_2 are similar, then they have the same canonical form.
   b) Show conversely that if N_1 and N_2 have the same canonical form, then they are similar.

14. a) Show that if $N_1, N_2 \in M_{n\times n}(F)$ are nilpotent matrices with $\text{rank}(N_1) = \text{rank}(N_2) = n - 1$, then N_1 and N_2 are similar.
   b) Show that if $\text{rank}(N_1) = \text{rank}(N_2) = 1$, then N_1 and N_2 are similar.
   c) What can be said if $1 < \text{rank}(N_i) < n - 1$?

15. a) Show that if $T: V \to V$ is a linear mapping on a finite-dimensional space V with the property that for each $\mathbf{x} \in V$ there exists an integer $k \geqslant 1$ (depending on the vector $\mathbf{x}$) such that $T^k(\mathbf{x}) = \mathbf{0}$, then T is nilpotent.
   b) Is the same true if V is not finite-dimensional? (*Hint:* Consider the mapping $T: P(F) \to P(F)$ defined by $T(p) = p'$.)

16. Let $W \subset M_{n\times n}(F)$ be the subset $W = \{A \mid A \text{ is nilpotent—that is, } A^k = 0 \text{ for some } k \geqslant 1\}$.
   a) Is W a vector subspace of $M_{n\times n}(F)$?
   b) Show that if $N_1, N_2 \in W$ and $N_1N_2 = N_2N_1$, then $\text{Span}\{N_1, N_2\} \subset W$.
   c) What is the dimension of the largest vector subspace of $M_{n\times n}(F)$ contained in W?

§6.3. JORDAN CANONICAL FORM

We are now ready to describe and prove the existence of our promised canonical form for general linear mappings. Our method will be to combine the results of

Sections 6.1 and 6.2. To begin, we consider the following refinement of Theorem (6.1.8) and its Corollary (6.1.11). Although the proof of this result is somewhat complicated, the result itself is of fundamental importance, and it will be beneficial to the reader to study it carefully.

(6.3.1) Proposition. Let $T: V \to V$ be a linear mapping whose characteristic polynomial has $\dim(V)$ roots (λ_i with respective multiplicities m_i, $1 \leq i \leq k$) in the field F over which V is defined.

a) There exist subspaces $V_i' \subset V$ ($1 \leq i \leq k$) such that

(i) Each V_i' is invariant under T,

(ii) $T|_{V_i'}$ has exactly one distinct eigenvalue λ_i, and

(iii) $V = V_1' \oplus \cdots \oplus V_k'$

b) There exists a basis β for V such that $[T]_\beta^\beta$ has a direct sum decomposition into upper-triangular blocks of the form $\begin{bmatrix} \lambda & & & * \\ 0 & \lambda & & \\ \vdots & \ddots & \ddots & \\ 0 & \cdots & 0 & \lambda \end{bmatrix}$. (The entries above the diagonal are arbitrary and all entries in the matrix other than those in the diagonal blocks are zero.)

Before beginning the proof, we will indicate with a picture exactly what the relation is between this proposition and our previous results on triangularization. In Corollary (6.1.11) we saw that there is a basis α of V such that $[T]_\alpha^\alpha$ is upper-triangular of the form

$$\begin{bmatrix} \lambda_1 & & & & & & * \\ & \ddots & & & & & \\ & & \lambda_1 & & & & \\ & & & \lambda_2 & & & \\ & & & & \ddots & & \\ & & & & & \lambda_2 & \\ & 0 & & & & & \ddots \end{bmatrix} \tag{6.5}$$

(The $*$ above the diagonal represents the fact that those entries of the matrix are arbitrary.) For each i ($1 \leq i \leq k$) we draw in an $m_i \times m_i$ block on the main diagonal of the matrix in the rows and columns corresponding to the entries λ_i on the diagonal. Then, the content of this proposition is that by choosing another basis, we may obtain another upper-triangular matrix, but in which now all the entries outside these blocks on the diagonal are zero:

$$\begin{bmatrix} \begin{matrix} \lambda_1 & & * \\ & \ddots & \\ 0 & & \lambda_1 \end{matrix} & 0 & 0 \\ 0 & \boxed{\begin{matrix} \lambda_2 & & * \\ & \ddots & \\ 0 & & \lambda_2 \end{matrix}} & 0 \\ 0 & 0 & \ddots \end{bmatrix} \tag{6.6}$$

The entries within the blocks, above the main diagonal of the matrix, are still arbitrary, but note that in general there will be far fewer of these arbitrary entries in Eq. (6.6) than in the previous general form (6.5). In a sense, we are moving toward a simpler, canonical form for the matrix. In geometric terms, we can achieve the block form of the matrix in Eq. (6.6) if the first m_1 vectors in the basis (the ones corresponding to the diagonal entries λ_1), the second m_2 vectors (corresponding to the diagonal entries λ_2), and so on, all span subspaces that are invariant under T.

In order to replace the unwanted nonzero entries in Eq. (6.5) by zeros, we use an inductive procedure that adds a correction term to each member of the basis α to produce a corresponding vector in the new basis β. If $\mathbf{x} \in \beta$ and $T(\mathbf{x}) = \lambda_i\mathbf{x} + \mathbf{z}$, then the correction term will ensure that $\mathbf{z}$ depends only on the other vectors $\mathbf{y} \in \beta$ with $T(\mathbf{y}) = \lambda_i\mathbf{y} + \mathbf{w}$ for that eigenvalue λ_i. Thus the span of those vectors will be an invariant subspace of V.

Proof: a) By Corollary (6.1.11) there is a basis $\alpha = \{\mathbf{x}_{1,1}, \ldots, \mathbf{x}_{1,m_1}, \ldots, \mathbf{x}_{k,1}, \ldots, \mathbf{x}_{k,m_k}\}$ such that $[T]_\alpha^\alpha$ is upper-triangular with diagonal entries $\lambda_1, \ldots, \lambda_1, \ldots, \lambda_k, \ldots, \lambda_k$ in that order. Let $V_i = \text{Span}\{\mathbf{x}_{i,1}, \ldots, \mathbf{x}_{i,m_i}\}$ for each i $(1 \leq i \leq k)$. By construction the subspaces $V_1, V_1 + V_2, \ldots, V_1 + \cdots + V_k$ are all invariant under T.

The existence of our desired subspaces V_i' will follow if we can prove the existence of subspaces V_i' $(1 \leq i \leq k)$ such that

(i) Each V_i' is invariant under T,

(ii) $T \mid V_i'$ has exactly one eigenvalue λ_i, and

(iii′) For each i $(1 \leq i \leq k)$, $V_1 + \cdots + V_i = V_1' + \cdots + V_i'$ (Note that condition (ii) implies that $\dim(V_i') \leq m_i$, so that if condition (iii′) holds as well, then V will be the direct sum of the subspaces V_i'.)

We will show how to construct these subspaces and the desired basis $\beta = \{\mathbf{y}_{1,1}, \ldots, \mathbf{y}_{1,m_1}, \ldots, \mathbf{y}_{k,1}, \ldots, \mathbf{y}_{k,m_k}\}$ by an inductive process. The inductive construction will proceed on two levels, so to speak. The outer level produces the subspaces V_i', while the inner level produces the vectors in a basis for V_i'. First, let $V_1' = V_1$, and let $\mathbf{y}_{1,j} = \mathbf{x}_{1,j}$ for each j $(1 \leq j \leq m_1)$. Now, suppose we have constructed subspaces $V_1', \ldots, V_{\ell-1}'$ satisfying conditions (i) and (ii) and such that $V_1' + \cdots + V_{\ell-1}' = V_1 + \cdots + V_{\ell-1}$.

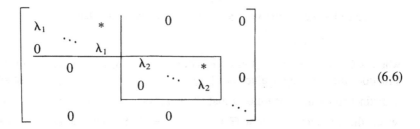

Consider the vector $\mathbf{x}_{\ell,1}$. Since $\mathbf{x}_{\ell,1} \in \alpha$ we know that

$$T(\mathbf{x}_{\ell,1}) = \lambda_\ell \mathbf{x}_{\ell,1} + \mathbf{z}$$

where $\mathbf{z} \in V_1 + \cdots + V_{\ell-1}$. Hence, $(T - \lambda_\ell I)(\mathbf{x}_{\ell,1}) = \mathbf{z}$. However, by construction, the only eigenvalues of $T|_{V_1 + \cdots + V_{\ell-1}}$ are $\lambda_1, \ldots, \lambda_{\ell-1}$ so since λ_ℓ is distinct from these, it follows that $(T - \lambda_\ell I) \mid_{V_1 + \cdots + V_{\ell-1}}$ is invertible. As a result, there is some vector $\mathbf{w} \in V_1 + \cdots + V_{\ell-1}$ such that $(T - \lambda_\ell I)(\mathbf{w}) = \mathbf{z}$. Hence, $(T - \lambda_\ell I)(\mathbf{x}_{1,\ell} - \mathbf{w}) = \mathbf{0}$. Let $\mathbf{v}_{\ell,1} = \mathbf{x}_{\ell,1} - \mathbf{w}$. (The vector $\mathbf{w}$ is the correction term mentioned before.)

Now, inductively again, assume we have constructed vectors $\mathbf{y}_{\ell,1}, \ldots, \mathbf{y}_{\ell,j-1}$ such that $T(\mathbf{y}_{\ell,n}) = \lambda_\ell \mathbf{y}_{\ell,n} + \mathbf{u}_n$, where $\mathbf{u}_n \in \text{Span}\{\mathbf{y}_{\ell,1}, \ldots, \mathbf{y}_{\ell,n-1}\}$ for each n $(1 \leqslant n \leqslant j - 1)$ and $\text{Span}\{\mathbf{y}_{\ell,1}, \ldots, \mathbf{y}_{\ell,j-1}\} + V_1' + \cdots + V_\ell' = \text{Span}\{\mathbf{x}_{\ell,1}, \ldots, \mathbf{x}_{\ell,j-1}\} + V_1' + \cdots + V_{\ell-1}'$. We know that $T(\mathbf{x}_{\ell,j}) = \lambda_\ell \mathbf{x}_{\ell,j} + \mathbf{u} + \mathbf{z}$, where $\mathbf{u} \in \text{Span}\{\mathbf{y}_{\ell,1}, \ldots, \mathbf{y}_{\ell,j-1}\}$ and $\mathbf{z} \in V_1' + \cdots + V_{\ell-1}'$ by the construction of the triangular basis α and induction. As before, we can rewrite this as $(T - \lambda_\ell I)(\mathbf{x}_{\ell,j}) = \mathbf{u} + \mathbf{z}$. Since the only eigenvalues of $T|V_1' + \cdots + V_{\ell-1}'$, are $\lambda_1, \ldots, \lambda_{\ell-1}$, as before, we can find some vector $\mathbf{w} \in V_1' + \cdots + V_{\ell-1}'$ such that $(T - \lambda_\ell I)(\mathbf{w}) = \mathbf{z}$. As a result, $(T - \lambda_\ell I)(\mathbf{x}_{\ell,j} - \mathbf{w}) = \mathbf{u} \in \text{Span}\{\mathbf{y}_{\ell,1}, \ldots, \mathbf{y}_{\ell,j-1}\}$. Therefore, we let $\mathbf{y}_{\ell,j} = \mathbf{x}_{\ell,j} - \mathbf{w}$. It can be seen that $\text{Span}\{\mathbf{y}_{\ell,1}, \ldots, \mathbf{y}_{\ell,j}\} + V_1' + \cdots + V_{\ell-1}' = \text{Span}\{\mathbf{x}_{\ell,1}, \ldots, \mathbf{x}_{\ell,j}\} + V_1' + \cdots + V_{\ell-1}'$, so the induction can continue, and we find the subspace $V_\ell' = \text{Span}\{\mathbf{y}_{\ell,1}, \ldots, \mathbf{v}_{\ell,m_\ell}\}$ after m_ℓ steps in all. By construction, V_ℓ' is invariant under T, since it is invariant under $T - \lambda_\ell I$. Thus, we find the desired basis for V.

b) This statement follows immediately from part a. ■

The subspaces V_i' constructed in the proof are extremely important. They are called the *generalized eigenspaces* of T. Note that since the only eigenvalue of $T|_{V_i'}$ is λ_i, if $\beta_i = \{\mathbf{v}_{i,1}, \ldots, \mathbf{v}_{i,m_i}\}$ is the basis of V_i' constructed in the proof of the proposition, then $[(T - \lambda_i I)|V_i']_{\beta_i}^{\beta_i}$ is upper-triangular with diagonal entries all equal to zero. As a result, the linear mapping $N_i = (T - \lambda_i I)|V_i'$ is nilpotent. Now, since $m_i = \dim(V_i')$, it is easy to see that $N_i^{m_i} = 0$ (the zero mapping on V_i') and that, furthermore, $V_i' = \text{Ker}((T - \lambda_i I)^{m_i})$ (as a mapping on all of V). This is true because of the observation we used several times in the proof of Proposition (6.3.1). Since the λ_i are distinct, on the sum of the V_j' *for* $j \neq i$, $T - \lambda_i I$ is invertible.

These observations motivate the following definition, which gives a more intrinsic way to view the subspaces V_i'.

(6.3.2) Definitions. Let $T: V \to V$ be a linear mapping on a finite-dimensional vector space V. Let λ be an eigenvalue of T with multiplicity m.

a) The λ-*generalized eigenspace*, denoted by K_λ, is the kernel of the mapping $(T - \lambda I)^m$ on V.

b) The nonzero elements of K_λ are called *generalized eigenvectors* of T.

In other words, a generalized eigenvector of T is any vector $\mathbf{x} \neq \mathbf{0}$ such that $(T - \lambda I)^m(\mathbf{x}) = \mathbf{0}$. Note that this will certainly be the case if $(T - \lambda I)^k(\mathbf{x}) = \mathbf{0}$ for some $k \leqslant m$. In particular, the eigenvectors with eigenvalue λ are also generalized eigenvectors, and we have $E_\lambda \subset K_\lambda$ for each eigenvalue.

(6.3.3) Examples

a) Let $T: \mathbf{C}^5 \to \mathbf{C}^5$ be the linear mapping whose matrix with respect to the standard basis is

$$\begin{bmatrix} 2 & 1 & 0 & 0 & 0 \\ 0 & 2 & 1 & 0 & 0 \\ 0 & 0 & 2 & 0 & 0 \\ 0 & 0 & 0 & 3-i & 1 \\ 0 & 0 & 0 & 0 & 3-i \end{bmatrix}$$

Since $\lambda = 2$ is an eigenvalue of multiplicity $m = 3$, the generalized eigenspace K_2 is by definition $K_2 = \text{Ker}((T - 2I)^3)$. It is easily checked that $K_2 = \text{Span}\{\mathbf{e}_1, \mathbf{e}_2, \mathbf{e}_3\}$. Each of these vectors is a generalized eigenvector:

$$(T - 2I)(\mathbf{e}_1) = \mathbf{0}, \quad \text{so} \quad (T - 2I)^3(\mathbf{e}_1) = \mathbf{0}$$

$$(T - 2I)^2(\mathbf{e}_2) = \mathbf{0}, \quad \text{so} \quad (T - 2I)^3(\mathbf{e}_2) = \mathbf{0}$$

and

$$(T - 2I)^3(\mathbf{e}_3) = \mathbf{0}$$

Similarly, the generalized eigenspace K_{3-i} has dimension two and is spanned by $\{\mathbf{e}_4, \mathbf{e}_5\}$.

b) Consider the mapping $T: \mathbf{C}^4 \to \mathbf{C}^4$ whose matrix with respect to the standard basis is

$$A = \begin{bmatrix} 1 & 1 & 0 & 0 \\ 0 & 0 & 0 & -8 \\ 0 & 1 & 0 & -12 \\ 0 & 0 & 1 & -6 \end{bmatrix}$$

Computing, we find that the characteristic polynomial of T is $\det(T - \lambda I) = (\lambda - 1)(\lambda + 2)^3$. Hence we have two generalized eigenspaces, K_1 and K_{-2}. By the definition we have $K_1 = \text{Ker}(T - I)$, which is the set of solutions of

$$\begin{bmatrix} 0 & 1 & 0 & 0 \\ 0 & -1 & 0 & -8 \\ 0 & 1 & -1 & -12 \\ 0 & 0 & 1 & -7 \end{bmatrix} \cdot \mathbf{x} = \mathbf{0}$$

After reducing to echelon form, it may be seen that the first component of $\mathbf{x}$ is the only free variable, and hence we have $K_1 = \text{Span}\{\mathbf{e}_1\}$.

Similarly, we have $K_{-2} = \text{Ker}((T + 2I)^3)$, which is the set of solutions of

$$(A + 2I)^3\mathbf{x} = \begin{bmatrix} 27 & 19 & -8 & 8 \\ 0 & 0 & 0 & 0 \\ 0 & 0 & 0 & 0 \\ 0 & 0 & 0 & 0 \end{bmatrix} \cdot \mathbf{x} = \mathbf{0}$$

Thus, $K_{-2} = \text{Span}\{(-19/27,1,0,0), (8/27,0,1,0), (-8/27,0,0,1)\}$.

(6.3.4) Proposition

a) For each eigenvalue λ of T, K_λ is an invariant subspace of V.

b) If λ_i $(1 \leq i \leq k)$ are the distinct eigenvalues of T, then $V = K_{\lambda_1} \oplus \cdots \oplus K_{\lambda_k}$.

c) If λ is an eigenvalue of multiplicity m, then $\dim(K_\lambda) = m$.

Proof: All three of these statements follow from Proposition (6.3.1) and the remarks before the Definitions (6.3.2). ■

Our canonical form for general linear mapping comes from combining the decomposition of V into the invariant subspaces K_{λ_i} for the different eigenvalues with our canonical form for nilpotent mappings [applied to the nilpotent mappings $N_i = (T - \lambda_i I)|_{K_{\lambda_i}}$]. Since N_i is nilpotent, by Theorem (6.2.8) there exists a basis γ_i for K_{λ_i} such that $[N_i]_{\gamma_i}^{\gamma_i}$ has a direct sum decomposition into blocks of the

$$\begin{bmatrix} 0 & 1 & 0 & \cdots & 0 \\ 0 & 0 & 1 & \cdots & 0 \\ \vdots & & & \ddots & \\ 0 & \cdots & & & 1 \\ & & & & 0 \end{bmatrix}$$

As a result, $[T|_{K_{\lambda_i}}]_{\gamma_i}^{\gamma_i} = [N_i]_{\gamma_i}^{\gamma_i} + \lambda_i I$ has a direct sum decomposition into diagonal blocks of the form

$$\begin{bmatrix} \lambda_i & 1 & 0 & \cdots & 0 \\ 0 & \lambda_i & 1 & \cdots & 0 \\ \vdots & \ddots & \ddots & \ddots & \\ \vdots & & & & 1 \\ 0 & & \cdots & 0 & \lambda_i \end{bmatrix} \qquad (6.7)$$

In this form the blocks each have the same number in each diagonal entry, 1's immediately above the diagonal, and 0's everywhere else.

(6.3.5) Definitions

a) A matrix of the form (6.7) is called a *Jordan block matrix*.

b) A matrix $A \in M_{n \times n}(F)$ is said to be *in Jordan canonical form* if A is a direct sum of Jordan block matrices.

The major result of this chapter is the following theorem.

(6.3.6) Theorem. (Jordan Canonical Form) Let $T: V \to V$ be a linear mapping on a finite-dimensional vector space V whose characteristic polynomial has $\dim(V)$ roots in the field F over which V is defined.

a) There exists a basis γ (called a *canonical basis*) of V such that $[T]_\gamma^\gamma$ has a direct sum decomposition into Jordan block matrices.

b) In this decomposition the number of Jordan blocks and their sizes are uniquely determined by T. (The order in which the blocks appear in the matrix may be different for different canonical bases, however.)

Proof:

a) In each generalized eigenspace K_{λ_i}, as we have noted, by Theorem (6.2.8) there is a basis γ_i such that $[T|_{K_{\lambda_i}}]_{\gamma_i}^{\gamma_i}$ has a direct sum decomposition into Jordan blocks with diagonal entries λ_i. In addition, we know that V is the direct sum of the subspaces K_{λ_i}. Let $\gamma = \bigcup_{i=1}^{k} \gamma_i$. Then γ is a basis for V, and by Exercise 6 of Section 6.1, we see that $[T]_\gamma^\gamma$ has the desired form.

b) This statement follows from Corollary (6.2.11). ■

We will give several examples of computing the Jordan canonical form in the next section.

EXERCISES

1. Show that the given vector is a generalized eigenvector of the given mapping for the given eigenvalue.

a) $\mathbf{x} = (3, 1, 0, 0)$ for $T: \mathbf{R}^4 \to \mathbf{R}^4$ defined by $\begin{bmatrix} 1 & 2 & 2 & 4 \\ 0 & 1 & -1 & 1 \\ 0 & 0 & 1 & 0 \\ 0 & 0 & 0 & 3 \end{bmatrix}$, $\lambda = 1$.

b) $\mathbf{x} = (3, 7, -1, 2)$ for $T: \mathbf{C}^4 \to \mathbf{C}^4$ defined by

$\begin{bmatrix} 0 & 0 & 0 & -1 \\ 1 & 0 & 0 & -4 \\ 0 & 1 & 0 & -6 \\ 0 & 0 & 1 & -4 \end{bmatrix}$, $\lambda = -1$.

c) $p(x) = x^4 + 2x^3 - 1$ for $T: P_4(\mathbf{R}) \rightarrow P_4(\mathbf{R})$ defined by $T(p) = p''$, $\lambda = 0$.
d) $f(x) = 2e^{3x} + 4xe^{3x}$ for $T: V \rightarrow V$ defined by $T(f) = f'$, where $V = \text{Span}\{e^{3x}, xe^{3x}, x^2e^{3x}, e^{-x}, xe^{-x}\}$, $\lambda = 3$.

2. Find the generalized eigenspace for each of the mappings of Exercise 1 for the given eigenvalue.

3. Let $A \in M_{n\times n}(F)$, and assume that all the eigenvalues of A lie in F. Show that there exists an invertible matrix $Q \in M_{n\times n}(F)$ such that $Q^{-1}AQ$ is in Jordan form.

4. a) Show that every matrix in $M_{2\times 2}(\mathbf{C})$ is similar to a matrix of one of the following forms

$$\begin{bmatrix} \lambda_1 & 0 \\ 0 & \lambda_2 \end{bmatrix} \quad \begin{bmatrix} \lambda & 1 \\ 0 & \lambda \end{bmatrix}$$

b) Show that every matrix in $M_{3\times 3}(\mathbf{C})$ is similar to a matrix of one of the following forms:

$$\begin{bmatrix} \lambda_1 & 0 & 0 \\ 0 & \lambda_2 & 0 \\ 0 & 0 & \lambda_3 \end{bmatrix} \quad \begin{bmatrix} \lambda_1 & 1 & 0 \\ 0 & \lambda_1 & 0 \\ 0 & 0 & \lambda_2 \end{bmatrix} \quad \begin{bmatrix} \lambda & 1 & 0 \\ 0 & \lambda & 1 \\ 0 & 0 & \lambda \end{bmatrix}$$

(*Hint:* Use Exercise 3 and reduce to Jordan form. Note: in the preceding matrices the λ_i are not necessarily distinct.)

c) Find a similar list of possible Jordan forms for matrices in $M_{4\times 4}(\mathbf{C})$.

5. a) Show that a linear mapping $T: V \rightarrow V$ on a finite-dimensional space is diagonalizable if and only if for each eigenvalue λ, $E_\lambda = K_\lambda$.
b) Show that if $T: V \rightarrow V$ is diagonalizable, then the Jordan canonical form of T is the diagonal form.

6. Let $A \in M_{n\times n}(\mathbf{C})$ be invertible.
a) Show that there exist a diagonalizable matrix D and a unipotent matrix U (see Exercise 8 of Section 6.2) such that $A = DU = UD$.
b) Show that D and U are uniquely determined by A.

7. Let $A \in M_{n\times n}(\mathbf{C})$.
a) Show that there exist a diagonalizable matrix D and a nilpotent matrix N such that $A = D + N$, and $ND = DN$.
b) Show that D and N are uniquely determined by A.

8. Show that if $p \in P(F)$ is any polynomial and B is a $k \times k$ Jordan block matrix with diagonal entries λ, then $p(B)$ is the matrix

$$\begin{bmatrix} p(\lambda) & p_1(\lambda) & p_2(\lambda) & \cdots & p_{k-1}(\lambda) \\ 0 & p(\lambda) & p_1(\lambda) & \cdots & p_{k-2}(\lambda) \\ 0 & 0 & p(\lambda) & \cdots & p_{k-3}(\lambda) \\ & & & \vdots & \\ 0 & 0 & & \cdots & p(\lambda) \end{bmatrix}$$

where $p_n(x)$ is the polynomial $p_n(x) = (1/n!)\, p^{(n)}(x)$. ($p^{(n)}(x)$ is the nth derivative of $p(x)$.)

9. Show that if $A \in M_{n\times n}(F)$ has rank 1, then A is diagonalizable or A is nilpotent. [*Hint:* The Jordan canonical form matrix similar to A (see Exercise 3) also has rank 1.]

10. a) Show that if B is a $k \times k$ Jordan block matrix with diagonal entries $\lambda \neq 0$, then B is invertible, and compute B^{-1}.
    b) Show that the Jordan form of B^{-1} also has exactly one $k \times k$ block.
    c) Using part b, show that if $A \in M_{n\times n}(\mathbf{R})$ is any invertible matrix, the number of Jordan blocks and their sizes in the Jordan form of A^{-1} are the same as the number of Jordan blocks and their sizes in the Jordan form of A.

11. One simple way used to describe Jordan form matrices is the *Segre characteristic* of the matrix. If J is a matrix in Jordan form, λ_i $(1 \leq i \leq k)$ are the eigenvalues, and J has Jordan blocks of sizes $n_{11}, \ldots, n_{1\ell_1}$ with diagonal entries λ_1, blocks of sizes $n_{21}, \ldots, n_{2\ell_2}$ with diagonal entries λ_2, and so on, then the Segre characteristic of the matrix is defined to be the symbol

$$\{(n_{11}, \ldots, n_{1\ell_1}), \ldots, (n_{k1}, \ldots, n_{k\ell_k})\}$$

    a) Show that in the Segre characteristic $\sum_{j=1}^{\ell_i} n_{ij} = m_i$ (the multiplicity of λ_i as an eigenvalue of J) and $\sum_{i=1}^{k}\left[\sum_{j=1}^{\ell_i} n_{ij}\right] = \dim(V)$.
    b) Write down the Jordan form matrix with eigenvalues $\lambda_1 = 4$, $\lambda_2 = 2 + i$, $\lambda_3 = -1 - i$, and Segre characteristic $\{(3, 2, 2), (2, 1), (3, 1)\}$.
    c) What does the Segre characteristic of a diagonal matrix look like?

§6.4. COMPUTING JORDAN FORM

In this section we will give several examples of computing the Jordan canonical form of a linear mapping. The process we use is summarized in the following outline. One interesting observation is that it is somewhat easier to determine the canonical form itself than to produce a canonical basis. Finding a canonical basis will in general involve solving many additional systems of linear equations.

(6.4.1) Computation of Jordan Canonical Form

a) Find all the eigenvalues of T and their multiplicities by factoring the characteristic polynomial completely. (We will usually work over an algebraically closed field such as $\mathbf{C}$ to ensure that this is always possible.)

b) For each distinct eigenvalue λ_i in turn, construct the cycle tableau for a canonical basis of K_{λ_i} with respect to the mapping $N_i = (T - \lambda_i I)|K_{\lambda_i}$ using the method of Lemma (6.2.9), in the following modified form. For each j, the number of boxes in the jth column of the tableau for λ_i will be

$$\dim(\mathrm{Ker}(T - \lambda_i I)^j) - \dim(\mathrm{Ker}(T - \lambda_i I)^{j-1})$$

(computed on V). This is the same as $\dim(\mathrm{Ker}(N_i^j) - \dim(\mathrm{Ker}(N_i^{j-1}))$, since $\mathrm{Ker}((T - \lambda_i I)^j)$ is contained in K_{λ_i} for all $j \geq 1$. As a result it is never actually necessary to construct the matrix of the restricted mapping N_i.

c) Form the corresponding Jordan blocks and assemble the matrix of T.

It should be emphasized at this point that this process is exactly the same as the process we described before for nilpotent mappings. We just apply the techniques to each $T - \lambda_i I$, in turn. Note, however, that if T has several different eigenvalues, then it will *not* be the case that each $T - \lambda_i I$ in nilpotent on all of V. (This only happens when we restrict $T - \lambda_i I$ to the generalized eigenspace K_{λ_i}.) Hence, we should not expect to obtain the zero matrix by computing powers of $T - \lambda_i I$. Rather, we should look for the smallest integer j such that $\dim(\mathrm{Ker}(T - \lambda_i I)^j) = m_i$, because when this occurs we will have found all the vectors in K_{λ_i}, hence the relevant cycle tableau.

(6.4.2) Example. For our first example, let us assume that we have a linear mapping $T: \mathbf{C}^8 \to \mathbf{C}^8$ with distinct eigenvalues $\lambda_1 = 2$, $\lambda_2 = i$, $\lambda_3 = -i$ having multiplicities 4, 3, and 1, respectively. Assume that $N_1 = (T - 2I)|_{K_2}$ satisfies:

$$\begin{aligned} \dim(\mathrm{Ker}(N_1)) &= 2 \\ \dim(\mathrm{Ker}(N_1^2)) &= 3 \\ \dim(\mathrm{Ker}(N_1^3)) &= 4 \end{aligned} \tag{6.8}$$

and that $N_2 = (T - iI)|_{K_i}$ satisfies

$$\begin{aligned} \dim(\mathrm{Ker}(N_2)) &= 1 \\ \dim(\mathrm{Ker}(N_2^2)) &= 2 \\ \dim(\mathrm{Ker}(N_2^3)) &= 3 \end{aligned} \tag{6.9}$$

(If we had started from an explicit matrix, this information would have been obtained in the usual way by reducing the matrices of the powers of the mappings $T - \lambda I$ to echelon form and counting the number of free variables.)

From Eq. (6.8) we see that the cycle tableau for N_1 is

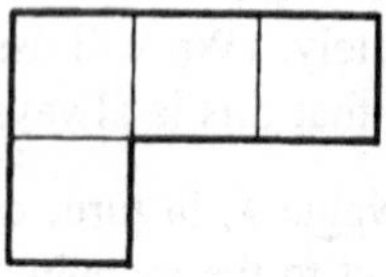

As a result, the canonical form of the nilpotent mapping N_1 is

$$\begin{bmatrix} 0 & 1 & 0 & 0 \\ 0 & 0 & 1 & 0 \\ 0 & 0 & 0 & 0 \\ 0 & 0 & 0 & 0 \end{bmatrix}$$

Hence, the canonical form for $T|_{K_2}$ is

$$\begin{bmatrix} 2 & 1 & 0 & 0 \\ 0 & 2 & 1 & 0 \\ 0 & 0 & 2 & 0 \\ 0 & 0 & 0 & 2 \end{bmatrix}$$

(Jordan blocks of sizes 3 and 1, respectively).

Similarly, the cycle tableau for a canonical basis of $N_2 = (T - iI)|_{K_i}$ is

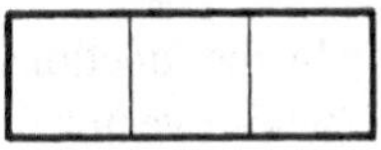

, which corresponds to a canonical form containing one Jordan block:

$$\begin{bmatrix} i & 1 & 0 \\ 0 & i & 1 \\ 0 & 0 & i \end{bmatrix}$$

The one eigenvector for $\lambda = -i$ gives us a Jordan block of size 1 (i.e., a single diagonal entry). Hence, putting this all together, the canonical form of T on $\mathbf{C}^8$ is

$$C = \begin{bmatrix} 2 & 1 & 0 & & & & & \\ 0 & 2 & 1 & & & & & \\ 0 & 0 & 2 & & & & & \\ & & & 2 & & & & \\ & & & & i & 1 & 0 & \\ & & & & 0 & i & 1 & \\ & & & & 0 & 0 & i & \\ & & & & & & & -i \end{bmatrix}$$

(All the other entries in the matrix are zero.) The union of the two nonoverlapping cycles for $\lambda = 2$, the cycle for $\lambda = i$, and the eigenvector for $\lambda = -i$ would form a canonical basis γ, and $[T]_\gamma^\gamma = C$.

(6.4.3) Example. Let $V = \text{Span}\{e^x, xe^x, x^2e^x, x^3e^x, e^{-x}, xe^{-x}\}$. Let $T: V \to V$ be defined by $T(f) = 2f + f''$. We begin by computing $[T]_\beta^\beta$, where β is the given basis of V. [The reader should convince him- or herself that these six functions are indeed linearly independent in $C^\infty(\mathbf{R})$.] We find that

$$[T]_{\beta}^{\beta} = \begin{bmatrix} 3 & 2 & 2 & 0 & 0 & 0 \\ 0 & 3 & 4 & 6 & 0 & 0 \\ 0 & 0 & 3 & 6 & 0 & 0 \\ 0 & 0 & 0 & 3 & 0 & 0 \\ 0 & 0 & 0 & 0 & 3 & -2 \\ 0 & 0 & 0 & 0 & 0 & 3 \end{bmatrix}$$

In this case, since $[T]_{\beta}^{\beta}$ is upper-triangular, we see that T has only one eigenvalue, $\lambda = 3$, with multiplicity $m = 6$. As a result, the generalized eigenspace K_3 is all of V. With respect to the same basis β, the nilpotent mapping $N = T - 3I$ has matrix

$$[N]_{\beta}^{\beta} = \begin{bmatrix} 0 & 2 & 2 & 0 & 0 & 0 \\ 0 & 0 & 4 & 6 & 0 & 0 \\ 0 & 0 & 0 & 6 & 0 & 0 \\ 0 & 0 & 0 & 0 & 0 & 0 \\ 0 & 0 & 0 & 0 & 0 & -2 \\ 0 & 0 & 0 & 0 & 0 & 0 \end{bmatrix}$$

From this we see that $\dim(\mathrm{Ker}(N)) = 2$. A basis for $\mathrm{Ker}(N)$ is given by the functions e^x (whose vector of coordinates with respect to β is $\mathbf{e}_1$), and e^{-x} (whose vector of coordinates is $\mathbf{e}_5$).

Now, continuing on to the powers of N, we find

$$[N^2]_{\beta}^{\beta} = \begin{bmatrix} 0 & 0 & 8 & 24 & 0 & 0 \\ 0 & 0 & 0 & 24 & 0 & 0 \\ 0 & 0 & 0 & 0 & 0 & 0 \\ 0 & 0 & 0 & 0 & 0 & 0 \\ 0 & 0 & 0 & 0 & 0 & 0 \\ 0 & 0 & 0 & 0 & 0 & 0 \end{bmatrix}$$

so that $\dim(\mathrm{Ker}(N^2)) = 4$. In the same way, it is easily checked that $\dim(\mathrm{Ker}(N^3)) = 5$ and $\dim(\mathrm{Ker}(N^4)) = 6$. Hence, the cycle tableau for N is

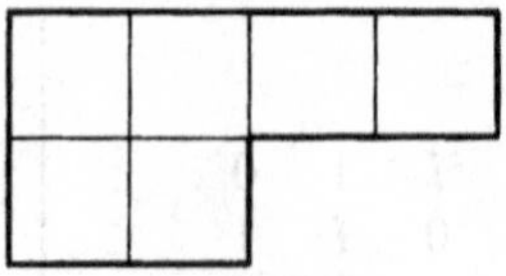

and the canonical form of T is the Jordan form matrix

$$C = \begin{bmatrix} 3 & 1 & 0 & 0 & & \\ 0 & 3 & 1 & 0 & & \\ 0 & 0 & 3 & 1 & & \\ 0 & 0 & 0 & 3 & & \\ & & & & 3 & 1 \\ & & & & 0 & 3 \end{bmatrix} \tag{6.10}$$

To find a canonical basis for T, we can proceed as in Section 6.2. Since $[N^3]_{\beta}^{\beta}$ is the matrix with a 48 in the first row and the fourth column and zeros in every other entry (check!), it follows that $e^x \in \text{Ker}(N) \cap \text{Im}(N^3)$, so we can take e^x as the final vector for the cycle of length 4. In this case it is also relatively easy to find a vector that will be the initial vector of the cycle. By the form of the matrix previously described, it follows that $N^3((1/48)x^3e^x) = e^x$, so that $\gamma_1 = \{e^x, (1/2)(e^x + xe^x), (1/8)(xe^x + x^2e^x), (1/48)x^3e^x\}$ is a cycle of length 4. (The "middle vectors" in the cycle may be found by computing $N((1/48)x^3e^x)$ and $N^2((1/48)x^3e^x)$. Similarly, $e^{-x} \in \text{Ker}(N) \cap \text{Im}(N)$ and $\gamma_2 = \{e^{-x}, -(1/2)xe^{-x}\}$ is a cycle of length 2. Since $\{e^x, e^{-x}\}$ is linearly independent, the two cycles are nonoverlapping and it follows that $\gamma = \gamma_1 \cup \gamma_2$ is a canonical basis for T. We have $[T]_\gamma^\gamma = C$ from Eq. (6.10).

We now turn to a linear mapping with several distinct eigenvalues. In this case we must apply the method of the previous example to $T - \lambda_i I$ for each eigenvalue λ_i separately.

(6.4.4) Example. Let $T: \mathbf{C}^5 \to \mathbf{C}^5$ be the linear mapping whose matrix with respect to the standard basis is

$$A = \begin{bmatrix} 1-i & 1 & 0 & 0 & 0 \\ 0 & 1-i & 0 & 0 & 0 \\ i & -1 & 1 & 0 & 0 \\ 0 & i & 1 & 1 & 1 \\ -i & 1 & 0 & 0 & 1 \end{bmatrix}$$

To begin, we compute the characteristic polynomial of T, and we find $\det(T - \lambda I) = (1 - \lambda)^3((1 - i) - \lambda)^2$. Hence, $\lambda = 1$ is an eigenvalue of multiplicity 3 and $\lambda = 1 - i$ is an eigenvalue of multiplicity 2.

Consider $\lambda = 1 - i$ first. We have $\dim(\text{Ker}(T - (1 - i)I)) = 1$, since there is only one free variable when we reduce the matrix $A - (1 - i)I$ to echelon form. At this point we could continue and compute $(A - (1 - i)I)^2$, however, since $\dim(K_{1-i}) =$ (multiplicity of $1 - i$) $= 2$, it can only be the case that $\dim(\text{Ker}((T - (1 - i)I)^2)) = 2$. Hence, the corresponding cycle tableau for this eigenvalue is

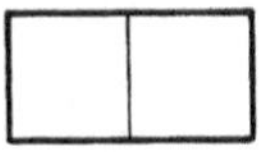

There will be one Jordan block in the canonical form: $\begin{bmatrix} 1-i & 1 \\ 0 & 1-i \end{bmatrix}$.

Now, for $\lambda = 1$, we consider $T - I$. The reader should check that there are two free variables in the system $(T - I)(\mathbf{x}) = \mathbf{0}$, so $\dim(\text{Ker}(T - I)) = 2$. Similarly, $\dim(\text{Ker}((T - I)^2)) = 3$. Hence, $K_1 = \text{Ker}((T - I)^2)$ already in this case. The cycle tableau for $\lambda = 1$ and the corresponding canonical form are

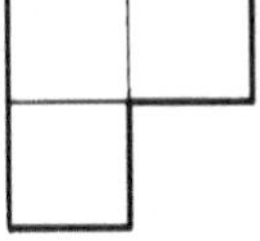

and

$$\begin{bmatrix} 1 & 1 & \\ 0 & 1 & \\ & & 1 \end{bmatrix}$$

Hence, the canonical form of T on $\mathbf{C}^5$ is

$$C = \begin{bmatrix} 1-i & 1 & & & \\ 0 & 1-i & & & \\ & & 1 & 1 & \\ & & 0 & 1 & \\ & & & & 1 \end{bmatrix}$$

We leave it as an exercise for the reader to find a canonical basis in this example.

EXERCISES

1. Complete the computations of Example (6.4.4) by finding a canonical basis for the linear mapping T.

2. a) Let $T: \mathbf{C}^9 \to \mathbf{C}^9$ be a linear mapping and assume that the characteristic polynomial of T is $((2 - i) - \lambda)^4(3 - \lambda)^5$. Assume that

$$\dim(\mathrm{Ker}(T - (2 - i)I)) = 3$$

and $$\dim(\mathrm{Ker}((T - (2 - i)I)^2)) = 4$$

$$\dim(\mathrm{Ker}(T - 3I)) = 2$$

$$\dim(\mathrm{Ker}((T - 3I)^2)) = 4$$

$$\dim(\mathrm{Ker}((T - 3I)^3)) = 5$$

Find the Jordan canonical form for T.

b) How many different canonical forms are there for linear mappings with characteristic polynomial as in part a? Compute the dimensions of the kernels of the $T - \lambda I$ in each case.

3. Let $T: \mathbf{C}^8 \to \mathbf{C}^8$ be a linear mapping with characteristic polynomial $(i - \lambda)^4(5 - \lambda)^4$. Assume that

$$\dim(\mathrm{Im}(T - iI)) = 6$$
$$\dim(\mathrm{Im})(T - iI)^2)) = 4$$
and $$\dim(\mathrm{Im}(T - 5I)) = 6$$
$$\dim(\mathrm{Im}((T - 5I)^2)) = 5$$
$$\dim(\mathrm{Im}((T - 5I)^3)) = 4$$

What is the Jordan canonical form of T?

4. Find the Jordan canonical form and a canonical basis for each of the mappings in Exercise 1 of Section 6.3 of this Chapter.

5. Find the Jordan canonical form and a canonical basis for each of the following mappings.

a) $T: \mathbf{C}^4 \to \mathbf{C}^4$ defined by the matrix $\begin{bmatrix} 1 & -3 & 5 & 3 \\ -2 & -6 & 0 & 13 \\ 0 & 0 & 1 & 0 \\ -1 & -4 & 7 & 8 \end{bmatrix}$.

b) $T: \mathbf{C}^4 \to \mathbf{C}^4$ defined by the matrix $\begin{bmatrix} 0 & 1 & 0 & 0 \\ -1 & 0 & 0 & -1 \\ 0 & 0 & i & 1 \\ 0 & 0 & 0 & i \end{bmatrix}$.

c) Let $V = \text{Span}\{\sin(x), \cos(x), e^{4x}, xe^{4x}, x^2e^{4x}\}$ (using coefficients in $\mathbf{C}$), and let $T: V \to V$ be the mapping defined by $T(f) = f' - 2f$.

§6.5. THE CHARACTERISTIC POLYNOMIAL AND THE MINIMAL POLYNOMIAL

In this section we consider some further properties of linear mappings involving their characteristic polynomials and relating to the Jordan canonical form. Recall that by the Cayley-Hamilton theorem, if $T: V \to V$ is a linear mapping on a finite-dimensional space and $p(t) = \det(T - tI)$ is its characteristic polynomial, then $p(T) = 0$ (the zero mapping on V). For example, the mapping of Example (6.4.3) has characteristic polynomial $p(t) = (t - 3)^6$, so we know that $(T - 3I)^6 = 0$. However, in that example, we saw that actually a lower power of $(T - 3I)$, namely, $(T - 3I)^4$ is already the zero mapping. In other words, there is a factor $q(t) = (t - 3)^4$ of $p(t)$ such that $q(T) = 0$. Note that $q(t)$ has strictly smaller degree than $p(t)$.

This example suggests the following general questions:

a) Given a linear mapping $T: V \to V$, defined over a field F, which polynomials $q(t) \in P(F)$ satisfy $q(T) = 0$?

b) In particular, when are there polynomials $q(t)$ of a degree less than $\dim(V)$ such that $q(T) = 0$?

c) How does the set of polynomials $q(t)$ such that $q(T) = 0$ reflect other properties of T such as diagonalizability, the number and sizes of the Jordan blocks in the canonical form of T, and so on?

We begin with some new terminology. We will always use our usual notation $P(F)$ for the set of polynomials $p(t) = a_nt^n + \cdots + a_0$ with coefficients a_i in the field F. As we saw in Chapter 1, $P(F)$ is an infinite-dimensional vector space under the usual operations of polynomial sum and multiplication of a polynomial by a scalar in F. The elements of $P(F)$ can also be multiplied in the usual fashion.

(6.5.1) Example. In $P(\mathbf{C})$, let $p(t) = (1 + i)t^3 + 2$ and $q(t) = 2t^2 - it$. Then $p(t)q(t)$ is the polynomial $((1 + i)t^3 + 2)(2t^2 - it) = (2 + 2i)t^5 + (1 - i)t^4 + 4t^2 - 2it$.

(6.5.2) Definition. Let $T: V \to V$ be a linear mapping on a vector space defined over F. The *ideal of* T, denoted I_T, is the set of all polynomials $q(t) \in P(F)$ such that $q(T) = 0$.

Note that with this new terminology, the Cayley-Hamilton theorem may be rephrased as saying that $p(t) = \det(T - tI) \in I_T$. The most important properties of I_T are the following.

(6.5.3) Proposition

a) I_T is a vector subspace of $P(F)$.

b) If $p(t) \in I_T$ and $q(t) \in P(F)$, then the product polynomial $p(t)q(t) \in I_T$.

Proof: a) Let $p_1(t)$ and $p_2(t) \in I_T$. Then for any scalar $c \in F$ we have $(cp_1 + p_2)(T) = cp_1(T) + p_2(T) = c\,0 + 0 = 0$. Hence, $(cp_1 + p_2)(t) \in I_T$ and by our general criterion for subspaces, I_T is a subspace of $P(F)$.

b) Let $p(t) \in I_T$, and let $q(t) \in P(F)$ be any other polynomial. Then we have $(pq)(T) = p(T)q(T) = 0q(T) = 0$. (The product here is the composition of linear mappings on V.) Hence, $p(t)q(t) \in I_T$, as claimed. ■

Note that because of part b of the proposition, if we have any polynomial $p(t) \in I_T$ with degree $\geqslant 1$ ($p(t) = \det(T - tI)$ is one such polynomial), then the polynomials $tp(t)$, $t^2p(t)$, $t^3p(t)$, and so on are all in I_T. Hence, I_T contains polynomials of arbitrarily high degree in t, and it follows that, as a vector space, I_T is infinite-dimensional. Nevertheless, there is a very nice description of I_T, which arises from the *division algorithm* for polynomials. For completeness, we include a statement of this basic result from high school algebra.

(6.5.4) Proposition. [Division Algorithm in $P(F)$] Let $p(t)$ and $s(t)$ be any polynomials in $P(F)$. There are unique polynomials $q(t)$ and $r(t)$, such that $s(t) = q(t)p(t) + r(t)$ and either $r(t) = 0$, or $\deg(r(t)) < \deg(p(t))$. (The degree of the zero polynomial is not defined.)

(6.5.5) Example. Given $p(t)$ and $s(t)$, the polynomials $q(t)$ and $r(t)$ may be computed by a sort of long division process. For instance, if $p(t) = t^2 + 2t$ and $s(t) = t^4 - 3t^3 + 2$, then we compute

$$
\begin{array}{r|l}
 & t^2 - 5t + 10 \\ \hline
t^2 + 2t & t^4 - 3t^3 + 2 \\
 & \underline{t^4 + 2t^3} \\
 & \quad - 5t^3 + 2 \\
 & \quad \underline{- 5t^3 - 10t^2} \\
 & \qquad 10t^2 + 2 \\
 & \qquad \underline{10t^2 + 20t} \\
 & \qquad\quad - 20t + 2
\end{array}
$$

Hence, $q(t) = t^2 - 5t + 10$ and $r(t) = -20t + 2$ in this case.

In the division algorithm $q(t)$ is called the *quotient* of $s(t)$ by $p(t)$ and $r(t)$ is called the *remainder*. Note that $p(t)$ divides $s(t)$ exactly if and only if the remainder $r(t) = 0$. [This a consequence of the uniqueness of the polynomials $q(t)$ and $r(t)$.]

To study the ideal of T in $P(F)$, we will look at a slightly more general situation—we will consider all subsets of $P(F)$ that satisfy the two properties we proved for I_T in Proposition (6.5.3).

(6.5.6) Proposition. Let $I \subset P(F)$ be any subset such that I is a vector subspace of $P(F)$ and such that the product of any $p(t) \in I$ and any $q(t) \in P(F)$ satisfies $p(t)q(t) \in I$. Then there exists a unique polynomial $g(t)$ in I such that

(i) $g(t)$ divides $f(t)$ for all $f(t) \in I$, and

(ii) $g(t)$ is a *monic* polynomial [i.e., the coefficient of the highest power term in $g(t)$ is 1].

Proof: existence: Since if $g(t)$ divides $f(t)$, then $\deg(g(t)) \leqslant \deg(f(t))$, to find the polynomial $g(t)$, we should look at the polynomials in I of the smallest nonnegative degree. Let $\tilde{g}(t)$ be any such polynomial and multiply $\tilde{g}(t)$ by a scalar $c \in F$ if necessary to obtain a monic polynomial. By our hypothesis on I, it follows that this monic polynomial is also in I, and we will call it $g(t)$. Now if $f(t)$ is any other polynomial in I, applying the division algorithm, we find polynomials $q(t)$ and $r(t)$ such that $f(t) = q(t)g(t) + r(t)$ and either $r(t) = 0$, or $\deg(r) < \deg(g)$. Rearranging this equation we have $r(t) = f(t) - q(t)g(t)$. Since $g(t) \in I$, $q(t)g(t) \in I$ as well. Furthermore, since I is a vector subspace of $P(F)$ and $f(t) \in I$ and $q(t)g(t) \in I$, it follows that $r(t) \in I$ as well. But $g(t)$ was chosen as a polynomial of the smallest nonnegative degree in I, so $\deg(r) \geqslant \deg(g)$ if r is not the zero polynomial. Thus, it can only be that $r(t) = 0$, so by our preceding remark, $g(t)$ divides $f(t)$. The proof of the uniqueness of $g(t)$ is left as an exercise for the reader. ■

The unique polynomial $g(t)$ whose existence is guaranteed by the proposition is called the *generator* of I. Applying this proposition to the ideal of a linear mapping $T: V \to V$, we obtain a uniquely determined polynomial called the minimal polynomial of T.

(6.5.7) Definition. The *minimal polynomial* of $T: V \to V$ is the unique polynomial $g(t) \in I_T$ such that

(i) $g(t)$ divides every $f(t) \in I_T$, and

(ii) $g(t)$ is monic.

The proof of Proposition (6.5.6) shows that the minimal polynomial of a mapping T may also be characterized as the monic polynomial of the smallest degree in I_T.

(6.5.8) Examples

a) The minimal polynomial of the mapping of Example (6.4.2) is $g(t) = (t - 2)^2(t - i)^3(t + i)$.

b) The minimal polynomial of the mapping of Example (6.4.3) is $g(t) = (t - 3)^4$.

c) The minimal polynomial of the mapping of Example (6.4.4) is $g(t) = (t - (2 - i))^2(t - 1)^2$.

Each of these facts may be derived directly, but as the reader will no doubt have noticed, there is a general pattern that allows us to compute the minimal polynomial in a very simple manner, once we know the Jordan canonical form of the mapping in question. First, since the characteristic polynomial is an element of I_T, by definition the minimal polynomial must divide the characteristic polynomial. Furthermore, if $J \in M_{n\times n}(F)$ is the Jordan canonical form matrix of T and $g(t)$ is the minimal polynomial of T, then $g(J) = 0$ (why?). By the fact that J has a direct sum decomposition into Jordan block matrices, it follows that $g(B) = 0$ for all the blocks B on the diagonal of J as well. This observation leads to the following result.

(6.5.9) Proposition. Let $T: V \to V$ be a linear mapping whose characteristic polynomial has dim(V) roots in the field F over which V is defined. Let λ_i $(1 \leq i \leq k)$ be the distinct eigenvalues with respective multiplicities m_i. Then

a) The minimal polynomial of T factors as

$$g(t) = (t - \lambda_1)^{\ell_1} \cdots (t - \lambda_k)^{\ell_k}$$

where $1 \leq \ell_i \leq m_i$ for each i.

b) In this factorization, the integer ℓ_i is the length of the longest cycle for the nilpotent mapping $N_i = (T - \lambda_i I)|_{K_{\lambda_i}}$ (or equivalently, the size of the largest Jordan block with diagonal entries λ_i in the Jordan canonical form of T.)

c) $K_{\lambda_i} = \text{Ker}((T - \lambda_i I)^{\ell_i})$ for each i.

Proof: a) If we use the fact that $g(t)$ divides $\det(T - tI) = (t - \lambda_1)^{m_1} \cdots (t - \lambda_k)^{m_k}$, then it follows immediately from the uniqueness of factorizations for polynomials that $g(t)$ can contain only the factors $(t - \lambda_i)$ and that each $(t - \lambda_i)$ appears to a power ℓ_i with $0 \leq \ell_i \leq m_i$. In fact, each $\ell_i \geq 1$ since it must be the case that $g(B) = 0$ for each block B in the Jordan canonical form of T: If B is any one of the blocks with diagonal entries λ_i, then there must be a factor in $g(t)$ to annihilate B. Since $(B - \lambda_j I)$ is invertible if $j \neq i$, it follows that $g(t)$ must contain a factor $(t - \lambda_i)^{\ell_i}$ with $\ell_i \geq 1$.

Parts b and c are left as exercises for the reader. They may both be proved by a careful use of the fact that if B is a Jordan block of size s with diagonal entries λ, then $(B - \lambda I)^s = 0$. ∎

This proposition says in particular that the minimal polynomial of a linear mapping T may be computed easily if we already know the canonical form of T.

(6.5.10) Example. Let $T: \mathbf{C}^9 \to \mathbf{C}^9$ be the linear mapping defined by the Jordan form matrix

$$\begin{bmatrix} 2 & 1 & & & & & & & \\ 0 & 2 & & & & & & & \\ & & 2 & 1 & & & & & \\ & & 0 & 2 & & & & & \\ & & & & 3 & 1 & 0 & & \\ & & & & 0 & 3 & 1 & & \\ & & & & 0 & 0 & 3 & & \\ & & & & & & & 3 & \\ & & & & & & & & 3 \end{bmatrix}$$

Then the minimal polynomial is $g(t) = (t - 2)^2(t - 3)^3$.

The following corollary of Proposition (6.5.9) contains some further observations about the minimal polynomial of a linear mapping and its relation with other properties of the mapping in the two extreme cases.

(6.5.11) Corollary. With notation as before,

a) T is diagonalizable if and only if the minimal polynomial of T is $g(t) = (t - \lambda_1) \cdots (t - \lambda_k)$ (all factors appearing to the first power).

b) The Jordan canonical form contains exactly one Jordan block for each eigenvalue λ_i if and only if the minimal polynomial of T is equal to the characteristic polynomial of T (up to sign). [That is, $g(t) = (t - \lambda_1)^{m_1} \cdots (t - \lambda_k)^{m_k}$]

Proof: Exercise. ■

To conclude this chapter, we now consider the following general question. With all the information and different techniques we have built up for studying linear mappings, starting in Chapter 4 and continuing in this chapter, it is reasonable to ask whether we now have a complete description of all the different possible types of behavior a linear mapping can exhibit. Put another way, can we now *classify* all linear mappings?

From our study of the process of changing basis in Section 2.7, we know that this question is equivalent to asking for a classification of matrices $A \in M_{n\times n}(F)$ up to *similarity*. [Recall that two matrices $A, B \in M_{n\times n}(F)$ are said to be *similar* if and only if there exists an invertible matrix Q such that $B = Q^{-1}AQ$.] Two matrices represent the same linear mapping (but with respect to different bases) exactly when they are similar.

Hence, our general question may be rephrased as follows: Given two matrices, can we determine whether they are similar? It is interesting to note that the characteristic polynomial and the minimal polynomial do not alone give us enough information to answer these questions.

(6.5.12) Example. The characteristic and minimal polynomials of the matrices

$$A_1 = \begin{bmatrix} 2 & 1 & 0 & 0 \\ 0 & 2 & 0 & 0 \\ 0 & 0 & 2 & 1 \\ 0 & 0 & 0 & 2 \end{bmatrix} \quad \text{and} \quad A_2 = \begin{bmatrix} 2 & 1 & 0 & 0 \\ 0 & 2 & 0 & 0 \\ 0 & 0 & 2 & 0 \\ 0 & 0 & 0 & 2 \end{bmatrix}$$

are equal. [They are, respectively, $p(t) = (t - 2)^4$ and $g(t) = (t - 2)^2$.] Nevertheless, these matrices are not similar. The easiest way to see this is to note that if two matrices are similar, then the dimensions of the corresponding eigenspaces must be the same (why?). Here, however, $\dim(\text{Ker}(A_1 - 2I)) = 2$, whereas $\dim(\text{Ker}(A_2 - 2I)) = 3$.

Hence, examples like this tell us that we must look for more detailed information in order to distinguish matrices that are not similar (and to be able to conclude that matrices are similar). It is precisely the *Jordan canonical form* that allows us to answer this sort of question. In what follows, we say that matrices A and B have *the same* Jordan canonical form if the number, the sizes, and the diagonal entries in the Jordan block matrices making up the canonical forms are the same (the blocks *may* appear in different orders, however). For simplicity, we also consider matrices $A, B \in M_{n \times n}(\mathbf{C})$ and work over $F = \mathbf{C}$ so that the existence of the Jordan canonical forms is guaranteed by Theorem (6.3.6).

(6.5.13) Theorem

a) Let A, $B \in M_{n \times n}(\mathbf{C})$ have the same canonical form. Then A and B are similar.

b) Conversely, if A and B are similar, then they have the same canonical form.

Proof:

a) If A and B have the same Jordan canonical form J, then we know there are invertible matrices P and Q such that $P^{-1}AP = J = Q^{-1}BQ$. Hence, $B = (QP^{-1})A(PQ^{-1})$, using the associativity of matrix multiplication to group the terms. Since $(PQ^{-1})^{-1} = (Q^{-1})^{-1}P^{-1} = QP^{-1}$, we see that $B = (PQ^{-1})^{-1}A(PQ^{-1})$. Hence, by definition B and A are similar.

b) If A and B are similar matrices, then we can view them as the matrices of a fixed linear mapping on an n-dimensional vector space V with respect to two different bases. This implies that the eigenvalues of A and B and their respective multiplicities are the same. Furthermore, it implies that $\dim(\text{Ker}(A - \lambda I)^j) = \dim(\text{Ker}(B - \lambda I)^j)$ for each eigenvalue λ and each integer j (these dimensions depend on the linear mapping, not on the matrix chosen to represent it). Hence, the cycle tableaux and the Jordan forms are the same as well. ■

EXERCISES

1. Prove Proposition (6.5.9b, c). (See the proof in the text for a hint.)

2. Prove Corollary (6.5.11).

3. Compute the minimal polynomial of each of the mappings of Exercise 4 in Section 6.4.

4. Show that if $A, B \in M_{n \times n}(F)$ are similar, then A and B have the same characteristic polynomial and the same minimal polynomial. [The converse of this statement is *false*—see Example (6.5.12).]

5. Which of the pairs of the following matrices are similar?

$$\begin{bmatrix} -3 & 3 & -2 \\ -7 & 6 & -3 \\ 1 & -1 & 2 \end{bmatrix} \quad \begin{bmatrix} 0 & 1 & -1 \\ -4 & 4 & -2 \\ -2 & 1 & 1 \end{bmatrix} \quad \begin{bmatrix} 0 & -1 & -1 \\ -3 & -1 & -2 \\ 7 & 5 & 6 \end{bmatrix}$$

6. a) Let $f(t), g(t) \in P(F)$ be two polynomials. Let

$$I_1 = \{q_1(t)f(t) \mid q_1(t) \in P(F)\} \quad \text{and} \quad I_2 = \{q_2(t)g(t) \mid q_2(t) \in P(F)\}$$

and let $I = I_1 \cap I_2$. Show that I_1, I_2, and I all satisfy the hypotheses of Proposition (6.5.6). Hence, there is a unique monic polynomial $m(t)$ such that I consists of all the multiples of $m(t)$. [The polynomial $m(t)$ is called the *least common multiple* of $f(t)$ and $g(t)$.]

b) Show that if $f(t) = (t - \lambda_1)^{e_1} \cdots (t - \lambda_k)^{e_k}$ and $g(t) = (t - \lambda_1)^{f_1} \cdots (t - \lambda_k)^{f_k}$, where $e_i, f_i \geq 0$, then the least common multiple of $f(t)$ and $g(t)$ is the polynomial $m(t) = (t - \lambda_1)^{M_1} \cdots (t - \lambda_k)^{M_k}$, where for each i, M_i is the larger of e_i and f_i.

c) Show that if $A \in M_{n \times n}(\mathbf{C})$ has a direct sum decomposition

$$A = \begin{bmatrix} A_1 & 0 \\ 0 & A_2 \end{bmatrix}$$

with diagonal blocks A_1 and A_2, then the minimal polynomial of A is the least common multiple of the minimal polynomials of A_1 and A_2.

7. Show that if $A \in M_{n \times n}(\mathbf{C})$, then A and A^t are always similar. [*Hint:* To compute the cycle tableaux for the different generalized eigenspaces, you would need to look at $\dim(\text{Ker}(A - \lambda I)^k)$ and $\dim(\text{Ker}(A^t - \lambda I)^k)$ for the different eigen values λ and the integers $k \leq m =$ multiplicity of λ. Show that these dimensions are always equal.]

8. Prove the division algorithm in $P(F)$, Proposition (6.5.4). [*Hint:* induction on the degree of the polynomial $s(t)$ gives the cleanest proof.]

9. a) If $A \in M_{8 \times 8}(\mathbf{C})$ and the minimal polynomial of A is $g(t) = (t - 2)^4(t - 3i)$, what are the possible Jordan forms for A?

b) Same question for $A \in M_{9 \times 9}(\mathbf{C})$ with minimal polynomial $g(t) = (t - i)^3(t - 2)(t - 3)^2$.

10. Show that $A \in M_{n \times n}(\mathbf{C})$ is diagonalizable if and only if the minimal polynomial of A has no root of multiplicity > 1.

11. Let $A \in M_{n \times n}(\mathbf{C})$, and consider the subspace $W = \text{Span}\{A^k | \ k \geq 0\}$ in $M_{n \times n}(\mathbf{C})$. Show that the dimension of W is equal to the degree of the minimal polynomial of A.

CHAPTER SUMMARY

The Jordan canonical form is the single most powerful technique known for describing the structure of linear mappings on finite-dimensional vector spaces. In this chapter we have seen one way to derive the Jordan form, and some of its applications.

The underlying idea in our development has been that in order to understand a general linear mapping $T: V \to V$, which may not have enough eigenvectors to form a basis of V, we need to look for subspaces of V that are *invariant* under T [Definition (6.1.2)]. These invariant subspaces have some of the same properties as the eigenspaces of T, in that they give us a way to break the space V down into smaller pieces that are preserved by the mapping T. By analyzing the action of T on invariant subspaces, we can reconstruct the action of T on the whole space.

Our approach to the Jordan form consisted of three steps:

1) *Reduction to upper-triangular form* [Theorem (6.1.8)]. We saw that if all the eigenvalues of $T: V \to V$ lie in the field F over which V is defined, then it is *always* possible to find a basis $\beta = \{\mathbf{x}_1, \ldots, \mathbf{x}_n\}$ such that $[T]_\beta^\beta$ is upper-triangular, or equivalently, such that each subspace $W_i = \text{Span}\{\mathbf{x}_1, \ldots, \mathbf{x}_i\}$ is invariant under T.

2) *Analysis of nilpotent mappings* [Theorem (6.2.8)]. By studying the invariant *cyclic subspaces* [Definition (6.2.1)] for a nilpotent mapping and how they fit together in V, we were able to prove the existence of a *canonical basis* [Definition (6.2.7)] for every nilpotent mapping—a basis that is a union of nonoverlapping cycles. We also gave a convenient method for computing what these canonical bases, and the corresponding matrix of the mapping, should look like, starting from the *cycle tableau* [introduced in the proof of Proposition (6.2.4)].

3) *The general case*. Finally, we saw that for a general linear mapping $T: V \to V$, if all the eigenvalues of T lie in the field of scalars F for V, then letting λ be an eigenvalue of multiplicity m as a root of the characteristic polynomial:

a) $K_\lambda = \text{Ker}((T - \lambda I)^m)$ is an invariant subspace of V, called the λ-*generalized eigenspace* of T [Definition (6.3.2)],

b) $V = \oplus K_{\lambda_i}$ [Proposition (6.3.4)], and

c) $N = (T - \lambda I)|_{K_\lambda}$ is a nilpotent mapping on K_λ.
Thus, by applying our results on nilpotent mappings to each of the $N_i = (T - \lambda_i I)|_{K_{\lambda_i}}$, we obtained the Jordan canonical form of T, a matrix with a direct sum decomposition into *Jordan block matrices* [Definition (6.3.5).

In the final section of the chapter we saw how the Jordan canonical form of the matrix of a mapping T gives insights into the set of polynomials $p(t) \in P(F)$ such that $p(T) = 0$, whose importance is hinted at by the Cayley-Hamilton theorem [Theorem (6.1.12)]. Given the Jordan form matrix of T, we can find the *minimal polynomial* of T [Definition (6.5.7)] very easily. Furthermore, we saw that two matrices are similar if and only if they have the same Jordan form [Theorem (6.5.13)]. It is this fact that justifies our claim that the Jordan canonical form gives a complete catalog of all possible linear mappings on finite-dimensional vector spaces (over an algebraically closed field such as $\mathbf{C}$).

In Chapter 7 we will see how the Jordan canonical form may be applied in the study of linear differential equations. Other applications are given in the Supplementary Exercises.

SUPPLEMENTARY EXERCISES

1. **True-False**. For each true statement, give a brief proof. For each false statement, give a counterexample.
   a) If V is finite-dimensional vector space over a field F, and $T: V \to V$ is linear, then there always exist subspaces of V invariant under T that are different from $\{\mathbf{0}\}$ and V.
   b) Let $N: V \to V$ be nilpotent, and let $W \subset V$ be the subspace spanned by a cycle for N. Then W is invariant under N.
   c) Every $A \in M_{n \times n}(\mathbf{R})$ is similar to an upper-triangular matrix U in $M_{n \times n}(\mathbf{R})$.
   d) If $\dim(V) = 5$, $N: V \to V$ is nilpotent, and $\dim(\text{Ker}(N)) = 2$, then every canonical basis for N contains a cycle of length 3.
   e) For each eigenvalue λ of T, $\dim(E_\lambda) \leq \dim(K_\lambda)$.
   f) If $A \in M_{n \times n}(F)$ and $\det(A - tI) = \pm t^n$, then A is nilpotent.
   g) If $A, B \in M_{n \times n}(F)$ are nilpotent and $AB = BA$, then $cA + B$ is also nilpotent.
   h) If $A \in M_{n \times n}(\mathbf{R})$, all entries in the Jordan form matrix similar to A are also real.
   i) Let $T: V \to V$ be linear. The ideal of T, $I_T \subset P(F)$, is a finite-dimensional subspace of $P(F)$.
   j) If the Jordan form of a matrix A is a "scalar matrix" cI for some $c \in F$, then $A = cI$.
   k) If A, B are matrices with the same minimal polynomial, they have the same characteristic polynomial as well.

2. Which of the following matrices are triangularizable over $F = \mathbf{R}$? over $F = \mathbf{C}$? over $F =$ the field of rational numbers?

a) $\begin{bmatrix} 1 & 2 \\ -2 & 7 \end{bmatrix}$ b) $\begin{bmatrix} 1 & 0 & 3 \\ 0 & -2 & -4 \\ 3 & -4 & 7 \end{bmatrix}$ c) $\begin{bmatrix} 0 & 0 & 0 & -16 \\ 1 & 0 & 0 & -32 \\ 0 & 1 & 0 & -24 \\ 0 & 0 & 1 & -8 \end{bmatrix}$

3. a) Show that if $T: V \rightarrow V$ is linear, and λ is an eigenvalue of T, then for each $i \geq 0$, $\text{Ker}((T - \lambda I)^i) \subset \text{Ker}((T - \lambda I)^{i+1})$.

b) How is this reflected in the cycle tableau for a canonical basis for the nilpotent mapping $N = (T - \lambda I)|_{K_\lambda}$?

4. For each linear mapping, find the generalized eigenspace for the given eigenvalue.

a) $T: \mathbf{C}^4 \rightarrow \mathbf{C}^4$ defined by $A = \begin{bmatrix} 0 & 4 & 0 & i \\ 0 & 0 & 0 & 0 \\ 0 & 0 & 0 & 3 \\ 0 & 0 & 1 & 0 \end{bmatrix}$, $\lambda = 0$.

b) $T: \mathbf{C}^4 \rightarrow \mathbf{C}^4$ defined by $A = \begin{bmatrix} 1 & 3 & 3 & -1 \\ 0 & 1 & 1 & 0 \\ 0 & 0 & 4 & 9 \\ 0 & 0 & -1 & -2 \end{bmatrix}$, $\lambda = 1$.

c) $V = \text{Span}\{\cos(x), \sin(x), \cos(2x), \sin(2x), \cos(3x), \sin(3x)\}$, $T: V \rightarrow V$ defined by $T(f) = f'$, $\lambda = i$. (Take $F = \mathbf{C}$).

d) $V = \text{Span}\{1, x, y, x^2, xy, y^2\}$, $T: V \rightarrow V$ defined by $T(f)(x, y) = f(y, x)$, $\lambda = 1$.

5. Compute the Jordan canonical form of each of the mappings in Exercise 4.

6. Compute the minimal polynomial of each of the mappings in Exercise 4.

Let $S \subset L(V, V)$ be a set of linear mappings on V. Let $W \subset V$ be a subspace of V. We say W is *invariant under* S if W is invariant under each of the mappings $T \in S$.

7. a) Let S be the set of mappings on $\mathbf{C}^2$ defined by the matrices in $\left\{\begin{bmatrix} 1 & 1 \\ -1 & 1 \end{bmatrix}, \begin{bmatrix} 4 & -4 \\ 4 & 4 \end{bmatrix}\right\}$. Find a subspace $W \neq \{\mathbf{0}\}$, $\mathbf{C}^2$, which is invariant under S.

b) Show that if W is invariant under S, then W is invariant under $\text{Span}(S) \subset L(V, V)$.

c) Show that if $S = \{T, U\}$ and $TU = UT$ in $L(V, V)$ (V a finite-dimensional vector space over $F = \mathbf{C}$), then there exists a one-dimensional subspace $W \subset V$ invariant under S.

A set of linear mappings $S \subset L(V, V)$ is said to be *irreducible* if the only subspaces of V that are invariant under S are $\{\mathbf{0}\}$ and V.

8. Show that the set of mappings defined by the matrices in

$$S = \left\{ \begin{bmatrix} 0 & 1 & 0 & 0 \\ -1 & 0 & 0 & 0 \\ 0 & 0 & 0 & 1 \\ 0 & 0 & -1 & 0 \end{bmatrix}, \begin{bmatrix} 0 & 0 & 0 & 1 \\ 0 & 0 & 1 & 0 \\ 0 & -1 & 0 & 0 \\ -1 & 0 & 0 & 0 \end{bmatrix} \right\} \quad \text{in } L(\mathbf{R}^4, \mathbf{R}^4)$$

is irreducible.

9. Show that if $S \subset L(V, V)$ (V finite-dimensional, if you like) is irreducible and $C(S) = \{U \in L(V, V) | UT = TU \text{ for all } T \in S\}$, then every mapping in $C(S)$ other than the zero mapping is invertible. [*Hint:* Show that if $U \in C(S)$, then $\text{Ker}(U)$ and $\text{Im}(U)$ are invariant under S. This result is a special case of a general result known as Schur's lemma.]

10. a) Show that if $A \in M_{n \times n}(\mathbf{C})$ and $J = Q^{-1}AQ$ is the Jordan form of A, then $J^k = Q^{-1}A^kQ$ for all k.
    b) Show that if $p(t) \in P(\mathbf{C})$, then $p(J) = Q^{-1}p(A)Q$.

11. Show that if all the eigenvalues λ of $A \in M_{n \times n}(\mathbf{C})$ satisfy $|\lambda| < 1$, then $\lim_{k \to \infty} A^k = 0$, in the sense that all the entries of A^k go to zero as k increases without bound. (*Hint:* Use Exercise 10. What happens to J^k?)

12. Two competing supermarkets in the small town of Mansfield, PA (Super Duper and Honchell's Market) divide all the customers in the town between them (the number remains constant). In each month each customer shops in only one of the two stores but may switch to the other store at the beginning of the following month. By observing the numbers of the customers who shop in the two stores over a period of many months, the manager of the Super Duper store notices that in each month, 80 percent of the Super Duper customers in that month remain Super Duper customers in the next month, whereas 20 percent switch to Honchell's. On the other hand, of the Honchell's customers, 90 percent stay Honchell's customers, and 10 percent switch to Super Duper. The same pattern is observed in every month.
    a) Show that if there are SD_0 Super Duper customers and H_0 Honchell's customers at the beginning of one month, then the number SD_n of Super Duper customers and the number H_n of Honchell's customers after n months may be computed via the matrix product

$$\begin{bmatrix} SD_n \\ H_n \end{bmatrix} = \begin{bmatrix} 0.8 & 0.1 \\ 0.2 & 0.9 \end{bmatrix}^n \begin{bmatrix} SD_0 \\ H_0 \end{bmatrix}$$

    b) To determine what will happen *in the long run*, the manager sees that she needs to compute $L = \lim_{n \to \infty} T^n$, where $T = \begin{bmatrix} 0.8 & 0.1 \\ 0.2 & 0.9 \end{bmatrix}$ is the matrix from part a. Use the result of Exercise 10a to compute this limit. What is the limit of J^n, where J is the Jordan form of T?

c) Let SD_∞ and H_∞ be the eventual numbers of customers, computed as

$$\begin{bmatrix} SD_\infty \\ H_\infty \end{bmatrix} = L \begin{bmatrix} SD_0 \\ H_0 \end{bmatrix}$$

where $L = \lim_{n \to \infty} T^n$ as in part b. Show that no matter what SD_0 and H_0 are, the limiting market *fractions* $SD_\infty/(SD_\infty + H_\infty)$ and $H_\infty/(SD_\infty + H_\infty)$ are always the same. (They are 1/3, 2/3, respectively.)

This two-competitor system is a simple example of a *Markov chain*. The matrix T is called the *transition matrix* of the Markov chain. Note that the entries in each column sum to 1. In general, the long-term behavior of a Markov chain may be studied by analyzing the powers of the transition matrix, T^k as $k \to \infty$. For readers who may wish to explore this topic further, we recommend the book *Finite Markov Chains*, by J. Kemeny and J. Snell. (Princeton: D. Van Nostrand, 1960).

CHAPTER 7

Differential Equations

Introduction

The applications of linear algebra are numerous and diverse both in and out of mathematics. One reason is that linear mappings that depend on several variables arise naturally in a variety of areas. Among the major mathematical applications of linear algebra are the theory of linear differential equations, the theory of Markov chains, and the theory of linear programming. Since each of these topics deserves a lengthy treatment, we have chosen to present only one of them in this, our last, chapter. We have chosen to consider linear differential equations for several reasons, not the least of which is that we like the theory. However, one of the best reasons is that at various times, to study these equations, we must use all the theory that we have developed so far in this text. Moreover, the material we will see naturally points the way to further study in both the theory of differential equations and linear algebra.

We begin this chapter by describing some typical problems informally in order to convey the flavor of the theory. The most important and most interesting examples of differential equations arise in trying to solve real-world problems. With this goal in mind, we present several examples beginning with those you have seen in calculus—the exponential growth and radioactive decay equations—and ending with the problem of describing the motion of a vibrating string. As one goes on in the theory of differential equations, it is hard not to be astonished by the great variety of physical problems that can be modeled by differential equations.

Naturally, we will be able to discuss only a small part of such a large subject. We focus on linear differential equations and begin with a geometric approach to solving systems of first-order linear differential equations with constant coefficients in one variable. In order to obtain solutions in all instances, we need to develop the algebraic side of the theory. We can apply these techniques to a single higher-order equation in one variable with constant coefficients by replacing the one higher-order equation by a related system of first-order equations.

The last section discusses the wave equation in one space variable, an extremely important second-order partial differential equation. This is the equation that is used to model wavelike phenomena, although the more space variables and the more complicated the wave phenomenon under study are, the more involved the model and the equation become.

At various times in the text we have referred to examples of infinite-dimensional vector spaces or have commented on the necessity of having a finite-dimensional vector space in order to apply certain results. In this chapter we will see our first serious use of infinite-dimensional vector spaces. These vector spaces will be vector spaces of functions and the linear transformations on these vector spaces will be defined by the linear differential equations. All of our work ranging from questions of linear dependence, independence, and bases to Hermitian inner products will be applied in this context.

---

§7.1. TWO MOTIVATING EXAMPLES

First, let us say what we mean by a differential equation. In rough terms, a differential equation is a relation between a function and its derivatives. More precisely, let $u: \mathbf{R}^n \to \mathbf{R}^k$ be a mapping (which need not be linear!). We say u satisfies a differential equation (or system of differential equations) if there is an equation (or system of equations) relating the function u and its derivatives. If $n = 1$, the equations are called ordinary differential equations since they involve only ordinary derivatives of u, that is, du/dt, d^2u/dt^2, and so on. If $n > 1$, the equations are called partial differential equations since they involve partial derivatives $\partial u/\partial x_j$, $\partial^2 u/\partial x_j \partial x_k$, and so forth.

In the study of differential equations we usually begin with an equation that is determined by a mathematical analysis of a problem. These equations are often determined empirically by observing relationships between quantities and their rates of change in some real-world situation or physical system. For example, the well-known exponential growth equation [Example (7.1.1)] is based on the observation of Malthus circa 1800 that the rate of growth of certain populations is proportional to their size.

Then we look to find a solution, a function that satisfies the equation subject to given *initial conditions,* for example, the number of individuals in a population at $t = 0$. If the problem has been posed correctly, the resulting solution should accurately model the original physical system. It is hoped that the solutions will yield further insight into the physical system.

At this point it would be helpful to give two examples.

(7.1.1) Examples

a) This first example should be familiar from one variable calculus. Let $u(t)$ be a function $u: \mathbf{R} \to \mathbf{R}$. Here $t \in \mathbf{R}$ is thought of as time. The ordinary differential equation $du/dt = \lambda u$, for $\lambda \in \mathbf{R}$ a constant, is a mathematical model for both exponential growth ($\lambda > 0$), and radioactive decay ($\lambda < 0$). If $u(t)$ is interpreted as population size, or the amount of a radioactive substance in a sample, then du/dt is the rate of growth of the population, or the rate of decay of the radioactive substance. The equation is the mathematical translation of the statement that the rate of change of u is proportional to the value of u at all times. A solution in either case is given by the function $u(t) = u(0)e^{\lambda t}$.

b) The motion of a particle under the influence of gravity is governed by Newton's equation. Let $\mathbf{x}(t) = (x_1(t), x_2(t), x_3(t))$ denote the path of a particle moving through three-dimensional space. In its most general form Newton's equation is written

$$(\text{mass}) \cdot d^2\mathbf{x}/dt^2 = F(t, \mathbf{x}(t), d\mathbf{x}/dt)$$

where F is a function depending on the physical situation under study. In words, the motion of a particle depends only on time t, the velocity $d\mathbf{x}/dt$, and the acceleration $d^2\mathbf{x}/dt^2$, not on any higher derivatives. Such qualitative insights are crucial in correctly formulating differential equations.

In the particular case of a particle moving vertically and only undergoing acceleration due to the force of gravity Newton's equation becomes the familiar equation $d^2x/dt^2 = x''(t) = -32$ ft/sec^2.

In the remainder of this section, we consider two further problems that lead to differential equations. The first problem of interest to us involves systems of equations with one time derivative. Let a particle move in Euclidean space $\mathbf{R}^n$ (or $\mathbf{C}^n$—the mathematics involved is the same.) As time evolves the particle traces a path in $\mathbf{R}^n$. This movement can be described by a function $\mathbf{u}: \mathbf{R} \to \mathbf{R}^n$, $\mathbf{u}(t) = (u_1(t), \ldots, u_n(t))$, giving the position of the particle at each time t. For simplicity, we assume that the function $\mathbf{u}(t)$ is infinitely differentiable. Then the velocity vector of the path $\mathbf{u}(t)$ is the derivative of $\mathbf{u}$ with respect to t. We denote this by $\mathbf{u}'(t) = (u_1'(t), \ldots, u_n'(t))$. See Figure 7.1.

To formulate our problem, let us assume that the movement of the particle is governed by a fluid flow in $\mathbf{R}^n$. (We can also think of a charged particle in a magnetic field or a body moving in a gravitational field.) That is, we think of $\mathbf{R}^n$

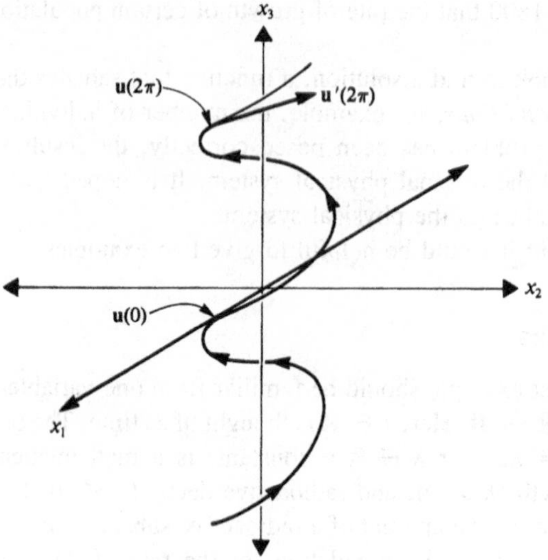

$\mathbf{u}(t) = (\cos(t), \sin(t), t)$ and
the velocity vector $\mathbf{u}'(2\pi)$.

Figure 7.1

as being filled with a moving fluid. Assume we have dropped our test particle into the fluid and wish to determine its path. To write a system of equations governing the fluid flow, we must make assumptions of a mathematical nature.

We assume that the velocity of the fluid flow at a particular point, and thus of a particle moving with the fluid at that point, is independent of time and depends only on the location in $\mathbf{R}^n$. Thus, each component $u_i'(t)$ of the velocity vector is a function $f_i(x_1, \ldots, x_n)$ of the position $x_1 = u_1(t), \ldots, x_n = u_n(t)$ of the particle. Further, we assume that the functions f_i are linear functions of the u_i:

$$u_i' = a_{i1}u_1 + \cdots + a_{in}u_n \qquad \text{for } i = 1, \ldots, n \tag{7.1}$$

(This last assumption is quite restrictive—for a general fluid flow, the components of the velocity need not be linear functions of the position.)

If we introduce matrix notation, the system (7.1) can be rewritten as

$$\mathbf{u}' = A\mathbf{u} \tag{7.2}$$

where A is the matrix of coefficients a_{ij}. In order to obtain a particular solution to this system of equations, we must specify the initial state of the system, that is, the initial position of the particle, $\mathbf{u}(0) = \mathbf{u}_0$, which is where we place our particle in the fluid flow. Since the fluid flow is not changing with time, and the velocity of the particle depends only on the flow, we expect that this initial condition will determine the position of the particle at *all later times*. In other words, we expect that there should be only one solution to the equations $\mathbf{u}' = A\mathbf{u}$, subject to $\mathbf{u}(0) = \mathbf{u}_0$.

Let us consider a simple example that will indicate the geometry of the situation.

(7.1.2) Example. Assume that our system of equations is

$$u_1' = \quad -u_2$$
$$u_2' = u_1$$

and $\mathbf{u}_0 = (1, 0)$. At each point in $\mathbf{R}^2$ we should imagine having a vector that indicates the direction and magnitude of the fluid flow. So, for example, at the point (2, 1) the velocity vector is $(-1, 2)$. In Figure 7.2 we illustrate this.

In this particular case, it is easy to see that $\mathbf{u}$ and $\mathbf{u}'$ are orthogonal:

$$\langle \mathbf{u}, \mathbf{u}' \rangle = u_1 \cdot (-u_2) + u_2 \cdot u_1 = 0$$

This is reflected in the figure. An example of a path in $\mathbf{R}^2$ whose velocity vector is orthogonal to all lines through the origin is a circle. To see this, we consider the derivative of the square of the length $\mathbf{u}(t)$.

$$\begin{aligned}\frac{d}{dt}(\|\mathbf{u}(t)\|^2) &= \frac{d}{dt}(\langle \mathbf{u}(t), \mathbf{u}(t)\rangle) \\ &= \frac{d}{dt}((u_1(t))^2 + (u_2(t))^2) \\ &= 2u_1(t)u_1'(t) + 2u_2(t)u_2'(t) \\ &= 2\langle \mathbf{u}(t), \mathbf{u}'(t)\rangle\end{aligned}$$

But this last inner product is zero in our case. Thus $\frac{d}{dt}(\|\mathbf{u}(t)\|^2) = 0$. This implies that $\|\mathbf{u}(t)\|^2$ is a constant and therefore $\|\mathbf{u}(t)\|$ is also a constant. If the vector $\mathbf{u}(t)$ has constant length r, then $\mathbf{u}(t)$ must lie on a circle of radius r centered at the origin for all values of t. A circle of radius r through the origin can be parametrized by $\mathbf{u}(t) = (r\cos(\omega t + b), r\sin(\omega t + b))$. For $\mathbf{u}(t)$ to pass through (1, 0) we must

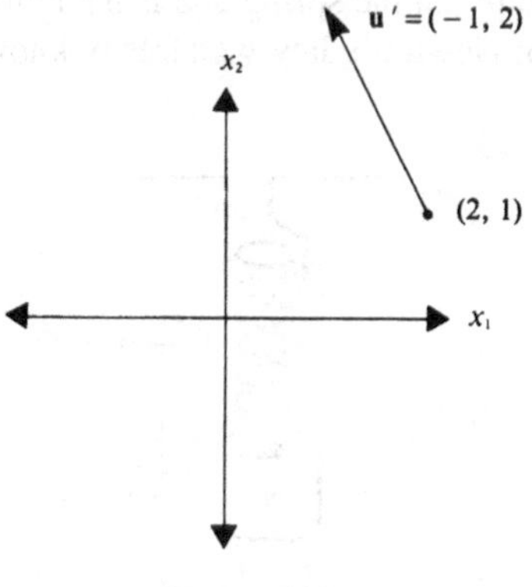

Figure 7.2

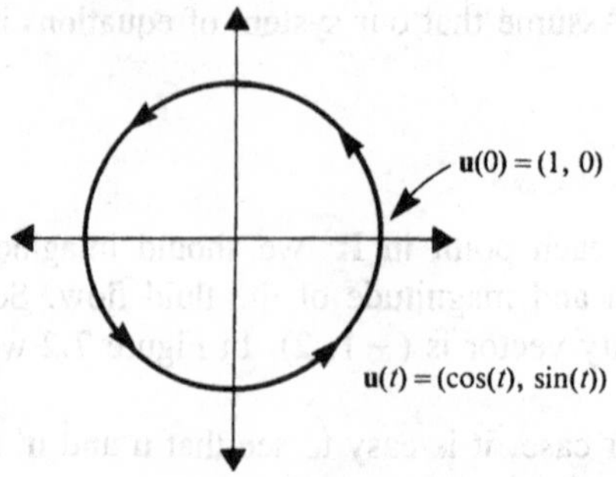

Figure 7.3

of course have the radius $r = 1$ so that $\cos(\omega \cdot 0 + b)$ must equal 1. This implies that $b = 0$ (up to a multiple of 2π). Further $\mathbf{u}'(0) = (0, 1)$ implies that $\omega \cdot \cos(0) = \omega = 1$. Therefore, we expect that $\mathbf{u}(t) = (\cos(t), \sin(t))$ is a solution satisfying the initial condition, which indeed it is. See Figure 7.3.

Of course, this geometric technique does not succeed in all cases. In Sections 7.2 and 7.3 we will give two methods for solving systems of the form of equation (7.1). Note, however, that a sketch of the velocity vectors of the fluid flow as in Figure 7.2 can aid in understanding the geometry of the solution.

Our second problem of interest also involves determining the motion of a particle. However, in this case the particle of fixed mass m is attached to one end of a spring, while the other end is held fixed. See Figure 7.4. The mass-spring system is often called a harmonic oscillator.

Assume our coordinate system is chosen so that the equilibrium, or rest position of the mass is at the origin. Then Hooke's law governs the motion of the particle on the end of the spring: The force F required to move the particle a fixed distance x, that is, stretch or compress the spring a distance x, is proportional to x. In mathematical terms we have $F = kx$ for some constant of proportionality k. The constant k depends only on the spring and is naturally enough called the spring constant. However, from Newton's law we already know that force is mass times

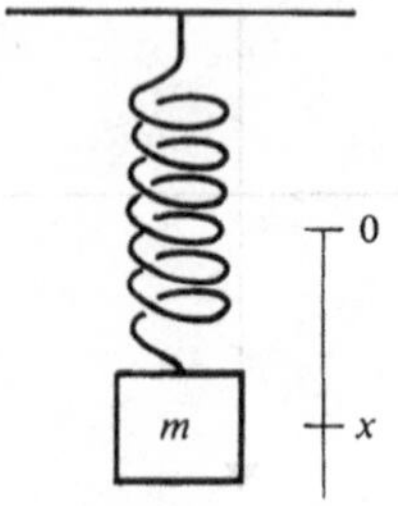

Figure 7.4

acceleration. Since the force of the spring and the force of gravity act in opposite directions, we obtain $F = mx'' = -kx$ or $mx'' + kx = 0$. [The position function $x(t)$ is regarded as a function of time.]

The equation $mx'' + kx = 0$ is called a second-order ordinary differential equation since the highest-order derivative of the unknown function x that appears is the second derivative. Like our first equation, this one too has constant coefficients.

As it stands this equation is a somewhat simplified model of a physical system since it neglects air resistance or friction within the spring. Suppose we now take into account air resistance. Assume that the resistance due to movement through the air is proportional to velocity and equals rx' where r is constant. The new equation is

$$mx'' + rx' + kx = 0$$

In both cases to specify initial conditions, we must specify not only the initial position, but the initial velocity with which we release the mass. Thus, we assume that we know $x(0)$ and $x'(0)$.

In Section 7.4 we will show how to exhibit explicit solutions to equations of the form $mx'' + kx = 0$ and $mx'' + rx' + kx = 0$. To solve these equations, we use a substitution to reduce the single second-order equation to two first-order linear equations and apply the techniques of Sections 7.2 and 7.3. Finally, in Section 7.5 we will turn this around and study such equations from the point of view of a vector space V of functions and a linear transformation $T: V \to V$. In the preceding case V would be the vector space of real- or complex-valued differentiable functions on $\mathbf{R}$ and T would be given by $m \cdot d^2/dt^2 + k$ or $m \cdot d^2/dt^2 + r d/dt + k$.

§7.2. CONSTANT COEFFICIENT LINEAR DIFFERENTIAL EQUATIONS: THE DIAGONALIZABLE CASE

In Section 7.1 we showed how to rewrite a certain type of system of differential equations using matrices. To be precise, we had a system of equations with constant coefficients, a_{ij}, of the form

$$du_i/dt(t) = a_{i1}u_1(t) + \cdots + a_{in}u_n(t), \qquad u_i(0) = b_i$$

where each $u_i(t)$ is a real-valued function of a real variable, that is, $u_i: \mathbf{R} \to \mathbf{R}$, for $i = 1, \ldots, n$. Writing $\mathbf{u}(t) = (u_1(t), \ldots, u_n(t))$, a vector-valued function of $t \in \mathbf{R}$, and A for the matrix of coefficients, our system became

$$d\mathbf{u}/dt = \mathbf{u}'(t) = A\mathbf{u}, \ \mathbf{u}(0) = \mathbf{u}_0$$

In this section we will give a solution in the case that A is diagonalizable over $\mathbf{R}$. As with our initial discussions of diagonalization in Chapter 4, this will have a geometric flavor. Although the results of this section can be formulated over the complex numbers, that is, when $u_i: \mathbf{R} \to \mathbf{C}$ is a complex-valued function of a real variable, so that $\mathbf{u}: \mathbf{R} \to \mathbf{C}^n$, and A is a diagonalizable $n \times n$ matrix with complex entries, we consider this case in Section 7.3, where we give a more general algebraic construction of solutions.

First, let us consider a simple example.

(7.2.1) Example. Let A be a 2×2 diagonal matrix.

$$A = \begin{bmatrix} \lambda_1 & 0 \\ 0 & \lambda_2 \end{bmatrix}, \qquad \lambda_1 \neq \lambda_2,\ \lambda_i \neq 0$$

so that our system is

$$du_1/dt = \lambda_1 u_1, \qquad u_1(0) = b_1$$

$$du_2/dt = \lambda_2 u_2, \qquad u_2(0) = b_2$$

Because each equation involves only one of the functions, we can solve for u_1 and u_2 separately:

$$u_1(t) = b_1 e^{\lambda_1 t} \quad \text{and} \quad u_2(t) = b_2 e^{\lambda_2 t}$$

so that

$$\mathbf{u}(t) = \begin{bmatrix} e^{\lambda_1 t} & 0 \\ 0 & e^{\lambda_2 t} \end{bmatrix} \begin{bmatrix} b_1 \\ b_2 \end{bmatrix}$$

In order to analyze this example, we consider solutions for particular choices of $\mathbf{u}_0 = (b_1, b_2)$. For fixed $\mathbf{u}_0$, $\mathbf{u}(t) = (e^{\lambda_1 t}b_1, e^{\lambda_2 t}b_2)$ traces a path in $\mathbf{R}^2$ as t varies. The initial conditions $\mathbf{u}_0 = (b_1, 0)$ and $\mathbf{u}_0 = (0, b_2)$ yield "special" solutions. For if $\mathbf{u}_0 = (b_1, 0)$, $\mathbf{u}(t) = (e^{\lambda_1 t}b_1, 0)$ is always a positive multiple of the initial position. In other words, the solution path $\mathbf{u}(t)$ always lies in the line spanned by the initial position vector (in fact, in the same half-line or ray). Similarly, the solution curve corresponding to the initial condition $\mathbf{u}_0 = (0, b_2)$ also lies in a line. However, if both b_1 and b_2 are nonzero, the solution curve lies in a straight line only if $e^{\lambda_1 t} = e^{\lambda_2 t}$ for all values of t (why?). Since $\lambda_1 \neq \lambda_2$, this is clearly false. Thus, the only solution curves that lie in straight lines are those whose initial position vectors lie on the axes.

Let us pursue this example further. Using x-y coordinates in the plane, the path $\mathbf{u}(t) = (e^{\lambda_1 t}b_1, e^{\lambda_2 t}b_2)$ intersects any vertical line $x =$ constant at most once. Thus, we may express the set of points $\{\mathbf{u}(t) | \ t \in \mathbf{R}\}$ as the graph of a function $y = f(x)$ for an appropriate domain for x. Assume that $b_1 > 0$ and $b_2 > 0$. Since $x = e^{\lambda_1 t}b_1$, we can express t as $t = (1/\lambda_1)\ln(x/b_1)$. Then $y = e^{\lambda_2 t}b_2$ is expressed in terms of x as

$$y = b_2\, e^{(\lambda_2/\lambda_1)\ln(x/b_1)}$$

$$= b_2\, e^{\ln[(x/b_1)^{(\lambda_2/\lambda_1)}]}$$

$$= b_2\,(x/b_1)^{(\lambda_2/\lambda_1)}$$

For example, if $\lambda_1 = 1$ and $\lambda_2 = 2$ and $b_1 = b_2 = 1$, then $\mathbf{u}(t)$ lies in the graph of $y = x^2$, $x > 0$. If $b_1 = 2$ and $b_2 = 1$, $\mathbf{u}(t)$ lies in the graph of $y = x^2/4$. In

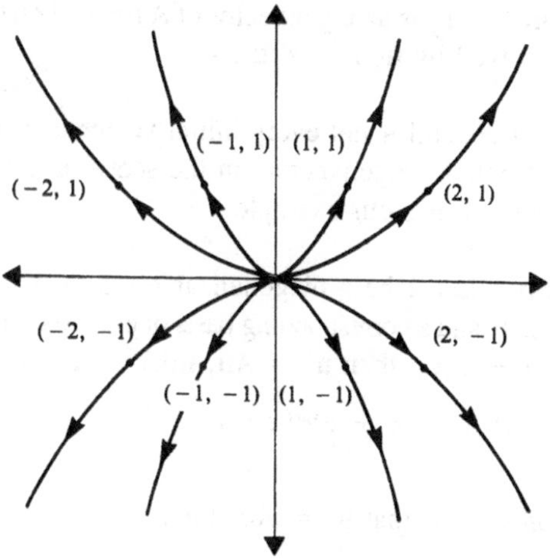

Figure 7.5

Figure 7.5 we illustrate these and several other graphs and solutions for $\lambda_1 = 1$ and $\lambda_2 = 2$.

By examining the possible values of λ_1, λ_2 and $\mathbf{u}_0$, we can obtain reasonable geometric insight into the behavior of all solutions (see Exercise 6).

For a general diagonalizable matrix A, the pattern we noticed in this example can be exploited to solve the system $\mathbf{u}' = A\mathbf{u}$. Note that the special initial position vectors we considered in Example (7.2.1) are just the eigenvectors of the matrix of coefficients. This observation is the starting point for the solution in the diagonalizable case.

Let A be an $n \times n$ matrix, and let $\mathbf{u}_0$ be an eigenvector of A for the eigenvalue λ, so that for any multiple $a\mathbf{u}_0$ we have $A(a\mathbf{u}_0) = \lambda(a\mathbf{u}_0)$. Thus, at any point on the line spanned by $\mathbf{u}_0$, the velocity vector of the solution curve through the point satisfies $d\mathbf{u}/dt|_{t=0} = A(a\mathbf{u}_0) = \lambda(a\mathbf{u}_0)$, and therefore, also lies in the line. This leads us to expect that the solution curve corresponding to the initial condition $\mathbf{u}_0$ lies in the line spanned by $\mathbf{u}_0$. Examining Example (7.2.1), we are led to conjecture a solution of the form

$$\mathbf{u}(t) = e^{\lambda t}\mathbf{u}_0$$

This is a path lying in the line spanned by $\mathbf{u}_0$ satisfying the initial condition $\mathbf{u}(0) = \mathbf{u}_0$. Of course, we must verify that this satisfies the differential equation, but this is easily done:

$$d/dt(e^{\lambda t}\mathbf{u}_0) = \lambda e^{\lambda t}\mathbf{u}_0 = \lambda(e^{\lambda t}\mathbf{u}_0) = A(e^{\lambda t}\mathbf{u}_0) = A\mathbf{u}(t)$$

Therefore, we have proven Proposition (7.2.2).

(7.2.2) Proposition. If $\mathbf{u}_0$ is an eigenvector of A for λ, then the system $\mathbf{u}' = A\mathbf{u}$, with $\mathbf{u}(0) = \mathbf{u}_0$, is solved by $\mathbf{u}(t) = e^{\lambda t}\mathbf{u}_0$.

Of course, for a general A not every initial vector $\mathbf{u}_0$ is an eigenvector of A. However, if A has "enough" eigenvectors in the sense of Chapter 4, we can give a general solution that mimics our example.

(7.2.3) Proposition. Let A be a diagonalizable real $n \times n$ matrix with eigenvalues $\lambda_1, \ldots, \lambda_n \in \mathbf{R}$ and a corresponding basis of eigenvectors $\mathbf{v}_1, \ldots, \mathbf{v}_n \in \mathbf{R}^n$. If $\mathbf{u}_0 = a_1\mathbf{v}_1 + \cdots + a_n\mathbf{v}_n$, then $\mathbf{u}' = A\mathbf{u}$, $\mathbf{u}(0) = \mathbf{u}_0$, is solved by

$$\mathbf{u}(t) = e^{\lambda_1 t}(a_1\mathbf{v}_1) + \cdots + e^{\lambda_n t}(a_n\mathbf{v}_n)$$

Proof: We need only show that $\mathbf{u}' = A\mathbf{u}$. First

$$\mathbf{u}'(t) = \sum_{i=1}^{n} \lambda_i\, e^{\lambda_i t} a_i \mathbf{v}_i$$

On the other hand

$$\begin{aligned} A\mathbf{u}(t) &= \sum_{i=1}^{n} A(e^{\lambda_i t} a_i \mathbf{v}_i) \\ &= \sum_{i=1}^{n} e^{\lambda_i t} a_i\, A\mathbf{v}_i \\ &= \sum_{i=1}^{n} e^{\lambda_i t} a_i \lambda_i \mathbf{v}_i \end{aligned}$$

Therefore, $\mathbf{u}' = A\mathbf{u}$ as claimed. ■

As we noted before, from the physical interpretation of the solutions of the system as the paths followed by particles in a fluid flow, we expect that this is the only solution satisfying the given initial condition. You will derive a proof of this in Exercise 5.

If $\mathbf{u}_0 = (b_1, \ldots, b_n)$ in standard coordinates, then

$$Q^{-1}\mathbf{u}_0 = \begin{bmatrix} a_1 \\ \vdots \\ a_n \end{bmatrix}$$

where Q is the matrix whose columns are the eigenvectors $\mathbf{v}_1, \ldots, \mathbf{v}_n$. Thus, $\mathbf{u}(t)$ can be rewritten as

$$\mathbf{u}(t) = Q \begin{bmatrix} e^{\lambda_1 t} & & \\ & \ddots & \\ & & e^{\lambda_n t} \end{bmatrix} Q^{-1}\mathbf{u}_0$$

so that expressing our solutions in terms of matrices involves the eigenvectors of A and the "exponential" of the diagonal form for A. In the next section we will be quite specific about the exponential of a matrix. For now, let us say only that if D is a diagonal matrix, the exponential of D is obtained by exponentiating each of the diagonal entries of D. Thus, if Λ is the diagonal form of A, we have exponentiated Λt above.

(7.2.4) Example. Let $A = \begin{bmatrix} 2 & 2 & 0 \\ 0 & 1 & 0 \\ 0 & 1 & 2 \end{bmatrix}$. [See Example (4.2.3b).] The eigenvalues of A are 2, 2, and 1 and the corresponding eigenvectors are (1, 0, 0), (0, 0, 1), and (2, -1, 1), that is, A is diagonalizable.

A solution $\mathbf{u}(t)$ for any initial position $\mathbf{u}_0$ is given by

$$\begin{bmatrix} 1 & 0 & 2 \\ 0 & 0 & -1 \\ 0 & 1 & 1 \end{bmatrix} \begin{bmatrix} e^{2t} & 0 & 0 \\ 0 & e^{2t} & 0 \\ 0 & 0 & e^{t} \end{bmatrix} \begin{bmatrix} 1 & 0 & 2 \\ 0 & 0 & -1 \\ 0 & 1 & 1 \end{bmatrix}^{-1} \mathbf{u}_0$$

$$= \begin{bmatrix} e^{2t} & 2(e^{2t} - e^{t}) & 0 \\ 0 & e^{t} & 0 \\ 0 & e^{2t} - e^{t} & e^{2t} \end{bmatrix} \mathbf{u}_0$$

Once the solutions have been found, it is often useful to obtain a qualitative understanding of the set of all solutions. A particular question of interest concerns the behavior of solution curves for large values of the parameter t (both negative and positive).

For example, let A be a diagonalizable real matrix that satisfies the additional condition that its largest eigenvalue is positive and has multiplicity one. (This assumption will considerably simplify the analysis.) That is, $\lambda_n > 0$ and $|\lambda_i| < \lambda_n$ for $i < n$, and $A\mathbf{v}_i = \lambda_i \mathbf{v}_i$ for some $\mathbf{v}_i \neq \mathbf{0}$. We have already shown that for the initial condition $\mathbf{u}(0) = a_1\mathbf{v}_1 + \cdots + a_n\mathbf{v}_n$,

$$\mathbf{u}(t) = e^{\lambda_1 t}a_1\mathbf{v}_1 + \cdots + e^{\lambda_n t}a_n\mathbf{v}_n$$

Clearly, if $\lambda_i > 0$, $\lim_{t\to\infty} e^{\lambda_i t} = \infty$ and $\lim_{t\to\infty} \mathbf{u}(t)$ is not defined. Notice, however, that $\lim_{t\to\infty} e^{-\lambda_n t}\,\mathbf{u}(t)$ is defined:

$$\lim_{t\to\infty} e^{-\lambda_n t}\mathbf{u}(t) = \lim_{t\to\infty} e^{(\lambda_1-\lambda_n)t}a_1\mathbf{v}_1 + \cdots + e^{(\lambda_{n-1}-\lambda_n)t}a_{n-1}\mathbf{v}_{n-1} + a_n\mathbf{v}_n$$

$$= a_n\mathbf{v}_n$$

since $\lim_{t\to\infty} e^{(\lambda_i-\lambda_n)t} = 0, \qquad$ for $i < n$.

Since $e^{-\lambda_n t}\,\mathbf{u}(t)$ is a scalar multiple of $\mathbf{u}(t)$, this limit yields the limiting direction of $\mathbf{u}(t)$ as $t \to \infty$. Figure 7.6 illustrates the case $\lambda_1 < 0 < \lambda_2$.

Exercises 4 and 6 will indicate how to graph the set of solutions for the case $n = 2$ and A diagonalizable.

If the matrix A is not diagonalizable, a different approach is required. The solutions we produce will resemble those obtained by the preceding construction

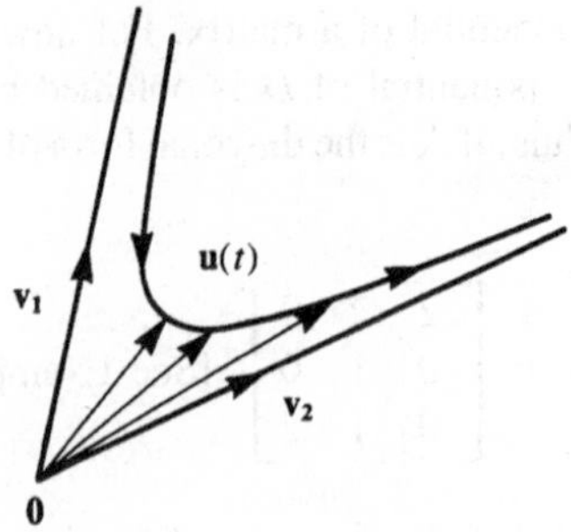

Figure 7.6

and will, quite naturally, utilize the Jordan canonical form of the matrix as a substitute for the diagonal form of a diagonalizable matrix. And, of course, if A is diagonalizable, this new solution will coincide with the one we have already established.

EXERCISES

1. Using the technique of Example (7.2.1), sketch the solutions to the system of equations of the form of Example (7.2.1) for the following choices of λ. (Choose initial values in each quadrant of the x-y plane.)
   a) $\lambda_1 = \lambda_2 = 1$
   b) $\lambda_1 = 1, \quad \lambda_2 = -1$
   c) $\lambda_1 = 0, \quad \lambda_2 = 1$
   d) $\lambda_1 = 2, \quad \lambda_1 = 1$
   e) $\lambda_1 = -1, \quad \lambda_2 = -2$

2. Solve the system $\mathbf{u}' = A\mathbf{u}$ for a general $\mathbf{u}_0$ and the given A.

   a) $A = \begin{bmatrix} 2 & 1 \\ 1 & 2 \end{bmatrix}$ b) $A = \begin{bmatrix} 1 & 1 \\ 0 & -1 \end{bmatrix}$ c) $A = \begin{bmatrix} 1 & 1 & 0 \\ & 2 & 1 \\ & & 3 \end{bmatrix}$.

3. Find all the straight-line solutions to the systems in Exercise 2.

4. Let Λ be the diagonal form of the diagonalizable matrix A. That is, $A = Q\Lambda Q^{-1}$ where Q is the matrix of eigenvectors of A. Prove that if $\mathbf{v}(t)$ is a solution to the system of equations $\mathbf{v}' = \Lambda\mathbf{v}$ with $\mathbf{v}_0 = Q^{-1}\mathbf{u}_0$, then $\mathbf{u}(t) = Q\mathbf{v}(t)$ is a solution to $\mathbf{u}' = A\mathbf{u}$ with $\mathbf{u}(0) = \mathbf{u}_0$.

5. Let A be a diagonalizable $n \times n$ matrix.
   a) Show that the only solution of the system $\mathbf{u}' = A\mathbf{u}$ with $\mathbf{u}(0) = \mathbf{0}$ is the constant solution $\mathbf{u}(t) = \mathbf{0}$.
   b) From part a deduce that if $\mathbf{u}(t)$ and $\hat{\mathbf{u}}(t)$ are two solutions of $\mathbf{u}' = A\mathbf{u}$ subject to $\mathbf{u}(0) = \mathbf{u}_0$, then $\mathbf{u}(t) = \hat{\mathbf{u}}(t)$.

6. Exercise 4 can be used to sketch solutions to $\mathbf{u}' = A\mathbf{u}$ for any diagonalizable A. First solve $\mathbf{v}' = \Lambda\mathbf{v}$ and sketch the resulting $\mathbf{v}(t)$. (Since Λ is diagonal, the techniques we have developed so far yield a solution.) Then sketch $Q\mathbf{v}(t)$ by applying the mapping defined by the matrix Q to the solution curves $\mathbf{v}(t)$. Apply this technique to the systems of Exercise 2, parts a and b.

7. Let A be a diagonalizable $n \times n$ matrix with real eigenvalues. Let E_+, E_-, and E_0, respectively denote the spans of the sets of eigenvectors of A corresponding to positive, negative, and zero eigenvalues. Prove that
   a) If $\mathbf{u}_0 \in E_+$, then $\lim_{t\to+\infty} \|\mathbf{u}(t)\| = \infty$ and $\lim_{t\to-\infty} \mathbf{u}(t) = \mathbf{0}$.
   b) If $\mathbf{u}_0 \in E_-$, then $\lim_{t\to+\infty} \mathbf{u}(t) = \mathbf{0}$ and $\lim_{t\to-\infty} \|\mathbf{u}(t)\| = \infty$.
   c) If $\mathbf{u}_0 \in E_0$, then $\mathbf{u}(t) = \mathbf{u}_0$ for all t.
   Can you formulate a general result for $\lim_{t\to\infty} \|\mathbf{u}(t)\|$ if $\mathbf{u}_0 \notin E_+$, E_-, or E_0?

In the text we have dealt with differential equations, in which the unknown function u depends on a continuous parameter $t \in \mathbf{R}$. In some applications another type of equation, called a *difference equation,* also arises. In a difference equation we express a relationship between the values of a function (or functions) $u: N \to \mathbf{C}$ for different values of $n \in N$ (the set of natural numbers). For example, we can think of u as the sequence of measured values of some quantity at a discrete sequence of times (e.g., measurements made once each hour). A difference equation might then relate the value of u at each hour to the value at the next hour. In the following exercises we will study some elementary facts about difference equations, and we will see that there are many similarities between these equations and the differential equations we have already seen.

Let $u_i: N \to \mathbf{C}$, $1 \leqslant i \leqslant p$ be functions (or equivalently, sequences of complex numbers). An example of a system of first-order difference equations on $\mathbf{u} = (u_1, \ldots, u_p)$ is a system of equations of the form

$$\begin{aligned} u_1(n+1) &= a_{11}u_1(n) + \cdots + a_{1p}u_p(n) \\ &\ \vdots \\ u_p(n+1) &= a_{p1}u_1(n) + \cdots + a_{pp}u_p(n) \end{aligned} \tag{7.3}$$

($a_{ij} \in \mathbf{C}$ constant), which are assumed to hold for all natural numbers n. Usually we will want to solve the system subject to some *initial conditions* $u_i(1) = b_i$ $(1 \leqslant i \leqslant p)$.

8. In the case $p = 1$, show that there is exactly one solution of the equation $u(n+1) = au(n)$ satisfying $u(1) = b$, namely, $u(n) = b \cdot a^{n-1}$.

9. Show that in general the system (7.3) with the preceding initial conditions can be written in matrix form as

$$\mathbf{u}(n+1) = A\mathbf{u}(n), \qquad \mathbf{u}(1) = \mathbf{b}$$

   where $\mathbf{u} = (u_1, \ldots, u_p)$, $\mathbf{b} = (b_1, \ldots, b_p)$, and $A = (a_{ij})$ is the $p \times p$ matrix of coefficients in Eq. (7.3).

10. Using the result of Exercise 8, solve the system $\mathbf{u}(n+1) = A\mathbf{u}(n)$, subject to $\mathbf{u}(1) = \mathbf{b}$ in the case that A is a diagonal matrix.

11. More generally, if A is a diagonalizable matrix with diagonal form Λ (so that $A = Q\Lambda Q^{-1}$ for some invertible matrix Q), then show that if $\mathbf{v}(n+1) = \Lambda\mathbf{v}(n)$, $\mathbf{v}(1) = \mathbf{b}$, then $\mathbf{u}(n) = Q\mathbf{v}(n)$ solves the system $\mathbf{u}(n+1) = A\mathbf{u}(n)$, $\mathbf{u}(1) = Q\mathbf{b}$.

12. Use the method outlined in Exercises 10 and 11 to solve:

a)
$$\begin{aligned} u_1(n+1) &= 2u_1(n) + u_2(n) \\ u_2(n+1) &= u_1(n) + 2u_2(n) \\ (u_1(1), u_2(1)) &= (3, 4) \end{aligned}$$

b)
$$\begin{aligned} u_1(n+1) &= 4u_1(n) + 2u_2(n) \\ u_2(n+1) &= 3u_2(n) - u_3(n) \\ u_3(n+1) &= 2u_3(n) \\ (u_1(1), u_2(1), u_3(1)) &= (0, 1, 0) \end{aligned}$$

§7.3. CONSTANT COEFFICIENT LINEAR DIFFERENTIAL EQUATIONS: THE GENERAL CASE

To derive the general solution to the equation $\mathbf{u}' = A\mathbf{u}$ for an arbitrary matrix A, we need to develop several new concepts that combine linear algebra, calculus, and analysis. Recall that in Section 7.2 we saw that the solutions of $\mathbf{u}' = A\mathbf{u}$ in the case that A is diagonalizable could be expressed in terms of the "exponential" of the diagonal form of A. It is this observation that we use to derive our solutions in general. First, we define the exponential of a matrix, then we show that this exponential allows us to produce a solution, and, finally, we show how to compute the exponential of a general matrix.

At first sight it is not clear how the exponential of a matrix should be defined. However, if we recall from calculus that the ordinary exponential function e^x can be defined using a MacLaurin series, $e^x = \sum_{k=0}^{\infty} (1/k!)x^k$, which converges absolutely for all values of x, then we are on the right track. The advantage of this definition is that, even though it involves the sum of an infinite series, the partial sums of the series involve only the operations of multiplication and addition. Since we can also multiply and add square matrices, we can try the following provisional method to define e^A for a matrix A.

(7.3.1) Definition. Let A be an $n \times n$ matrix. The exponential of A, denoted e^A, is the $n \times n$ matrix defined by the series

$$e^A = \sum_{k=0}^{\infty} (1/k!)A^k$$

Of course, unless the series in Definition (7.3.1) converges in some sense, this definition is of no use to us. Before we prove that the series converges, let us consider an example.

(7.3.2) Example. Let A be a diagonal matrix with diagonal entries a_{ii}, $i = 1, \ldots, n$. A^k is also a diagonal matrix and its diagonal entries are a_{ii}^k, $i = 1, \ldots, n$. Therefore, the mth partial sum of the series e^A is the matrix whose diagonal entries are $\sum_{k=0}^{m} (1/k!)a_{ii}^k$ and whose off-diagonal entries are zero. The ith diagonal entry of e^A is the series $\sum_{k=0}^{\infty} (1/k!)a_{ii}^k$, which is $e^{a_{ii}}$. Therefore,

$$e^A = \begin{bmatrix} e^{a_{11}} & & & \\ & \cdot & & \\ & & \cdot & \\ & & & \cdot \\ & & & & e^{a_{nn}} \end{bmatrix}$$

Thus, in this case the series in Definition (7.3.1) *does* converge. An easy calculation now shows that

$$e^{At} = \begin{bmatrix} e^{a_{11}t} & & & \\ & \cdot & & \\ & & \cdot & \\ & & & \cdot \\ & & & & e^{a_{nn}t} \end{bmatrix}$$

can be used to produce a solution to $\mathbf{u}' = A\mathbf{u}$, $\mathbf{u}(0) = \mathbf{b}$. The promised solution is the function $\mathbf{u}(t) = e^{At}\mathbf{b}$.

In order to show that the series e^A converges for an arbitrary matrix, we need to be able to say that the partial sums of the series converge to some limiting matrix as we include more and more terms. To do this, we need to be able to measure the "size" of matrices, so we will introduce a norm on the set of $n \times n$ matrices, that is, a real-valued function on the set of $n \times n$ matrices that we denote by $\| A \|$. The function $\| \quad \|$ will behave much like the norm—or length—of a vector, $\| \mathbf{v} \| = \langle \mathbf{v}, \mathbf{v} \rangle^{1/2}$, although the norm of a matrix will be defined not in terms of the particular entries of the matrix, but rather, in terms of its "geometric behavior."

(7.3.3) Definition. Let A be an $n \times n$ matrix. The norm of A, denoted $\| A \|$, is defined by

$$\| A \| = \max_{\substack{\mathbf{x} \in \mathbf{R}^n \\ \mathbf{x} \neq \mathbf{0}}} \| A\mathbf{x} \| / \| \mathbf{x} \|$$

Notice first that $\| A\mathbf{x} \|$ and $\| \mathbf{x} \|$ are the norms of vectors in $\mathbf{R}^n$. Since we have ruled out $\mathbf{x} = \mathbf{0}$, the quotient is well defined. The norm is the maximum over all such $\mathbf{x}$ of this quotient.

(7.3.4) Remark. In this section we will also want to apply Definition (7.3.3) to matrices with complex entries. In this case we must replace the usual inner product on $\mathbf{R}^n$, $\langle \mathbf{x}, \mathbf{y} \rangle = \sum_{i=1}^{n} x_i y_i$ by the Hermitian inner product on $\mathbf{C}^n$, $\langle \mathbf{z}, \mathbf{w} \rangle = \sum_{i=1}^{n} z_i \overline{w}_i$ (see Section 5.3). Thus, (7.3.3) would read

$$\| A \| = \max_{\substack{\mathbf{z} \in \mathbf{C}^n \\ \mathbf{z} \neq \mathbf{0}}} \| A\mathbf{z} \| / \| \mathbf{z} \|$$

Since all the results of this section hold for $\mathbf{R}^n$ or $\mathbf{C}^n$ interchangeably, we will usually state our results for $\mathbf{R}^n$.

This quotient can be simplified somewhat algebraically:

$$\| A\mathbf{x} \| / \| \mathbf{x} \| = \| (1/ \| \mathbf{x} \|)A\mathbf{x} \| = \| A(\mathbf{x}/ \| \mathbf{x} \|) \|$$

Since $\mathbf{x}/ \| \mathbf{x} \|$ is a unit vector, we see that

$$\| A \| = \max_{\| \mathbf{x} \| = 1} \| A\mathbf{x} \| \tag{7.4}$$

that is, the maximum value of $\| A\mathbf{x} \|$ over all $\mathbf{x}$ for which $\| \mathbf{x} \| = 1$.

For the moment, assume that $\| A \|$ is finite (we will see shortly that this is always the case). If $\mathbf{y} \neq \mathbf{0}$ is a particular vector, we must have

$$\| A\mathbf{y} \| / \| \mathbf{y} \| \leqslant \| A \|$$

since $\| A \|$ is the maximum value of such a quotient. But then $\| A\mathbf{y} \| \leqslant \| A \| \cdot \| \mathbf{y} \|$, so that $\| A \|$ gives an upper bound on the amount by which A can stretch (or shrink) a vector.

With this geometric intuition in mind, let us compute $\| A \|$ explicitly taking equation (7.4) as the definition. Recall that if $\mathbf{x}$ is regarded as a column vector, $\| \mathbf{x} \| = \langle \mathbf{x},\mathbf{x} \rangle^{1/2} = (\mathbf{x}^t\mathbf{x})^{1/2}$. Thus, to find the maximum value of $\| A\mathbf{x} \|$, it suffices to find the maximum value of $(A\mathbf{x})^t (A\mathbf{x}) = \mathbf{x}^t A^t A\mathbf{x}$. Now A^tA is a symmetric matrix so we may apply the spectral theorem [Theorem (4.6.1)] to A^tA. Let $\mathbf{v}_1, \ldots, \mathbf{v}_n$ be an orthonormal basis for $\mathbf{R}^n$ consisting of eigenvectors of A^tA, and let $\lambda_1, \ldots, \lambda_n$ be the corresponding eigenvalues.

If $\mathbf{x} = x_1\mathbf{v}_1 + \cdots + x_n\mathbf{v}_n$, then

$$\begin{aligned}
\mathbf{x}^t(A^tA)\mathbf{x} &= (x_1\mathbf{v}_1 + \cdots + x_n\mathbf{v}_n)^t A^tA(x_1\mathbf{v}_1 + \cdots + x_n\mathbf{v}_n) \\
&= (x_1\mathbf{v}_1^t + \cdots + x_n\mathbf{v}_n^t)(\lambda_1 x_1\mathbf{v}_1 + \cdots + \lambda_n x_n\mathbf{v}_n) \\
&= \sum_{j=1}^{n} \left[\sum_{i=1}^{n} \lambda_j x_i x_j \langle \mathbf{v}_i, \mathbf{v}_j \rangle \right] \\
&= \sum_{j=1}^{n} \lambda_j x_j^2
\end{aligned}$$

since $\{\mathbf{v}_1, \ldots, \mathbf{v}_n\}$ is an orthonormal set. Since $\langle A\mathbf{x}, A\mathbf{x} \rangle \geq 0$, each λ_j must be nonnegative. Furthermore, if $A \neq 0$, then A^tA has some nonzero, hence strictly positive eigenvalue (why?). Assume that $\lambda_n \geq \lambda_i$ for $i \leqslant n$. Then $\sum_{j=1}^{n} \lambda_j x_j^2 \leqslant$

$\sum_{j=1}^{n} \lambda_n x_j^2 = \lambda_n \cdot \langle \mathbf{x}, \mathbf{x} \rangle$. If $\langle \mathbf{x}, \mathbf{x} \rangle = 1$, we are left with λ_n. Therefore, for all unit vectors $\mathbf{x}$, $\| A\mathbf{x} \|^2 \leq \lambda_n$, and hence, $\| A \| \leq \sqrt{\lambda_n}$. However, if $\mathbf{x}$ is chosen to be $\mathbf{v}_n$ the preceding calculation shows $\| A\mathbf{v}_n \| = \sqrt{\lambda_n}$. Therefore, we have proven Proposition (7.3.5).

(7.3.5) Proposition. The norm of a nonzero $n \times n$ matrix $A \in M_{n\times n}(\mathbf{R})$ is the square root of the largest eigenvalue of A^tA, which is necessarily positive.

(7.3.6) Example. We compute the norm of $A = \begin{bmatrix} 2 & 1 \\ 0 & 2 \end{bmatrix}$. $A^tA = \begin{bmatrix} 4 & 2 \\ 2 & 5 \end{bmatrix}$, so the largest eigenvalue of A^tA is $(9 + \sqrt{17})/2$ and $\| A \| = \sqrt{(9 + \sqrt{17})/2}$.

The matrix norm has properties similar to those of the vector norm in $\mathbf{R}^n$ [compare this with Proposition (4.3.4)]:

(7.3.7) Proposition. Let $A \in M_{n\times n}(\mathbf{R})$ be any matrix.
(i) For all $a \in \mathbf{R}$, $\| aA \| = |a| \cdot \| A \|$.
(ii) $\| A \| \geq 0$ and $\| A \| = 0$ implies $A = 0$.
(iii) $\| A + B \| \leq \| A \| + \| B \|$.
(iv) $\| AB \| \leq \| A \| \cdot \| B \|$.

Proof: See Exercise 1. ■

In order to show that the series e^A converges, we require a result from analysis that we will phrase in the language of matrices and the matrix norm.

(7.3.8) Proposition. Let $\sum_{j=0}^{\infty} B_j$ be a series of matrices. If the series converges absolutely in norm, that is, $\sum_{j=0}^{\infty} \| B_j \|$ converges, then the series $\sum_{j=0}^{\infty} B_j$ converges.

Note that $\sum_{j=0}^{\infty} \| B_j \|$ is simply a series of real numbers.

In order to apply this to the series e^A, we must show the series $\sum_{j=0}^{\infty} \| (1/j!)A^j \|$ converges. The partial sums S_k of this series satisfy

$$S_k = \sum_{j=0}^{k} \| (1/j!)A^j \| \leq \sum_{j=0}^{k} (1/j!) \| A \|^j$$

[by Proposition (7.3.7i and iv)].

But the last term here is just the kth partial sum of the series of real numbers $\sum_{j=0}^{\infty} (1/j!) \| A \|^j = e^{\| A \|}$. Thus, the sequence of partial sums S_k, which is an increasing sequence (all the terms in the sum are nonnegative), is bounded above by $e^{\|A\|}$. Therefore, the series $\sum_{j=0}^{\infty} \|(1/j!)A^j\|$ converges. Applying Proposition (7.3.8) we conclude with Proposition (7.3.9).

(7.3.9) Proposition. The series e^A converges for every matrix A.

We now want to show that if A is an $n \times n$ matrix, then $e^{At}\mathbf{u}_0$ is a solution to $\mathbf{u}' = A\mathbf{u}$ with $\mathbf{u}(0) = \mathbf{u}_0$. Once we have proven the following lemma, we will be able to apply the definition of the derivative of a vector function directly in order to evaluate $d/dt(e^{At}\mathbf{u}_0)$.

As with the usual exponential function e^x that satisfies $e^{x+y} = e^xe^y$, we want to show that the exponential e^A satisfies $e^{A+B} = e^Ae^B$. At this point a difference between multiplication in $\mathbf{R}$ and multiplication of matrices intervenes: matrix multiplication is not commutative. Thus, we are left with Lemma (7.3.10).

(7.3.10) Lemma. If A and B are commuting $n \times n$ matrices, $AB = BA$, then $e^{A+B} = e^Ae^B$.

Proof: Since $AB = BA$, the binomial theorem can be applied to expand the jth term of the series defining e^{A+B}:

$$(1/j!)(A + B)^j = (1/j!) \sum_{i=0}^{j} \begin{bmatrix} j \\ i \end{bmatrix} A^{j-i}B^i \tag{7.5}$$

We now want to evaluate the jth term of the product series e^Ae^B. Since both series, $e^A = \sum_{j=0}^{\infty} (1/j!)A^j$ and $e^B = \sum_{i=0}^{\infty} (1/j!)B^i$, converge absolutely in norm, to compute the product e^Ae^B, we can multiply the two power series together formally. The terms of total degree j in the product are

$$\sum_{i=0}^{j} (1/(j-i)!)A^{j-i} \cdot (1/i!)B^i \tag{7.6}$$

Since $(1/j!)\begin{bmatrix} j \\ i \end{bmatrix} = (1/j!)(j!/(j-i)!i!) = 1/((j-i)!i!)$ we see that Eqs. (7.5) and (7.6) are the same. Hence, it is also true that $e^{A+B} = e^Ae^B$, since e^{A+B} is the sum over j of the terms in Eq. (7.5), whereas e^Ae^B is the sum over j of the terms in Eq. (7.6). ■

(7.3.11) Proposition. $d/dt(e^{At}) = Ae^{At}$ for any $n \times n$ matrix A.

Proof: Using the definition of the derivative, we must evaluate the limit

$$d/dt(e^{At}) = \lim_{h\to 0} (e^{A(h+t)} - e^{At})/h$$

Applying the lemma, we see that $e^{A(h+t)} = e^{Ah+At} = e^{Ah}e^{At}$ since $(Ah)(At) = (At)(Ah)$. Then we may factor e^{At} from the numerator of the quotient to obtain

$$\lim_{h\to 0} ((e^{Ah} - I)/h)e^{At} = [\lim_{h\to 0} (e^{Ah} - I)/h]e^{At} \tag{7.7}$$

Expanding $e^{Ah} - I$, we obtain

$$(I + (Ah) + (Ah)^2/2! + \cdots) - I = \sum_{j=1}^{\infty} (1/j!)(Ah)^j$$

Dividing by h we see that Eq. (7.7) is

$$[\lim_{h\to 0} \sum_{j=1}^{\infty} (1/j!)\, A^j h^{j-1}]\, e^{At} \tag{7.8}$$

The limit in Eq. (7.8) is the term of degree zero in h, that is, the term with $j = 1$, multiplied by e^{At}. This is just Ae^{At}. ■

(7.3.12) Corollary. $\mathbf{u}(t) = e^{At}\,\mathbf{u}_0$ is a solution to the system of equations $\mathbf{u}' = A\mathbf{u}$ subject to $\mathbf{u}(0) = \mathbf{u}_0$.

Proof: This is an immediate consequence of Proposition (7.3.11). ■

In order to evaluate e^{At} for an arbitrary matrix A, we will require one more lemma whose proof will be left as an exercise (see Exercise 8 of Section 6.1)

(7.3.13) Lemma. Let A and B be similar $n \times n$ matrices: $A = QBQ^{-1}$, for Q an invertible matrix. Then

$$e^A = Qe^BQ^{-1}$$

We can now formulate a procedure for calculating e^{At}. Suppose that A is an $n \times n$ matrix with complex entries. Of course, this includes the case that A has real entries. As a consequence of Theorem (6.3.6), there is an invertible matrix Q whose columns are a Jordan canonical basis for A, so that $A = QJQ^{-1}$, where J is the Jordan canonical form of A. Applying Lemma (7.3.13), we obtain $e^{At} = Qe^{Jt}Q^{-1}$. Thus, it suffices to be able to calculate e^{Jt} for a matrix J in Jordan canonical form.

Before we proceed with calculating e^{Jt} for a general J, notice that if A is diagonalizable, J is diagonal and the columns of Q are a basis of $\mathbf{R}^n$ or $\mathbf{C}^n$ consisting of eigenvectors of A. Using Example (7.3.2), we see that this is exactly the result we obtained in Section 7.2 using geometric considerations.

Let us return to the general situation. J is a matrix of the form

$$J = \begin{bmatrix} J_1 & & & \\ & \cdot & & \\ & & \cdot & \\ & & & \cdot \\ & & & & J_m \end{bmatrix}$$

where each J_i is a Jordan block matrix

$$J_i = \begin{bmatrix} \lambda_i & 1 & & \\ & \cdot & & \\ & & \cdot & \\ & & \cdot & 1 \\ & & & \lambda_i \end{bmatrix}$$

Applying Exercise 4, we ultimately need only calculate $e^{J_i t}$.

Each Jordan block can be written $J_i = D + N$ with $DN = ND$. $D = \lambda_i I$ and $N = J - \lambda_i I$, that is,

$$J_i = \lambda_i I + \begin{bmatrix} 0 & 1 & & \\ & \cdot & & \\ & & \cdot & \\ & & \cdot & 1 \\ & & & 0 \end{bmatrix}$$

From Lemma (7.3.10) we see that

$$e^{J_i t} = e^{\lambda_i I t}\, e^{Nt} \tag{7.9}$$

We calculated $e^{\lambda_i I t}$ in Example (7.3.2). It remains to calculate e^{Nt}. Since N is nilpotent, if N is an $(l + 1) \times (l + 1)$ matrix, we have $N^{l+1} = 0$; and it is an easy calculation to show that

$$e^{Nt} = \begin{bmatrix} 1 & t & t^2/(2!) & \cdots & t^l/(l!) \\ & 1 & t & \cdots & \\ & & \cdot & & \\ & & & \cdot & \\ & & & \cdot & \\ & & & 1 & t \\ & & & & 1 \end{bmatrix} \tag{7.10}$$

If we can produce the Jordan form and corresponding Jordan basis for A (which as we know is a rather large assumption), we can use (7.3.12), (7.3.13), Exercise 4, and equations (7.9) and (7.10) to solve any system of equations $\mathbf{u}' = A\mathbf{u}$, $\mathbf{u}(0) = \mathbf{u}_0$, explicitly.

Let's put this to work in two examples.

(7.3.14) Examples

a) Let $A = \begin{bmatrix} 0 & 5 & -9 \\ 1 & 5 & -9 \\ 1 & 4 & -8 \end{bmatrix}$. We use the techniques of this section to compute e^{At}. The eigenvalues of A are -2, -2, and 1. Since $\dim(\text{Ker}(A + 2I)) = 1$, the Jordan form of A is

$$J = \begin{bmatrix} -2 & 1 & 0 \\ 0 & -2 & 0 \\ 0 & 0 & 1 \end{bmatrix}$$

A corresponding Jordan canonical basis is $\{(2, 1, 1), (3, 1, 1), (1, 2, 1)\}$. Thus

$$e^{At} = Qe^{Jt}Q^{-1}$$

$$= Q\begin{bmatrix} e^{-2t} & & \\ & e^{-2t} & \\ & & e^{t} \end{bmatrix}\begin{bmatrix} 1 & t & 0 \\ & 1 & 0 \\ & & 1 \end{bmatrix} Q^{-1}$$

$$= \begin{bmatrix} 2 & 3 & 1 \\ 1 & 1 & 2 \\ 1 & 1 & 1 \end{bmatrix}\begin{bmatrix} e^{-2t} & te^{-2t} & 0 \\ 0 & e^{-2t} & 0 \\ 0 & 0 & e^{t} \end{bmatrix}\begin{bmatrix} -1 & -2 & 5 \\ 1 & 1 & -3 \\ 0 & 1 & -1 \end{bmatrix}$$

$$= \begin{bmatrix} (2t + 1)e^{-2t} & (2t - 1)e^{-2t} + e^{t} & (-6t + 1)e^{-2t} - e^{t} \\ te^{-2t} & (t - 1)e^{-2t} + 2e^{t} & (-3t + 2)e^{-2t} - 2e^{t} \\ te^{-2t} & (t - 1)e^{-2t} + e^{t} & (-3t + 2)e^{-2t} - e^{t} \end{bmatrix}$$

b) Let us consider an example with nonreal eigenvalues:

$$A = \begin{bmatrix} 0 & -1 \\ 1 & 0 \end{bmatrix}$$

The eigenvalues of A are i and $-i$. The corresponding eigenvectors are $(i, 1)$ and $(-i, 1)$. Therefore

$$e^{At} = \begin{bmatrix} i & -i \\ 1 & 1 \end{bmatrix}\begin{bmatrix} e^{it} & 0 \\ 0 & e^{-it} \end{bmatrix}\begin{bmatrix} i & -i \\ 1 & 1 \end{bmatrix}^{-1}$$

$$= (1/2)\begin{bmatrix} (e^{it} + e^{-it}) & i(e^{it} - e^{-it}) \\ -i(e^{it} - e^{-it}) & (e^{it} + e^{-it}) \end{bmatrix}$$

$$= \begin{bmatrix} \cos(t) & -\sin(t) \\ \sin(t) & \cos(t) \end{bmatrix}$$

Obtaining the final form of our answer requires a basic calculation with complex exponentials. Using the fact that $i^2 = -1$, we can show that e^{it} is expressed in the form $a + ib$ as $\cos(t) + i\sin(t)$ (see Exercise 5). Since $e^{-it} = \overline{e^{it}}$, it follows that $e^{it} + e^{-it}$ is twice the real part of e^{it}; thus, $(1/2)(e^{it} + e^{-it}) = \cos(t)$. In a similar manner, $(-i/2)(e^{it} - e^{-it}) = \sin(t)$.

Notice that even though our calculations involved complex numbers and complex vectors, our final result involves only real-valued functions. Why should we have anticipated this?

In the exercises we explore the importance of the Jordan form of A in determining the qualitative behavior of the solutions.

EXERCISES

1. Prove Proposition (7.3.7).

2. Prove Lemma (7.3.13).

3. a) Prove that if $A \in M_{n\times n}(\mathbf{R})$ is a symmetric matrix and λ_n is the eigenvalue of A with the largest absolute value, then $\| A \| = |\lambda_n|$.
   b) Show that the same statement is true if $A \in M_{n\times n}(\mathbf{C})$ is a Hermitian matrix ($\bar{A}^t = A$). [*Hint:* Try to prove a complex analog of Proposition (7.3.5) first. The relevant matrix in the complex case is $\bar{A}^t A$.]

4. Let A be a block diagonal matrix,

$$A = \begin{bmatrix} A_1 & & & \\ & \cdot & & \\ & & \cdot & \\ & & & \cdot \\ & & & & A_k \end{bmatrix}$$

Prove that

$$e^A = \begin{bmatrix} e^{A_1} & & & \\ & \cdot & & \\ & & \cdot & \\ & & & \cdot \\ & & & & e^{A_k} \end{bmatrix}$$

5. The complex exponential function e^z is defined by the MacLaurin series $e^z = \sum_{n=0}^{\infty} z^n/n!$ for $z \in \mathbf{C}$. This series converges absolutely for all $z \in \mathbf{C}$. This exercise develops basic properties of e^z.
   a) Let $z = a + ib$, use the idea of the proof of Lemma (7.3.10) to show that $e^z = e^a e^{ib}$.
   b) Use the power-series definition of the exponential function to show that $e^{ib} = \cos(b) + i\sin(b)$.
   c) Show that $\overline{e^{ib}} = e^{-ib}$.

6. Let $A = \begin{bmatrix} \lambda & 1 \\ 0 & \lambda \end{bmatrix}$. Sketch the solutions to $\mathbf{u}' = A\mathbf{u}$ for $\mathbf{u}(0) = (1, 0)$ and $\mathbf{u}(0) = (0, 1)$ in each of the three cases $\lambda > 0$, $\lambda < 0$, and $\lambda = 0$.

7. Use the techniques of Exercise 6 of Section 7.2 and Exercise 6 of this section to sketch solutions to $\mathbf{u}' = A\mathbf{u}$ for $A =$

   a) $\begin{bmatrix} 3 & 1 \\ -1 & 1 \end{bmatrix}$ b) $\begin{bmatrix} 1 & 1 \\ -1 & -1 \end{bmatrix}$ c) $\begin{bmatrix} 2 & 1 \\ -1 & 0 \end{bmatrix}$

8. Calculate e^{At} for $A =$

a) $\begin{bmatrix} 1 & -1 & 1 \\ & 1 & -1 \\ & & 1 \end{bmatrix}$ b) $\begin{bmatrix} 0 & 1 & 0 & 0 \\ -1 & 0 & 0 & -1 \\ & & i & 1 \\ & & 0 & i \end{bmatrix}$

9. Show that if N is a nilpotent matrix, then all components of solutions of the system $\mathbf{u}' = N\mathbf{u}$ are polynomials in t.

The techniques we have developed so far do not allow us to sketch solutions to $\mathbf{u}' = A\mathbf{u}$ if A is a real matrix with complex eigenvalues that are not real. The following problems will develop a technique for accomplishing this for 2×2 matrices A.

10. a) [See Example (7.3.14b)]. Calculate e^{At} for a matrix A of the form

$$A = \begin{bmatrix} 0 & b \\ -b & 0 \end{bmatrix}, b \in \mathbf{R}.$$

b) Use (7.3.10) to calculate e^{At} for a matrix of the form $A = \begin{bmatrix} a & b \\ -b & a \end{bmatrix}$.

11. a) Using Exercise 10a, sketch the solution curves of $\mathbf{u}' = A\mathbf{u}$ for $A = \begin{bmatrix} 0 & b \\ -b & 0 \end{bmatrix}$, where b is positive.
Also sketch the solution curves for b negative.

b) Using Exercise 10b, sketch the solution curves of $\mathbf{u}' = A\mathbf{u}$ for $A = \begin{bmatrix} a & b \\ -b & a \end{bmatrix}$ for $b > 0$ and $a > 0$.
Repeat this for $b > 0$ and $a < 0$, $b < 0$ and $a > 0$, and $b < 0$ and $a < 0$.

12. Let B be a real matrix similar to a matrix A of the form $\begin{bmatrix} a & b \\ -b & a \end{bmatrix}$; that is, $B = Q^{-1}AQ$. How would you sketch solution curves to $\mathbf{u}' = B\mathbf{u}$?

13. Let B be a 2×2 real matrix with eigenvalues $a + ib$ and $a - ib$, $b \neq 0$, with corresponding complex eigenvectors $\mathbf{z}$ and $\bar{\mathbf{z}}$. Let $\mathbf{z} = \mathbf{x} + i\mathbf{y}$, where $\mathbf{x}$ and $\mathbf{y} \in \mathbf{R}^2$, and let Q be the matrix whose columns are $\mathbf{x}$ and $\mathbf{y}$. Show that

$$Q^{-1}BQ = \begin{bmatrix} a & b \\ -b & a \end{bmatrix}$$

by direct calculation. [*Hint:* Use the facts that $(\mathbf{z} + \bar{\mathbf{z}})/2 = \mathbf{x}$ and $(\mathbf{z} - \bar{\mathbf{z}})/2i = \mathbf{y}$.]

14. Use the preceding problems to sketch the solutions to $\mathbf{u}' = A\mathbf{u}$ for $A =$

a) $\begin{bmatrix} 2 & 4 \\ -2 & -2 \end{bmatrix}$ b) $\begin{bmatrix} -2 & 4 \\ -8 & 6 \end{bmatrix}$

15. a) Let A be a 2×2 real matrix with eigenvalues $a \pm ib$, $b \neq 0$. Determine the general behavior of solutions to $\mathbf{u}' = A\mathbf{u}$ for $\mathbf{u}_0 \neq \mathbf{0}$ as $t \to \pm\infty$; that is, evaluate $\lim_{t\to\infty} \| \mathbf{u}(t) \|$ for $\mathbf{u}_0 \neq \mathbf{0}$. (*Hint:* Consider the cases $a < 0$, $a = 0$, and $a > 0$ separately.)

    b) Let A be an $n \times n$ real matrix which is diagonalizable over $\mathbf{C}$. Determine the general behavior of solutions to $\mathbf{u}' = A\mathbf{u}$ for $\mathbf{u}_0 \neq \mathbf{0}$ as $t \to \pm\infty$; that is, evaluate $\lim_{t\to\infty} \| \mathbf{u}(t) \|$ for $\mathbf{u}_0 \neq 0$. (*Hint:* Compare with Exercise 7 of Section 7.2. As in part a of this problem, consider different cases based upon the signs of the real parts of the eigenvalues.)

The following exercises continue the study of difference equations that began in the Exercises from Section 7.2

16. Let B be a $p \times p$ Jordan block matrix with diagonal entries λ. Show that if $\lambda \neq 0$ the system of difference equations $\mathbf{u}(n + 1) = B\mathbf{u}(n)$ has solutions of the form

$$u_1(n) = (c_{10} + c_{11}n + \cdots + c_{1,p-1}n^{p-1})\,\lambda^{n-1}$$

$$u_2(n) = (c_{20} + c_{21}n + \cdots + c_{2,p-2}n^{p-2})\lambda^{n-1}$$

$$\vdots$$

$$u_p(n) = c_{p0}\lambda^{n-1}$$

where the $c_{ij} \in \mathbf{C}$ are determined by the initial conditions $\mathbf{u}(1)$. If $\lambda = 0$, show that these formulas are valid for $n \geq p$. (What happens for $n < p$?)

17. Show that if $A \in M_{n\times n}(\mathbf{C})$ has Jordan canonical form J (so $A = QJQ^{-1}$ for some invertible matrix Q), and if $\mathbf{v}(n + 1) = J\mathbf{v}(n)$ with $\mathbf{v}(1) = \mathbf{b}$, then $\mathbf{u}(n) = Q\mathbf{v}(n)$ solves $\mathbf{u}(n + 1) = A\mathbf{u}(n)$, $\mathbf{u}(1) = Q\mathbf{b}$.

Hence, we can solve $\mathbf{u}(n + 1) = A\mathbf{u}(n)$ by

1. Finding the Jordan form J of A,
2. Using Exercise 16 to solve the system $\mathbf{v}(n + 1) = J\mathbf{v}(n)$, and
3. Changing coordinates to find $\mathbf{u} = Q\mathbf{v}$.

18. Use this method to solve $\mathbf{u}(n + 1) = A\mathbf{u}(n)$ for

    a) $A = \begin{bmatrix} 1 & 1 \\ -1 & -1 \end{bmatrix}$, $\mathbf{u}(1) = (1, 0)$

    b) $A = \begin{bmatrix} 0 & 3 & -7 \\ 1 & 1 & -5 \\ 1 & 2 & -6 \end{bmatrix}$, $\mathbf{u}(1) = (0, 1, 1)$

§7.4. ONE ORDINARY DIFFERENTIAL EQUATION WITH CONSTANT COEFFICIENTS

In this section we apparently switch our focus by considering a single differential equation in one variable of the special form

$$u^{(n)}(t) + a_{n-1}u^{(n-1)}(t) + \cdots + a_1u^{(1)}(t) + a_0u^{(0)}(t) = 0 \tag{7.11}$$

Here $u: \mathbf{R} \to \mathbf{C}$ is a complex-valued function of a real variable, $u^{(i)}$ denotes the ith derivative of $u(t)$ with respect to t [so that $u^{(0)} = u$] and the coefficients $a_{n-1}, \ldots, a_0$ are complex numbers. Notice that if $a_{n-1}, \ldots, a_0 \in \mathbf{R}$ it makes sense to look for solutions of the form $u: \mathbf{R} \to \mathbf{R}$, however, we will remain in the more general context. As with systems of equations, we will need to specify initial conditions in order to determine $u(t)$ uniquely. These conditions usually take the form $u^{(i)}(0) = b_i$, $i < n$. That is, we specify the value of u and all its derivatives up to the order of the equation minus one at time $t = 0$.

Since the right-hand side of such an equation is zero, Eq. (7.11) is called a homogeneous nth-order ordinary differential equation with constant coefficients.

We have already encountered the simplest of these equations, the case $n = 1$. Suppose we wish to solve

$$u^{(1)}(t) + a_0u(t) = 0, \quad u(0) = b$$

First, let us rewrite this in a more familiar form

$$du/dt = u^{(1)}(t) = -a_0u(t), \qquad u(0) = b$$

Of course, this is just the exponential growth equation again. By now, it should be immediate that the solution is $u(t) = be^{-a_0t}$.

One general method of solution is to reduce a single equation in one variable and higher-order derivatives to a *system* of equations in several variables with no derivatives higher than first order. (Once we have used this method to derive the general form of the solutions, we will also see a more economical way to produce the solutions.) Before presenting the general case, let us consider a second-order example.

(7.4.1) Example. Consider the equation

$$u^{(2)} + 4u^{(1)} + 3u^{(0)} = 0, \qquad u(0) = 1, \qquad u^{(1)}(0) = 1 \tag{7.12}$$

Notice that if we make a substitution $x(t) = u(t)$ and $y(t) = u^{(1)}(t)$, our equation becomes

$$y' + 4y + 3x = 0 \quad \text{or} \quad y' = -3x - 4y$$

The functions x and y are also related by

$$x' = y$$

Our initial conditions become $x(0) = 1$ and $y(0) = 1$. Therefore, solving the original second-order equation (7.12) is the same as solving the system of first-order equations

$$\begin{aligned} x' &= y \\ y' &= -3x - 4y \end{aligned} \tag{7.13}$$

subject to the initial conditions $(x(0), y(0)) = (1, 1)$.

The equations (7.13) can be rewritten in the form

$$\begin{bmatrix} x \\ y \end{bmatrix}' = \begin{bmatrix} 0 & 1 \\ -3 & -4 \end{bmatrix} \begin{bmatrix} x \\ y \end{bmatrix} = A \begin{bmatrix} x \\ y \end{bmatrix}$$

From Section 7.3 we know that we can solve for x and y

$$\begin{bmatrix} x \\ y \end{bmatrix} = e^{At} \cdot \begin{bmatrix} x(0) \\ y(0) \end{bmatrix}$$

Let us carry out the calculation of e^{At}. The characteristic polynomial of A is $\lambda^2 + 4\lambda + 3$ and the eigenvalues are -1 and -3. The corresponding eigenvectors are $(-1, 1)$ and $(1, -3)$, respectively. Thus

$$\begin{aligned} e^{At} &= \begin{bmatrix} -1 & 1 \\ 1 & -3 \end{bmatrix} \begin{bmatrix} e^{-t} & 0 \\ 0 & e^{-3t} \end{bmatrix} \begin{bmatrix} -1 & 1 \\ 1 & -3 \end{bmatrix}^{-1} \\ &= 1/2 \begin{bmatrix} 3e^{-t} - e^{-3t} & e^{-t} - e^{-3t} \\ -3e^{-t} + 3e^{-3t} & -e^{-t} + 3e^{-3t} \end{bmatrix} \end{aligned}$$

Therefore, $x(t) = 1/2(3e^{-t} - e^{-3t})x(0) + 1/2(e^{-t} - e^{-3t})y(0)$. Substituting $(1, 1)$ for $(x(0), y(0))$, we see that

$$x(t) = 2e^{-t} - e^{-3t}$$

It is a simple calculation to verify that this is, in fact, a solution satisfying the given initial condition.

It is interesting to note the relationship between the matrix A and the original equation. If we rewrite Eq. (7.12) viewing differentiation as a linear transformation between vector spaces of functions, we see that Eq. (7.12) becomes

$$(d^2/dt^2 + 4\,d/dt + 3)u = 0 \tag{7.14}$$

Let D denote differentiation with respect to t and denote the second derivative by D^2. Then Eq. (7.14) becomes

$$(D^2 + 4D + 3)u = 0 \tag{7.15}$$

The polynomial $D^2 + 4D + 3$ is precisely the characteristic polynomial of A, with D substituted for the variable λ. Thus, the eigenvalues of A can be determined directly from the equation itself.

Let us return to the problem at hand. Consider a general nth-order differential equation of the form (7.11). Let $x_0(t) = u(t)$, $x_1(t) = u^{(1)}(t), \ldots, x_{n-1}(t) = u^{(n-1)}(t)$. Then the functions $x_i(t)$ satisfy the system of equations

$$\begin{array}{rcl}
x_0' &=& x_1 \\
x_1' &=& \quad x_2 \\
x_2' &=& \qquad x_3 \\
\vdots && \qquad\qquad \ddots \\
x_{n-2}' &=& \qquad\qquad\qquad x_{n-1} \\
x_{n-1}' &=& -a_0x_0 - a_1x_1 - a_2x_2 - a_3x_3 - \cdots - a_{n-1}x_{n-1}
\end{array}$$

Solving Eq. (7.11) is equivalent to solving the system $\mathbf{x}' = A\mathbf{x}$ for

$$A = \begin{bmatrix} 0 & 1 & & & & \\ & 0 & 1 & & & \\ & & & \ddots & & \\ & & & & 0 & 1 \\ -a_0 & -a_1 & -a_2 & \cdots & -a_{n-2} & -a_{n-1} \end{bmatrix} \tag{7.16}$$

By a general pattern that can be seen in the previous example, the characteristic polynomial of A, $\det(A - \lambda I)$, is $\pm(\lambda^n + a_{n-1}\lambda^{n-1} + \cdots + a_1\lambda + a_0)$ (see Exercise 1). From the results of Sections 7.2 and 7.3 we have Proposition (7.4.2).

(7.4.2) Proposition. Equation (7.11) is solved by the first entry of $e^{At}\mathbf{b}$, where A is the matrix of Eq. (7.16) and $\mathbf{b} = (u(0), u^{(1)}(0), \ldots, u^{n-1}(0))$.

Proof: We know $x_0(t)$ is the first entry $e^{At}\mathbf{b}$. Since we have set $u(t) = x_0(t)$, $u(t)$ is also the first entry of $e^{At}\mathbf{b}$. ■

The method of solving equations of the form (7.11) suggested by the proposition is fairly reasonable if n is small. Nevertheless, as the following examples will show, the calculations can become quite complicated. For this reason, we will also try to use these examples to deduce and explain an easier method for producing the solutions.

(7.4.3) Examples

a) Consider the equation $u^{(3)}(t) + 3u^{(2)}(t) + 3u^{(1)}(t) + u(t) = 0$. If we convert this to a first-order system of equations as before, we obtain the equations $\mathbf{x}' = A\mathbf{x}$, where A is the 3×3 matrix

$$A = \begin{bmatrix} 0 & 1 & 0 \\ 0 & 0 & 1 \\ -1 & -3 & -3 \end{bmatrix}$$

As we have noted, the characteristic polynomial of A can be read off from the equation: $\det(A - \lambda I) = -(\lambda^3 + 3\lambda^2 + 3\lambda + 1) = -(\lambda + 1)^3$. It is easily checked that $\dim(\mathrm{Ker}(A + I)) = 1$, so the Jordan canonical form of A is a single 3×3 Jordan block, B, with $\lambda = -1$ on the diagonal. Let Q be the invertible matrix such that $A = QBQ^{-1}$ (the exact entries of Q are not important for the conclusion we want to draw from this example, so we will omit the computation of Q).

We know that $e^{At} = Qe^{B^t}Q^{-1}$ from Lemma (7.3.13). Hence

$$e^{At} = Q \cdot \begin{bmatrix} e^{-t} & te^{-t} & (t^2/2)e^{-t} \\ 0 & e^{-t} & te^{-t} \\ 0 & 0 & e^{-t} \end{bmatrix} \cdot Q^{-1}$$

Therefore, from Proposition (7.4.2), we see that the solution of our equation for given initial conditions **b** will be the first component of $e^{At}\mathbf{b}$. From the form of the entries in e^{At} we see that $u(t) \in \text{Span}\{e^{-t}, te^{-t}, t^2e^{-t}\}$. The coefficients in the linear combination will depend on both the initial conditions **b** and the entries of the matrix Q, and could be found in a straightforward way by computations in any particular case.

b) Consider the equation $u^{(4)}(t) - 6u^{(3)}(t) + 12u^{(2)}(t) - 8u^{(1)}(t) = 0$. We will determine the Jordan form of the corresponding matrix and the associated Jordan basis. Given initial conditions for the function u we could then explicitly solve the equation. The corresponding matrix A is

$$A = \begin{bmatrix} 0 & 1 & 0 & 0 \\ 0 & 0 & 1 & 0 \\ 0 & 0 & 0 & 1 \\ 0 & 8 & -12 & 6 \end{bmatrix}$$

The characteristic polynomial of A is $\lambda(\lambda - 2)^3$ and the eigenvalues are 0 and 2 with multiplicities 1 and 3. The kernel of $(A - 2I)$ has dimension one. Thus, the Jordan canonical form of A is

$$J = \begin{bmatrix} 0 & & & \\ & 2 & 1 & \\ & & 2 & 1 \\ & & & 2 \end{bmatrix}$$

To find a Jordan canonical basis for A, we must find a cycle of vectors for the eigenvalue 0 and for the eigenvalue 2. For $\lambda = 0$ we need only produce an eigenvector, and $\mathbf{e}_1$ will suffice. To find a cycle for $\lambda = 2$, we must find vectors $\mathbf{v}_1$, $\mathbf{v}_2$ and $\mathbf{v}_3$ so that $\mathbf{v}_1$ is an eigenvector for $\lambda = 2$, $\mathbf{v}_1 = N\mathbf{v}_2$, and $N\mathbf{v}_3 = \mathbf{v}_2$ for $N = A - 2I$. We have

$$N = \begin{bmatrix} -2 & 1 & & \\ & -2 & 1 & \\ & & -2 & 1 \\ 0 & 8 & -12 & 4 \end{bmatrix} \quad \text{and} \quad N^2 = \begin{bmatrix} 4 & -4 & 1 & 0 \\ 0 & 4 & -4 & 1 \\ 0 & 8 & -8 & 2 \\ 0 & 16 & -16 & 4 \end{bmatrix}$$

A simple calculation shows $\mathbf{v}_1 = (1, 2, 4, 8)$ is an eigenvector for 2. Solving $N^2\mathbf{v}_3 = \mathbf{v}_1$, we see $\mathbf{v}_3 = (1/4, 0, 0, 2)$. Finally, $\mathbf{v}_2 = (-1/2, 0, 2, 8)$. The Jordan canonical basis is $\{(1, 0, 0, 0), (1, 2, 4, 8), (-1/2, 0, 2, 8), (1/4, 0, 0, 2)\}$. The solution is the first component of $Qe^{J^t}Q^{-1}\mathbf{b}$, where Q is the matrix whose columns are the Jordan canonical basis of the matrix A.

As in Example (7.4.3a), we can see what the solution will look like by considering the functions that appear in the exponential of the Jordan form matrix

J. In this case we will have $u(t) \in \text{Span}\{1\ (= e^{0t}), e^{2t}, te^{2t}, t^2e^{2t}\}$. Once again, the coefficients in the linear combination making up a particular solution may be computed from the entries of the change of basis matrix Q and the initial conditions. See Exercise 2.

If A has real entries, but some of the roots of the characteristic polynomial of A are not real, then the exponential functions in the solutions may be rewritten in terms of sines and cosines (see Exercise 5 of Section 7.3).

(7.4.4) Example. Let u satisfy the equation $u^{(2)} + 4u = 0$ subject to $u(0) = 0$ and $u'(0) = 2$.

The corresponding 2×2 matrix is $A = \begin{bmatrix} 0 & 1 \\ -4 & 0 \end{bmatrix}$ whose characteristic polynomial is $\lambda^2 + 4$. The eigenvalues of A are $\pm 2i$. The corresponding eigenvectors are $(1, \pm 2i)$. Therefore, e^{At} is

$$\begin{aligned} e^{At} &= \begin{bmatrix} 1 & 1 \\ 2i & -2i \end{bmatrix} \begin{bmatrix} e^{2it} & 0 \\ 0 & e^{-2it} \end{bmatrix} \begin{bmatrix} 1 & 1 \\ 2i & -2i \end{bmatrix}^{-1} \\ &= \begin{bmatrix} \cos(2t) & (1/2)\sin(2t) \\ -2\sin(2t) & \cos(2t) \end{bmatrix} \end{aligned}$$

(See Exercise 5 of Section 7.3 in order to simplify $e^{\pm 2it}$.) Then $u(t)$ is the first component of $e^{At}\begin{bmatrix} 0 \\ 2 \end{bmatrix}$. That is, $u(t) = \sin(2t)$.

In Section 7.3 we indicated how to sketch a solution to $\mathbf{x}' = A\mathbf{x}$. Suppose we do this for the current example, what does it tell us? The solution $\mathbf{x}(t) = (u(t), u'(t))$ simultaneously gives us information about $u(t)$ and $u'(t)$. For example, when u' changes sign, $u(t)$ reverses direction. This can be seen in the accompanying figure that was obtained by the technique developed in the Exercises of Section 7.3. See Figure 7.7. The ellipse is traversed clockwise and when $t = \pi$, we have completed one revolution.

We will now take the patterns noticed in the previous examples and derive a much easier method for solving equations of the form (7.11). It is this method (rather than the conversion to first-order systems) that is used in practice to solve equations of this kind.

As in Eqs. (7.14) and (7.15), we can rewrite equation (7.11) in the form $p(D)u = 0$, where $p(\lambda) = \lambda^n + a_{n-1}\lambda^{n-1} + \cdots + a_0$. (By Exercise 1, this is also the characteristic polynomial of the matrix of the corresponding first-order system, up to sign.)

(7.4.5) Theorem. Let $\lambda_1, \ldots, \lambda_k$ be the distinct roots of $p(\lambda) = 0$ in $\mathbf{C}$, and assume $p(\lambda)$ factors as

$$p(\lambda) = (\lambda - \lambda_1)^{m_1} \cdots (\lambda - \lambda_k)^{m_k}$$

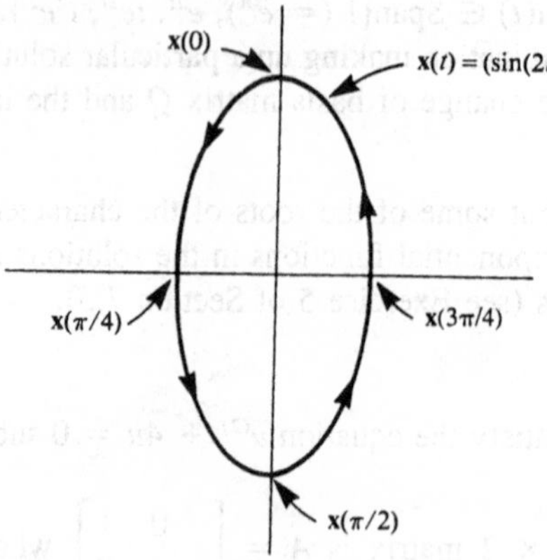

Figure 7.7

Then every solution of the equation $p(D)u = 0$ is in Span$\{e^{\lambda_1 t}, te^{\lambda_1 t}, \ldots, t^{m_1-1}e^{\lambda_1 t}, \ldots, e^{\lambda_k t}, te^{\lambda_k t}, \ldots, t^{m_k-1}e^{\lambda_k t}\}$. Moreover, every function in this set is a solution of the equation $p(D)u = 0$.

Proof: The first statement follows from Proposition (7.4.2) and the computation of e^{Jt} for a Jordan form matrix J from Section 7.3. [See also Examples (7.4.3)]. The scalars in the linear combination making up any particular solution may be computed if we specify initial conditions $u^{(i)}(0) = b_i$ for i, $0 \leqslant i \leqslant n-1$. The last statement follows by a direct computation (Exercise 4). ■

For explicitly solving equations of the form (7.11), this result is clearly far superior to the ad hoc method we used in Examples (7.4.3), since it requires far less work.

(7.4.6) Example. Solve $u^{(3)} + 2u^{(2)} + u^{(1)} = 0$ subject to $u^{(2)}(0) = 1$, $u^{(1)}(0) = -1$, and $u(0) = 0$. Using Theorem (7.4.5), we first write the equation in the form $(D^3 + 2D^2 + D)u = 0$. Hence, $p(\lambda) = \lambda^3 + 2\lambda^2 + \lambda$ in this case. Factoring, we find $p(\lambda) = \lambda(\lambda + 1)^2$. Hence, by the theorem our solution will have the form

$$u(t) = c_{10}e^{0t} + (c_{20} + c_{21}t)e^{-t}$$

for some scalars c_{ij}. These coefficients can be computed from the initial conditions as follows. We have

$$u^{(1)}(t) = (-c_{20} + c_{21})e^{-t} - c_{21}te^{-t}$$

$$u^{(2)}(t) = (c_{20} - 2c_{21})e^{-t} + c_{21}te^{-t}$$

Hence, the c_{il} may be found by substituting $t = 0$ in the three equations and solving the resulting system of linear equations. The final result is

$$c_{10} = -1, \qquad c_{20} = 1, \qquad c_{21} = 1$$

Therefore, our solution is $u(t) = -1 + e^{-t}$.

We conclude this section with one final observation about the equations of the form (7.11). Namely, we can also study these equations from a different point of view. If $p(\lambda)$ is any polynomial, we can view $L = p(D)$ as a linear mapping $L: V \rightarrow V$, where V is the vector space of functions $u: \mathbf{R} \rightarrow \mathbf{C}$ with derivatives of all orders. The solutions of the differential equation $L(u) = p(D)u = 0$ are precisely the functions in Ker(L). By Theorem (7.4.5) we have the following interesting pattern.

(7.4.7) Corollary. Let $p(\lambda)$ be a polynomial of degree n. Then the kernel of the linear mapping $L = p(D)$ has dimension n in V.

Proof: There are $m_1 + \cdots + m_k = n$ functions in the spanning set for Ker(L) given in the theorem. Furthermore, it is easy to show these functions are linearly independent (Exercise 3). Therefore, dim(Ker(L)) $= n$. ■

We may now ask for a description of the image of L, for a description of the set $L^{-1}(v)$ for any $v \in$ Im(L), and so forth. In general, can we apply the machinery we developed in previous chapters to $L: V \rightarrow V$? These questions are a point of departure for a much broader study of linear algebra and analysis. In the next section we will explore a few of these questions in a particular case.

EXERCISES

1. Let A be the matrix of equation (7.16). Prove that the characteristic polynomial of A is $(-1)^n(\lambda^n + a_{n-1}\lambda^{n-1} + \cdots + a_1\lambda + a_0)$, where n is the size of the matrix.

2. Complete the computations of Example (7.4.3b) by solving the given equation with the initial conditions $u^{(3)}(0) = 1$, $u^{(2)}(0) = -2$, $u^{(1)}(0) = 0$, and $u(0) = 0$.

3. Show that for any distinct $\lambda_1, \ldots, \lambda_k \in \mathbf{C}$ and any positive integers $m_1, \ldots, m_k$, $\{e^{\lambda_1 t}, te^{\lambda_1 t}, \ldots, t^{m_1-1}e^{\lambda_1 t}, \ldots, e^{\lambda_k t}, te^{\lambda_k t}, \ldots, t^{m_k-1}e^{\lambda_k t}\}$ is a linearly independent set of functions.

4. a) Show that if l is an integer and $0 \leq l \leq m - 1$, then $(D - \lambda)^m(t^l e^{\lambda t}) = 0$.
   b) Conclude that if $p(D) = (D - \lambda_1)^{m_1} \cdot \cdots \cdot (D - \lambda_k)^{m_k}$, then any $u \in$ Span$\{e^{\lambda_1 t}, te^{\lambda_1 t}, \ldots, t^{m_1-1}e^{\lambda_1 t}, \ldots, e^{\lambda_k t}, te^{\lambda_k t}, \ldots, t^{m_k-1}e^{\lambda_k t}\}$ solves $p(D)u = 0$.

5. Solve the following equations subject to the given initial conditions: using Theorem (7.4.5)
   a) $u^{(2)} + 4u^{(1)} + 4u = 0, \quad u(0) = 0, u^{(1)}(0) = 1$
   b) $u^{(4)} + 6u^{(2)} + 8u = 0, \quad u(0) = 1, u^{(1)}(0) = 0, u^{(2)}(0) = 0, \quad u^{(3)}(0) = 0$
   c) $u^{(3)} - u = 0, u(0) = 1, u^{(1)}(0) = 1, u^{(2)}(0) = 0$
   d) $u^{(2)} + iu = 0, u(0) = i, u^{(1)}(0) = 1$

6. Consider a second-order equation $u^{(2)} + bu^{(1)} + cu = 0$. Determine the values of b and c and the corresponding initial conditions on $u(0)$, $u'(0)$ so that
   a) $\lim_{t\to\infty} | u(t) | = +\infty$
   b) $\lim_{t\to\infty} u(t) = 0$
   c) There exists a constant k so that $|u(t)| \leqslant k$ for all t.

7. Solve the equations for a mass suspended from a spring of Section 7.1.
   a) $mx'' + kx = 0$, subject to $x(0) = 0, x'(0) = 1$
   b) $mx'' + rx' + kx = 0, \quad m = 2, r = 1, x(0) = 0, x'(0) = 1$

8. In this problem we will investigate some more general equations similar to those of the form (7.11). Consider the equation $u^{(1)}(t) + au(t) = w(t)$. This is an example of an inhomogeneous first-order equation if $w \neq 0$.
   a) Show that the associated homogenous equation $u^{(1)}(t) + au(t) = 0$ is solved by any constant multiple of e^{-at}.
   b) Prove that if the solution $u(t)$ to the inhomogeneous equation is written in the form $u(t) = v(t)e^{-at}$ (a nonconstant multiple of e^{-at}), then $v(t)$ must satisfy $v' = w(t)e^{at}$.
   c) Show that $u(t) = ce^{-at} + e^{-at} \int w(t)e^{at}dt$ is a solution of the equation, where c is an arbitrary constant of integration.
   d) Use integration by parts to show that the integral $\int w(t)e^{at}dt$ can be evaluated in closed form (i.e., there is an antiderivative in the usual calculus sense) if $w(t)$ is a function of the form $w(t) = p(t)e^{bt}$, where $p(t)$ is a polynomial in t and b is a constant.

9. In this exercise we outline an alternate, inductive process that is sometimes used to solve equations of the form (7.11). We write Eq. (7.11) as $p(D)u = 0$ [see Eqs. (7.14) and (7.15) and Exercise 5]. Assume that $p(D)$ factors as $p(D) = (D - \lambda_n) \cdots (D - \lambda_1)$ so that

$$p(D)u = (D - \lambda_n) \cdots (D - \lambda_2)(D - \lambda_1)u = 0$$

   a) Show that solving the inhomogeneous equation $p(D)u = w$ is equivalent to solving the two equations $(D - \lambda_n)v = w$ and $(D - \lambda_{n-1})\cdot \cdots \cdot(D - \lambda_1)u = v$.
   b) Show how Exercise 8 can be used to solve the first equation in part a.
   c) Then conclude by induction that the second equation in part a can be solved for u once v has been computed.

10. Use the inductive procedure of Exercise 9 to solve the equations of Exercise

5. [Is this method any easier than the method discussed in the text, derived from Theorem (7.4.5)?]

11. Consider any differential equation of the form

$$L(u) = (D^n + c_{n-1}(t)D^{n-1} + \cdots + c_0(t))u(t) = v(t)$$

and the associated homogeneous equation

$$L(u) = 0$$

(These are called *linear differential equations.*)
a) Show that L defines a linear mapping on the vector space V of functions u: $\mathbf{R} \to \mathbf{C}$ with derivatives of all orders.
b) Show that the solutions of the homogeneous equation $L(u) = 0$ are precisely the elements of $\mathrm{Ker}(L) \subset V$.
c) Show that any two solutions of the inhomogeneous equation $L(u) = v$ differ by an element of $\mathrm{Ker}(L)$.
d) Conclude that if $v \in \mathrm{Im}(L)$, u_p is any particular solution, that is $L(u_p) = v$, then every solution of the equation $L(u) = v$ has the form $u = u_p + u_h$, where $u_h \in \mathrm{Ker}(L)$. [*Hint:* See Proposition (2.4.11).]

12. The inhomogeneous equations considered in Exercise 8 are special cases of the linear equations that were studied in Exercise 11. The general first-order linear equation has the form

$$u^{(1)}(t) + p(t)u(t) = q(t) \tag{7.17}$$

where p and q are given functions of t. As in Exercise 11, the equation

$$u^{(1)}(t) + p(t)u(t) = 0 \tag{7.18}$$

is known as the homogeneous equation associated to (7.17).
a) Show that the homogeneous equation (7.18) may always be solved by separating variables, yielding the solution $u_h(t) = ce^{-\int p(t)dt}$, where c is an arbitrary constant of integration.
b) Show that $u_p(t) = e^{-\int p(t)dt} \cdot \int e^{\int p(t)dt} q(t)dt$ is a particular solution of the inhomogeneous equation (7.17).
c) Conclude that the general solution of (7.17) has the form $u_p(t) + u_h(t)$.

13. Using the technique developed in Exercise 12
a) Solve the equation $u^{(1)}(t) + tu(t) = t, \quad u(0) = 2$.
b) Solve the equation $u^{(1)}(t) + (1/t)u(t) = \sin(t)$, $u(\pi) = 0$.

The following exercises continue the study of difference equations begun in Sections 7.2 and 7.3. There are also higher-order difference equations. A kth-order difference equation is an equation on $u: N \to \mathbf{C}$ relating the values of $u(n)$, $u(n + 1), \ldots,$ $u(n + k)$ for all $n \in N$. For example, consider an equation of the form

$$u(n + k) + a_{k-1}u(n + k - 1) + \cdots$$
$$+ a_1u(n + 1) + a_0u(n) = 0 \quad (7.19)$$

where the $a_i \in \mathbf{C}$ are constant.

14. Show that a kth-order difference equation of the form (7.19) is equivalent to a system of k first-order difference equations of the kind we studied in Sections 7.2 and 7.3.

15. Formulate and prove a result parallel to Theorem (7.4.5) giving the general form for a solution of (7.19). (*Hint:* Use the Jordan form of the matrix of the equivalent first-order system and the result of Exercise 16 of Section 7.3).

§7.5. AN EIGENVALUE PROBLEM

In the previous section, following Example (7.4.6), we indicated how to rewrite a single ordinary differential equation as a linear transformation on a vector space of functions. In this section we pursue this idea further and discuss the eigenvalues and eigenfunctions of such a linear transformation. We begin with a partial differential equation and then reduce the problem of finding a solution to finding the solution of a related ordinary differential equation.

An eigenvalue problem for a linear transformation acting on functions arises from attempting to describe the motion of a vibrating string. Suppose we have a thin flexible string that is anchored at the origin and the point $(\pi, 0)$ in the x-y plane. Label a point on the string (x, y). See Figure 7.8.

Assume that there is exactly one point on the string with a given x-value for $0 \leq x \leq \pi$. Thus, for each time t, y is the graph of a function of x. We write $y = y(x, t)$. By analyzing the physics of the moving string (see, for example, Ritger and Rose, *Differential Equations with Applications* [New York: McGraw-Hill, 1968], Section 14-2), it can be shown that $y(x, t)$ satisfies

$$\frac{\partial^2 y}{\partial x^2} = c^{-2}\frac{\partial^2 y}{\partial t^2} \quad (7.20)$$

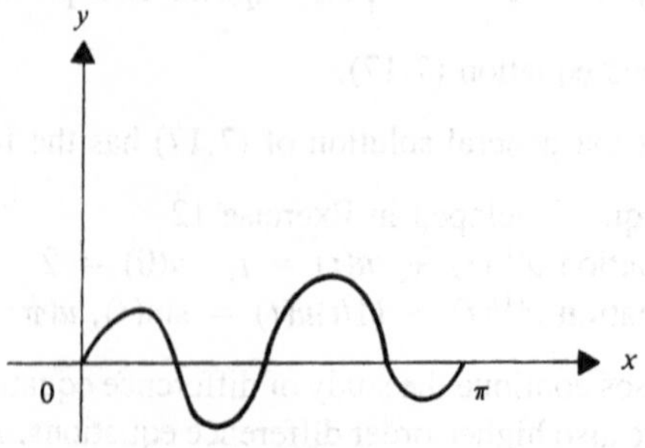

Figure 7.8

where c is a constant depending on the length, mass, and tension of the string. Equation (7.20) is called the one-dimensional wave equation (it has one spatial derivative $\partial^2/\partial x^2$). This is a second-order partial differential equation.

Any function y satisfying this equation should also satisfy $y(0, t) = y(\pi, t) = 0$ if y is to be a model for the vibrating string. Further, we might assume that we have "plucked" the string initially so that $y(x, 0) = f(x)$ for some function f, and the string is not moving until it is released: $\partial y/\partial t(x, 0) = 0$.

A first approach to a solution is to guess that the string vibrates by simple harmonic motion. That is, given the initial position of any point on the string $f(x)$, its resulting motion is described by $y(x, t) = f(x) \cdot \cos(\omega t)$, $\omega \in \mathbf{R}$. Notice that $y(x, 0) = f(x)$ and $\partial y/\partial t(x, 0) = -\omega f(x) \sin(\omega \cdot 0) = 0$. This initial guess leads us to consider solutions that are products of a function of x and a function of t:

$$y(x, t) = f(x)h(t)$$

Then Eq. (7.20) implies

$$f''(x)h(t) = 1/c^2\, f(x)h''(t)$$

or

$$f''(x)/f(x) = 1/c^2\, h''(t)/h(t)$$

Since the left-hand side depends only on x and the right-hand side depends only on t, both sides must equal a constant, which we write as $-k$. Then we can *separate variables* and solve

$$\frac{f''(x)}{f(x)} = -k \quad \text{and} \quad \frac{h''(t)}{h(t)} = -kc^2$$

independently. In this section we concentrate on solving the first equation. The second equation can be solved using the same techniques. In finding such f, we develop several new ideas.

First, rewrite the equation as $f''(x) = -kf(x)$. If we denote d^2/dx^2 by L, the equation reads

$$L(f) = -kf, \qquad f(0) = f(\pi) = 0 \tag{7.21}$$

L is a linear transformation on a vector space of functions that we will specify subsequently. To solve Eq. (7.21) then, we are looking for both a function f and a scalar k. That is, we must solve the eigenvalue-eigenvector problem for L!

Let V be the vector space of infinitely differentiable functions $f(x)$ on the interval $[0, \pi] \subset \mathbf{R}$, which take values in the complex numbers $\mathbf{C}$. [See Example (5.2.5c).] By infinitely differentiable we mean that f has continuous derivatives of arbitrarily high order. Since f: $[0, \pi] \to \mathbf{C}$, we may write $f(x) = u(x) + iv(x)$, where u and v are real-valued functions u, v: $[0, \pi] \to \mathbf{R}$. The differentiability of f is equivalent to the differentiability of u and v. From the point of view of analysis it is preferable to replace V by a larger vector space called a completion of V. A rigorous discussion of this larger vector space is beyond the realm of linear algebra. We, therefore, will proceed in a somewhat ad hoc manner.

Rather than use the results of this chapter directly, we would like to use the concepts of Chapter 4 and Section 5.3 to simplify our search for the eigenvalues $-k$ of L. In these chapters we developed the theory of real and complex vector spaces with inner products. The major result was the spectral theorem that not only said that a symmetric or self-adjoint transformation is diagonalizable, but also that the eigenvalues are real and the eigenvectors may be chosen to be orthonormal. It turns out that many of these properties are true for L as well. To apply these ideas to L, we first must introduce an inner product on the vector space V defined earlier:

(7.5.1) Definition. Let $f, g \in V$. Define the *Hermitian inner product* of f and g by

$$\langle f, g\rangle = \int_0^\pi f(x)\overline{g(x)}\, dx.$$

To evaluate such an integral of complex-valued functions, we can separate the integral into real and imaginary parts. If $f(x) = u_1(x) + iv_1(x)$ and $g(x) = u_2(x) + iv_2(x)$, then

$$\langle f, g\rangle = \int_0^\pi (u_1(x) + iv_1(x))(u_2(x) - iv_2(x))\, dx$$

$$= \int_0^\pi (u_1(x)u_2(x) + v_1(x)v_2(x))\, dx$$

$$+ i\int_0^\pi (v_1(x)u_2(x) - v_2(x)u_1(x))\, dx$$

Notice since f and g are differentiable, they are integrable and all the usual rules of integration apply, for example, the integral of a sum is the sum of the integrals. Each of the resulting two integrals is an integral of a real-valued function so that the usual techniques of evaluating integrals of real-valued functions can be used.

Of course, we must show that $\langle\ ,\ \rangle$ is, in fact, an inner product.

(7.5.2) Proposition. The complex-valued function on pairs of functions in V, $\langle f, g\rangle = \int_0^\pi f\overline{g}\, dx$ is a Hermitian inner product.

Proof: Exercise 2. ■

(7.5.3) Example. Let $f(x) = e^{i\alpha x}$ and $g(x) = e^{i\beta x}$ with $\alpha \neq \beta$ real numbers. Then

$$\langle f, g\rangle = \int_0^\pi e^{i\alpha x}\, \overline{e^{i\beta x}}\, dx$$

$$= \int_0^\pi e^{i\alpha x}\, e^{-i\beta x}\, dx$$

$$= \int_0^\pi e^{i(\alpha-\beta)x}\, dx$$

$$= (i(\alpha - \beta))^{-1} e^{i(\alpha-\beta)x} \Big|_0^\pi$$

$$= -i(\alpha - \beta)^{-1} (e^{i(\alpha-\beta)\pi} - 1)$$

Let us analyze this further. If $\alpha - \beta$ is an even integer, then $e^{i(\alpha-\beta)\pi} = 1$ and $\langle f, g\rangle = 0$; that is, f and g are orthogonal. Now consider $\langle f, f\rangle = \int_0^\pi e^{i\alpha t}\overline{e^{i\alpha t}}\, dt = \int_0^\pi dt = \pi$. Thus, $\pi^{-1/2} e^{i\alpha t}$ is also a unit vector. Already we can see the makings of an orthonormal collection of functions. Momentarily we will see how to choose the α.

To apply the results of Chapters 4 and 5, we must have a symmetric or self-adjoint transformation. Recall the following definition.

(7.5.4) Definition. Let $L: V \to V$ be a linear transformation of a complex vector space with Hermitian inner product $\langle\ ,\ \rangle$. L is said to be *self-adjoint* or *Hermitian* if for all f and $g \in V$, $\langle Lf, g\rangle = \langle f, Lg\rangle$.

To show that our transformation $L = d^2/dx^2$ is Hermitian, we have to impose a restriction on the functions f and g. Let $W \subset V$ be the subspace of V consisting of those f such that $f(0) = f(\pi) = 0$. This is the side condition of Equation (7.21), which amounts to the string being attached at $x = 0$ and $x = \pi$. (See Exercise 3.) We can then show Proposition (7.5.5)

(7.5.5) Proposition. $L: W \to V$ is a Hermitian linear transformation, that is, for f and g in W, $\langle Lf, g\rangle = \langle f, Lg\rangle$.

Proof: Expand the left-hand side, $\langle Lf, g\rangle = \int_0^\pi f''\overline{g}\, dx$, and apply integration by parts:

$$\langle Lf, g\rangle = f'\overline{g}\,\Big|_0^\pi - \int_0^\pi f'\overline{g}'\, dx$$

Since $g \in W$ the first term is zero. Applying parts one more time, we obtain

$$-\int_0^\pi f'\overline{g}'\, dx = -\left[f\overline{g}'\,\Big|_0^\pi - \int_0^\pi f\overline{g}''\, dx\right]$$

Since $f \in W$ the first term is zero, and we are left with $\int_0^\pi f\overline{g}''\, dx = \langle f, Lg\rangle$. Therefore, L is Hermitian. ■

Since L is Hermitian, its eigenvalues are all real [see Proposition (5.3.10) of Chapter 5—the same proof applies here].

Returning to our original problem, we are looking for an f that satisfies

$$Lf = -kf,\ f \in W, \quad \text{and} \quad k \in \mathbf{R},\ k \neq 0 \tag{7.22}$$

Using the techniques of Section 7.4 (see Exercise 4), we can easily see that f is a linear combination of $e^{\pm i\sqrt{k}x}$. Can we choose k so that such a linear combination is in W?

If $c_1e^{i\sqrt{k}x} + c_2e^{-i\sqrt{k}x} \in W$, the "boundary" conditions that $f(0) = f(\pi) = 0$ imply that

$$c_1 + \qquad c_2 = 0$$

$$e^{i\pi\sqrt{k}}c_1 + e^{-i\pi\sqrt{k}}c_2 = 0$$

Therefore, (c_1, c_2) is the kernel of the matrix

$$\begin{bmatrix} 1 & 1 \\ e^{i\pi\sqrt{k}} & e^{-i\pi\sqrt{k}} \end{bmatrix}$$

The matrix has a non-zero kernel if and only if its determinant, $e^{-i\pi\sqrt{k}} - e^{i\pi\sqrt{k}} = 0$, or $e^{-i\pi\sqrt{k}} = e^{i\pi\sqrt{k}}$. This is true if and only if $\sqrt{k}$ is an integer (why?). In this case if we choose $c_1 = 1$ and $c_2 = -1$, the corresponding f will be an eigenvector. Notice that

$$f(x) = e^{i\sqrt{k}x} - e^{-i\sqrt{k}x} = 2i \cdot \sin(\sqrt{k}x)$$

Since $\sqrt{k}$ is an integer, k is the square of an integer. Replace k by m^2, $m \in \mathbf{Z}$, $m > 0$. Thus, we can write $f(x) = 2i \cdot \sin(mx)$ for the eigenfunction corresponding to m. As in Chapter 4, we can replace $f(x)$ by a "unit" eigenfunction g; that is, $\langle g, g\rangle = 1$. It is an easy calculation to show that $\langle f, f\rangle = 2\pi$. It follows that $g_m(x) = \sqrt{2/\pi}\,\sin(mx)$ is an eigenfunction of unit length for the eigenvalue $-m^2$. [Notice that since $|-i| = 1$, we may multiply the unit eigenfunction $i\sqrt{2/\pi}\cdot\sin(mx)$ by $-i$ to obtain the unit eigenfunction $\sqrt{2/\pi}\cdot\sin(mx)$, which has the same span. Compare with Example (5.3.4).] From Proposition (5.3.11) we know that the eigenfunctions for distinct eigenvalues must be orthogonal; that is, $\langle g_m, g_n\rangle = 0$ for $m \neq n$. [This could also be shown directly using Example (7.5.3).] Thus, we have shown Proposition (7.5.6).

(7.5.6) Proposition. The linear transformation $L: W \to V$ defined by $L(f) = d^2f/dx^2$ is a Hermitian linear transformation with eigenvalues $-m^2$, $m \in \mathbf{Z}$, and corresponding orthonormal eigenfunctions $g_m(x) = \sqrt{2/\pi}\cdot\sin(mx)$.

The next step in the program would be to show that $\alpha = \{g_1(x), g_2(x), \ldots\}$ forms a basis for the subspace W of V. Consequently, L would be diagonalizable. Here again we run into a problem in analysis disguised as linear algebra, whose solution is beyond the scope of this course. Briefly, the difficulty is this: Not every element of W is a finite linear combination of the eigenfunctions in α. More generally, there is no infinite list $\alpha = \{f_1(x), f_2(x), \ldots\}$ with the property that every element of V is a finite linear combination of elements of α. By now the reader should have noticed our emphasis on *finite* linear combinations. It is natural then to allow infinite linear combinations of elements α. In fact, this is the correct line of approach providing one is very careful about which infinite linear combinations are allowed. The correct notion is that if α is an orthonormal set of functions, that is, $\langle f_i, f_j\rangle = 0$ if $i \neq j$ and $\langle f_i, f_i\rangle = 1$, then the linear combination

$$a_1 f_1 + a_2 f_2 + \cdots, \quad a_i \in \mathbf{C}$$

is permitted if the sequence of finite linear combinations

$$s_k = a_1 f_1 + \cdots + a_k f_k, \quad k = 1, 2, \ldots, \infty$$

converges to a well-defined limit. This is where linear algebra and analysis merge and we are forced to leave our discussion incomplete.

Before we illustrate these ideas with a simple example, notice that another problem has arisen while we sketched the solution to the first. Not every convergent sequence $\{s_k\}$, $k = 1, \ldots, \infty$ converges to an element of W! Thus, in order to write every element of W as a linear combination of our proposed basis, we were forced to introduce infinite linear combinations of the basis that in turn will force us to expand W.

(7.5.7) Example. Consider the function $f(x) = \pi x - x^2$. f is an element of W since $f(0) = f(\pi) = 0$. Is f a linear combination of the eigenfunctions g_m of the Hermitian transformation L? If there are coefficients a_m so that $f = a_1 g_1 + a_2 g_2 + \cdots$, then $\langle f, g_m\rangle = a_m$ since the g_m form an orthonormal set. This inner product can be evaluated directly.

$$\begin{aligned}\langle f, g_m\rangle &= \sqrt{2/\pi}\int_0^\pi (\pi x - x^2)\cdot \sin(mx)\,dx \\ &= \sqrt{2/\pi}\cdot (2/m^3)(1 - \cos(m\pi))\end{aligned} \tag{7.23}$$

Notice, if m is even $\cos(m\pi) = 1$, so that $\langle f, g_{2n}\rangle = 0$. If m is odd, $m = (2n + 1)$, $\cos((2n + 1)\pi) = -1$, so that $\langle f, g_{2n+1}\rangle = \sqrt{2/\pi}\, 4/(2n + 1)^3$. This leads us to claim, without proof, that

$$f(x) = (4\sqrt{2/\pi}) \sum_{n=0}^{\infty} 1/(2n + 1)^3\, g_{2n+1}(x)$$

that is, f can be expressed as an infinite linear combination of the eigenfunctions. This series expansion of f is called a Fourier expansion of f and the coefficients are called the Fourier coefficients of f.

Returning to the original string problem, it can be shown that for $k = m^2$ the equation

$$-m^2 = 1/c^2\, h''(t)/h(t)$$

or equivalently

$$h'' = -m^2 c^2 h$$

subject to $h(0) = 1$, $h'(0) = 0$ (why?) is solved by $h_m(t) = \cos(mct)$.

Our conclusion is that if the initial shape of the string is given by $f(x) = g_m(x)$, then each point on the string *is* undergoing simple harmonic motion [although with different amplitudes at different points, determined by the value of $f(x)$ at time $t = 0$] and this motion is described by

$$y(x, t) = \sqrt{2/\pi}\,\sin(mx)\cos(mct)$$

This solution is called the mth normal mode of vibration of the string. In Figure 7.9 we graph the first three modes of a string for $t = 0$. The solution $y(x, t)$ for a fixed choice of m has the initial shape $y(x, 0) = \sqrt{2/\pi}\,\sin(mx)$.

If the initial shape of the string $y(x,0) = f(x)$ is a general continuous function, the solution to the wave equation can be expressed as an "infinite linear combination" or Fourier series in the normal modes. It is shown in analysis that if $f(x)$ is a continuous function satisfying $f(\pi) = f(0)$, then f can be expressed as a series in the eigenfunctions $g_m(x)$: $f(x) = \sum_{m=0}^{\infty} a_m g_m(x)$ with $a_m \in \mathbf{R}$. We can construct the corresponding solution $y(x, t)$ directly by letting $y(x, t)$ be the series in the normal modes with the same coefficients:

$$y(x, t) = \sum_{m=0}^{\infty} a_m g_m(x) h_m(t) \tag{7.24}$$

To show that this $y(x, t)$ satisfies the wave equation, we must differentiate $y(x, t)$ and compute $\partial^2 y/\partial x^2$ and $\partial^2 y/\partial t^2$. In analysis, we prove that series of this form may be differentiated term by term, so that, for example

$$\partial y/\partial x = \sum_{m=0}^{\infty} \frac{\partial}{\partial x}\left[a_m g_m(x) h_m(t)\right]$$

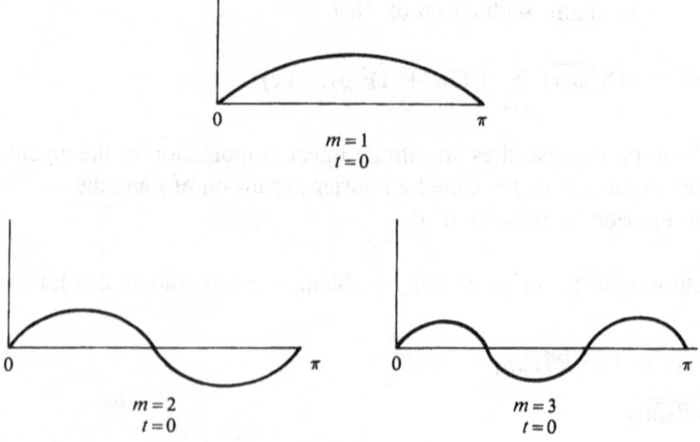

Figure 7.9

In Exercise 7 you will use this fact to show that every $y(x, t)$ as in Eq. (7.24) is a solution of the wave equation.

For example, if $f(x) = \pi x - x^2$, the function of Example (7.5.7), then

$$y(x, t) = (4\sqrt{2/\pi}) \sum_{n=0}^{\infty} (2n + 1)^{-3} g_{2n+1}(x) h_{2n+1}(t)$$

Thus, providing that certain analytic problems can be solved, such as showing that the eigenfunctions span the vector space of possible initial shapes and showing that all the series in question converge, we have constructed the general solution to the wave equation.

EXERCISES

1. Use the real and imaginary parts of $e^{i\alpha t}$ and $e^{i\beta t}$ to evaluate the integral of Example (7.5.3).
2. Prove Proposition (7.5.2). (Remember that a Hermitian inner product is conjugate symmetric, $\langle f, g\rangle = \overline{\langle g, f\rangle}$, complex linear in the first function, and positive-definite.)
3. Show that $W = \{f \in V: f(0) = f(\pi) = 0\}$ is a vector subspace of the space of differentiable functions $f: [0,\pi] \to \mathbf{C}$.
4. Use the techniques of Section 7.4 to show that $Lf = f'' = -kf$ has a solution of the form $f(t) = c_1 e^{i\sqrt{k}t} + c_2 e^{-i\sqrt{k}t}$.
5. Verify equation (7.23), $\langle f, g_m\rangle = \sqrt{2/\pi} \cdot (2/m^3)(1 - \cos(m\pi))$.
6. Show that $h'' = -m^2c^2h$, $h(0) = 1$, $h'(0) = 0$ is solved by $h_m(t) = \cos(mct)$.
7. Prove that if $f(x) = \sum_{m=0}^{\infty} a_m g_m(x)$, then $y(x, t) = \sum_{m=0}^{\infty} a_m g_m(x) h_m(t)$ is a solution to the wave equation with $y(x,0) = f(x)$, by differentiating the series term by term (g_m and h_m are defined as in the text).
8. How do the results of this section change if the interval $[0, \pi]$ is replaced by an interval of the form $[0. L]$ for some $L > 0$? of the form $[-L, L]$?
9. Let V be the vector space of differentiable functions $f: [0, \pi] \to \mathbf{C}$. Define $T: V \to V$ by $T(f) = x \cdot f(x)$.
   a) Prove that T is Hermitian.
   b) Can you find eigenvalues and eigenvectors for T?
10. Let W be the vector space of differentiable functions $f{:}[0, \pi] \to \mathbf{C}$ with $f(0) = f(\pi) = 0$. Define $T: W \to V$ (V defined as in Exercise 9) by $T(f)(x) = i\partial f/\partial x$.

a) Prove that T is Hermitian.
b) Find the eigenvalues and eigenvectors of T. Are the eigenvectors orthogonal? (Are they even in W?)

11. a) Repeat the calculations of Example (7.5.7) for the string initial position given by

$$u(x, 0) = f(x) = \begin{cases} x/\pi & \text{if } 0 \leq x \leq \pi/2 \\ (1 - x/\pi) & \text{if } \pi/2 \leq x \leq \pi \end{cases}$$

(Note that the graph of this function gives the shape of a string plucked at its midpoint.)
b) Find the solution $u(x, t)$ to the wave equation that satisfies the initial condition given in part a.

12. In the physics of heat conduction it is shown that if a thin rod of a homogeneous material is placed in a surrounding medium of constant temperature, and the temperature $u(x, t)$ of the rod at position $x \in [0, \pi]$ and time $t = 0$ is given by some initial temperature function $u(x, 0) = T(x)$, then as time goes on the temperature $u(x, t)$ satisfies the *heat equation*: $\partial u/\partial t = a \cdot \partial^2 u/\partial x^2$, where a is a constant depending on the thermal properties of the material of the rod.
a) Show that if $u(x, t) = f(x)h(t)$ solves the heat equation, then $f''(x) + kf(x) = 0$ and $h'(t) + akh(t) = 0$ for some constant k. (*Hint:* Use the same separation of variables idea used in the text for the wave equation.)
b) Assume $u(0, t) = u(\pi, t) = 0$ for all t (blocks of ice are placed at the ends of the rod). Then show that the constant $k = m^2$ for some integer m.
c) Deduce from a) and b) that $u(x, t) = e^{-am^2t} \sin(mx)$ is a solution of the heat equation subject to the boundary conditions $u(0, t) = u(\pi, t) = 0$. (*Hint:* Solve the two ordinary differential equations in part a.) These solutions are somewhat analogous to the normal modes of vibration of a string.

CHAPTER SUMMARY

We considered three different types of linear differential equations: *systems of constant coefficient first-order linear ordinary differential equations*, a single *constant coefficient higher-order linear ordinary differential equation*, and a *second-order linear partial differential equation*.

The systems of constant coefficient equations were expressed in matrix form by $\mathbf{u}' = A\mathbf{u}$, with initial condition $\mathbf{u}(0) = \mathbf{u}_0$, where A is an $n \times n$ real or complex matrix and $\mathbf{u} = \mathbf{u}(t)$ is a function $\mathbf{u}: \mathbf{R} \to \mathbf{R}^n$ (or $\mathbf{C}^n$). In the special case that A is diagonalizable, we used the corresponding basis of eigenvectors to produce a solution [Proposition (7.2.3)]. This particular form of the solution is advantageous both because it is conceptually clear to understand and because it yields information about the limiting values of solutions as $t \to \pm\infty$.

In the general case, the solution to $\mathbf{u}' = A\mathbf{u}$ is considerably more intricate. Formally, we saw that the solution is always given by $\mathbf{u}(t) = e^{At}\mathbf{u}_0$. Although this statement is quite concise, we necessarily had to develop several new concepts and methods—the *exponential of a matrix* [Definition (7.3.1)], the *norm of a matrix* [Definition (7.3.3)], and the subsequent propositions implying convergence—in order to make sense of this statement. But with this preparatory work in hand, we were able to express e^{At} directly in terms of the *Jordan form of* A and the *Jordan canonical basis for* A.

We applied our method of producing solutions to $\mathbf{u}' = A\mathbf{u}$ to solve a single higher-order constant coefficient linear differential equation. Given the single equation, we saw how to *reduce the order of the equation* to first order and simultaneously increase the number of variables. Using the coefficients of the single equation, we produced the corresponding A [Equation (7.16)]. Proposition (7.4.2) gave the solution to the original equation. We then saw how to construct a solution directly without passing to the system of equations [Theorem (7.4.5)]. It is sufficient to be able to factor the polynomial $p(\lambda)$ in order to solve $P(D)u = 0$.

In Section 7.5 we concentrated on a single second-order partial differential equation, the *wave equation*. It is hard to overemphasize the fundamental importance of this equation. Our world abounds with wavelike phenomena, from water waves to sound waves to electromagnetic waves. Although it is possible to construct solutions in a more direct manner, our emphasis on *linear transformations of an infinite-dimensional vector space* allowed us to concentrate on the relationship between *eigenvectors* and the solutions to the wave equation. Our first step was to *separate variables,* treating the space variable x and the time variable t independently. Working with the x variable alone, we obtained a *self-adjoint linear transformation* $L(f) = f''$ acting on an infinite dimensional *Hermitian inner product space*. We produced a collection of eigenvectors for this transformation. Each of the eigenvectors corresponded to an individual solution to the wave equation for a different initial condition. These solutions are called the *normal modes* of the string. Finally, we indicated how to produce a solution to the wave equation given any continuous function as the initial data. This solution was a *linear combination of the normal modes*.

SUPPLEMENTARY EXERCISES

1. **True-False.** For each true statement give a brief proof. For each false statement give a counterexample.
   a) If a solution to $\mathbf{u}' = A\mathbf{u}$, satisfies $\mathbf{u}(2\pi) = a\mathbf{u}(0)$ for $a \in \mathbf{R}$, $a \neq 0$, then the initial condition $\mathbf{u}(0) = \mathbf{u}_0$ yields a straight-line solution.
   b) If $\mathbf{u}(t)$ is a straight-line solution to $\mathbf{u}' = A\mathbf{u}$, then $\mathbf{u}(0)$ is an eigenvector of A.
   c) Let $\mathbf{u}(t)$ be a solution to $\mathbf{u}' = A\mathbf{u}$. If λ and μ are distinct eigenvalues of A and $\mathbf{u}(0) \in E_\lambda + E_\mu$, then $\mathbf{u}(t) \subset E_\lambda + E_\mu$.
   d) Let A and B be $n \times n$ matrices, then $e^{A+B} = e^A e^B$.
   e) The matrix e^{At} is invertible for any $n \times n$ matrix A.
   f) If A is a real $n \times n$ matrix, then e^{At} is a real $n \times n$ matrix for any $t \in \mathbf{R}$ whether or not the eigenvalues of A are real numbers.

g) If the polynomial $p(\lambda) = a_2\lambda^2 + a_1\lambda + a_0$ has roots in $\mathbf{C}\setminus\mathbf{R}$, then no nonzero polynomial $f(t)$ is a solution to $p(D)(f(t)) = 0$.

h) If $f(t)$ and $g(t)$ are differentiable functions on the interval $[0, \pi]$, with $f''(t) = \lambda f(t)$ and $g''(t) = \mu g(t)$, for $\lambda \neq \mu$, then $\int_0^{2\pi} f(t)g(t)\, dt = 0$.

i) Let $f(t)$ be a unit vector in the vector space of differentiable functions f: $[0, \pi] \to \mathbf{C}$, that is, $\| f \| = (\int_0^{\pi} f(t)\bar{f}(t)\, dt)^{1/2} = 1$, then $f(t)$ is not zero for any $t \in [0, \pi]$.

2. Solve the system $\mathbf{u}' = A\mathbf{u}$ for a general $\mathbf{u}_0$ and the given A:

a) $A = \begin{bmatrix} 1 & -1 \\ 1 & 1 \end{bmatrix}$ b) $A = \begin{bmatrix} 1 & 0 & -1 \\ 0 & 1 & 0 \\ 1 & 0 & 1 \end{bmatrix}$

c) $A = \begin{bmatrix} 2 & 1 & 0 & 0 \\ 1 & 2 & 0 & 0 \\ 0 & 0 & 1 & -1 \\ 0 & 0 & 1 & 1 \end{bmatrix}$

3. Solve the following constant coefficient equations subject to the given conditions:

a) $u^{(2)} + au = 0$, $a \in \mathbf{R}$, $a > 0$, $u(0) = 0$, $u'(0) = 1$

b) $u^{(2)} - au = 0$, $a \in \mathbf{R}$, $a > 0$, $u(0) = 0$, $u'(0) = 1$

c) $u^{(2)} - a^2u = 0$, $a \in \mathbf{R}$, $a > 0$, $u(0) = 0$, $u'(0) = 1$

4. a) Let A be an $n \times n$ matrix (real or complex). Prove that if $\mathbf{v}$ is an eigenvector of A for the eigenvalue λ, then $\mathbf{v}$ is an eigenvector of e^{At} for the eigenvalue $e^{\lambda t}$.

b) Prove that if A is a diagonalizable $n \times n$ matrix, then e^{At} is diagonalizable.

5. a) Let A be a real diagonalizable $n \times n$ matrix with eigenvalues $\mu_1, \ldots, \mu_n$. Under what conditions on the eigenvalues is there a real $n \times n$ matrix B so that $e^B = A$?

b) If the matrix B is allowed to be complex, how does this change your answer?

6. Prove that if A is a skew-adjoint complex $n \times n$ matrix, then e^{At} is unitary. (See Exercises 7 and 8 of Section 5.3.)

In Proposition (7.5.6) we produced an orthonormal collection of vectors in the vector space V consisting of infinitely differentiable functions f: $[0, \pi] \to \mathbf{C}$. In the following problems we develop a different orthonormal collection in a similar space. Let V denote the vector space of continuous functions f: $[-1, 1] \to \mathbf{C}$ and define a Hermitian inner product on V by

$$\langle f, g \rangle = \int_{-1}^{1} f(t)\overline{g(t)}\, dt.$$

7. Let $q_n(t)$ denote the polynomial $\frac{d^n}{dt^n}(t^2 - 1)^n$.
   a) Prove that $(d^k/dt^k(t^2 - 1)^n)(\pm 1) = 0$ for $k = 1, \ldots, n - 1$.
   b) Prove that $\langle q_n(t), t^k \rangle = 0$ for $k = 1, \ldots, n - 1$. (*Hint:* Use induction on k and integration by parts.)
   c) Prove that $\langle q_n(t), q_m(t) \rangle = 0$ for $n \neq m$ (thus, the polynomials $(q_1, q_2, \ldots)$ form an orthogonal set of vectors in V).

8. Prove that $\|q_n\| = (n!)\, 2^n \sqrt{2/(2n + 1)}$ as follows:
   a) By repeated use of integration by parts, prove that
   $\langle q_n(t), q_n(t) \rangle = (2n)! \langle (1 - t)^n, (1 + t)^n \rangle$
   b) Prove that $\langle (1 - t)^n, (1 + t)^n \rangle = \dfrac{(n!)^2}{(2n)!(2n + 1)} 2^{2n+1}$
   c) Use parts a and b to prove the result.

In Exercises 7 and 8 we have produced an orthonormal sequence of polynomials $p_n(t) = q_n(t)/\|q_n(t)\|$ in V. It is sometimes preferable to replace these polynomials by different scalar multiples of the $q_n(t)$. Define $L_n(t) = \sqrt{2/(2n + 1)}\, p_n(t)$. These polynomials are called the *Legendre polynomials*.

9. a) Write out the explicit formulas for the first five Legendre polynomials $L_0, \ldots, L_4$.
   b) Sketch the graphs of $L_0, \ldots, L_4$.

10. If $f(t) \in V$ can be expressed as $f(t) = \sum_{n=0}^{\infty} a_n L_n(t)$, the a_n are called the Legendre coefficients of $f(t)$. Find the Legendre coefficients of the following functions $f(t)$. [*Hint:* see Example (7.5.7)].
   a) $f(t) = t^2 + t + 1$
   b) $f(t) = t^3$
   c) $f(t) = |t|$.

11. The Legendre polynomials can also be defined using the Gram-Schmidt process.
   a) Prove that $\text{Span}\{p_0(t), \ldots, p_n(t)\} = \text{Span}\{1, t, t^2, \ldots, t^n\}$ in V for all n.
   b) Let $r_0(t), \ldots, r_n(t)$ denote the orthonormal set of vectors obtained from applying the Gram-Schmidt process to $\{1, t, t^2, \ldots, t^n\}$. Prove that $r_i(t) = \pm p_i(t)$ for all $i \geq 1$.

12. Prove that the Legendre polynomial L_n $(n \geq 0)$ satisfies the second-order linear differential equation

$$(1 - t^2)u^{(2)} - 2tu^{(1)} + n(n + 1)u = 0$$

where $u = u(t)$. (This equation does not have constant coefficients, so the results of Section 7.4 do not apply.) Notice that if we denote by T the linear transformation $T(u) = (1 - t^2)u^{(2)} - 2tu^{(1)}$, then the Legendre polynomial $L_n(t)$ is an eigenvector of T with eigenvalue $-n(n + 1)$.

APPENDIX 1

Some Basic Logic and Set Theory

In this appendix we bring together in a very informal and sketchy way some of the basic notions from logic and set theory that are used frequently in mathematics and in the text. This material is, in a sense, much more basic than the subject matter of the text. Indeed, these ideas can be viewed as part of the language of all branches of mathematics.

§A1.1. SETS

The set concept is basic to all of mathematics. However, in our treatment, we will not make any formal definition of what a set is. Rather, we rely on the experience and intuition the reader has developed in dealing with sets, and concentrate, instead, on the operations that can be performed on sets. For us, a set is any collection of objects, for example, the names in a list, or the collection of all whole numbers.

The objects contained in a set are called its *elements* or *members*. We denote the statement "x is an element of the set A" by $x \in A$. Two sets A, B are *equal*, denoted $A = B$, if and only if they have the same elements. If every element of a set A is also contained in a set B, then we say A is a *subset* of B, and we write $A \subseteq B$. Note that if $A \subseteq B$ and $B \subseteq A$, then $A = B$, and conversely. We sometimes write $A \subset B$ if A is a subset of B, but $A \neq B$.

The set with no elements is called the *empty set*, and denoted ϕ. By definition $\phi \subseteq A$ for all sets A.

To illustrate these ideas, let Z be the set of all integers, and let P be the set of all prime numbers. We have $5 \in P$, and $P \subset Z$. However, $P \neq Z$, since there are integers that are not prime.

§A1.2. STATEMENTS AND LOGICAL OPERATORS

Mathematical discourse consists of statements or assertions about mathematical objects (i.e., sets), and operations performed on mathematical objects. The two simplest forms of statements are two that we have already encountered, namely, "$x \in A$" ("x is an element of the set A"), and "$A = B$" ("the sets A and B are equal").

As we know from everyday language, assertions may be combined to produce new ("compound") assertions. The five most basic ways this can be done are described here. We let p and q represent any statements. Then we can combine p and q to form new statements using *logical operators*. We have

1. "*p and q*," called the *conjunction* of p, q. This statement is often denoted $p \wedge q$ in symbolic logic.
2. "*p or q*," called the *disjunction* of p, q. This statement is often denoted $p \vee q$ in symbolic logic.
3. "*not p*," or "*it is false that p*," called the *negation* of p. This statement is often denoted $\sim p$ in symbolic logic.
4. "*if p, then q*," or "*p implies q*," or "*q only if p*," often denoted $p \rightarrow q$ in symbolic logic.
5. "*p if and only if q*," or "*p is equivalent to q*," often denoted $p \leftrightarrow q$ in symbolic logic.

In most written mathematics it is customary to use the English words (or their French, German, Russian, . . . equivalents) rather than the symbolic forms for the logical operators. We usually follow this practice in what follows, except when we feel it is clearer or more concise to use the symbols.

If we assume that the statements used to form a compound statement have a well-defined "truth-value" true or false (denoted for brevity by T and F), then we can express the truth value of a compound statement as a function of the truth values of its component parts by means of "truth tables."

For example, the statement "p and q" is true if and only if both p and q are true, so the truth table for conjunction is

p	q	p and q
T	T	T
T	F	F
F	T	F
F	F	F

The logical disjunction is the "inclusive" or. That is "p or q" is true if p, q, or both are true:

p	q	p or q
T	T	T
T	F	T
F	T	T
F	F	F

Negation simply reverses the truth value of a proposition:

p	not p
T	F
F	T

The logical implication operator does not quite conform to the natural language sense of "implication." By definition, when p is false, "p implies q" is true, regardless of the truth value of q. (We have already seen an instance of this in action, in the statement $\phi \subset A$ for all sets A. Do you see why?)

p	q	p implies q
T	T	T
T	F	F
F	T	T
F	F	T

Finally, logical equivalence is described as follows:

p	q	p if and only if q
T	T	T
T	F	F
F	T	F
F	F	T

In mathematics, propositions of the forms "p implies q" and "p if and only if q" are especially common and important. For statements of the form "p implies q" we can consider two related statements: "q implies p," called the *converse* of the original statement, and "not q implies not p," called the *contrapositive* of the original statement. For instance, if we consider the statement "if it is Tuesday, then this is Belgium," then the converse is "if this is Belgium, then it is Tuesday." The contrapositive is the statement "if this is not Belgium, then it is not Tuesday."

A frequent source of error for beginning students of mathematics is confusing statements and their converses. In general, the converse of a statement is not equivalent to, and does not follow from, the statement itself. The student should take pains to keep this distinction clear in his or her mind. For example, in the hypothetical European tour described by our previous examples, the converse *does not* follow from the original statement! (The tour could be spending several days in Belgium.)

When a statement is always true, independent of the truth values of its component statements, and solely by reason of the definitions of the logical operators, it is called a *tautology*. Truth tables may be used to demonstrate that a given proposition is a tautology. For example, we will see that the contrapositive "not q implies not p" *is* equivalent to the statement "p implies q." To do this, we construct a truth table for the (symbolic) statement $(p \to q) \leftrightarrow (\sim q \to \sim p)$. First evaluate the subexpressions $(p \to q)$ and $(\sim q \to \sim p)$ for all possible combinations of truth values of p and q. Then combine the results using the operator "$\leftrightarrow$". The following table gives the result of this:

p	q	$(p \to q)$	$\leftrightarrow$	$(\sim q \to \sim p)$
T	T	T	T	T
T	F	F	T	F
F	T	T	T	T
F	F	T	T	T

Since the value of the whole statement is always T, it is a tautology. In proofs it is sometimes easier to prove the contrapositive of a statement, so this equivalence can come in quite handy.

Some other useful tautologies are as follows:

(A1.2.1) Tautologies

T1) a) $(p \wedge q) \leftrightarrow (q \wedge p)$ (commutative laws)
b) $(p \vee q) \leftrightarrow (q \vee p)$

T2) a) $(p \wedge (q \wedge r)) \leftrightarrow ((p \wedge q) \wedge r)$ (associative laws)
b) $(p \vee (q \vee r)) \leftrightarrow ((p \vee q) \vee r)$

T3) a) $(p \wedge (q \vee r)) \leftrightarrow ((p \wedge q) \vee (p \wedge r))$ (distributive laws)
b) $(p \vee (q \wedge r)) \leftrightarrow ((p \vee q) \wedge (p \vee r))$

T4) $\sim(\sim p) \leftrightarrow p$ (double negation)

T5) $(p \to q) \leftrightarrow (\sim p \vee q)$

T6) $(p \leftrightarrow q) \leftrightarrow ((p \rightarrow q) \wedge (q \rightarrow p))$
T7) $((p \vee q) \rightarrow r) \leftrightarrow ((p \rightarrow r) \wedge (q \rightarrow r))$
T8) $(p \rightarrow (q \vee r)) \leftrightarrow ((p \wedge \sim q) \rightarrow r)$
T9) a) $\sim (p \wedge q) \leftrightarrow (\sim p \vee \sim q)$ (DeMorgan's laws)
b) $\sim (p \vee q) \leftrightarrow (\sim p \wedge \sim q)$

DeMorgan's laws show the effect of negating a conjunction or disjunction. For example, the negation of the statement "$x > 0$ and x is an integer" is "$x \leqslant 0$ *or* x is not an integer."

In general, these tautologies are used most often by mathematicians as *models* or *patterns* for arguments or parts of arguments. For example, to prove a statement of the form "p if and only if q," we frequently use T6 (usually implicitly) and construct a proof by showing that "p implies q" and "q implies p" are both true.

Similarly, T7 is used when we break a proof into several *cases:* if each one of two or more separate hypotheses implies a result r, then their disjunction (which, we hope, covers *all* the possibilities) implies r.

T8 is used in several ways. First, to prove a statement of the form "p implies q or r," we can apply T8 directly and replace the assertion to be proved by one of the form "p and not q implies r." This form is often easier to prove.

T8 is also the basis of one form of the method of *proof by contradiction* (or *reductio ad absurdum*, to give its fancier Latin name.) Since T8 is a tautology, it remains true for all statements p, q, and r. In particular, if we let r be the same statement as q, then T8 shows

$$(p \rightarrow q) \leftrightarrow (p \rightarrow (q \vee q)) \leftrightarrow ((p \wedge \sim q) \rightarrow q)$$

Thus, if by assuming p and the negation of q we arrive at q (a patent contradiction), it must be that "p implies q" is true.

§A1.3. STATEMENTS WITH QUANTIFIERS

In mathematics, two further logical operators are frequently used in forming statements. These are the so-called *quantifiers* "for all," and "for some." "For all" is called the *universal quantifier* (in symbols: $\forall$). The statement "for all x in some specified set S, $p(x)$" (where $p(x)$ is some assertion involving x) is sometimes written symbolically as $(\forall x \in S)(p(x))$. Similarly, the operator "for some" is called the *existential quantifier* (in symbols: $\exists$). The statement "for some x in some specified set S, $p(x)$" is written as $(\exists x \in S)(p(x))$.

In fact, most statements in mathematics involve these quantifiers in one way or another. For example

1. The statement $A \subseteq B$ for sets is equivalent to the statement "for all $x \in A$, $x \in B$" (or $(\forall x \in A)(x \in B)$).

2. The statement "r is a rational number" is equivalent to "there exists an integer $m \neq 0$ such that $m \cdot r$ is an integer," or $(\exists m \in Z)((m \neq 0) \wedge (m \cdot r \in Z))$

3. For a slightly more involved example, recall the definition of continuity at a point $a \in \mathbf{R}$ for functions $f: \mathbf{R} \to \mathbf{R}$. Let $\mathbf{R}^+$ denote the set of positive real numbers. In symbols, this definition comes out as

$$(\forall \varepsilon \in \mathbf{R}^+)(\exists \delta \in \mathbf{R}^+)((0 < |x - a| < \delta) \to (|f(x) - f(a)| < \varepsilon))$$

(Whew!)

To conclude our sketch of the elements of logic that are used frequently in mathematics, we indicate how the logical operator "not" that was introduced previously interacts with quantifiers.

(A1.3.1) Negating Statements with Quantifiers

a) $\sim (\forall x \in S)(p(x)) \leftrightarrow (\exists x \in S)(\sim p(x))$

b) $\sim (\exists x \in S)(p(x)) \leftrightarrow (\forall x \in S)(\sim p(x))$

A moment's thought will convince you of the validity of these rules. For a down-to-earth example, the negation of the statement "all cows are white" is "some cow is not white." For another example, the negation of the statement "r is a rational number" is equivalent to "for all $m \in Z$, $m \neq 0$ implies $m \cdot r \notin Z$." It would be a good exercise to work out the negation of the statement "f is continuous at a."

Many mathematical statements effectively contain quantifiers implicitly, even though this may not be evident from the form of the statement itself. This is especially true of statements involving *variables* or undetermined terms of various kinds, and is usually the result of a slightly informal way of saying things. For example, the true theorem of calculus: "if f is differentiable at a, then f is continuous at a," when analyzed closely, is seen to be the assertion "*for all* functions f, f is differentiable at a implies f is continuous at a." Note the implicit universal quantifier connected to f.

This type of statement is another frequent source of error for students of mathematics. *We repeat*, a statement $p(x)$ involving a "variable" x (such as the undetermined function f in our example) often means ($\forall x \in$ some specified set) $(p(x))$. To prove such a statement, we must show that $p(x)$ is true for all possible "values" of x. By the same token, the negation of such a statement is ($\exists x \in$ the specified set) (not $p(x)$). To disprove the assertion, we need only find *one* counterexample.

Thus, the "converse" of our example statement: "If f is continuous at a, then f is differentiable at a" is *false*. The function $f(x) = |x|$ is a counterexample to this assertion, since f is continuous everywhere, but not differentiable at $a = 0$.

§A1.4. FURTHER NOTIONS FROM SET THEORY

The most common method used to describe sets in mathematics is to specify the set, say, B, as the subset of a given set, say, A, for which some proposition is true. The usual notation for this is

$$B = \{x \in A | \; p(x)\}$$

where p is the proposition. For example

1. $[0, 1] \subset \mathbf{R}$ is the subset $\{x \in \mathbf{R}|\ (0 \leqslant x) \text{ and } (x \leqslant 1)\}$
2. The unit circle in $\mathbf{R}^2$ is the set $\{(x, y) \in \mathbf{R}^2|\ x^2 + y^2 = 1\}$.

If the statement $p(x)$ is false for all x in the set A, then we obtain the empty set. For example

$$\phi = \{x \in \mathbf{R}|\ (x < 3) \quad \text{and} \quad (x > 5)\}$$

Sets may be combined with other sets to form new sets in several ways. The most important among these are as follows:

1. *union:* The union of two sets A, B, denoted $A \cup B$, is the set

$$A \cup B = \{x|x \in A \quad \text{or} \quad x \in B\}$$

Unions of collections of more than two sets may be defined in the same way.

2. *intersection:* The intersection of two sets A, B, denoted $A \cap B$, is the set

$$A \cap B = \{x|\ x \in A \quad \text{and} \quad x \in B\}$$

Intersections of collections of more than two sets may be defined in the same way.

3. *difference:* The difference of two sets A, B, denoted $A \setminus B$, is the set

$$A \setminus B = \{x|x \in A \quad \text{and} \quad \text{not } x \in B\}.$$

$B \setminus A$ may be defined by the same process, but, in general, $B \setminus A \neq A \setminus B$. When $B \subseteq A$, $A \setminus B$ is also called the *complement* of B (in A).

For example, if $A = \{x \in \mathbf{R}|\ x \geqslant 1\}$ and $B = \{x \in \mathbf{R}|\ x < 3\}$, then

$$A \cup B = \{x \in \mathbf{R}|\ (x \geqslant 1) \quad \text{or} \quad (x < 3)\} = \mathbf{R}$$

$$A \cap B = \{x \in \mathbf{R}|\ (x \geqslant 1) \quad \text{and} \quad (x < 3)\} = [1, 3)$$

$$A \setminus B = \{x \in \mathbf{R}|\ (x \geqslant 1) \quad \text{and} \quad (x \geqslant 3)\} = [3, +\infty)$$

$$B \setminus A = \{x \in \mathbf{R}|\ (x < 3) \quad \text{and} \quad (x < 1)\} = (-\infty, 1)$$

Another important operation that can be used to produce new sets is the *Cartesian product*. If A and B are sets, the Cartesian product of A and B, denoted $A \times B$, is the set of *ordered pairs* (a, b), where $a \in A$ and $b \in B$. These are called *ordered pairs* since the element $a \in A$ is always the first component of the pair, whereas the element $b \in B$ is the second. We have

$$A \times B = \{(a, b)|\ a \in A \quad \text{and} \quad b \in B\}$$

For example, if $A = \{1, 2\}$ and $B = \{3, 4, 5\}$ then

$$A \times B = \{(1, 3), (1, 4), (1, 5), (2, 3), (2, 4), (2, 5)\}$$

Note: It is this construction that is used to produce the coordinate plane in analytic geometry: $\mathbf{R}^2 = \mathbf{R} \times \mathbf{R}$.

If we have general sets $A_1, \ldots, A_n$, we may define their Cartesian product in a similar fashion. The set $A_1 \times \cdots \times A_n$ is the set of all *ordered n-tuples*:

$$A_1 \times \cdots \times A_n = \{(a_1, \ldots, a_n) | \; a_i \in A_i \quad \text{for all } i\}$$

Some of the important properties of these operators on sets are as follows:

(A1.4.1) Proposition. For all sets A, B, C

S1) $A \cup \phi = A$ and $A \cap \phi = \phi$
S2) $A \cup A = A$ and $A \cap A = A$
S3) a) $A \cap B = B \cap A$ (commutative laws)
b) $A \cup B = B \cup A$
S4) a) $(A \cap B) \cap C = A \cap (B \cap C)$ (associative laws)
b) $(A \cup B) \cup C = A \cup (B \cup C)$
S5) a) $A \cap (B \cup C) = (A \cap B) \cup (A \cap C)$ (distributive laws)
b) $A \cup (B \cap C) = (A \cup B) \cap (A \cup C)$
S6) a) $A \setminus (B \cap C) = (A \setminus B) \cup (A \setminus C)$ (DeMorgan's laws)
b) $A \setminus (B \cup C) = (A \setminus B) \cap (A \setminus C)$

Each of these follows from the definitions and the properties of the logical operators appearing there. For example, we will show how the second DeMorgan law for sets follows from the second DeMorgan law for logical operators:

$$\begin{aligned}
A \setminus (B \cup C) &= \{x | \; x \in A \text{ and } \text{not } x \in B \cup C\} \\
&= \{x | \; x \in A \text{ and } \text{not } (x \in B \text{ or } x \in C)\} \\
&= \{x | \; x \in A \text{ and } \text{not } (x \in B) \text{ and } \text{not } (x \in C)\} \\
&\qquad \text{[by DeMorgan's law (T9b)]} \\
&= \{x | \; (x \in A \text{ and } \text{not } x \in B) \text{ and } (x \in A \text{ and } \text{not } x \in C)\} \\
&= \{x | \; x \in A \text{ and } \text{not } x \in B\} \cap \{x | \; x \in A \text{ and } \text{not } x \in C\} \\
&= (A \setminus B) \cap (A \setminus C)
\end{aligned}$$

The other derivations are similar (and mostly easier). ■

§A1.5. RELATIONS AND FUNCTIONS

In everyday life you have run into many examples of relations, or correspondences, between sets. For instance, if we are dealing with sets of people, an example of a relation would be *parenthood* (i.e., the relation for two people X and Y, X is a parent of Y).

Mathematically, a relation (between elements of a set A and a set B) is a *subset* R of the Cartesian product $A \times B$. For example, if A and B are two sets of people, we could define the parenthood relation as

$R = \{(X, Y) \in A \times B | X \text{ is a parent of } Y\}$

If $a \in A$, $b \in B$ and $(a, b) \in R$, then we say that a is *related* to b (for this relation R). The proposition "a is related to b" is sometimes written $a\, R\, b$ [i.e., $a\, R\, b$ means $(a, b) \in R \subseteq A \times B$]. Note that in a general relation, each $a \in A$ can be related to one, many, or no elements of B, and similarly, for $b \in B$.

Some important examples of mathematical relations that you have seen are the relations $=$, $<$, $\leqslant$, $>$, $\geqslant$ (for sets of real numbers), congruence of geometric figures, equality of sets, and so forth.

A relation $R \subseteq A \times A$ (between elements of a single set A) is said to be

1. *reflexive* if $a\, R\, a$ for all $a \in A$,
2. *symmetric* if $a\, R\, a'$ implies $a'\, R\, a$ for all $a, a' \in A$, and
3. *transitive* if $a\, R\, a'$ and $a'\, R\, a''$ imply $a\, R\, a''$ for all $a, a', a'' \in A$.

A relation with all three of these properties is called an *equivalence relation*. Examples include equality of numbers, equality of sets, and congruence of geometric figures. On the other hand, the relation $\leqslant$ is reflexive and transitive, but not symmetric. The parenthood relation has none of these properties.

Perhaps the most important examples of relations in mathematics are *functions*. Up to this point in your mathematical training, you may have seen functions defined only in the relatively naive way as "rules" that assign elements of one set to elements of another set. A more satisfactory and precise definition is the following.

A function F (from A to B) is defined by giving

1. the set A, called the *domain* of the function
2. the set B, called the *target* of the function, and
3. a relation $F \subset A \times B$ such that for each $a \in A$ there is *exactly one* $b \in B$ such that $(a, b) \in F$.

We usually think of F as both the "name" of the function and as the set of ordered pairs. The exact interpretation we have in mind should be clear from the context. If $a \in A$, the unique $b \in B$ such that $(a, b) \in F$ is called the *image* of a under F, and is sometimes denoted $b = F(a)$. This is the connection with the definition of functions with which you are probably more familiar. If we want to emphasize the quality of F as a *mapping*, taking elements of A to elements of B, we write $F: A \rightarrow B$.

In effect, our definition of a function amounts to giving the function by specifying the set of ordered pairs in its "graph" $\{(a, b) | b = F(a)\} \subset A \times B$, just as functions $f: \mathbf{R} \rightarrow \mathbf{R}$ are studied via their graphs in calculus.

For two functions to be equal, they must have the same domain, the same target, and the same image at each element of the domain.

Some typical examples of functions are

1. Given any set A, we can define the *identity function* $id_A: A \to A$. For each $a \in A$, $id_A(a) = a$ itself.
2. From calculus, you are familiar with functions $f: \mathbf{R} \to \mathbf{R}$, for example, the function defined by $f(x) = x^3 + \sin(x) - e^{3x}$.
3. Let T be the set of triangles in the plane. We can define the area function on T: Area: $T \to \mathbf{R}$. For each $t \in T$, Area$(t) \in \mathbf{R}$ is the area of that triangle.

For every function $F: A \to B$, the set $\{b \in B|\ b = F(a)$ for some $a \in A\}$ is called the *image* of F, and denoted Im(F) or $F(A)$. $F(A)$ need not equal the target, B. More generally, if $C \subset A$ is any subset, we define $F(C) = \{b \in B|\ b = F(c)$ for some $c \in C\}$. For example, if $F: \mathbf{R} \to \mathbf{R}$ is the function defined by $F(x) = x^2$ [i.e., the set of ordered pairs $F = \{(x, y) \in \mathbf{R} \times \mathbf{R}|\ y = x^2\}$], then $F(\mathbf{R}) = \{y \in \mathbf{R}|\ y \geqslant 0\} \neq \mathbf{R}$.

For a function $F: A \to B$, for each $a \in A$, there is exactly one $b \in B$ with $b = F(a)$. However, given $b \in B$, there may be one, many, or no $a \in A$ such that $F(a) = b$. In general, given a subset $C \subset B$, we define $F^{-1}(C)$, called the *inverse image* of C to be the set

$$F^{-1}(C) = \{a \in A \mid F(a) \in C\}$$

the set of all $a \in A$ that "map into" C. For example, if $F: \mathbf{R} \to \mathbf{R}$ is the function defined by $F(x) = x^2$ again, then

$$F^{-1}(\{9\}) = \{x|\ x^2 = 9\} = \{-3, 3\}$$

$$F^{-1}([0, 4)) = \{x|\ 0 \leqslant x^2 \text{ and } x^2 < 4\} = (-2, 2)$$

$$F^{-1}([16, +\infty)) = \{x|\ x^2 \geqslant 16\} = (-\infty, -4] \cup [4, +\infty)$$

If $F: A \to B$ is a function, and $C, D \subset A$, $E, G \subset B$, then we can ask how images and inverse images interact with unions, intersections, and differences. The facts are as follows.

(A1.5.1) Proposition

1. a) $F(C \cup D) = F(C) \cup F(D)$
   b) $F(C \cap D) \subset F(C) \cap F(D)$ (but they are not always equal)
2. a) $F^{-1}(E \cup G) = F^{-1}(E) \cup F^{-1}(G)$
   b) $F^{-1}(E \cap G) = F^{-1}(E) \cap F^{-1}(G)$
3. $F^{-1}(B - E) = A - F^{-1}(E)$
4. $F(F^{-1}(E)) \subset E$ (but not always equal)
5. $F^{-1}(F(C)) \supset C$ (but not always equal).

These follow directly from the definitions.

§A1.6. INJECTIVITY, SURJECTIVITY, AND BIJECTIVITY

If we consider why some of these inclusions are not equalities, we are led to look at the following notions. A function $F: A \to B$ is said to be *injective* (or one-to-one) if, for each $b \in F(A)$, there is exactly one $a \in A$ with $F(a) = b$. Another, more useful, way of saying the same thing is: F is injective if for all $a, a' \in A$, $F(a) = F(a')$ implies $a = a'$. For example, the function $F: \mathbf{R} \to \mathbf{R}$ defined by $F(x) = x^2$ is *not* injective, since, for example, $F(2) = F(-2) = 4$. On the other hand, the function $G: \mathbf{R} \to \mathbf{R}$ defined by $G(x) = x^3$ is injective. We see this since, if $x^3 = x'^3$, then $0 = x^3 - x'^3 = (x - x')(x^2 + xx' + x'^2)$. Now, $x^2 + xx' + x'^2 = (x + (1/2)x')^2 + (3/4)\,x'^2$ after completing the square. We see that this is > 0 unless $x = x' = 0$. Thus, the original equality is true if and only if $x = x'$. Hence, F is injective. (We have avoided the "obvious" approach of simply taking cube roots in the original equation, since the existence of a *unique* real cube root for every $x \in \mathbf{R}$ follows from what we are trying to show!)

A function $F: A \to B$ is said to be *surjective* (or onto) if $F(A) = B$. For example, $F: \mathbf{R} \to \mathbf{R}$ defined by $F(x) = x^2$ is not surjective, since $F(\mathbf{R}) = [0, +\infty)$. However, the function defined by $F(x) = x^3 + 1$ is surjective.

A function that is both injective and surjective is said to be *bijective*. If $F: A \to B$ is bijective, then by definition, F sets up a one-to-one correspondence between the elements of A and those of B.

Hence, we can relate those notions to the previously noted properties of images and inverse images.

(A1.6.1) Proposition. Let $F: A \to B$ be a function, and let $C \subset A$, and $E \subset B$ as before. Then

1. $F(F^{-1}(E)) = E$ for all subsets of B if and only if F is surjective.
2. $F^{-1}(F(C)) = C$ for all subsets of A if and only if F is injective.

For example, to see part 1, note that if F is surjective, then for each $e \in E$, there is some $a \in A$ with $F(a) = e$. Hence, this $a \in F^{-1}(E)$, so $e \in F(F^{-1}(E))$, and consequently, $E \subset F(F^{-1}(E))$. Since the opposite inclusion is always true, we have $E = F(F^{-1}(E))$. Conversely, if $F(F^{-1}(E)) = E$ for all subsets E of B, then in particular we have, for each $b \in B$, $F(F^{-1}(\{b\})) = \{b\}$. In particular, $F^{-1}(\{b\}) \neq \phi$ for all $b \in B$. Hence, F is surjective. The proof of part 2 is left as an exercise for the reader. ■

§A1.7. COMPOSITES AND INVERSE MAPPINGS

If $F: A \to B$ and $G: B \to C$ are functions, the *composite* of F and G, denoted by $G \circ F: A \to C$, is the function defined by $(G \circ F)(a) = G(F(a))$ for all $a \in A$. For example, if $F: \mathbf{R} \to \mathbf{R}$ is the function defined by $F(x) = x + 5$, and $G: \mathbf{R} \to \mathbf{R}$ is the function defined by $G(x) = x^3 + 7\sin(x)$, then for each $x \in \mathbf{R}$, $(G \circ F)(x) = (x + 5)^3 + 7\sin(x + 5)$. On the other hand, $(F \circ G)(x) = x^3$

$+ 7\sin(x) + 5$. In general, though, we will not be able to form the composite $F \circ G$ if the target of G is not a subset of the domain of F.

The notions of injectivity, surjectivity, and bijectivity introduced in the last section can also be interpreted in terms of compositions:

(A1.7.1) Proposition. Let $F: A \to B$ be a function.

1. F is injective if and only if there exists a function $L: B \to A$ with $L \circ F = id_A$ [i.e., the function $id_A: A \to A$ with $id_A(a) = a$ for all $a \in A$]. The function L is called a *left inverse* of F if it exists.
2. F is surjective if and only if there exists a function $R: B \to A$ with $F \circ R = id_B$. The function R is called a *right inverse* of F if it exists.
3. F is bijective if and only if there exists a function $G: B \to A$ with $F \circ G = id_B$ and $G \circ F = id_A$. The function G is necessarily unique if it exists and if so, it is called the *inverse function* of F. The usual notation for the inverse function is F^{-1}.

Of course, part 3 follows from parts 1 and 2. Part 1 is true since, if F is injective, we can construct a function L as follows. For each $b \in F(A)$, define $L(b) = a$, the unique $a \in A$ such that $F(a) = b$. For all other $b \in B - F(A)$, we can pick an arbitrary element $x \in A$ and define $L(b) = x$. Then for any $a \in A$ $(L \circ F)(a) = L(F(a)) = a$, so $L \circ F = id_A$ as required. Conversely, if such a mapping L exists, given a and $a' \in A$ with $F(a) = F(a')$, we apply L to both sides of the equation, yielding $L(F(a)) = L(F(a'))$. But, $L \circ F = id_A$, so that $L(F(a)) = a$ and $L(F(a')) = a'$. Hence $a = a'$ and F is injective.

The proof of part 2 is similar. ■

For example, $F: \mathbf{R} \to \mathbf{R}$ defined by $F(x) = x^3$ is bijective as we have seen. Hence, F has an inverse function, defined by $F^{-1}(x) = x^{1/3}$, the cube root function.

The reader should be careful to keep in mind the distinction between the inverse function of a bijective function $F: A \to B$ and the inverse image $F^{-1}(C)$ for $C \subset B$. Even though the notation used is similar, the ideas are quite different. In particular, $F^{-1}(C)$ is a *set*, defined for *all* functions F (not just bijective ones) and all subsets C of B. Of course, if F *is* bijective, then the set $F^{-1}(C)$ does coincide with the image of C under the inverse function.

§A1.8. SOME (OPTIONAL) REMARKS ON MATHEMATICS AND LOGIC

The question of the precise relation between mathematics and logic, which may have occurred to you in our previous discussion, is one that has received much attention from both philosophers and mathematicians. In fact, there is today no single, universally held opinion in this matter. Some logicians (chiefly G. Frege, B. Russell, and W. Quine) have argued that mathematics *is* logic—that, in effect, every mathematical statement is equivalent to, and can be translated into, a statement in logic. To get a feeling for what this might be like, look at our translation of the statement "f is continuous at a" and *then* realize that we have not rigorously defined the real numbers, absolute values, arithmetic operations, and so on, in terms of logic! Indeed, it took Russell and Whitehead several hundred pages of logic to define the number 1 in their monumental *Principia Mathematica* (Cambridge: Cambridge University Press, 1950). Needless to say, no one does mathematics in this way.

Others (following Hilbert's work on the axiomatic foundations of geometry) have argued that mathematics is the study of *formal systems*, in which the rules of logic are used as tools, or rules of inference, to deduce statements ("theorems") from certain axioms or basic assumptions. The axioms are set down once and for all as the starting points for the formal system.

It is our experience that most professional mathematicians present their ideas in the second, "formalist" way, but really hold yet a third (more romantic and subjective, if you will) view. Namely, they feel that the axiomatic approach is itself a tool, and that in "good" mathematics (whatever that may be) the axiom system used reflects (or at least closely approximates) "the real nature of the universe," or "the way things really are." For them, mathematical statements are statements about this (idealized) reality, and logic is a tool for deducing true statements.

The whole notion of truth or falsity of a statement is actually another extremely subtle point, and the accepted meanings of these terms vary, depending on which of the viewpoints described above we adopt. For example, the intuitive idea of truth and falsity certainly includes the so-called *principle of the excluded middle*—every statement is either true or false (this is the *exclusive* or). Most classical logical systems have included this as a basic assumption. (Note that in our propositional logic, the statement "p or not p" is, in fact, a tautology, whereas "p and not p" is never true.)

Nevertheless, most "formalists" would say that, within a formal system, a statement is "true" *if and only if it can be deduced from the axioms* of the system by applying the legal rules of logical reasoning. (Saying that a statement is "false" means that its negation is "true.") This lends to the intuitively rather unsettling possibility that there can be *undecidable* statements in a given system—those that are neither "true" nor "false" in this sense. The logician Gödel showed that this is the case in certain systems describing even so simple a part of mathematics as the arithmetic of natural numbers!

Without taking sides or dwelling on these issues, we have implicitly assumed that all the statements we encounter are either true or false, in an unspecified, but (we hope) intuitively clear sense. The ultimate criterion for deciding if a statement in mathematics is true, however, is that a convincing proof can be constructed.

EXERCISES

1. a) Verify tautology T5 of (A1.2.1) by constructing an appropriate truth table.
   b) Do the same for T7.
   c) Do the same for T8.
   d) Do the same for T9.

2. For each of the following statements of the form $p \to q$, identify p and q, and give the converse and contrapositive.
   a) If Jim Rice bats .300, then the Red Sox will win the pennant.
   b) If the sun is shining, then it is not raining.
   c) If the function f is odd, then $f(0) = 0$.
   d) If the function f is periodic with period 2π, then f assumes each real value infinitely often.

3. Give the negation of each of the following statements with quantifiers.
   a) There exists $x \in \mathbf{R}$ such that $f(x) = 0$.
   b) For all $m \in N$, x^m is an integer.
   c) There exists an $x \in \mathbf{R}$ such that for all $y \in \mathbf{R}$, $x + y > 0$.
   d) For all $x \in \mathbf{R}$ and all $y \in \mathbf{R}$, $f(x,y) = f(y,x)$.

4. Let $A = \{1,4,7,9,10\}$ and $B = \{3,4,8,9,12\}$. Find $A \cup B$, $A \cap B$, $A \setminus B$ and $B \setminus A$.

5. Repeat Exercise 4 with $A = [2,4]$ and $B = [1,5/2] \cup [3,5]$ in $\mathbf{R}$.

6. Repeat Exercise 4 with $A = \{(x,y) \in \mathbf{R}^2 \mid x > y\}$ and $B = \{(x,y) \mid x^2 + y^2 \geqslant 1\}$.

7. a) Prove the distributive laws for set union and set intersection—S5 in (A1.4.1).
   b) Prove the first DeMorgan law—S6a.

8. Which of the following relations are equivalence relations. If not, which properties fail to hold?
   a) Let n be a fixed integer, and let $R \subset Z \times Z$ be defined by $(m,m') \in R$ if $m - m'$ is divisible by n.
   b) $R \subset \mathbf{R} \times \mathbf{R}$ defined by $(x,y) \in R$ if $x \geqslant y$.
   c) The relation of similarity for triangles in the plane.
   d) $R \subset \mathbf{R} \times \mathbf{R}$ defined by $(x,y) \in R$ if $x - y$ is an integer.

9. Which of the following functions $f\colon \mathbf{R} \to \mathbf{R}$ are injective, surjective, bijective, neither? If f is bijective, find its inverse function f^{-1}.
   a) $f(x) = |x|$
   b) $f(x) = x^3 + 5$
   c) $f(x) = x^3 - 3x + 1$
   d) $f(x) = \begin{cases} (x^2 + 1)^{-1} \text{ if } x > 0 \\ x^3 \text{ if } x \leqslant 0 \end{cases}$

10. For each of the functions in Exercise 9, determine $f^{-1}([1,\infty))$.

11. Prove the statements in Proposition (A1.5.1).

12. Give an example to show that, in general, $(\forall x \in S)(\exists\, y \in S)(p(x,y))$ does not imply $(\exists y \in S)(\forall x \in S)\ (p(x,y))$.

13. Let X be a finite set, and let $P(X)$ denote the set of all subsets of X. For example, $P(\{1,2\}) = \{\phi, \{1\}, \{2\}, \{1,2\}\}$.
    a) Write out the elements of $P(\{1,2,3\})$ and $P(\{1,2,3,4\})$.
    b) Show that $P(X) \subset P(Y)$ if and only if $X \subset Y$.

14. Show that the set of functions $S = \{x,\ 1/x,\ 1 - x,\ (x - 1)/x,\ x/(x - 1),\ 1/(1 - x)\}$ is closed under composition—$f \circ g \in S$ for all $f, g \in S$.

15. Let S be a set, and let $R \subset S \times S$ be an equivalence relation. If $s \in S$, we define $[s] = \{t \in S \mid (s,t) \in R\}$, called the equivalence class of s.
   a) Show that if $(s,s') \notin R$, then $[s] \cap [s'] = \phi$,
   b) Show that if $(s,s') \in R$, then $[s] = [s']$.
   c) Show that S is the disjoint union of the distinct equivalence classes for R.
   d) What are the equivalence classes for the relation from Exercise 8a?

APPENDIX 2

Mathematical Induction

The technique of proof by mathematical induction is used frequently both in linear algebra and in other branches of mathematics. Often we wish to prove a statement about every natural number or systems of any number of equations or matrices of any size. In each of these cases we are really dealing with a *collection* of statements $p(n)$, indexed by the set of natural numbers $N = \{n \mid n \text{ is an integer} \geqslant 1\}$ (Sometimes collections of statements indexed by the set $\{n \mid n \geqslant 0\}$ or by the set $\{n \mid n \geqslant \text{k}\}$ for some other integer k also appear.)

Mathematical induction gives a concise and powerful method for proving the whole collection of statements $p(n)$ in two steps.

(A2.1) Principle of Mathematical Induction. Let $p(n)$, $n \in N$, be a collection of statements, and assume that

(i) $p(1)$ is true, and

(ii) if $p(k)$ is true for a general integer k, then $p(k + 1)$ is true.

Then $p(n)$ is true for all $n \geqslant 1$.

To justify the claim that the truth of $p(n)$ for all $n \geqslant 1$ follows from the two statements (i) and (ii) of (A2.1), we use a more intuitive property of sets of integers.

Well-Ordering Property: Every nonempty set of natural numbers has a smallest element.

For example, the smallest natural number is 1; the smallest natural number divisible by both 15 and 6 is 30, and so forth. We will not prove the well-ordering property, rather, we will treat it as an axiom describing the behavior of the set of integers.

To see that the well-ordering property justifies the principle of mathematical induction (A2.1), consider a collection of statements $p(n)$ for which (i) and (ii) hold. Let S be the set $\{n \in N \mid p(n) \text{ is true}\}$. We claim that $S = N$, or equivalently, $p(n)$ is true for all $n \in N$. If not, then there exist $n \in N$ such that $p(n)$ is false, and hence $N \setminus S$ is nonempty. Hence, by the well-ordering property, there is a smallest element $n_0 \in N \setminus S$. [Note that $n_0 > 1$, since by our assumption (i), $p(1)$ is true.]

Since n_0 is the smallest element of $N \setminus S$, $n_0 - 1 \in S$ [i.e., $p(n_0 - 1)$ *is* true]. But then, by assumption (ii), $p((n_0 - 1) + 1) = p(n_0)$ must also be true, so $n_0 \in S$. This contradicts the way we chose n_0, so it can only be the case that $S = N$ and $N \setminus S$ is empty. ■

We now give several kinds of examples to show how induction is used in practice.

(A2.2) Example. Suppose we wish to prove the formula for the sum of the first n natural numbers:

$$\sum_{m=1}^{n} m = (1/2)n(n + 1)$$

[For example, $1 = (1/2) \cdot 1 \cdot 2$, $1 + 2 = (1/2) \cdot 2 \cdot 3$, $1 + 2 + 3 = (1/2) \cdot 3 \cdot 4$, and so on.] Here our statement $p(n)$ is the formula itself. To prove it by induction, we proceed using (A2.1).

(i) $p(1)$ is true by the computations done previously.

(ii) Suppose we know that $p(k)$ is true—that is, $\sum_{m=1}^{k} m = (1/2)k(k + 1)$.

We must show that this implies that $p(k + 1)$ is true—that is, $\sum_{m=1}^{k+1} m = (1/2)(k + 1)(k + 2)$. To prove this, note that

$$\sum_{m=1}^{k+1} m = \sum_{m=1}^{k} m + (k + 1)$$

$$= (1/2)k(k + 1) + (k + 1)$$

since we assume that $p(k)$ is true. [This kind of step occurs in all induction proofs.

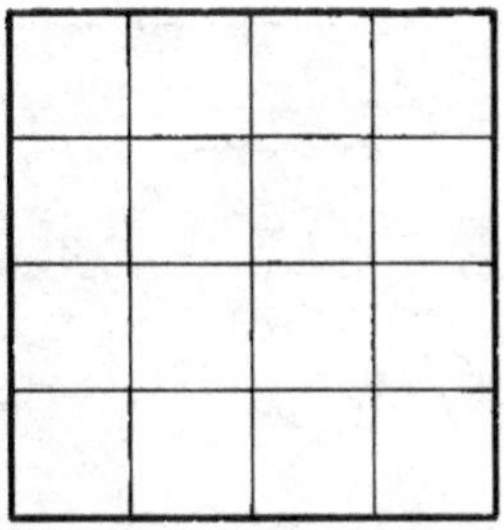

Figure A2.1

We sometimes refer to the assumption $p(k)$ as the *induction hypothesis*.] Now by elementary algebraic manipulations, we continue

$$\sum_{m=1}^{k+1} m = (1/2)k(k + 1) + (k + 1)$$

$$= (1/2)(k^2 + 3k + 2)$$

$$= (1/2)(k + 1)(k + 2)$$

So $p(k)$ implies $p(k + 1)$ for all k, and the formula is proved in general. ∎

Our second example gives a geometric example of induction in action and illustrates the real power of the method to simplify proofs that might otherwise require long and complicated analyses.

(A2.3) Example. Consider a square $2^n \times 2^n$ array of square floor tiles. For instance, with $n = 2$. See Figure A2.1.

Assume that all but one of the tiles are colored black and the remaining tile is colored white. The white tile may be located anywhere in the array.

Question: Is it possible to tile the black portion of the design using only groups of three black tiles of the form in Figure A2.2?

For $n = 1$ (a 2×2 pattern) the answer to our question is clearly yes! With some trial-and-error work, the reader should also see that the answer to our question is yes when $n = 2$ and $n = 3$, no matter where the single white tile is placed.

In fact, the answer to our question is always yes, and the easiest proof of this fact uses induction. To begin, note that the really important thing about the design we are trying to cover with our groups of three tiles is that there is one tile

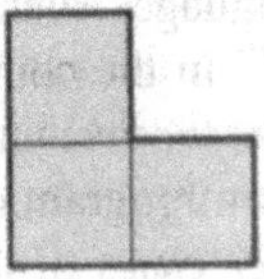

Figure A2.2

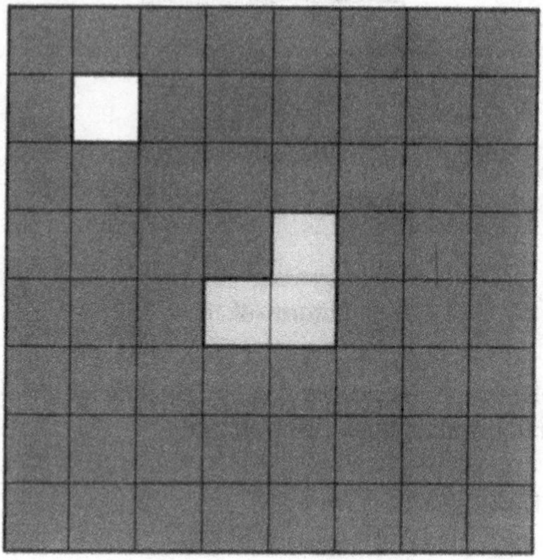

Figure A2.3

missing from the part of the array to be tiled. Hence, we take our statement $p(n)$ to be the assertion that all $2^n \times 2^n$ arrays with *one tile removed* (e.g., the white tile) can be tiled using our groups of three black tiles. Following (A2.1), we have already seen that $p(1)$ is true. Now assume $p(k)$ is true (all $2^k \times 2^k$ arrays with one tile removed can be tiled as described earlier) and consider a $2^{k+1} \times 2^{k+1}$ array. See Figure A2.3.

If we divide the array into quadrants by dividing the sides in halves, we obtain four $2^k \times 2^k$ subarrays. The single white tile will lie in one of these quadrants (say, for the purposes of illustration, in the upper left quadrant). At the point where the four quadrants meet, we can start our tiling by using one of our groups of three black tiles, aligned so that the "notch" in the group of three falls on the quadrant containing the white tile (see Figure A2.3). Then the remaining portion of the design to be tiled consists of four $2^k \times 2^k$ arrays, *each* with exactly one tile removed! By our induction hypothesis, we know that we can tile each of these arrays using the group of three black tiles, and our proof that the answer to our question is yes for all n is complete. ■

(A2.4) Remark (for readers with programming experience). There is a very close connection between mathematical induction and the notion of *recursion* in computer programs.

In modern programming languages such as Pascal, it is possible for a procedure or a function to "call itself" in the course of carrying out its intended job. We say that such subprograms are *recursive,* and this feature is a powerful problem-solving tool. When we realize that a programming task can be solved by reducing the problem to one or more *smaller cases* of the *same* task, recursion is a natural technique to use.

A recursive *algorithm* for *doing* the tiling described in Example (A2.3)—not just proving that it is possible—can be constructed along lines very similar to those of the proof:

To tile a $2^k \times 2^k$ array as before:
  If $k = 1$, then
    Finish tiling using one group of three black tiles.
  Otherwise
    Subdivide the array into four $2^{k-1} \times 2^{k-1}$ quadrants.
    Place the "middle tile" as in the proof.
    Tile the remaining parts of each of the four quadrants, using this same procedure.

The last step in the else block consists of four recursive calls to our tiling procedure. Note that after enough subdivisions, we always reach the base case $k = 1$, so the recursion will stop.

In some induction proofs, a variation of Principle (A2.1) is also used. We may assume that the truth of *all* the statements $p(1)$ through $p(k)$ as our induction hypothesis to prove the statement $p(k + 1)$ if we desire.

(A2.5) Principle of Mathematical Induction (Modified Form). Let $p(n)$, $n \in N$ be a collection of statements, and assume that

(i) $p(1)$ is true, and

(ii) the conjunction of the statements $p(1)$ through $p(k)$ implies $p(k + 1)$ for all k.

Then $p(n)$ is true for all natural numbers $n \geqslant 1$.

(A2.6) Example. As in the text, we use the notation $P(F)$ for the set of all polynomials with coefficients in the field F. (For the reader who has not seen the definition of a field in Chapter 5, nothing essential will be lost in the following by taking $F = \mathbf{R}$, or $F =$ the rational numbers.)

A nonconstant polynomial $q(t) \in P(F)$ is said to be *irreducible* in $P(F)$ if whenever we factor $q(t) = r(t)s(t)$ as the product of two other polynomials $r(t)$, $s(t) \in P(F)$, either $r(t)$ or $s(t)$ is a constant polynomial. For example, all the linear polynomials $q(t) = at + b$ in $P(\mathbf{R})$ are irreducible. The polynomial $q(t) = t^2 + 1$ is also irreducible. This is true since if we could factor $t^2 + 1 = r(t)s(t)$ in $P(\mathbf{R})$, where neither $r(t)$ nor $s(t)$ is a constant polynomial, then by looking at the possible degrees of $r(t)$ and $s(t)$, it is clear that both $r(t)$ and $s(t)$ would have to have degree 1. This would imply that $t^2 + 1 = 0$ has real roots, which, of course, is not the case. Hence, $t^2 + 1$ is irreducible in $P(\mathbf{R})$.

Claim: Every nonconstant polynomial $q(t) \in P(F)$ can be factored as a product of irreducible polynomials.

Sometimes, to make it clear which quantity we are using to index our collection of statements in an induction proof, we say we are using *induction on that quantity*.

We will prove this claim by induction on the degree of the polynomial $q(t)$. If $q(t)$ has degree 1, then $q(t)$ is itself irreducible and we are finished. Now assume, following Principle (A2.5), that our claim is true for polynomials of all degrees less than $k + 1$, and consider any polynomial $q(t)$ of degree $k + 1$. If $q(t)$ is itself irreducible, we are done. If not, there exist polynomials $r(t)$ and $s(t) \in P(F)$ such that neither $r(t)$ nor $s(t)$ is a constant polynomial, and $q(t) = r(t)s(t)$. But since neither $r(t)$ nor $s(t)$ is constant, both have a degree less than $k + 1$. By our induction hypothesis, both can be factored as a product of irreducible polynomials, and hence, the same is true of $q(t)$. ■

In linear algebra induction proofs are quite common. Some important examples in the text are as follows:

1. The proof of Theorem (1.5.8) (reduction to echelon form systems of linear equations). This proof proceeds by induction on the number of equations in the system.

2. The proof of Theorem (4.6.1) (the spectral theorem in $\mathbf{R}^n$). Here we use induction on the dimension of the vector space.

3. The proof of Theorem (6.2.8) (existence of canonical bases for nilpotent linear mappings). Here again, we use induction on the dimension of the domain vector space, in the modified form (A2.5).

Indeed, induction on the dimension of a vector space (or on the size of a matrix) is probably the most common kind of induction argument encountered in linear algebra.

EXERCISES

1. Show that $\sum_{m=1}^{n} m^2 = (1/6)n(n + 1)(2n + 1)$ for all natural numbers n.

2. Show that $\sum_{m=1}^{n} m^3 = \left[\sum_{m=1}^{n} m\right]^2 = (1/4)n^2(n + 1)^2$ for all natural numbers n.

3. Let $\begin{bmatrix} n \\ k \end{bmatrix}$ denote the coefficient of x^k in the binomial expansion of $(x + 1)^n$.

We have $\begin{bmatrix} n \\ k \end{bmatrix} = n(n - 1) \cdots (n - k + 1)/k!$

a) Show that for all n and all j with $0 \leq j \leq n + 1$

$$\begin{bmatrix} n \\ j \end{bmatrix} + \begin{bmatrix} n \\ j + 1 \end{bmatrix} = \begin{bmatrix} n + 1 \\ j + 1 \end{bmatrix}.$$

(Pascal's triangle—the proof of this statement does not require induction.)

b) Using part a and induction, show that if $f,\ g: \mathbf{R} \to \mathbf{R}$ are functions with derivatives of all orders, then the following product rule for higher derivatives is valid:

$$(fg)^{(n)}(x) = \sum_{j=0}^{n} \begin{bmatrix} n \\ j \end{bmatrix} f^{(n-j)}(x)\, g^{(j)}(x)$$

[As usual, the superscript (l) denotes the lth derivative. By definition, $f^{(0)} = f$ for all functions.]

4. A natural number $p > 1$ is said to be prime if whenever we factor p as the product of two natural numbers $p = rs$, either $r = 1$ or $s = 1$. Show that every natural number $n > 1$ can be written as a product of prime numbers.

5. Show that the prime factorization of a natural number n is unique, up to the order of the factors. (*Hint:* If a prime number p divides a product of integers rs, then p divides r or p divides s).

6. Consider a $2^n \times 2^n \times 2^n$ three-dimensional cubical array of cubical blocks. Assume that all but one of the blocks are black and the remaining block is white. Show that the black portion of the array can be constructed using groups of seven black blocks as in Figure A2.4.

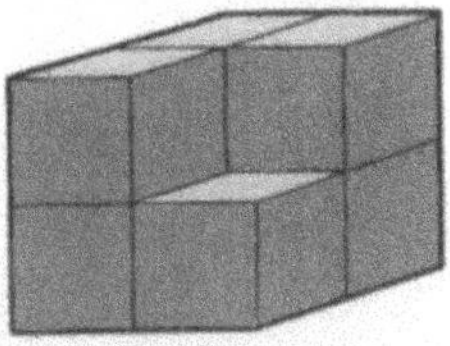

Figure A2.4

(This is a $2 \times 2 \times 2$ cube with one corner removed.) How can this statement be generalized to higher dimensions?

7. Let $P(X)$ be the set of subsets of a finite set X as defined in Exercise 13 of Appendix 1. Show that if X contains n members, then $P(X)$ contains 2^n elements.

8. Let $m \in N$ be any nonnegative integer, written in the standard decimal notation:

$$m = d_k d_{k-1} \cdots d_1 d_0$$

where $0 \leqslant d_i \leqslant 9$ for all i. (That is, d_0 is the ones digit of m, d_1 is the tens digit of m, and so on.) Define a function $T: N \to N$ by $T(m) = (d_k + 1)(d_{k-1} + 1) \cdot \cdots \cdot (d_1 + 1)(d_0 + 1)$, using the decimal expansion of m as above.

a) Show that if $m = 10n + d_0$, where $0 \leqslant d_0 \leqslant 9$, then $T(m) = (d_0 + 1) \cdot T(n)$. (You do not need induction for this part.)

b) Show that for all natural numbers k, $\sum_{n=0}^{10^k - 1} T(n) = 55^k$.

9. Let $a_n = ((1 + \sqrt{5})/2)^n + ((1 - \sqrt{5})/2)^n$ for all $n \geqslant 0$.
   a) Show that $a_{n+1} = a_n + a_{n-1}$.
   b) From part a, using induction, deduce that a_n is always an integer.

10. Generalize the results of Exercise 9 to the sequences defined by

$$a_n = ((\alpha + \sqrt{\beta})/2)^n + ((\alpha - \sqrt{\beta}/2)^n$$

where α and β are integers such that $\alpha^2 - \beta$ is divisible by 4.

Solutions

Chapter 1, Section 1

1. a) (3,9,6)
   b) 4(1,3,2) − (−2,3,4) = (6,9,4)
   c) −(1,3,2) + (−2,3,4) + 3(−3,0,3) = (−12,0,11)

3. Let $\mathbf{x} = (x_1, \ldots, x_n)$, $\mathbf{y} = (y_1, \ldots, y_n)$, $\mathbf{z} = (z_1, \ldots, z_n) \in \mathbf{R}^n$ and $c, d \in \mathbf{R}$.

   1)
$$\begin{aligned}(\mathbf{x} + \mathbf{y}) + \mathbf{z} &= (x_1 + y_1, \ldots, x_n + y_n) + (z_1, \ldots, z_n)\\ &= ((x_1 + y_1) + z_1, \ldots, (x_n + y_n) + z_n)\\ &= (x_1 + (y_1 + z_1), \ldots, x_n + (y_n + z_n))\\ &= (x_1, \ldots, x_n) + (y_1 + z_1, \ldots, y_n + z_n)\\ &= \mathbf{x} + (\mathbf{y} + \mathbf{z}).\end{aligned}$$

   2)
$$\begin{aligned}\mathbf{x} + \mathbf{y} &= (x_1 + y_1, \ldots, x_n + y_n)\\ &= (y_1 + x_1, \ldots, y_n + x_n)\\ &= \mathbf{y} + \mathbf{x}.\end{aligned}$$

3) The additive identity is $\mathbf{0} = (0, \ldots, 0)$, since $\mathbf{x} + \mathbf{0} = (x_1 + 0, \ldots, x_n + 0) = (x_1, \ldots, x_n) = \mathbf{x}$.

4) The additive inverse of $\mathbf{x}$ is $-\mathbf{x} = (-x_1, \ldots, -x_n)$, since

$$\begin{aligned}\mathbf{x} + (-\mathbf{x}) &= (x_1 + (-x_1), \ldots, x_n + (-x_n))\\ &= (0, \ldots, 0)\\ &= \mathbf{0}.\end{aligned}$$

5) $$\begin{aligned}c(\mathbf{x} + \mathbf{y}) &= c(x_1 + y_1, \ldots, x_n + y_n)\\ &= (c(x_1 + y_1), \ldots, c(x_n + y_n))\\ &= (cx_1 + cy_1, \ldots, cx_n + cy_n)\\ &= (cx_1, \ldots, cx_n) + (cy_1, \ldots, cy_n)\\ &= c\mathbf{x} + c\mathbf{y}.\end{aligned}$$

6) $$\begin{aligned}(c + d)\mathbf{x} &= ((c + d)x_1, \ldots, (c + d)x_n)\\ &= (cx_1 + dx_1, \ldots, cx_n + dx_n)\\ &= (cx_1, \ldots, cx_n) + (dx_1, \ldots, dx_n)\\ &= c\mathbf{x} + d\mathbf{x}.\end{aligned}$$

7) $$\begin{aligned}(cd)\mathbf{x} &= ((cd)x_1, \ldots, (cd)x_n)\\ &= (c(dx_1), \ldots, c(dx_n))\\ &= c(dx_1, \ldots, dx_n)\\ &= c(d\mathbf{x}).\end{aligned}$$

8) $$\begin{aligned}1 \cdot \mathbf{x} &= (1 \cdot x_1, \ldots, 1 \cdot x_n)\\ &= (x_1, \ldots, x_n)\\ &= \mathbf{x}.\end{aligned}$$

5. No, it is possible to add two elements of V and obtain a polynomial of degree less than n—for example with $n = 2$, $(x^2 + 2x) + (-x^2 - 1) = 2x - 1$. The sum operation does not always take pairs of elements of V to elements of V.

6. a) No. 1, 2, 3, 4, and 6.

8. a) If $\mathbf{x} + \mathbf{y} = \mathbf{x} + \mathbf{z}$ then adding $-\mathbf{x}$ to both sides, we have $-\mathbf{x} + (\mathbf{x} + \mathbf{y}) = -\mathbf{x} + (\mathbf{x} + \mathbf{z})$. By associativity, this shows $(-\mathbf{x} + \mathbf{x}) + \mathbf{y} = (-\mathbf{x} + \mathbf{x}) + \mathbf{z}$, so $\mathbf{0} + \mathbf{y} = \mathbf{0} + \mathbf{z}$, or $\mathbf{y} = \mathbf{z}$.

   b) By the distributive property of axiom 5, $(a + b)(\mathbf{x} + \mathbf{y}) = (a + b)\mathbf{x} + (a + b)\mathbf{y}$. Then, using axiom 6, we can rewrite this as $a\mathbf{x} + b\mathbf{x} + a\mathbf{y} + b\mathbf{y}$.

9. a) $\mathbf{R}^6$.
   b) $\mathbf{R}^9$.

Chapter 1, Section 2

1. If $\mathbf{x}, \mathbf{y} \in \{\mathbf{0}\}$, then $\mathbf{x} = \mathbf{y} = \mathbf{0}$. By Theorem (1.2.8), since $c\mathbf{x} + \mathbf{y} = c\mathbf{0} + \mathbf{0} = \mathbf{0}$, $\{\mathbf{0}\}$ is a subspace of V.

3. a) Yes. This follows from Corollary (1.2.14).
   b) No, W is neither closed under sums nor closed under scalar multiples. For example, $(\pi,0) \in W$, but $(1/2) \cdot (\pi,0) = (\pi/2,0)$ is not.
   c) Yes. $(a_1 + a_2 + a_3)^2 = 0$ if and only if $a_1 + a_2 + a_3 = 0$, so Corollary (1.2.14) applies.
   d) No, W is not closed under scalar multiplication. If $\mathbf{a} \in W$ with $a_3 > 0$ and $c \in \mathbf{R}$, $c < 0$, then $c\mathbf{a} \notin W$.
   e) No, W is not closed under scalar multiplication. For example, $(1,1,1) \in W$ and $(1/2)(1,1,1) \notin W$.
   f) Yes. If f, $g \in W$, then $(cf + g)'(x) + 4(cf + g)(x) = cf'(x) + g'(x) + c \cdot 4f(x) + 4g(x) = c(f'(x) + 4f(x)) + c(g'(x) + 4g(x)) = c \cdot 0 + 0 = 0$. Hence, $cf + g \in W$, so W is a subspace of $C^1(\mathbf{R})$.
   g) No, the zero function is not contained in W.
   h) Yes, if p, $q \in W$, then $(cp + q)(\sqrt{2}) = cp(\sqrt{2}) + q(\sqrt{2}) = c \cdot 0 + 0 = 0$. Hence $cp + q \in W$, and W is a subspace.
   i) No, the zero polynomial is not contained in W.
   j) Yes. $W = P_2(\mathbf{R})$. Or, apply Theorem (1.2.8). $p(x)$ and $q(x) \in W$ imply that $(cp(x) + q(x))' = cp'(x) + q'(x)$. Since $P_1(\mathbf{R})$ is a vector space and $p'(x)$, $q'(x) \in P_1(\mathbf{R})$, $c \cdot p'(x) + q'(x) \in P_1(\mathbf{R})$.
   k) Yes. If $f(x)$, $g(x) \in W$ and $c \in \mathbf{R}$, $(cf + g)(x + 2\pi) = c \cdot f(x + 2\pi) + g(x + 2\pi) = c \cdot f(x) + g(x) = (cf + g)(x)$. Thus, $cf + g$ is periodic with period 2π, that is, $cf + g \in W$.

5. →: If $\mathbf{v} + W$ is a subspace of V, then $\mathbf{0} \in \mathbf{v} + W$, so there is some vector $\mathbf{y} \in W$ with $\mathbf{0} = \mathbf{v} + \mathbf{y}$. This shows $\mathbf{y} = -\mathbf{v} \in W$, since additive inverses are unique (Proposition (1.1.6c)). Thus $(-1)(-\mathbf{y}) = \mathbf{v} \in W$, since W is a subspace.

   ←: If $\mathbf{v} \in W$, then let $\mathbf{v} + \mathbf{x}$ and $\mathbf{v} + \mathbf{y}$ be any two elements of $\mathbf{v} + W$. Then $c(\mathbf{v} + \mathbf{x}) + (\mathbf{v} + \mathbf{y}) = c\mathbf{v} + c\mathbf{x} + \mathbf{v} + \mathbf{y} \in W$.

7. a) Each line W through the origin can be defined by an equation $mx_1 - x_2 = 0$ (or $x_1 = 0$ if the line is vertical). In either case Corollary (1.2.14) shows that W is a subspace of $\mathbf{R}^2$.
   b) Suppose W is a subspace of $\mathbf{R}^2$, $W \neq \{\mathbf{0}\}$, and W is not a line through the origin. Then W contains two nonzero vectors (a,b) and (c,d) which lie along different lines through the origin, so (c,d) is not a scalar multiple of (a,b). If $(e,f) \in \mathbf{R}^2$ is any vector, we claim there are scalars $x_1,x_2 \in \mathbf{R}$ such that $x_1(a,b) + x_2(c,d) = (e,f)$. This vector equation yields two scalar equations

$$ax_1 + cx_2 = e$$
$$bx_1 + dx_2 = f$$

   which can be uniquely solved for x_1 and x_2. Thus $W = \mathbf{R}^2$.

9. If $f,g \in C^\infty(\mathbf{R})$, then for any $c \in \mathbf{R}$, and any integer $n \geqslant 1$, $cf + g$ is n-times differentiable, and $(d^n/dx^n)(cf + g) = c(d^n/dx^n)f + (d^n/dx^n)g$. Hence $cf + g \in C^\infty(\mathbf{R})$.

11. a) $\begin{bmatrix} 1 & 5 & -2 \\ 10 & 2 & 13 \end{bmatrix}$ b) $\begin{bmatrix} 0 & -37 \\ 17 & 14 \end{bmatrix}$.

13. If $A = \begin{bmatrix} a_{11} & a_{12} \\ a_{21} & a_{22} \end{bmatrix}$, $B = \begin{bmatrix} b_{11} & b_{12} \\ b_{21} & b_{22} \end{bmatrix} \in W$, then

$cA + B = \begin{bmatrix} ca_{11} + b_{11} & ca_{12} + b_{12} \\ ca_{21} + b_{21} & ca_{22} + b_{22} \end{bmatrix}$ and $3(ca_{11} + b_{11}) - 2(ca_{22} + b_{22}) = c(3a_{11} - 2a_{22}) + (3b_{11} - 2b_{22}) = c \cdot 0 + 0 = 0$. Hence $cA + B \in W$, and W is a subspace of $M_{2\times 2}(\mathbf{R})$.

16. A 1-1 correspondence can be constructed between the elements of $M_{m\times n}(\mathbf{R})$ and $\mathbf{R}^{mn}$ as follows. Define $T: M_{m\times n}(\mathbf{R}) \to \mathbf{R}^{mn}$ by

$$T\begin{bmatrix} a_{11} & \cdots & a_{1n} \\ \vdots & & \vdots \\ a_{m1} & \cdots & a_{mn} \end{bmatrix} = (a_{11}, \ldots, a_{1n}, a_{21}, \ldots, a_{2n}, \ldots, a_{m1}, \ldots, a_{mn}).$$ Note that T also "preserves" operations of sum and scalar multiplication.

Chapter 1, Section 3

1. a) $\text{Span}(S) = \{(a_1,0,a_2) \in \mathbf{R}^3 | a_1, a_2 \in \mathbf{R}\}$. This is the $x_1 - x_3$ plane in $\mathbf{R}^3$.
   b) $\text{Span}(S) = \{(a_1,a_2,0,0) \in \mathbf{R}^4 | a_1,a_2 \in \mathbf{R}\}$. This is the $x_1 - x_2$ plane in $\mathbf{R}^4$.
   c) $\text{Span}(S) = \{(a,a,a) \in \mathbf{R}^3 | a \in \mathbf{R}\}$. This is the line through $(1,1,1)$ and $(0,0,0)$ in $\mathbf{R}^3$.
   d) $\text{Span}(S) = \{a_0 + a_1x + a_2x^2 | a_0, a_1, a_2 \in \mathbf{R}\}$. This is $P_2(\mathbf{R}) \subset P_4(\mathbf{R})$.

3. $\text{Span}(S) \subseteq P_2(\mathbf{R})$, so it suffices to show that $P_2(\mathbf{R}) \subseteq \text{Span}(S)$. Let $a_0 + a_1x + a_2x^2 \in P_2(\mathbf{R})$. Then $a_0 + a_1x + a_2x^2 = (a_0 - a_1) + (a_1 - a_2) \cdot (1 + x) + a_2(x^2 + x + 1)$. So $a_0 + a_1x + a_2x^2 \in \text{Span}(S)$. Hence $P_2(\mathbf{R}) = \text{Span}(S)$.

5. a) Let $\mathbf{u}$ and $\mathbf{v} \in W_1 + \cdots + W_n$, with $\mathbf{u} = \mathbf{w}_1 + \cdots + \mathbf{w}_n$ and $\mathbf{v} = \mathbf{x}_1 + \cdots + \mathbf{x}_n$, where $\mathbf{w}_i$ and $\mathbf{x}_i \in W_i$. For all $c \in \mathbf{R}$, $c\mathbf{u} + \mathbf{v} = (c\mathbf{w}_1 + \cdots + c\mathbf{w}_n) + (\mathbf{x}_1 + \cdots + \mathbf{x}_n) = (c\mathbf{w}_1 + \mathbf{x}_1) + \cdots + (c\mathbf{w}_n + \mathbf{x}_n)$. Since each W_i is a subspace, $c\mathbf{w}_i + \mathbf{x}_i \in W_i$ and we conclude $c\mathbf{u} + \mathbf{v} \in W_1 + \cdots + W_n$.
   b) Use induction on the number of subspaces. Proposition (1.3.8) is the initial induction step. Assume the result holds for $n - 1$ subspaces. It follows from the definitions that $(W_1 + \cdots + W_{n-1}) + W_n = W_1 + \cdots + W_n$. By the induction hypothesis, $W_1 + \cdots + W_{n-1} = \text{Span}(S_1 \cup \ldots \cup S_{n-1})$. Applying (1.3.8) to the subspaces $W_1 + \cdots + W_{n-1}$ and W_n we have $W_1 + \cdots + W_n = (W_1 + \cdots + W_{n-1}) + W_n = \text{Span}((S_1 \cup \ldots \cup S_{n-1}) \cup S_n) = \text{Span}(S_1 \cup \ldots \cup S_n)$.

8. Suppose $\mathbf{x} = \mathbf{x}_1 + \mathbf{x}_2 = \mathbf{x}_1' + \mathbf{x}_2'$, where $\mathbf{x}_1, \mathbf{x}_1' \in W_1$ and $\mathbf{x}_2, \mathbf{x}_2' \in W_2$. Then $\mathbf{x}_1 - \mathbf{x}_1' = \mathbf{x}_2' - \mathbf{x}_2$. $\mathbf{x}_1 - \mathbf{x}_1'$ is in W_1, since W_1 is a subspace. Similarly $\mathbf{x}_2' - \mathbf{x}_2 \in W_2$. Hence, since $\mathbf{x}_1 - \mathbf{x}_1' = \mathbf{x}_2' - \mathbf{x}_2$, both vectors are in $W_1 \cap W_2 = \{\mathbf{0}\}$. This shows $\mathbf{x}_1 = \mathbf{x}_1'$ and $\mathbf{x}_2 = \mathbf{x}_2'$, so $\mathbf{x}_1$ and $\mathbf{x}_2$ are unique.

9. a) $\mathbf{R}^2 = W_1 + W_2$. Every $(x_1,x_2) \in \mathbf{R}^2$ can be written as $(x_1,x_2) = (x_1,0) + (0,x_2)$, where $(x_1,0) \in W_1$ and $(0,x_2) \in W_2$. Also, $W_1 \cap W_2 = \{(0,0)\}$, since if $(x_1,x_2) \in W_1 \cap W_2$, $x_1 = x_2 = 0$.
   c) i) For all $f \in F(\mathbf{R})$, we have $f(x) = (1/2)(f(x) + f(-x)) + (1/2)(f(x) - f(-x))$. Furthermore, $(1/2)(f(x) + f(-x))$ is even, since replacing x by $-x$ in this function yields $(1/2)(f(-x) + f(-(-x))) = (1/2)(f(-x) + f(x)) = (1/2)(f(x) + f(-x))$. Similarly, $(1/2)(f(x) - f(-x))$ is odd. Therefore $F(\mathbf{R}) = W_1 + W_2$.

ii) If $f \in W_1 \cap W_2$ then f is both even and odd so for all $x \in \mathbf{R}$ $f(x) = f(-x) = -f(x)$. Hence $f(x) = 0$ for all x.

11. a) $\text{Span}(S) = \left\{ \begin{bmatrix} a & b \\ c & d \end{bmatrix} \middle| c = d = 0 \right\}$, the set of all matrices with zeroes in the second row.

b) $\text{Span}(S) = \left\{ \begin{bmatrix} 2a & -a \\ -b & b \end{bmatrix} \middle| a, b \in \mathbf{R} \right\}$.

12. i) $M_{2\times2}(\mathbf{R}) = W_1 + W_2$, since

$$\begin{bmatrix} a & b \\ c & d \end{bmatrix} = \begin{bmatrix} a & (b + c)/2 \\ (b + c)/2 & d \end{bmatrix} + \begin{bmatrix} 0 & (b - c)/2 \\ (-b + c)/2 & 0 \end{bmatrix},$$

where the first matrix is in W_1 and the second matrix is W_2.

ii) If $\begin{bmatrix} a & b \\ c & d \end{bmatrix}$ is both symmetric and skew symmetric, then $a = d = 0$, and $b = c = -b$. Hence $b = c = 0$ as well, so $W_1 \cap W_2 = \left\{ \begin{bmatrix} 0 & 0 \\ 0 & 0 \end{bmatrix} \right\}$.

Chapter 1, Section 4

1. a) Dependent: $(1,1) - (1,3) + (0,2) = (0,0)$.
c) Dependent
e) Independent
g) Independent
i) Dependent
k) Dependent: $\sinh(x) + \cosh(x) - e^x = 0$ for all x.

2. a) No, a counterexample is $S_1 = \{(1,0),(0,1)\}$, $S_2 = \{(1,1)\}$ in $\mathbf{R}^2$.
b) Yes, $S_1 \cap S_2 \subset S_1$ which is linearly independent.
c) Yes, $S_1 \setminus S_2 \subset S_1$.

4. a) $\rightarrow$: If $\{\mathbf{v},\mathbf{w}\}$ is linearly dependent, $c_1\mathbf{v} + c_2\mathbf{w} = \mathbf{0}$ where $c_1 \neq 0$, or $c_2 \neq 0$. Say $c_1 \neq 0$, then $\mathbf{v} = (-c_2/c_1)\mathbf{w}$. Similarly, if $c_2 \neq 0$, $\mathbf{w} = (-c_1/c_2)\mathbf{v}$.

$\leftarrow$: If $\mathbf{v}$ is a scalar multiple of $\mathbf{w}$, then $\mathbf{v} = c\mathbf{w}$, for some $c \in \mathbf{R}$. Then $\mathbf{v} - c\mathbf{w} = \mathbf{0}$. Hence, $\{\mathbf{v},\mathbf{w}\}$ is linearly dependent. (This equation is actually a linear dependence since the coefficient of $\mathbf{v}$ is $1 \neq 0$.) Similarly, if $\mathbf{w}$ is a scalar multiple of $\mathbf{v}$, we obtain a different linear dependence.

b) One such example is $S = \{(1,1),(1,2),(1,3)\}$ in $\mathbf{R}^2$.

5. $\rightarrow$: If $\{\mathbf{v},\mathbf{w}\}$ is linearly independent and $c_1(\mathbf{v} + \mathbf{w}) + c_2(\mathbf{v} - \mathbf{w}) = \mathbf{0}$, then $(c_1 + c_2)\mathbf{v} + (c_1 - c_2)\mathbf{w} = \mathbf{0}$. Hence $c_1 + c_2 = 0$ and $c_1 - c_2 = 0$. So $c_1 = c_2 = 0$ and $\{\mathbf{v} + \mathbf{w},\mathbf{v} - \mathbf{w}\}$ is linearly independent.

$\leftarrow$: If $\{\mathbf{v} + \mathbf{w},\mathbf{v} - \mathbf{w}\}$ is linearly independent, and $c_1\mathbf{v} + c_2\mathbf{w} = \mathbf{0}$, then as $\mathbf{v} = (1/2)(\mathbf{v} + \mathbf{w}) + (1/2)(\mathbf{v} - \mathbf{w})$ and $\mathbf{w} = (1/2)(\mathbf{v} + \mathbf{w}) - (1/2)(\mathbf{v} - \mathbf{w})$, we have $\mathbf{0} = c_1((1/2)(\mathbf{v} + \mathbf{w}) + (1/2)(\mathbf{v} - \mathbf{w})) + c_2((1/2)(\mathbf{v} + \mathbf{w}) - (1/2)(\mathbf{v} - \mathbf{w})) = (c_1 + c_2)/2 \cdot (\mathbf{v} + \mathbf{w}) + (c_1 - c_2)/2 \cdot (\mathbf{v} + \mathbf{w})$. Since $\{\mathbf{v} + \mathbf{w}, \mathbf{v} - \mathbf{w}\}$ is linearly independent, $(c_1 + c_2)/2 = (c_1 - c_2)/2 = 0$. Hence $c_1 = c_2 = 0$, so $\{\mathbf{v},\mathbf{w}\}$ is linearly independent.

7. (By contradiction.) Suppose there were some $S' \subsetneq S$ with $\text{Span}(S') = \text{Span}(S)$. Since $S' \neq S$, there is some vector $\mathbf{x} \in S \setminus S'$. Since $\mathbf{x} \in S \subset \text{Span}(S')$, there exist scalars $c_i \in \mathbf{R}$ and vectors $\mathbf{x}_i \in S'$ such that $\mathbf{x} = c_1\mathbf{x}_1 + \cdots + c_n\mathbf{x}_n$. But then $\mathbf{0} = -\mathbf{x} + c_1\mathbf{x}_1 + \cdots + c_n\mathbf{x}_n$ is a linear dependence of vectors in S. Hence S is not linearly independent. This contradiction proves the claim.

9. a) Since $\text{Span}(S) = V$, given any $\mathbf{v} \in V$, we have $\mathbf{v} = c_1\mathbf{x}_1 + \cdots + c_n\mathbf{x}_n$ for some $\mathbf{x}_i \in S$ and $c_i \in \mathbf{R}$. Hence $-\mathbf{v} + c_1\mathbf{x}_1 + \cdots + c_n\mathbf{x}_n = \mathbf{0}$. This linear dependence shows $\{\mathbf{v}\} \cup S$ is linearly dependent.
   b) No, S itself could be linearly dependent, while $\text{Span}(S)$ could be strictly smaller than V.

11. If $c_1\mathbf{x}_1 + \cdots + c_n\mathbf{x}_n = \mathbf{0}$ for $\mathbf{x}_i \in S$, then we also have $0 \cdot \mathbf{x}_1 + \cdots + 0 \cdot \mathbf{x}_n = \mathbf{0}$. Since $\mathbf{0}$ can be written in only one way as a linear combination in $\text{Span}(S)$, $c_i = 0$ for all i.

Chapter 1, Section 5

1. a) $\{(1,0)\}$ c) $\{((-1/3)t, (11/9)t, t) | t \in \mathbf{R}\}$
   e) $\{((1/2)t_1 + (3/2)t_2 - 5/2, (-1/2)t_1 - (7/2)t_2 + 9/2, t_1, t_2) | t_1, t_2 \in \mathbf{R}\}$

2. a) $\{(1/3,1/3,1,0),(2/3,-5/6,0,1)\}$
   c) $\{(1/2,-1/4,1/2,1,0), (3/2,-5/4,1/2,0,1)\}$
   e) $\{-x^4 + 1, -x^3 + 1\}$

3. a) Yes
   c) No
   e) Yes

4. a)
$$\begin{aligned} (-11/6)x_1 + (1/3)x_2 + (1/2)x_4 &= 0 \\ -2x_1 + x_3 &= 0 \end{aligned}$$
   c) $x_1 + x_2 - x_3 + x_4 = 0$.

5. a) Yes
   c) No

Chapter 1, Section 6

1. Suppose the free variables are $x_{i_1}, \ldots, x_{i_k}$ and the spanning set is $\{\mathbf{v}_1, \ldots, \mathbf{v}_k\}$. Then the i_jth component of the vector $\mathbf{v}_j$ is 1, while the i_jth components of the remaining $\mathbf{v}_l$ are zero. Hence, in any linear combination $c_1\mathbf{v}_1 + \cdots + c_k\mathbf{v}_k = \mathbf{0}$, the equation coming from the i_jth components yields $c_j = 0, 1 \leqslant j \leqslant k$.

2. a) Basis $= \{(0,0,1,0),(-1,2,0,1)\}$. Dimension $= 2$.
   c) Basis $= \{(-1,0,1)\}$. Dimension $= 1$.
   e) Basis $= \{e^x - 2e^{2x} + e^{3x}\}$. Dimension $= 1$.

3. Let $W \subset V$ be a subspace, and let S_1 be a basis for W. By Theorem (1.6.6) there is a basis S for V such that $S_1 \subseteq S$. But then the number of vectors in S_1 must be less than or equal to the number of vectors in S. Hence $\dim(W) \leqslant \dim(V)$. If $\dim(W) = \dim(V)$ then $S_1 = S$, so $W = \text{Span}(S_1) = \text{Span}(S) = V$. If $W = V$ then $\dim(W) = \dim(V)$ follows from Corollary (1.6.11).

5. a) $W_1 \cap W_2 \subseteq W_1$ and $W_1 \cap W_2 \subseteq W_2$. Hence by Corollary (1.6.14) $\dim(W_1 \cap W_2) \leq \dim(W_1)$ and $\dim(W_1 \cap W_2) \leq \dim(W_2)$, so $\dim(W_1 \cap W_2) \leq$ the smaller of n_1 and n_2.
   b) If $W_1 = \text{Span}\{(1,0,0,0),(0,1,0,0)\}$ and $W_2 = \text{Span}\{(0,0,1,0),(0,0,0,1)\}$ in $\mathbf{R}^4$, then $W_1 \cap W_2 = \{\mathbf{0}\}$, so $\dim(W_1 \cap W_2) = 0$. If W_1 is as before, and $W_2 = \text{Span}\{(0,1,0,0), (0,0,1,0)\}$, then $W_1 \cap W_2 = \text{Span}\{(0,1,0,0)\}$, so $\dim(W_1 \cap W_2) = 1$. If $W_1 = W_2$, then $\dim(W_1 \cap W_2) = 2$.
   c) $\dim(W_1 \cap W_2) = n_1$ if and only if $W_1 \subseteq W_2$.
   Proof: If $W_1 \subseteq W_2$, then $W_1 \cap W_2 = W_1$, so $\dim(W_1) = \dim(W_1 \cap W_2) = n_1$. Conversely, if $\dim(W_1 \cap W_2) = n_1$, then since $W_1 \cap W_2 \subseteq W_1$, by Corollary (1.6.14), $W_1 = W_1 \cap W_2$. This is true if and only if $W_1 \subseteq W_2$.

7. a) $\{(1,2,3,4), (-1,0,0,0), (0,1,0,0), (0,0,1,0)\}$ using the standard basis as the spanning set:
   i) $\{(1,2,3,4),(-1,0,0,0)\} \cup \{(1,0,0,0)\}$ is linearly dependent so proceed
   ii) $\{(1,2,3,4),(-1,0,0,0)\} \cup \{(0,1,0,0)\}$ is independent so include that vector
   iii) $\{(1,2,3,4),(-1,0,0,0),(0,1,0,0)\} \cup \{(0,0,1,0)\}$ is linearly independent, hence a basis of $\mathbf{R}^4$
   c) $\{x^5 - 2x, x^4 + 3x^2, 6x^5 + 2x^3, x^5, x^4, 1\}$

9. a) $S' = \{(1,2,1),(-1,3,1),(1,1,1)\}$
   b) $S' = \{x^3 + x, 2x^3 + 3x, 3x^3 - x - 1, x^3 + x^2\}$

10. a) Let S be a basis for W. By Theorem (1.6.6), we know that there exists a set S' such that $S \cup S'$ is a basis for V. Let $W' = \text{Span}(S')$. We claim $V = W \oplus W'$. First, $V = \text{Span}(S \cup S') = \text{Span}(S) + \text{Span}(S') = W + W'$. Next, $W \cap W' = \{\mathbf{0}\}$ since if not, $S \cup S'$ would be linearly dependent.
    b) $\text{Span}\{(1,0,0),(0,1,0)\}$ is one possibility.

11. Suppose $\text{Span}(S_1) \cap \text{Span}(S_2)$ contains a non-zero vector $\mathbf{x}$. Then $\mathbf{x} = a_1\mathbf{y}_1 + \cdots + a_m\mathbf{y}_m$, where $\mathbf{y}_i \in S_1$, and $\mathbf{x} = b_1\mathbf{z}_1 + \cdots + b_n\mathbf{z}_n$, where $\mathbf{z}_j \in S_2$. Hence $(-a_1)\mathbf{y}_1 + \cdots + (-a_m)\mathbf{y}_m + b_1\mathbf{z}_1 + \cdots + b_n\mathbf{z}_n = \mathbf{0}$. Since $\mathbf{x} \neq \mathbf{0}$, at least one $a_i \neq 0$ (and similarly, at least one $b_j \neq 0$). Hence, there is a linear dependence among the vectors in $S_1 \cup S_2$. This contradiction shows $\text{Span}(S_1) \cap \text{Span}(S_2) = \{\mathbf{0}\}$.

13. In $\mathbf{R}^3$, every subspace W has dimension 0, 1, 2, or 3. If $\dim(W) = 0$, then $W = \{\mathbf{0}\}$. If $\dim(W) = 1$, then $W = \text{Span}\{\mathbf{x}\}$, a line through the origin. If $\dim(W) = 2$, $W = \text{Span}\{\mathbf{x},\mathbf{y}\}$, a plane through the origin. If $\dim(W) = 3$, then $W = \mathbf{R}^3$. Similarly, if W is a subspace in $\mathbf{R}^n$, $\dim(W) = 0,1,2, \ldots,$ or n.

14. a) The given set of matrices spans $M_{2\times 2}(\mathbf{R})$ and is linearly independent (see Exercise 12 of Section 1.4).
    b) One basis is
    $$\left\{\begin{bmatrix}1&0&0\\0&0&0\end{bmatrix}, \begin{bmatrix}0&1&0\\0&0&0\end{bmatrix}, \begin{bmatrix}0&0&1\\0&0&0\end{bmatrix}, \begin{bmatrix}0&0&0\\1&0&0\end{bmatrix}, \begin{bmatrix}0&0&0\\0&1&0\end{bmatrix}, \begin{bmatrix}0&0&0\\0&0&1\end{bmatrix}\right\}.$$
    $\dim(M_{2\times 3}(\mathbf{R})) = 6$.
    c) One basis is the set $\{E_{ij} | 1 \leq i \leq m, 1 \leq j \leq n\}$, where E_{ij} is the matrix with a 1 in the ith row and jth column and zeros in every other entry. There are $m \cdot n$ of these matrices, so $\dim(M_{m\times n}(\mathbf{R})) = m \cdot n$.

15. a) $\dim(W) = 2$. A basis is $\left\{\begin{bmatrix}-2&1\\0&0\end{bmatrix}, \begin{bmatrix}0&0\\3&1\end{bmatrix}\right\}$.

b) $\dim(W) = 3$. A basis is $\left\{\begin{bmatrix}1 & 0\\0 & 0\end{bmatrix}, \begin{bmatrix}0 & 0\\0 & 1\end{bmatrix}, \begin{bmatrix}0 & 1\\1 & 0\end{bmatrix}\right\}$.

c) $\dim(W) = 1$. A basis is $\left\{\begin{bmatrix}0 & 1\\-1 & 0\end{bmatrix}\right\}$.

Chapter 1, Supplementary Exercises

1. a) False. This is only true for homogeneous systems.
   b) True. Let α be a basis for W. We can extend α to a basis for V, so $\dim(W) \leq \dim(V)$.
   c) False. A counterexample is $S = \{(1,1,1),(2,2,2)\}$ in $\mathbf{R}^3$.
   d) True. $\{(1,1),(-1,3)\}$ is linearly independent, hence a basis for $\mathbf{R}^2$. $W_1 + W_2 = \text{Span}(\{(1,1), (-1,3)\}) = \mathbf{R}^2$.
   e) False. A counterexample is

$$x_1 + x_2 = 0$$

$$x_1 + 3x_2 = 0$$

   f) True. The zero vector is always a solution of a homogeneous system.
   g) False. A counterexample is

$$x_1 + x_2 + x_3 = 2$$

$$x_1 + x_2 + x_3 = 3$$

   h) True. It is closed under sums and scalar multiples.
   i) False. $\sin^2(x) - (1/2)\cdot 1 + (1/2)\cos(2x) = 0$ is a linear dependence.
   j) True. If $\mathbf{x} \in \text{Span}(S)$ and $\mathbf{x} = c_1\mathbf{x}_1 + \cdots + c_n\mathbf{x}_n = d_1\mathbf{x}_1 + \cdots + d_n\mathbf{x}_n$, then $\mathbf{0} = (c_1 - d_1)\mathbf{x}_1 + \cdots + (c_n - d_n)\mathbf{x}_n$, so $c_i = d_i$ for all i.
   k) False. There is no identity element for the $+'$ operation.

2. a) Basis $= \{(-1/2,1/4,1,0), (-1/2,3/4,0,1)\}$. Dimension $= 2$.
   b) Solution set $= \{((-1/2)t_1 - (1/2)t_2 + 1/2, (1/4)t_1 + (3/4)t_2 - 3/4, t_1, t_2) \mid t_1, t_2 \in \mathbf{R}\}$

4. Basis $= \left\{\begin{bmatrix}0 & 1\\0 & 0\end{bmatrix}, \begin{bmatrix}0 & 0\\1 & 0\end{bmatrix}, \begin{bmatrix}-1 & 0\\0 & 1\end{bmatrix}\right\}$. Dimension $= 3$.

6. a) No, it is not closed under vector sums. For example $(1,-1,0) \in W$ and $(1,0,-1) \in W$, but their sum $(2,-1,-1) \notin W$, since $2^3 + (-1)^3 + (-1)^3 = 6 \neq 0$. Yes, if $\mathbf{x} \in W$, then $\text{Span}(\{\mathbf{x}\}) \in W$.

8. $W = \{\mathbf{x} \in \mathbf{R}^4 \mid (-1/2)x_1 - (1/2)x_2 + x_3 - x_4 = 0\}$, so $(3,2,1,-1) \notin W$.

Chapter 2, Section 1

1. $\rightarrow$: If T is a linear transformation, $T(a\mathbf{u} + b\mathbf{v}) = T(a\mathbf{u}) + T(b\mathbf{v}) = aT(\mathbf{u}) + bT(\mathbf{v})$ using (i) and (ii) of Definition (2.1.1).
   $\leftarrow$: To obtain (i) of (2.1.1) let $a = b = 1$. To obtain (ii) of (2.1.1) let $\mathbf{v} = \mathbf{0}$.

2. $\rightarrow$: Assume T satisfies (2.1.2) and use induction on k. If the statement holds for k, then $T(\sum_{i=1}^{k+1} a_i\mathbf{v}_i) = T(\sum_{i=1}^{k} a_i\mathbf{v}_i) + a_{k+1}T(\mathbf{v}_{k+1})$ by (2.1.2). Apply the induction hypothesis to the first summand to obtain the result.
   $\leftarrow$: Let $k = 2$ and apply (2.1.2) to conclude T is a linear transformation.

3. The functions of parts a, c, e, and g are linear transformations. Verify (2.1.2) for these functions. The functions of parts d, f, and h are not linear transformations. Show (2.1.2) fails or show $T(\mathbf{0}) \neq \mathbf{0}$ for these functions.

5. a) Let f and $g \in V$. From the properties of derivatives, for a and $b \in \mathbf{R}$, $D(af + bg) = (af + bg)'(x) = af'(x) + bg'(x) = a \cdot D(f) + b \cdot D(g)$. Thus by (2.1.2), D is linear.
   b) Let $f, g \in V$ and $c_1, c_2 \in \mathbf{R}$. Then from the properties of integration $\text{Int}(c_1 f + c_2 g) = \int_a^b (c_1 f + c_2 g)\,dx = c_1 \int_a^b f dx + c_2 \int_a^b g dx = c_1 \cdot \text{Int}(f) + c_2 \cdot \text{Int}(g)$.
   c) Apply the calculations of part b in this case.

7. The pairs in parts a and c are perpendicular, the pair in part b is not perpendicular.

8. a) $R_{\pi/4}(\mathbf{x}) = (x_1/\sqrt{2} - x_2/\sqrt{2}, x_1/\sqrt{2} + x_2/\sqrt{2})$,
   $R_{-\pi/4}(\mathbf{x}) = (x_1/\sqrt{2} + x_2/\sqrt{2}, -x_1/\sqrt{2} + x_2/\sqrt{2})$.
   b) $R_{-\pi/4}(\mathbf{w}) = \mathbf{v}$.
   c) $R_{-\theta}(\mathbf{w}) = \mathbf{v}$.

10. a) Solve $\mathbf{x} = a_1\mathbf{v}_1 + a_2\mathbf{v}_2 + a_3\mathbf{v}_3$ for a_1, a_2, and a_3:
    $a_1 = (1/2)(x_1 - x_2 + x_3)$, $a_2 = (1/2)(-x_1 + x_2 + x_3)$, and
    $a_3 = (1/2)(x_1 + x_2 - x_3)$.

    b) $T(\mathbf{x}) = T(a_1\mathbf{v}_1 + a_2\mathbf{v}_2 + a_3\mathbf{v}_3) = a_1T(\mathbf{v}_1) + a_2T(\mathbf{v}_2) + a_3T(\mathbf{v}_3)$
    $= (1/2)(x_1 - x_2 + x_3)(1,0) + (1/2)(-x_1 + x_2 + x_3)(0,1) +$
    $(1/2)(x_1 + x_2 - x_3)(1,1)$
    $= (x_1, x_2)$.

12. a) Verify (2.1.2). Let $\mathbf{u}$, $\mathbf{v} \in V$ and $a, b \in \mathbf{R}$, then $R(a\mathbf{u} + b\mathbf{v}) = S(a\mathbf{u} + b\mathbf{v}) + T(a\mathbf{u} + b\mathbf{v}) = aS(\mathbf{u}) + bS(\mathbf{v}) + aT(\mathbf{u}) + bT(\mathbf{v})$, since S and T are linear. This equals $a(S(\mathbf{u}) + T(\mathbf{u})) + b(S(\mathbf{u}) + T(\mathbf{v})) = aR(\mathbf{u}) + bR(\mathbf{v})$.
    b) Verify (2.1.2). Let $\mathbf{u}$, $\mathbf{v} \in V$ and c, $d \in \mathbf{R}$, then $R(c\mathbf{u} + d\mathbf{v}) = aS(c\mathbf{u} + d\mathbf{v}) = acS(\mathbf{u}) + adS(\mathbf{v})$, since S is linear. This equals $cR(\mathbf{u}) + dR(\mathbf{v})$.
    c) Parts a and b show that $L(V,W)$ is closed under addition and scalar multiplication. It remains to verify axioms 1 through 8 of Definition (1.1.1). We verify 2 and 3 here. For all S, $T \in L(V,W)$ and for all $\mathbf{v} \in V$, $(S + T)(\mathbf{v}) = S(\mathbf{v}) + T(\mathbf{v}) = T(\mathbf{v}) + S(\mathbf{v}) = (T + S)(\mathbf{v})$. Therefore $T + S = S + T$. If 0 denotes the zero transformation: $0(\mathbf{v}) = \mathbf{0}$, then for all $S \in L(V,W)$, $(S + 0)(\mathbf{v}) = S(\mathbf{v}) + 0(\mathbf{v}) = S(\mathbf{v}) + \mathbf{0} = S(\mathbf{v})$. Therefore $S + 0 = S$.

15. $T|_U(x_1\mathbf{u}_1 + x_2\mathbf{u}_2) = T(x_1\mathbf{u}_1 + x_2\mathbf{u}_2) = x_1T(\mathbf{u}_1) + x_2T(\mathbf{u}_2)$
    $= x_1(3\cdot1 + 2\cdot1 + 1\cdot1) + x_2(3\cdot2 + 2\cdot1 + 1\cdot0) = 6x_1 + 8x_2$.

Chapter 2, Section 2

1. If $S,T: V \to W$ both satisfy $S(\mathbf{v}_j) = a_{1j}\mathbf{w}_1 + \cdots + a_{lj}\mathbf{w}_l = T(\mathbf{v}_j)$, then by Proposition (2.1.14), $S = T$.

2. It suffices to show that the two column vectors have the same entries. The ith entry of $A(a\mathbf{u} + b\mathbf{v})$ is $\sum_{j=1}^{k} a_{ij}(au_j + bv_j)$
   $= a(\sum_{j=1}^{k} a_{ij}u_j) + b(\sum_{j=1}^{k} a_{ij}v_j)$, which is the ith entry of $aA\mathbf{u} + bA\mathbf{v}$.

3. a) $[T] = \begin{bmatrix} 1 & -1 & 0 \\ 0 & 1 & -1 \\ 1 & 1 & -1 \\ -1 & 0 & 1 \end{bmatrix}$ b) $[T] = [a_1 \ldots a_n]$

c) $[D] = \begin{bmatrix} 0 & -1 \\ 1 & 0 \end{bmatrix}$

4. c) $\mathbf{R_a}$ reflects $\mathbf{R}^2$ in the line L, i.e., $\mathbf{R_a}(\mathbf{v})$ is the vector which is the "mirror image of $\mathbf{v}$" on the other side of L from $\mathbf{v}$. See the accompanying figure.

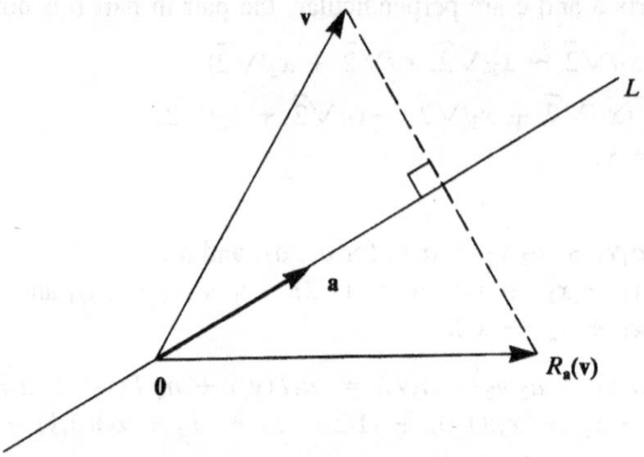

Figure for Exercise 4c.

5. a) $[T]^{\beta}_{\beta} = \begin{bmatrix} 1 & 0 & 0 \\ 1 & 1 & 0 \\ 1 & -1 & 1 \end{bmatrix}$

b) $S(1,1,0) = T(1,1,0) = T(\mathbf{e}_1) + T(\mathbf{e}_2) = (1,1,1) + (0,1,-1) = \mathbf{e}_1 + 2\mathbf{e}_2$ and

$S(0,0,1) = T(\mathbf{e}_3) = \mathbf{e}_3$. Therefore $[S]^{\beta}_{\alpha} = \begin{bmatrix} 1 & 0 \\ 2 & 0 \\ 0 & 1 \end{bmatrix}$

c) Since $(1,1,0) = \mathbf{e}_1 + \mathbf{e}_2$, the first column of $[S]^{\beta}_{\alpha}$ is $T(\mathbf{e}_1) + T(\mathbf{e}_2)$. Similarly, the second column is $T(\mathbf{e}_3)$.

7. a) $\begin{bmatrix} 5 \\ 7 \end{bmatrix}$ b) $[-10]$ c) $\begin{bmatrix} 2 \\ 14 \\ 9 \\ 8 \end{bmatrix}$

9. a) $[I]^{\beta}_{\alpha} = \begin{bmatrix} 1 & 3 \\ 2 & -4 \end{bmatrix}$ b) $[I]^{\beta}_{\alpha} = \begin{bmatrix} a_{11} & a_{12} \\ a_{21} & a_{22} \end{bmatrix}$

c) Let $\alpha = \{(a_{1i}, \ldots, a_{ni}) \mid i = 1, \ldots, n\}$. Then

$$[I]^{\beta}_{\alpha} = \begin{bmatrix} a_{11} & & a_{1n} \\ \cdot & & \cdot \\ \cdot & \cdots & \cdot \\ \cdot & & \cdot \\ a_{n1} & & a_{nn} \end{bmatrix}$$

11. a) If $S(\mathbf{x}_j) = \sum_{i=1}^{m} s_{ij}\mathbf{w}_i$ and $T(\mathbf{x}_j) = \sum_{i=1}^{m} t_{ij}\mathbf{w}_i$, then $(S + T)(\mathbf{x}_j) = \sum_{i=1}^{m} (s_{ij} + t_{ij})\mathbf{w}_i$. Thus, the i,j entry of $[S + T]_\alpha^\beta$, which is $s_{ij} + t_{ij}$, is the sum of the i,j entries of $[S]_\alpha^\beta$ and $[T]_\alpha^\beta$. Hence $[S + T]_\alpha^\beta = [S]_\alpha^\beta + [T]_\alpha^\beta$.

13. a) $T((1/2)(A + A^t)) = (1/2)(T(A) + T(A^t)) = (1/2)(A^t + (A^t)^t) = (1/2)(A^t + A) = (1/2)(A + A^t)$. Therefore $(1/2)(A + A^t) \in W_+$.
b) $T((1/2)(A - A^t)) = (1/2)(T(A) - T(A^t)) = (1/2)(A^t - (A^t)^t) = (1/2)(-A + A^t) = -(1/2)(A - A^t)$. Therefore $(1/2)(A - A^t) \in W_-$.
c) If $A \in W_+ \cap W_-$ then the entries a_{ij} of A satisfy both $a_{ij} = a_{ji}$ and $a_{ij} = -a_{ji}$. Therefore, $a_{ij} = 0$ for all i and j, so $A = 0$.
d) From parts a and b, $A = (1/2)(A + A^t) + (1/2)(A - A^t)$ so that $M_{n \times n}(\mathbf{R}) = W_+ + W_-$. From Exercise 9 of Section 1.3 and part c, $M_{n \times n}(\mathbf{R}) = W_+ \oplus W_-$.

14. a) $\dim M_{m \times n}(\mathbf{R}) = m \cdot n$, since a basis is given by the matrices E_{ij}, $1 \leqslant i \leqslant m$, $1 \leqslant j \leqslant n$, where E_{ij} is the matrix with a 1 in the ith row and the jth column and zeros in every other entry.

b) $\dim(W_+) = (1/2)\, n(n + 1)$, since a basis is given by E_{ii}, $1 \leqslant i \leqslant n$, and $E_{ij} + E_{ji}$, $i < j$. $\dim(W_-) = (1/2)n(n-1)$, since a basis is given by the matrices $E_{ij} - E_{ji}$, $i < j$.

Chapter 2, Section 3

1. a) Use elimination to show that $\text{Ker}(T) = \text{Span}\{(-1/2, -3/4, 1)\}$. Then one can verify that $\{(-1/2, -3/4, 1), \mathbf{e}_1, \mathbf{e}_2\}$ is of the desired form.
d) A basis for $\text{Ker}(T)$ is $\{(1,-1,1,-1)\}$. Then $\{(1,-1,1,-1), \mathbf{e}_1, \mathbf{e}_2, \mathbf{e}_3\}$ is a basis of the desired form.
f) A basis for $\text{Ker}(T)$ is given by the constant polynomial 1. Then $\{1, x, x^2, \ldots, x^n\}$ is a basis of the desired form.

2. If $\mathbf{a} = (a_1, a_2) \neq \mathbf{0}$, $(a_2, -a_1)$ is a basis for $\text{Ker}(P_\mathbf{a})$.

3. a) $\text{Ker}(T) = \{\mathbf{0}\}$ and $\{(1,2), (2,2)\}$ is a basis for $\text{Im}(T)$.
c) A basis for $\text{Ker}(T)$ is $\{(-1,-3,1,0,0,0), (-1/2,-1,0,-1/2,1,0), (-1/2,-1,0,1/2,0,1)\}$ and a basis for $\text{Im}(T)$ is $\{(1,-1,0),(0,1,1),(-1,1,2)\}$.

5. $\dim(\text{Ker}(T)) = 0$ and $\dim(\text{Im}(T)) = n$.

9. a) $\text{Ker}(\text{Mat}) = \{0\}$, that is, it consists of the zero transformation.
b) $\text{Im}(\text{Mat}) = M_{m \times n}(\mathbf{R})$.

10. a) Let A be an $m \times n$ matrix and denote the rows of A by $\mathbf{a}_1, \ldots, \mathbf{a}_m$. For operations of types (b) and (c) this is clear. For operations of type (a), if we add c times $\mathbf{a}_j$ to $\mathbf{a}_i$, the span of the rows of the resulting matrix is $\text{Span}\{\mathbf{a}_1, \ldots, \mathbf{a}_i + c\mathbf{a}_j, \ldots, \mathbf{a}_j, \ldots, \mathbf{a}_m\} \subseteq \text{Span}\{\mathbf{a}_1, \ldots, \mathbf{a}_m\}$. However, since $\mathbf{a}_i = (\mathbf{a}_i + c\mathbf{a}_j) - c\mathbf{a}_j$, it is also true that $\text{Span}\{\mathbf{a}_1, \ldots, \mathbf{a}_m\} \subseteq \text{Span}\{\mathbf{a}_1, \ldots, \mathbf{a}_i + c\mathbf{a}_j, \ldots, \mathbf{a}_j, \ldots, \mathbf{a}_m\}$. Thus the span of the rows of A is the same as the span of the rows of the transformed matrix.
b) Each non-zero row $\mathbf{a}_i$ from the echelon form matrix contains a leading term 1. Say the leading term in the ith row is in the $j(i)$th column. Let $c_1\mathbf{a}_1 + \cdots + c_k\mathbf{a}_k = \mathbf{0}$ be a linear relation among the $\mathbf{a}_i$, then the $j(i)$th components yield $c_i = 0$ for all i, $1 \leqslant i \leqslant k$.

c) If $\mathbf{a}_1, \ldots, \mathbf{a}_k$ are the non-zero rows of the echelon form matrix, then A has k basic variables, hence the rank of A is k. By part b, this is also the row rank of A, since the dimension of the span of the rows is k as well.

14. a) From our knowledge of derivatives from calculus, $T(f) = f'' = 0$ implies that $f(x) = ax + b$ for $a,b \in \mathbf{R}$. A basis for Ker(T) is $\{1,x\}$. No, $C^\infty(\mathbf{R})$ is not finite dimensional.

b) If $T(f) = f - f''$, $f - f'' = 0$ implies that $f = f''$. From our knowledge of calculus, $f(x) = e^x$ and $f(x) = e^{-x}$ satisfy this equation. Further, these are linearly independent.

Chapter 2, Section 4

1. a) both c) only surjective e) only surjective

2. a) both c) only injective e) both

3. a) $T^{-1}(\{\mathbf{w}\}) = \text{Ker}(T) = \text{Span}\{\mathbf{v}_2 + \mathbf{v}_4\}$.
b) $T^{-1}(\{\mathbf{w}\}) = \{\mathbf{v} \mid \mathbf{v} = (3\mathbf{v}_1 + 2\mathbf{v}_2) + s(\mathbf{v}_2 + \mathbf{v}_4),\ s \in \mathbf{R}\}$.

4. a) $\mathbf{x} = (1,2,0,0) + s(-1,-1,0,1) + t(-2,-7/2,1,0),\ s,t \in \mathbf{R}$.
c) $\mathbf{x} = (-1,1)$ e) $\mathbf{x} = (1,2)$

6. a) No. For example, let $U = V = \mathbf{R}^2$ and $S = I$ and $T = -I$. Then S and T are both injective and surjective, but $S + T$ is neither.

b) Yes. We have $\mathbf{v} \in \text{Ker}(S)$ if and only if $\mathbf{v} \in \text{Ker}(aS)$, and $\mathbf{w} \in \text{Im}(S)$ if and only if $\mathbf{w} \in \text{Im}(aS)$. Thus, S is injective (surjective) if and only if aS is injective (surjective).

7. a) We must show that for $\mathbf{a},\mathbf{b} \in \mathbf{R}^n$ and $\alpha,\beta \in \mathbf{R}$, we have that $*(\alpha\mathbf{a} + \beta\mathbf{b}) = \alpha*(\mathbf{a}) + \beta*(\mathbf{b})$, that is, we must show that each of these elements of $L(\mathbf{R}^n,\mathbf{R})$ agree on every $\mathbf{v} \in \mathbf{R}^n$. $*(\alpha\mathbf{a} + \beta\mathbf{b})(\mathbf{v}) = D_{\alpha a + \beta b}(\mathbf{v}) = \sum_{i=1}^{n} (\alpha a_i + \beta b_i)v_i = \alpha(\sum_{i=1}^{n} a_i v_i) + \beta(\sum_{i=1}^{n} b_i v_i) = \alpha D_a(\mathbf{v}) + \beta D_b(\mathbf{v}) = (\alpha D_a + \beta D_b)(\mathbf{v}) = (\alpha*(\mathbf{a}) + \beta*(\mathbf{b}))(\mathbf{v})$. Therefore $*$ is a linear transformation.

b) If $*(\mathbf{a}) = 0$ then $*(\mathbf{a})(\mathbf{e}_i) = a_i = 0$ for all i, thus $\mathbf{a} = \mathbf{0}$. Thus $*$ is injective. Since $\dim(\mathbf{R}^n) = \dim(L(\mathbf{R}^n,\mathbf{R})) = n$, $*$ is also surjective by (2.4.10).

9. a) $\text{Ker}(T|_U) = \{\mathbf{u} \in U \mid T(\mathbf{u}) = \mathbf{0}\} = U \cap \text{Ker}(T) = \{\mathbf{0}\}$. Hence $T|_U$ is injective.

b) Let $\mathbf{w} \in \text{Im}(T)$, so we can write $\mathbf{w} = T(\mathbf{v})$ for some $\mathbf{v} \in V$. Since $V = \text{Ker}(T) + U$, $\mathbf{v} = \mathbf{u}_0 + \mathbf{u}$ for some $\mathbf{u}_0 \in \text{Ker}(T)$, and $\mathbf{u} \in U$. But then $\mathbf{w} = T(\mathbf{v}) = T(\mathbf{u}_0) + T(\mathbf{u}) = \mathbf{0} + T(\mathbf{u}) = T(\mathbf{u})$. So every vector in Im($T$) is in Im($T|_U$). Hence $T|_U$ is surjective.

11. a) Let $a,b \in \mathbf{R}$ and $\mathbf{v} = \mathbf{u}_1 + \mathbf{u}_2$ and $\mathbf{w} = \mathbf{x}_1 + \mathbf{x}_2$ with $\mathbf{u}_1,\mathbf{x}_1 \in U_1$ and $\mathbf{u}_2,\mathbf{x}_2 \in U_2$. Then $T(a\mathbf{v} + b\mathbf{w}) = T((a\mathbf{u}_1 + b\mathbf{x}_1) + (a\mathbf{u}_2 + b\mathbf{x}_2)) = T_1(a\mathbf{u}_1 + b\mathbf{x}_1) + T_2(a\mathbf{u}_2 + b\mathbf{x}_2) = aT_1(\mathbf{u}_1) + bT_1(\mathbf{x}_1) + aT_2(\mathbf{u}_2) + bT_2(\mathbf{x}_2) = a(T_1(\mathbf{u}_1) + T_2(\mathbf{u}_2)) + b(T_1(\mathbf{x}_1) + T_2(\mathbf{x}_2)) = aT(\mathbf{v}) + bT(\mathbf{w})$. Therefore, T is linear.

b) If $T(\mathbf{v}) = T_1(\mathbf{u}_1) + T_2(\mathbf{u}_2) = \mathbf{0}$, then $T_1(\mathbf{u}_1) = -T_2(\mathbf{u}_2)$, so both vectors are in $U_1 \cap U_2$. However, $U_1 \cap U_2 = \{\mathbf{0}\}$ implies that $T(\mathbf{u}_1) = T_2(\mathbf{u}_2) = \mathbf{0}$. Since T_1 and T_2 are injective, $\mathbf{u}_1 = \mathbf{u}_2 = \mathbf{0}$ and $\mathbf{u} = \mathbf{0}$. Therefore T is injective.

c) If $\mathbf{v} = \mathbf{u}_1 + \mathbf{u}_2$, since T_1, T_2 are surjective there are vectors $\mathbf{x}_1 \in U_1$ and $\mathbf{x}_2 \in U_2$ so that $T_1(\mathbf{x}_1) = \mathbf{u}_1$ and $T_2(\mathbf{x}_2) = \mathbf{u}_2$. Thus $T(\mathbf{x}_1 + \mathbf{x}_2) = \mathbf{v}$, and T is surjective.

Chapter 2, Section 5

1. a) $(1/2)\begin{bmatrix} 1 & -1 \\ 1 & -1 \end{bmatrix}$ c) $\begin{bmatrix} \cos(\phi+\theta) & -\sin(\phi+\theta) \\ \sin(\phi+\theta) & \cos(\phi+\theta) \end{bmatrix}$

2. (ii) Let $\mathbf{u} \in U$. $(T(R+S))(\mathbf{u}) = T((R + S)(\mathbf{u})) = T(R(\mathbf{u}) + S(\mathbf{u})) = T(R(\mathbf{u})) + T(S(\mathbf{u}))$ since T is a linear transformation. This equals $(TR)(\mathbf{u}) + (TS)(\mathbf{u}) = (TR + TS)(\mathbf{u})$.
   (iii) The proof follows the same pattern as the proof of (ii).

3. (ii) Let $\mathbf{w} \in \text{Im}(TS)$. Then $\mathbf{w} = TS(\mathbf{u})$ for $\mathbf{u} \in U$. It follows that $\mathbf{w} = T(S(\mathbf{u}))$ and $\mathbf{w} \in \text{Im}(T)$. Therefore, $\text{Im}(TS) \subset \text{Im}(T)$.

5. a) $\begin{bmatrix} 7 & -3 \\ 2 & -2 \\ 19 & 5 \end{bmatrix}$ b) $\begin{bmatrix} -4 & 10 & 1 \\ 7 & -2 & 6 \\ -2 & 4 & 0 \\ 2 & 2 & 3 \end{bmatrix}$ c) $[33\ 9]$

6. a) Let $\mathbf{u} \in U$. $(IS)(\mathbf{u}) = I(S(\mathbf{u})) = S(\mathbf{u})$. Thus $IS = S$.

7. Let e_{ij} denote the entries of I: $e_{ii} = 1$, and $e_{ij} = 0$, $i \neq j$.
   a) Then the ijth entry of IA is $\sum_{k=1}^{n} e_{ik}a_{kj} = e_{ii}\, a_{ij} = a_{ij}$, the ijth entry of A. Thus $IA = A$.

9. a) It suffices to show that the ijth entry of AB is zero for $i > j$. The ijth entry of AB is $\sum_{k=1}^{n} a_{ik}b_{kj}$. Since $a_{ik} = 0$ for $k < i$ and $b_{kj} = 0$ for $k > j$, and since $i > j$ we have that one of $k < i$ and $k > j$ holds for all k. Hence for all k, one of b_{kj} and a_{ik} equals 0. Thus the ijth entry of AB is zero for $i > j$.
   b) Repeat the above argument replacing $i > j$ by $i < j$ and $k > j$ by $k < j$.

11. First Proof: The conditions are equivalent to saying that $a_{ij} = b_{ij} = 0$ if $(k+1) \leq i \leq (k+l)$ and $1 \leq j \leq k$. The ijth entry of AB is $\sum_{m=1}^{k+l} a_{im}b_{mj}$. If $(k+1) \leq i \leq (k+l)$ and $1 \leq m \leq k$ then $a_{im} = 0$. If $1 \leq j \leq k$, then $b_{mj} = 0$ for $k+1 \leq m \leq k+l$. Thus one of a_{im} and b_{mj} is zero for all m. Thus the sum is zero as we desired.
   Second Proof: This statement can also be proved by a more conceptual argument as follows. Let V be a $(k+l)$-dimensional vector space and let α be a basis for V, $\alpha = \{\mathbf{v}_1, \ldots, \mathbf{v}_k, \mathbf{v}_{k+1}, \ldots, \mathbf{v}_{k+l}\}$. Let T and $S \in L(V,W)$ be the linear mappings with $[T]_\alpha^\alpha = A$ and $[S]_\alpha^\alpha = B$. By the form of the matrices, for each i, $1 \leq i \leq k$, $T(\mathbf{v}_i) \in \text{Span}\{\mathbf{v}_1, \ldots, \mathbf{v}_k\}$ and $S(\mathbf{v}_i) \in \text{Span}\,\{\mathbf{v}_1, \ldots, \mathbf{v}_k\}$. The matrix product $AB = [TS]_\alpha^\alpha$. But if $i \leq k$,

$$\begin{aligned} TS(\mathbf{v}_i) &= T(S(\mathbf{v}_i)) \\ &= T(c_1\mathbf{v}_1 + \cdots + c_k\mathbf{v}_k), \text{ for some } c_i, \\ &= c_1T(\mathbf{v}_1) + \cdots + c_k\, T(\mathbf{v}_k). \end{aligned}$$

Since each $T(\mathbf{v}_i) \in \text{Span}\{\mathbf{v}_1, \ldots, \mathbf{v}_k\}$, $TS(\mathbf{v}_i) \in \text{Span}\{\mathbf{v}_1, \ldots, \mathbf{v}_k\}$ for each $1 \leqslant i \leqslant k$, so $[TS]_\alpha^\alpha$ will also have an $l \times k$ block of zeroes in the lower left corner.

14. The pqth entry of $E_{ij}(c)A$ is $\sum_{r=1}^{m} e_{pr}a_{rq}$. If $p \neq i$, then $e_{pr} = 0$ for $p \neq r$ and $e_{pp} = 1$. Thus the pqth entry is a_{pq} if $p \neq i$. If $p = i$, then $e_{pr} = 0$ for $p \neq r$ and $p \neq j$, $e_{pp} = 1$ and $e_{pj} = c$. Thus the pqth entry is $a_{pq} + ca_{jq}$ if $p = i$. Therefore $E_{ij}(c)A$ is the matrix obtained from A by adding c times the jth row to the ith row.

17. Note: the answers given for these problems are not the only correct ones.
   a) $E_{12}(-2)F_2(-1/5)E_{21}(-2)G_{12}$.
   b) $E_{12}(-1/2)F_1(1/2)G_{12}$.

19. The ijth entry of $(AB)^t$ is $\sum_{k=1}^{n} a_{jk}b_{ki} = \sum_{k=1}^{n} b_{ki}a_{jk}$. This is the same as the ijth entry of B^tA^t, since the ikth entry of B^t is b_{ki} and the kjth entry of A^t is a_{jk}.

Chapter 2, Section 6

1. a) No. $\text{Ker}(T) = \text{Span}\{1\}$.
   c) Yes. $T^{-1}(x_1,x_2,x_3) = x_1 + ((-3/2)x_1 + 2x_2 - (1/2)x_3)x + ((1/2)x_1 - x_2 + (1/2)x_3)x^2$.

2. b) No. $\dim(V) = 1$ and $\dim(W) = 2$.
   c) Yes. Let E_{ij} denote the 2×3 matrix all of whose entries are zero except the ijth entry, which is 1. Define T by $T(1) = E_{11}$, $T(x) = E_{12}$, $T(x^2) = E_{13}$, $T(x^3) = E_{21}$, $T(x^4) = E_{22}$, and $T(x^5) = E_{23}$.

3. a) $(-1/7)\begin{bmatrix} 2 & -3 \\ -3 & 1 \end{bmatrix}$ c) $\begin{bmatrix} 1/2 & 0 \\ -1/6 & 1/3 \end{bmatrix}$ e) $\begin{bmatrix} 3/2 & -1/2 & -1 \\ 1 & 0 & -1 \\ -1/2 & 1/2 & 1 \end{bmatrix}$
   h) The matrix is not invertible.

6. Since T is an isomorphism, T^{-1} exists and $TT^{-1} = I_W$ and $T^{-1}T = I_V$. These equations also say that T^{-1} is invertible.

9. a) Let $\{\mathbf{u}_1, \ldots, \mathbf{u}_k\}$ be a basis for U. Since every vector in $\text{Im}(T|_U)$ has the form $T(a_1\mathbf{u}_1 + \cdots + a_k\mathbf{u}_k) = a_1T(\mathbf{u}_1) + \cdots + a_kT(\mathbf{u}_k)$, it follows that $\{T(\mathbf{u}_1), \ldots, T(\mathbf{u}_k)\}$ spans $T(U)$. From Exercise 8, $T(\mathbf{u}_1), \ldots, T(\mathbf{u}_k)$ are also independent, thus $\dim(U) = \dim(T|_U)$and $T|_U$ is an isomorphism.
   b) By Exercise 6, T^{-1} is an isomorphism. Thus, by part a, $T^{-1}|_X$ is an isomorphism also.

10. a) This will follow from Exercise 9, if we show $\text{Ker}(S) = T(\text{Ker}(ST))$. If $\mathbf{v} \in \text{Ker}(ST)$, then $ST(\mathbf{v}) = \mathbf{0}$, which implies that $T(\mathbf{v}) \in \text{Ker}(S)$. Hence $T(\text{Ker}(ST)) \subset \text{Ker}(S)$. Conversely, if $S(\mathbf{w}) = \mathbf{0}$, by the surjectivity of T there is a $\mathbf{v} \in V$ with $T(\mathbf{v}) = \mathbf{w}$. But $ST(\mathbf{v}) = \mathbf{0}$. Thus, $\text{Ker}(S) = T(\text{Ker}(ST))$. If $\mathbf{x} \in \text{Im}(S)$, $\mathbf{x} = S(\mathbf{w})$ for some $\mathbf{w} \in W$. Since T is surjective, there is a $\mathbf{v} \in V$ with $T(\mathbf{v}) = \mathbf{w}$. Thus $\mathbf{x} = ST(\mathbf{v})$ and $\mathbf{x} \in \text{Im}(ST)$. The opposite inclusion follows from Exercise 3, Section 2.5. Thus, $\text{Im}(S) = \text{Im}(ST)$ and their dimensions are equal.

14. a) $A^{-1} = F_2(2)E_{12}(1)E_{21}(-3)F_1(1/2) = \begin{bmatrix} -1 & 1 \\ -3 & 2 \end{bmatrix}$

c) $A^{-1} = E_{12}(1)E_{13}(-2)E_{23}(1/3)F_3(-3/16)E_{32}(-2)F_2(1/3)E_{31}(-3)E_{21}(-1)$

$$= (1/16)\begin{bmatrix} -1 & 2 & 5 \\ -3 & 6 & -1 \\ 7 & 2 & -3 \end{bmatrix}$$

Chapter 2, Section 7

1. a) $\begin{bmatrix} \cos(\theta) & \sin(\theta) \\ -\sin(\theta) & \cos(\theta) \end{bmatrix}$ b) $\begin{bmatrix} \sin(\theta) + \cos(\theta) & \sin(\theta) \\ -2\sin(\theta) & \cos(\theta) - \sin(\theta) \end{bmatrix}$

2. $[T]_\beta^\beta = \begin{bmatrix} -1 & 4 & -1 \\ 0 & 3 & -1 \\ 0 & 0 & 2 \end{bmatrix}$

4. a) $\begin{bmatrix} 4 & 0 \\ 0 & -2 \end{bmatrix}$ c) $(1/2)\begin{bmatrix} 1 & -3 & 0 \\ 5 & 3 & 6 \\ 3 & 5 & 2 \end{bmatrix}$

6. a) If $A = Q^{-1}BQ$, then since Q is the matrix of an isomorphism it follows that when A and B are considered as the matrices of linear transformations, the dimensions of the images of A and B are the same, thus they have the same rank.
   b) A is invertible if and only if the rank of $A = n$, which is true if and only if the rank of $B = n$, which is true if and only if B is invertible. If $B = QAQ^{-1}$, then $B^{-1} = QA^{-1}Q^{-1}$.

8. Suppose that $A = QBQ^{-1}$. Let α be the standard basis for $\mathbf{R}^n$ and let T satisfy $[T]_\alpha^\alpha = A$. Then if β is given by the columns of Q, which necessarily form a basis since Q is invertible, it follows that $Q = [I]_\beta^\alpha$. Thus $B = Q^{-1}AQ = [I]_\beta^{\alpha\,-1}[T]_\alpha^\alpha[I]_\beta^\alpha = [T]_\beta^\beta$.

Chapter 2, Supplementary Exercises

1. a) False. A counterexample is $T(\mathbf{x}) = (x_1^2 + x_2^2, x_1x_2)$.
   b) False. $\begin{bmatrix} 1 & 1 \\ 0 & 1 \end{bmatrix}$ is injective and surjective but not a rotation.
   c) True. Since $\dim(\text{Im}(T)) \leq \dim(V)$ by the dimension theorem.
   d) False. Consider the zero transformation.
   e) True. Apply (2.4.8)
   f) True. A transformation is determined by its matrix with respect to a choice of basis and vice versa.
   g) False. $(TS)^{-1} = S^{-1}T^{-1}$.
   h) False. Let $V = W$. cI is an isomorphism for all $c \neq 0$.
   i) True. If $Q^{-1}AQ = C = P^{-1}BP$, then $A = (PQ^{-1})^{-1}B(PQ^{-1})$.
   j) False. Let $T = I$.

3. a) $\begin{bmatrix} 1 & 0 & -1 \\ -1 & 1 & 0 \\ 0 & -1 & 1 \end{bmatrix}$ b) $\begin{bmatrix} 0 & 1 & 0 \\ 0 & -1 & -5/2 \\ 0 & 3 & 15/2 \end{bmatrix}$

4. a) $\dim(\text{Ker}(T)) = 1$ and $\dim(\text{Im}(T)) = 2$.
   b) $\{(1,1,1)\}$ is a basis for $\text{Ker}(T)$ and $\{(1,-1,0), (0,1,-1)\}$ is a basis for $\text{Im}(T)$ in standard coordinates.

5. Neither injective nor surjective.

Chapter 3, Section 1

1. If $\mathbf{a}_1$ and $\mathbf{a}_2$ are dependent, say $\mathbf{a}_2 = c\mathbf{a}_1$, then $a_{11}a_{22} - a_{12}a_{21} = a_{11}(ca_{12}) - a_{12}(ca_{11}) = 0$, and the area is zero. If the vectors are independent, then $\mathbf{a}_1 \neq \mathbf{0}$ and $\| \mathbf{a}_1 \| \neq 0$. Further, $b \neq 0$, where b is the length of the perpendicular from $\mathbf{a}_2$ to $\mathbf{a}_1$. But the area is $b \| \mathbf{a}_1 \|$ and is therefore not zero.

2. (i) Area $(b\mathbf{a}_1 + c\mathbf{a}_1', \mathbf{a}_2) = (ba_{11} + ca_{11}')a_{22} - (ba_{12} + ca_{12}')a_{21}$
$= b(a_{11}a_{22} - a_{12}a_{21}) + c(a_{11}'a_{22} - a_{12}'a_{21})$
$= b\text{Area}(\mathbf{a}_1,\mathbf{a}_2) + c\text{Area}(\mathbf{a}_1',\mathbf{a}_2)$.
(i′) is similar to (i).
(ii) $\text{Area}(\mathbf{a}_1, \mathbf{a}_2) = a_{11}a_{22} - a_{12}a_{21} = -(a_{21}a_{12} - a_{22}a_{11})$
$= -\text{Area}(\mathbf{a}_2,\mathbf{a}_1)$.
(iii) $\text{Area}((1,0), (0,1)) = 1 \cdot 1 - 0 \cdot 0 = 1$.

3. a) 21 b) −4 c) 36.

6. $\det(A^{-1}) = (\det(A))^{-1}$.

7. $f(A) = f(\mathbf{a}_1,\mathbf{a}_2) = f(a_{11}\mathbf{e}_1 + a_{12}\mathbf{e}_2, a_{21}\mathbf{e}_1 + a_{22}\mathbf{e}_2)$
$= a_{11}f(\mathbf{e}_1, a_{21}\mathbf{e}_1 + a_{22}\mathbf{e}_2) + a_{12}f(\mathbf{e}_2, a_{21}\mathbf{e}_1 + a_{22}\mathbf{e}_2)$
$= a_{11}a_{22}f(\mathbf{e}_1,\mathbf{e}_2) + a_{12}a_{21}f(\mathbf{e}_2,\mathbf{e}_1)$
$= (a_{11}a_{22} - a_{12}a_{21})f(\mathbf{e}_1,\mathbf{e}_2)$
$= \det(A) \cdot f(I)$

Chapter 3, Section 2

1. a) 2 c) 2 e) 56

2. a) 0 c) 21

3. a) 0 c) 0

4. a) Any $a \in \mathbf{R}$ c) Any $a \in \mathbf{R}\setminus\{0\}$.

6. a) We will prove this by induction on n. This is trivially true for a 1×1 matrix. Now assume this holds for $(n-1) \times (n-1)$ diagonal matrices. Let A be an $n \times n$ diagonal matrix. Then expanding along the first row of A, $\det(A) = a_{11}\det(A_{11}) + 0 \cdot \det(A_{12}) + \cdots + 0 \cdot \det(A_{1n}) = a_{11}\det(A_{11})$. But A_{11} is an $(n-1) \times (n-1)$ diagonal matrix. Thus $\det(A_{11}) = a_{22} \ldots a_{nn}$. It follows that $\det(A) = a_{11} \cdot \ldots \cdot a_{nn}$.

7. By induction on k, the size of the upper left block. If $k = 1$, then $A = (a_{11})$ and we have a matrix of the form $\begin{bmatrix} a_{11} & 0 \ldots 0 \\ 0 & \\ \vdots & B \\ 0 & \end{bmatrix}$. Expanding along the first row,

$$\det \begin{bmatrix} a_{11} & 0 \cdots 0 \\ 0 & \\ \vdots & B \\ 0 & \end{bmatrix} = a_{11}\det(B) + 0 + \cdots + 0 = \det(A)\det(B).$$

Now suppose the result is true for $k = n$, and consider a matrix $\begin{bmatrix} A & 0 \\ 0 & B \end{bmatrix}$ where $A \in M_{(n+1)\times(n+1)}(\mathbf{R})$. As in the proof of Proposition (3.2.13), we can perform row operations of type (a) to simplify the matrix without changing its determinant. If all the entries in the first column of A are zero, then $0 = \det\begin{bmatrix} A & 0 \\ 0 & B \end{bmatrix} = \det(A) \cdot \det(B) = 0 \cdot \det(B)$. If there is some non-zero entry in the first column, then we can do row operations to eliminate all other entries from that column and we have

$$\det\begin{bmatrix} A & 0 \\ 0 & B \end{bmatrix} = \det\left[\begin{array}{cc|c} 0 & & \\ \vdots & & \\ a_{i1} & A' & 0 \\ \vdots & & \\ 0 & & \\ \hline & 0 & B \end{array}\right]$$

Expanding along the ith row we have

$$= (-1)^{1+i} a_{i1} \det\left[\begin{array}{c|c} A'_{i1} & 0 \\ \hline 0 & B \end{array}\right] + 0 + \cdots + 0$$

where A'_{i1} is the $n \times n$ matrix obtained by deleting the ith row of A', and the remaining terms are all zero since those minors all contain an entire column of zeroes (the 1st column). By induction we have

$$= (-1)^{1+i} a_{i1} \det(A'_{i1}) \cdot \det(B).$$

But $(-1)^{1+i} a_{i1} \det(A'_{i1}) = \det(A)$, because the same row operations could be performed on A directly. Hence $\det\begin{bmatrix} A & 0 \\ 0 & B \end{bmatrix} = \det(A)\det(B)$.

11. (Assuming $\mathbf{a}_1,\mathbf{a}_2$ lie in the $x_1 - x_2$ plane.)
The volume of the parallelepiped is the area of the face in the $x_1 - x_2$ plane multiplied by the height in the x_3 direction. This height is the x_3-coordinate. Thus, the volume is $(a_{11}a_{22} - a_{12}a_{21})a_{33}$, which is the determinant of $\begin{bmatrix} a_{11} & a_{12} & 0 \\ a_{21} & a_{22} & 0 \\ a_{31} & a_{32} & a_{33} \end{bmatrix}$.

12.

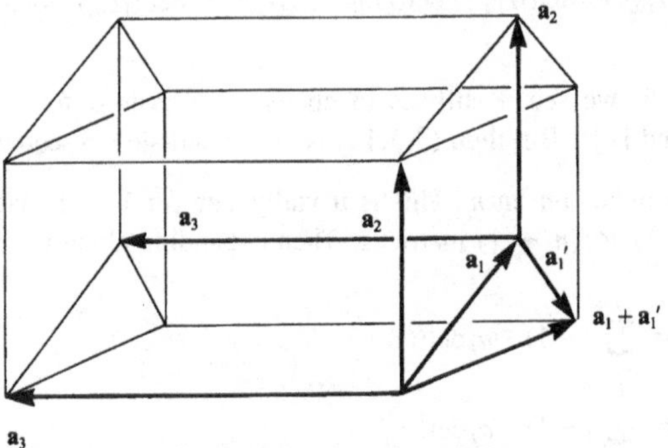

Figure for Exercise 12.

Volume $(\mathbf{a}_1 + \mathbf{a}'_1,\mathbf{a}_2,\mathbf{a}_3)$ = Volume$(\mathbf{a}_1,\mathbf{a}_2,\mathbf{a}_3)$ + Volume$(\mathbf{a}'_1,\mathbf{a}_2,\mathbf{a}_3) + V_1 - V_1$ = Volume$(\mathbf{a}_1,\mathbf{a}_2,\mathbf{a}_3)$ + Volume$(\mathbf{a}'_1,\mathbf{a}_2,\mathbf{a}_3)$, where V_1 is the volume of the prism-shaped region in the figure.

13. It suffices to prove that the ith entries of each equality are correct for each i. Note that the ith entry of $\mathbf{x} \times \mathbf{y}$ is $(-1)^{i+1}(x_j y_k - y_j x_k)$, where $j \neq i \neq k$ and $j < k$.

a) The ith entry of $(a\mathbf{x} + b\mathbf{y}) \times \mathbf{z}$ is $(-1)^{i+1}((ax_j + by_j)z_k - z_j(ax_k + by_k)) = a(-1)^{i+1}(x_j z_k - z_j x_k) + b(-1)^{i+1}(y_j z_k - z_j y_k)$, which is the ith entry of $a(\mathbf{x} \times \mathbf{z}) + b(\mathbf{y} \times \mathbf{z})$.

b) The ith entry of $\mathbf{x} \times \mathbf{y}$ is $(-1)^{i+1}(x_j y_k - y_j x_k) = (-1)(-1)^{i+1}(y_j x_k - x_j y_k)$, which is the ith entry of $-\mathbf{y} \times \mathbf{x}$.

15. It suffices to compute the determinant of the matrix whose rows are $\mathbf{x}, \mathbf{y}, \mathbf{x} \times \mathbf{y}$. This determinant is $\sum_{i=1}^{3} (x_j y_k - y_j x_k)^2$, $j \neq i \neq k$. This is a sum of squares not all of which are zero, thus it is not zero.

16. a) (2,6,2) b) (0,0,3) c) (0,0,0).

Chapter 3, Section 3

1. a) $(1/2)\begin{bmatrix} 0 & 2 \\ 1 & -3 \end{bmatrix}$ c) $(1/7)\begin{bmatrix} 18 & -11 & 4 & -1 \\ -11 & 11 & -4 & 1 \\ 4 & -4 & 4 & -1 \\ -1 & 1 & -1 & 2 \end{bmatrix}$

2. a) -2 b) 5 c) 7

3. a) Let $\mathbf{a}_i$, $i = 1, \ldots, n$ denote the rows of A. A calculation shows that if $\mathbf{a}_j$ is replaced by $c\mathbf{a}_j + d\mathbf{a}'_j$ to produce a matrix A'', then only the jth row of $A''B$ differs from the jth row of AB. The jth row of $A''B$ is $c \cdot$ (jth row of AB) $+ d \cdot$ (jth row of $A'B$), where A' is the matrix differing from A only in that its jth row is $\mathbf{a}'_j$. Then $f(A'') = \det(A''B) = c\det(AB) + d\det(A'B) = cf(A) + df(A')$. Thus f is a multilinear function of the rows of A.

b) Notice that interchanging the ith and jth rows of A interchanges the ith and jth rows of AB. Thus, if A' is A with the ith and jth rows interchanged, then $f(A') = \det(A'B) = -\det(AB) = -f(A)$. Thus f is an alternating function of the rows of A.

5. From Theorem (2.7.5), $[T]_\alpha^\alpha = [I]_\beta^\alpha[T]_\beta^\beta[I]_\alpha^\beta$ and $[I]_\beta^\alpha = ([I]_\alpha^\beta)^{-1}$. Thus $\det([T]_\alpha^\alpha) = \det([I]_\beta^\alpha[T]_\beta^\beta[I]_\alpha^\beta) = \det([I]_\alpha^{\beta-1})\det([T]_\beta^\beta)\det([I]_\alpha^\beta) = \det(([I]_\alpha^\beta)^{-1}\det([T]_\beta^\beta)\det([I]_\alpha^\beta) = \det([T]_\beta^\beta)$.

6. From Exercise 5, we see it suffices to choose any basis α for V and work with the matrices $[S]_\alpha^\alpha$ and $[T]_\alpha^\alpha$. But then (3.3.12) is an immediate consequence of (3.3.7).

9. We proceed by induction on n. This is trivially true for 1×1 matrices and assume it holds for $(n-1) \times (n-1)$ matrices. Then expanding along the jth column of A'

$$\begin{aligned} \det(A') &= \sum_{i=1}^{n} (-1)^{i+j} a_{ji} \det((A')_{ij}) \\ &= \sum_{i=1}^{n} (-1)^{i+j} a_{ji} \det(A_{ji}) \end{aligned}$$

by the induction hypothesis, since $(A')_{ij} = (A_{ji})^t$. The last expression is $\det(A)$ expanded along the jth row.

Chapter 3, Supplementary Exercises

1. a) True. The matrix is not invertible.
   b) True. Apply the multilinearity of det n times.
   c) False. cI, $c \neq 0$, is an isomorphism for any V. If V is finite dimensional and $c \neq 1$, then $\det(cI) = c^{\dim(V)} \neq 1$.
   d) False. Consider I and $\begin{bmatrix} 0 & 1 \\ 1 & 0 \end{bmatrix}$.
   e) True. Apply the alternating property of det.
   f) False. A must be invertible so that $\det(A) \neq 0$.
   g) True. Apply (3.3.7).
   h) False. For example, $\det\begin{bmatrix} 0 & 1 \\ 1 & 0 \end{bmatrix} = -1$.
   i) False. $\det(cA) = c^n\det(A)$.
   j) False. Let $A = I_{2\times 2}$ and $B = -I_{2\times 2}$.

2. a) 2 b) 6

4. a) 0 b) 120

8. If both A and B are invertible, then $\det(A) \neq 0$, and $\det(B) \neq 0$, so $\det(AB) = \det(A)\det(B) \neq 0$. This is a contradiction, thus one of $\det(A)$ or $\det(B)$ is zero; that is, A or B is not invertible.

9. a) $$\det(V) = (bc^2 - cb^2) - (ac^2 - ca^2) + (ab^2 - ba^2)$$
   $$= (c - b)(c - a)(b - a).$$

   V is invertible if and only if $\det(V) \neq 0$, if and only if $c \neq b$, $c \neq a$, and $b \neq a$. In other words, V is invertible if and only if a, b, and c are distinct real numbers.

   b) Let $a_1, \ldots, a_n$ be real numbers. Let V_n be the matrix whose ith column is $[1\ a_i\ a_i^2 \ldots a_i^{n-1}]^t$. We will prove that

   $$\det(V_n) = \prod_{i<j} (a_j - a_i).$$

   Consider $\det(V_n)$ as a polynomial in $a_1, \ldots, a_n$. From the properties of determinants we know that if $a_i = a_j$ for $i \neq j$, so that the ith column of V_n equals the jth column, then $\det(V_n) = 0$. Thus $(a_j - a_i)$ is a factor of $\det(V_n)$ for all pairs i and j with $i < j$. This gives $j - 1$ factors for each j, and hence $(n - 1) + \cdots + 2 + 1 = (1/2)n(n - 1)$ factors in all. However, the total degree of $\det(V_n)$ as a polynomial in the a_i is exactly $(n - 1) + \cdots + 2 + 1$. Thus, up to a scalar, the product of the factors is the determinant. The scalar may be determined, for example by comparing the coefficients of $a_n^{\,n-1} \cdot a_{n-1}^{\,n-2} \cdot \ldots \cdot a_2 \cdot 1$ in the two expressions. This coefficient is 1 in both cases, so $\det(V_n) = \prod_{i<j} (a_j - a_i)$, as claimed.

Chapter 4, Section 1

1. a) $\lambda = -5$ c) $\lambda = -1$

2. a) $\lambda^2 + 6\lambda + 5$ c) $\lambda^4 + 10\lambda^3 + 35\lambda^2 + 50\lambda + 24$

3. a) $\lambda = -2, 5$. $E_{-2} = \text{Span}\{(-4/3, 1)\}$ and $E_5 = \text{Span}\{(1,1)\}$.
   c) $\lambda = 0,1,2$. $E_0 = \text{Span}\{(-1,0,1)\}$, $E_1 = \text{Span}\{(0,1,0)\}$, and $E_2 = \text{Span}\{(1,0,1)\}$.

e) $\lambda = -1,1,3$. $E_{-1} = \text{Span}\{(-1,1,0,0)\}$, $E_1 = \text{Span}\{(5/2,-2,-1,1)\}$, and $E_3 = \text{Span}(\{(1,1,0,0),(0,0,1,1)\})$.

g) $\lambda = 1,4$. $E_1 = \text{Span}\{e^x\}$ and $E_4 = \text{Span}\{e^{2x}\}$.

4. Let $\mathbf{x} \in E_0$ then $T(\mathbf{x}) = 0 \cdot \mathbf{x} = \mathbf{0}$ so $\mathbf{x} \in \text{Ker}(T)$. Conversely, if $\mathbf{x} \in \text{Ker}(T)$, $T(\mathbf{x}) = \mathbf{0} = 0 \cdot \mathbf{x}$, so $\mathbf{x}$ is either an eigenvector with eigenvalue 0 or $\mathbf{x} = \mathbf{0}$. Thus, $E_0 = \text{Ker}(T)$.

5. a) The constant term in $\det(A - \lambda I)$ is the value of the polynomial with $\lambda = 0$. This is $\det(A)$.

6. a) If A is a 1×1 matrix, there is nothing to prove. If A is a 2×2 matrix, then $\det\begin{bmatrix} a_{11} - \lambda & a_{12} \\ a_{21} & a_{22} - \lambda \end{bmatrix} = (a_{11} - \lambda)(a_{22} - \lambda) - a_{12}a_{21}$. So the statement is true in this case as well. Now assume the result is true for $k \times k$ matrices, and consider

$$\det\begin{bmatrix} a_{11} - \lambda & a_{12} & \cdots & a_{1,k+1} \\ a_{21} & a_{22} - \lambda & \cdots & a_{2,k+1} \\ \vdots & \vdots & & \vdots \\ a_{k+1,1} & a_{k+1,2} & \cdots & a_{k+1,k+1} - \lambda \end{bmatrix}.$$

Expanding along the first row yields

$$= (a_{11} - \lambda)\det\begin{bmatrix} a_{22} - \lambda & & a_{2,k+1} \\ & \ddots & \\ a_{k+1,2} & & a_{k+1,k+1} - \lambda \end{bmatrix} + \sum_{j=2}^{k+1}(-1)^{1+j}a_{1j}\det((A - \lambda I)_{1j})$$

$((A - \lambda I)_{1j}$ is the $1j$ minor of $A - \lambda I)$. By induction, the first term is $(a_{11} - \lambda)[(a_{22} - \lambda)\cdots(a_{k+1,k+1} - \lambda) + \text{terms of degree} \leq k - 2 \text{ in } \lambda] = (a_{11} - \lambda)\cdots(a_{k+1,k+1} - \lambda) + \text{terms of degree} \leq k - 1$ in λ. The other terms contain at most terms of degree $k - 1$ in λ, since there are only $(k - 1)$ entries containing λ in each of those minors. Hence the sum has the form $(a_{11} - \lambda)\cdots(a_{k+1,k+1} - \lambda) + \text{terms of degree} \leq (k + 1) - 2$ as claimed.

10. a) This follows from the fact that every polynomial of degree 3 with real coefficients has at least one real root. (*Proof:* Consider the equation $p(x) = 0$. We can always divide through by the leading coefficient to put the equation in the form $x^3 + ax^2 + bx + c = 0$, $a,b,c \in \mathbf{R}$. Every polynomial defines a continuous function $p\colon \mathbf{R} \to \mathbf{R}$. For x sufficiently negative (say, $x < A$), $p(x) < 0$; on the other hand, for x sufficiently positive (say, $x > B$), $p(x) > 0$, since the degree is odd. Hence, by the intermediate value theorem for continuous functions, there is at least one solution x_0 of $p(x) = 0$ between A and B. This proves the existence of a root. The same argument shows that every polynomial of odd degree with real coefficients has at least one real root.) The matrix $A = \begin{bmatrix} 0 & -1 & 0 \\ 1 & 0 & 0 \\ 0 & 0 & 1 \end{bmatrix}$ has characteristic polynomial $p(t) = \det(A - tI) = -(t - 1)(t^2 + 1) = -(t - 1)(t^2 + 1)$, which has only one real root $t = 1$.

11. a) Since the determinant of an upper-triangular matrix is the product of the diagonal entries,

$$\det\begin{bmatrix} a_{11}-\lambda & & * \\ 0 & \ddots & \\ \vdots & \ddots & \\ 0 & \cdots \; 0 & a_{nn}-\lambda \end{bmatrix} = (a_{11}-\lambda)\cdots(a_{nn}-\lambda).$$

13. a) If A is an eigenvalue of A, then there exists a vector $\mathbf{x} \neq \mathbf{0}$ with $A\mathbf{x} = \lambda\mathbf{x}$. Hence, $A^2\mathbf{x} = A(\lambda\mathbf{x}) = \lambda A\mathbf{x} = \lambda^2\mathbf{x}$, $A^3\mathbf{x} = A(\lambda^2\mathbf{x}) = \lambda^2 A\mathbf{x} = \lambda^3\mathbf{x}$, and so (by induction) $A^m\mathbf{x} = \lambda^m\mathbf{x}$ for all $m \geq 1$. As a result $\mathbf{x}$ is an eigenvector of A^m with eigenvalue λ^m.

14. a) $\begin{bmatrix} 1 & -1 \\ 1 & -1 \end{bmatrix}$ is a nilpotent 2×2 matrix ($A^2 = 0$)

$\begin{bmatrix} 1 & -2 & 1 \\ 1 & -2 & 1 \\ 1 & -2 & 1 \end{bmatrix}$ is a nilpotent 3×3 matrix ($A^2 = 0$).

b) Let λ be an eigenvalue of A. Then by Exercise 13, λ^m is an eigenvalue of A^m for all $m \geq 1$. Take $m = k$ for which $A^k = 0$. Then λ^k is an eigenvalue of $A^k = 0$. But the only eigenvalue of the zero matrix is zero. Hence, $\lambda^k = 0$, so $\lambda = 0$.

15. a) Let λ be an eigenvalue of T. Then by Exercise 13, λ^2 is an eigenvalue of $T^2 = I$. But the only eigenvalue of I is 1. Hence $\lambda^2 = 1$, so $\lambda = 1$ or $\lambda = -1$.

b) Let $\mathbf{x} \in E_{+1} \cap E_{-1}$. Then we have $T(\mathbf{x}) = \mathbf{x}$ and $T(\mathbf{x}) = -\mathbf{x}$. Hence $\mathbf{x} = -\mathbf{x}$, so $\mathbf{x} = \mathbf{0}$, and $E_{+1} \cap E_{-1} = \{\mathbf{0}\}$.

c) For any $\mathbf{x} \in V$, $\mathbf{x} = (1/2)(\mathbf{x} + T(\mathbf{x})) + (1/2)(\mathbf{x} - T(\mathbf{x}))$. Notice that $T((1/2)\mathbf{x} + T(\mathbf{x})) = (1/2)(T(\mathbf{x}) + T^2(\mathbf{x})) = (1/2)(T(\mathbf{x}) + \mathbf{x})$, since $T^2 = I$. Hence $(1/2)(\mathbf{x} + T(\mathbf{x})) \in E_1$. Similarly, $T((1/2)(\mathbf{x} - T(\mathbf{x}))) = (1/2)(T(\mathbf{x}) - T^2(\mathbf{x})) = (1/2)(T(\mathbf{x}) - \mathbf{x})$, since $T^2 = I$ again, and this equals $-(1/2)(\mathbf{x} - T(\mathbf{x}))$. So $(1/2)(\mathbf{x} - T(\mathbf{x})) \in E_{-1}$.

Chapter 4, Section 2

1. a) $\lambda = 2,2$. $\dim(E_2) = 1 \neq m_2 = 2$. A is not diagonalizable.

c) $\lambda = -1,0,1$. $\dim(E_0) = 1 = m_1$ and $E_0 = \text{Span}\{(1,0,1)\}$. $\dim(E_{-1}) = 1 = m_{-1}$ and $E_{-1} = \text{Span}\{(3/2,-1/2,1)\}$. $\dim(E_1) = 1 = m_1$ and $E_1 = \text{Span}\{(3/2,1/2,1)\}$. A is diagonalizable and a basis of eigenvectors is $\{(1,0,1), (3/2,-1/2,1), (3/2,1/2,1)\}$.

e) $\lambda = 3,2$ and the remaining roots of the characteristic polynomial are not real numbers, so A is not diagonalizable.

g) $\lambda = 0,3,8$ and the remaining roots of the characteristic polynomial are not real, so A is not diagonalizable.

4. If T is diagonalizable, then we have a basis $\alpha = \{\mathbf{x}_1, \ldots, \mathbf{x}_n\}$ consisting of eigenvectors for T. Since $\dim(E_{\lambda_i}) \leq m_i$, if we let l_i be the number of eigenvectors for λ_i in α, we have $l_i \leq m_i$ for all i. But $\Sigma l_i = n$. Since $\Sigma m_i \leq n$, this implies $\Sigma m_i = n = \dim(V)$ and $m_i = l_i$ for each i. Therefore, for each i we have $l_i \leq \dim(E_{\lambda_i}) = m_i$, and hence $\dim(E_{\lambda_i}) = m_i$ for each i.

6. a) By Exercise 13 of Section 4.1, if $\mathbf{x}$ is an eigenvector of A with eigenvalue λ, then the same $\mathbf{x}$ is an eigenvector of A^k with eigenvalue λ^k. Hence if there is a basis α for $\mathbf{R}^n$ consisting of eigenvectors of A, the same basis will be a basis of eigenvectors for A^k. Hence, A^k is also diagonalizable.

8. By Exercise 15 of Section 4.1, if $T^2 = I$, then the eigenvalues of T can only be $\lambda = \pm 1$, and $V = E_1 \oplus E_{-1}$. Hence, if we let α_1 be a basis for E_1 and α_2 be a basis for E_{-1}, $\alpha_1 \cup \alpha_2$ will be a basis for V consisting of eigenvectors for T. Hence T is diagonalizable.

11. $T((1,0)) = (a,0)$ and $T((0,1)) = (1,a)$, thus T takes the unit square to a parallelogram as in the accompanying figure.

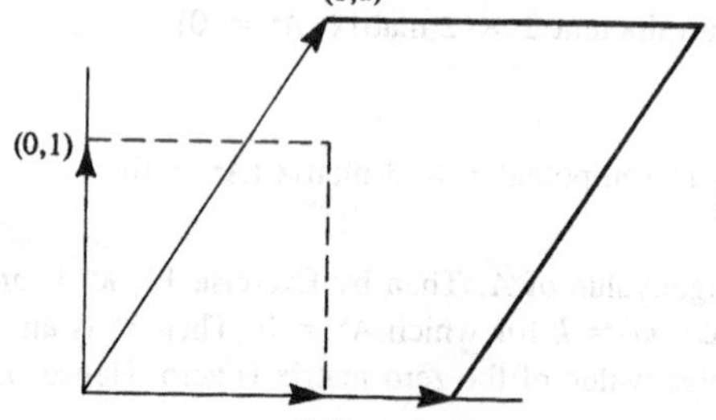

Figure for Exercise 11.

13. a) Let $\mathbf{x} \in E_{\lambda_i} \cap (\sum_{j \neq i} E_{\lambda_j})$

If $\alpha_l = \{\mathbf{x}_{l,1}, \ldots, \mathbf{x}_{l,d_l}\}$ is a basis for E_{λ_l}, $1 \leq l \leq k$, $d_l = \dim(E_{\lambda_l})$, then we can write $\mathbf{x} = c_{i,1}\mathbf{x}_{i,1} + \cdots + c_{i,d_i}\mathbf{x}_{i,d_i} = \sum_{j \neq i} (c_{j,1}\mathbf{x}_{j,1} + \cdots + c_{j,d_j}\mathbf{x}_{j,d_j})$. By Corollary (4.2.5), it follows that since the λ_i are distinct eigenvalues, $c_{i,m} = 0$ and $c_{j,m} = 0$ for all i, j, and m. Thus $\mathbf{x} = \mathbf{0}$.

15. Since $\det(A - \lambda I) = \det(A - \lambda I)^t = \det(A^t - \lambda I)$, the eigenvalues of A and A^t are the same. A is diagonalizable if and only if $\sum_{i=1}^{k} (n - \dim(\mathrm{Im}\,(A - \lambda_i I))) = n$. But by the fact that row rank $=$ column rank for all matrices (see Exercise 10 of Section 2.3), $\dim(\mathrm{Im}(A - \lambda_i I)) = \dim(\mathrm{Im}(A - \lambda_i I)^t) = \dim(\mathrm{Im}(A^t - \lambda_i I))$. Thus

$$\sum_{i=1}^{k} (n - \dim(\mathrm{Im}(A - \lambda_i I))) = \sum_{i=1}^{k} (n - \dim(\mathrm{Im}(A^t - \lambda_i I)))$$

so A is diagonalizable if λ and only if A^t is diagonalizable.

Chapter 4, Section 3

1. a) $\langle \mathbf{x},\mathbf{y}\rangle = \langle \mathbf{y},\mathbf{x}\rangle = 33$
   b) $\|\mathbf{x}\| = \sqrt{\langle \mathbf{x},\mathbf{x}\rangle} = \sqrt{26}$
   $\|\mathbf{y}\| = \sqrt{\langle \mathbf{y},\mathbf{y}\rangle} = \sqrt{54} = 3\sqrt{6}.$
   c) Cauchy-Schwarz: $|\langle \mathbf{x},\mathbf{y}\rangle| \leq \|\mathbf{x}\|\,\|\mathbf{y}\| \doteq 37.5.$
   Triangle: $\|\mathbf{x} + \mathbf{y}\| = \|(0,11,5)\| = \sqrt{146} \doteq 12.1 \leq \|\mathbf{x}\| + \|\mathbf{y}\| \doteq 12.4.$
   d) $\cos(\theta) \approx 33/37.5$, so $\theta \doteq .49$ radians or 28.4 degrees.

3. a) $\langle \mathbf{x} - c\mathbf{y}, \mathbf{x} - c\mathbf{y} \rangle = \|\mathbf{x} - c\mathbf{y}\|^2 \geq 0$. By the linearity in each variable

$$\begin{aligned}\langle \mathbf{x} - c\mathbf{y}, \mathbf{x} - c\mathbf{y} \rangle &= \langle \mathbf{x}, \mathbf{x} - c\mathbf{y} \rangle - c\,\langle \mathbf{y}, \mathbf{x} - c\mathbf{y} \rangle \\ &= \langle \mathbf{x}, \mathbf{x} \rangle - c\langle \mathbf{x}, \mathbf{y} \rangle - c\langle \mathbf{y}, \mathbf{x} \rangle + c^2\langle \mathbf{y}, \mathbf{y} \rangle.\end{aligned}$$

By the symmetry of the inner product

$$\begin{aligned}&= \langle \mathbf{x}, \mathbf{x} \rangle - 2c\langle \mathbf{x}, \mathbf{y} \rangle + c^2\,\langle \mathbf{y}, \mathbf{y} \rangle \\ &= \|\mathbf{x}\|^2 - 2c\langle \mathbf{x}, \mathbf{y} \rangle + c^2\|\mathbf{y}\|^2 \geq 0.\end{aligned}$$

b) Letting $c = \langle \mathbf{x}, \mathbf{y} \rangle / \|\mathbf{y}\|^2$, we have $\|\mathbf{x}\|^2 - \dfrac{2\langle \mathbf{x}, \mathbf{y} \rangle^2}{\|\mathbf{y}\|^2} + \dfrac{\langle \mathbf{x}, \mathbf{y} \rangle^2}{\|\mathbf{y}\|^2} \geq 0$

So $\|\mathbf{x}\|^2 \geq \langle \mathbf{x}, \mathbf{y} \rangle^2 / \|\mathbf{y}\|^2$ and $\langle \mathbf{x}, \mathbf{y} \rangle^2 \leq \|\mathbf{x}\|^2\,\|\mathbf{y}\|^2$.
Taking square roots,

$$|\langle \mathbf{x}, \mathbf{y} \rangle| \leq \|\mathbf{x}\|\,\|\mathbf{y}\|.$$

c) By part a with $c = -1$ we have

$$\langle \mathbf{x} + \mathbf{y}, \mathbf{x} + \mathbf{y} \rangle = \|\mathbf{x}\|^2 + 2\langle \mathbf{x}, \mathbf{y} \rangle + \|\mathbf{y}\|^2 \geq 0.$$

d) By part c

$$\|\mathbf{x} + \mathbf{y}\|^2 = \|\mathbf{x}\|^2 + 2\langle \mathbf{x}, \mathbf{y} \rangle + \|\mathbf{y}\|^2$$

By the Cauchy-Schwarz inequality $\langle \mathbf{x}, \mathbf{y} \rangle \leq \|\mathbf{x}\|\,\|\mathbf{y}\|$, so

$$\|\mathbf{x} + \mathbf{y}\|^2 \leq \|\mathbf{x}\|^2 + 2\|\mathbf{x}\|\,\|\mathbf{y}\| + \|\mathbf{y}\|^2 = (\|\mathbf{x}\| + \|\mathbf{y}\|)^2.$$

Taking square roots yields the triangle inequality:

$$\|\mathbf{x} + \mathbf{y}\| \leq \|\mathbf{x}\| + \|\mathbf{y}\|.$$

4. By 3a) with $c = 1$ we have

$$\|\mathbf{x} - \mathbf{y}\|^2 = \langle \mathbf{x} - \mathbf{y}, \mathbf{x} - \mathbf{y} \rangle = \|\mathbf{x}\|^2 - 2\langle \mathbf{x}, \mathbf{y} \rangle + \|\mathbf{y}\|^2.$$

If $\mathbf{x}$ and $\mathbf{y}$ are orthogonal, then $\langle \mathbf{x}, \mathbf{y} \rangle = 0$, so

$$\|\mathbf{x} - \mathbf{y}\|^2 = \|\mathbf{x}\|^2 + \|\mathbf{y}\|^2.$$

If $\mathbf{x}$ and $\mathbf{y}$ are non-zero, this is an algebraic form of the Pythagorean Theorem:

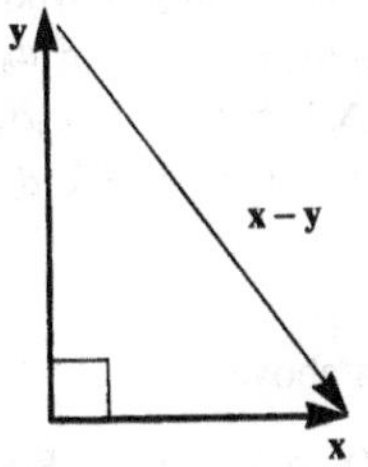

Figure for Exercise 4.

6. As in Exercise 5, use Exercise 3a) with $c = 1$ and $c = -1$. Subtracting those equations yields $\|\mathbf{x} + \mathbf{y}\|^2 - \|\mathbf{x} - \mathbf{y}\|^2 = 4\langle \mathbf{x}, \mathbf{y} \rangle$.

7. a) No, $\langle (1,4,2,0),(3,7,6,2) \rangle = 43 \neq 0$.

10. a) $\langle (x_1,x_2),(y_1,y_2) \rangle = [x_1\ x_2] \begin{bmatrix} 2 & 1 \\ 1 & 1 \end{bmatrix} \begin{bmatrix} y_1 \\ y_2 \end{bmatrix} = 2x_1y_1 + x_1y_2 + x_2y_1 + x_2y_2$

i) By the distributivity of matrix multiplication

$$\langle c\mathbf{x} + \mathbf{y},\mathbf{z}\rangle = (c[x_1\ x_2] + [y_1\ y_2])\begin{bmatrix}2 & 1\\ 1 & 1\end{bmatrix}\begin{bmatrix}z_1\\ z_2\end{bmatrix}$$

$$= c[x_1\ x_2]\begin{bmatrix}2 & 1\\ 1 & 1\end{bmatrix}\begin{bmatrix}z_1\\ z_2\end{bmatrix} + [y_1\ y_2]\begin{bmatrix}2 & 1\\ 1 & 1\end{bmatrix}\begin{bmatrix}z_1\\ z_2\end{bmatrix}$$

$$= c\langle \mathbf{x},\mathbf{z}\rangle + c\langle \mathbf{y},\mathbf{z}\rangle.$$

ii) $\langle \mathbf{x},\mathbf{y}\rangle = [x_1\ x_2]\begin{bmatrix}2 & 1\\ 1 & 1\end{bmatrix}\begin{bmatrix}y_1\\ y_2\end{bmatrix} = 2x_1y_1 + x_1y_2 + x_2y_1 + x_2y_2$

$\langle \mathbf{y},\mathbf{x}\rangle = [y_1\ y_2]\begin{bmatrix}2 & 1\\ 1 & 1\end{bmatrix}\begin{bmatrix}x_1\\ x_2\end{bmatrix} = 2y_1x_1 + y_1x_2 + y_2x_1 + x_2y_2$

which is the same.

iii) $\langle \mathbf{x},\mathbf{x}\rangle = [x_1\ x_2]\begin{bmatrix}2 & 1\\ 1 & 1\end{bmatrix}\begin{bmatrix}x_1\\ x_2\end{bmatrix} = 2x_1^2 + 2x_1x_2 + x_2^2$

$$= x_1^2 + (x_1 + x_2)^2.$$

This is zero if and only if $x_1 = 0$ and $x_1 + x_2 = 0$, which is true if and only if $x_1 = x_2 = 0$.

c) Let $A,B,C \in M_{n\times n}(\mathbf{R})$, $d \in \mathbf{R}$

i) $\langle dA + B,C\rangle = \mathrm{Tr}((dA + B)^tC) = \mathrm{Tr}((dA^t + B^t)C) = \mathrm{Tr}(dA^tC + B^tC) = d\mathrm{Tr}(A^tC) + \mathrm{Tr}(B^tC) = d\langle A,C\rangle + \langle B,C\rangle$.

ii) For any matrix M, $\mathrm{Tr}(M^t) = \mathrm{Tr}(M)$, since the diagonal entries are the same in each. Thus $\langle A,B\rangle = \mathrm{Tr}(A^tB) = \mathrm{Tr}((A^tB)^t) = \mathrm{Tr}(B^tA) = \langle B,A\rangle$.

iii) If $A = (a_{ij})$. The diagonal entries of A^tA are $\sum_{j=1}^{n} a_{ij}^2$ $(i = 1, \ldots, n)$. Thus $\langle A,A\rangle = \mathrm{Tr}(A^tA) = \sum_{i=1}^{n}(\sum_{j=1}^{n} a_{ij}^2) \geq 0$, and $\langle A,A\rangle = 0$ if and only if $a_{ij} = 0$ for all i and j.

13. a) $\langle(-4,5),(1,3)\rangle = [-4\quad 5]\begin{bmatrix}2 & 1\\ 1 & 1\end{bmatrix}\begin{bmatrix}1\\ 3\end{bmatrix} = [-4\quad 5]\begin{bmatrix}5\\ 4\end{bmatrix} = 0.$

15. Suppose that $\mathbf{x} = c_1\mathbf{x}_1 + \cdots + c_n\mathbf{x}_n$ and $\mathbf{y} = d_1\mathbf{x}_1 + \cdots + d_n\mathbf{x}_n$. Then in computing $B(\mathbf{x},\mathbf{y}) = B(c_1\mathbf{x}_1 + \cdots + c_n\mathbf{x}_n, d_1\mathbf{y}_1 + \cdots + d_n\mathbf{y}_n)$, by the bilinearity of B we have

$$= c_1B(\mathbf{x}_1,d_1\mathbf{x}_1 + \cdots + d_n\mathbf{x}_n) + \cdots + c_nB(\mathbf{x}_n,d_1\mathbf{x}_1 + \cdots + d_n\mathbf{x}_n)$$

$$= c_1d_1B(\mathbf{x}_1,\mathbf{x}_1) + c_1d_2B(\mathbf{x}_1,\mathbf{x}_2) + \cdots + c_1d_nB(\mathbf{x}_n,\mathbf{x}_1) + \cdots + c_nd_nB(\mathbf{x}_n,\mathbf{x}_n)$$

$$= \sum_{i=1}^{n}(\sum_{j=1}^{n} c_id_jB(\mathbf{x}_i,\mathbf{x}_j)).$$

On the other hand, if $\mathbf{x},\mathbf{y}$ are as above

$$[\mathbf{x}]_\alpha^t[B]_\alpha[\mathbf{y}]_\alpha = [c_1 \ \ldots \ c_n]\begin{bmatrix}B(\mathbf{x}_1,\mathbf{x}_1) & \cdots & B(\mathbf{x}_1,\mathbf{x}_n)\\ \vdots & & \vdots\\ B(\mathbf{x}_n,\mathbf{y}_1) & \cdots & B(\mathbf{x}_n,\mathbf{x}_n)\end{bmatrix}\begin{bmatrix}d_1\\ \vdots\\ d_n\end{bmatrix}$$

$$= \sum_{i=1}^{n}\left[\sum_{j=1}^{n} c_id_jB(\mathbf{x}_i,\mathbf{x}_j)\right]$$

So

$$B(\mathbf{x},\mathbf{y}) = [\mathbf{x}]_\alpha^t[B]_\alpha[\mathbf{y}]_\alpha \text{ as claimed.}$$

If $B(\mathbf{x},\mathbf{y}) = \langle \mathbf{x},\mathbf{y}\rangle$ on $\mathbf{R}^n$, then $[B]_\alpha = I \in M_{n\times n}(\mathbf{R})$.

17. Let $\alpha = \{\mathbf{x}_1, \ldots, \mathbf{x}_n\}$. We have $[\mathbf{x}_i]_\beta = [I]_\alpha^\beta[\mathbf{x}_i]_\alpha$, so $[\mathbf{x}_i]_\beta = Q[\mathbf{x}_i]_\alpha$ for all i, where $Q = [I]_\alpha^\beta$. Hence, by Exercise 15, (using β-coordinates), the (i,j) entry of $[B]_\alpha$, $B(\mathbf{x}_i,\mathbf{x}_j)$ can be computed as

$$\begin{aligned} B(\mathbf{x}_i,\mathbf{x}_j) &= [\mathbf{x}_i]_\beta^t[B]_\beta[\mathbf{x}_j]_\beta \\ &= (Q[\mathbf{x}_i]_\alpha)^t[B]_\beta(Q[\mathbf{x}_j]_\alpha) \\ &= [\mathbf{x}_i]_\alpha^t\,(Q^t[B]_\beta Q)[\mathbf{x}_j]_\alpha \end{aligned}$$

Of course, $[\mathbf{x}_j]_\alpha = \mathbf{e}_j \in \mathbf{R}^n$ and $[\mathbf{x}_i]_\alpha = \mathbf{e}_i \in \mathbf{R}^n$. So this is just the (i,j) entry of $Q^t[B]_\beta Q$. Thus, since all corresponding entries in the two matrices are the same, $[B]_\alpha = Q^t[B]_\beta Q$.

Chapter 4, Section 4

1. a) Basis: $\{(-1,-1,1,0), (-1,1,0,1)\}$ $\dim(W^\perp) = 2$

2. Let $\mathbf{x} \in W_2^\perp$, then $\langle \mathbf{x},\mathbf{y}\rangle = 0$ for all $\mathbf{y} \in W_2$. But $W_1 \subset W_2$, so $\langle \mathbf{x},\mathbf{z}\rangle = 0$ for all $\mathbf{z} \in W_1$. Therefore $\mathbf{x} \in W_1^\perp$. Hence $W_2^\perp \subset W_1^\perp$.

3. a) Let $\mathbf{x} \in (W_1 + W_2)^\perp$, then $\langle \mathbf{x},\mathbf{y} + \mathbf{z}\rangle = 0$ for all $\mathbf{y} \in W_1$ and $\mathbf{z} \in W_2$. Take $\mathbf{z} = \mathbf{0} \in W_2$, then $\langle \mathbf{x},\mathbf{y}\rangle = 0$ for all $\mathbf{y} \in W_1$. Hence $\mathbf{z} \in W_1^\perp$. Similarly, if $\mathbf{y} = 0 \in W_1$, then $\langle \mathbf{x},\mathbf{z}\rangle = 0$ for all $\mathbf{z} \in W_1$. Therefore $\mathbf{x} \in W_1^\perp \cap W_2^\perp$, so $(W_1 + W_2)^\perp \subseteq W_1^\perp \cap W_2^\perp$. On the other hand, if $\mathbf{x} \in W_1^\perp \cap W_2^\perp$, then for all $\mathbf{y} \in W_1$ and $\mathbf{z} \in W_2$, $\langle \mathbf{x},\mathbf{y} + \mathbf{z}\rangle = \langle \mathbf{x},\mathbf{y}\rangle + \langle \mathbf{x},\mathbf{z}\rangle = 0 + 0 = 0$. Hence $\mathbf{x} \in (W_1 + W_2)^\perp$, so $W_1^\perp \cap W_2^\perp \subseteq (W_1 + W_2)^\perp$. Combining the two inclusions gives the result.

4. a) $\begin{bmatrix} 4/5 & 0 & 0 & 2/5 \\ 0 & 0 & 0 & 0 \\ 0 & 0 & 0 & 0 \\ 2/5 & 0 & 0 & 4/5 \end{bmatrix}$

5. a) $\{(1/\sqrt{6})(1,2,1), (1/\sqrt{318})(-13,10,-7)\}$

7. a) Let $\mathbf{x},\mathbf{y} \in W^\perp$ and $c \in \mathbf{R}$. Then for all $\mathbf{z} \in W_1$, $\langle c\mathbf{x} + \mathbf{y},\mathbf{z}\rangle = c\langle \mathbf{x},\mathbf{z}\rangle = c \cdot 0 + 0 = 0$. Hence $c\mathbf{x} + \mathbf{y} \in W^\perp$, so $W^\perp$ is a subspace. (*Or:* $W^\perp$ is the set of solutions of a homogeneous system of linear equations, hence a subspace of $\mathbf{R}^n$.)
   b) Let $S = \{\mathbf{v}_1, \ldots, \mathbf{v}_k\}$ be a basis of W. Then $W^\perp$ is the set of solutions of

$$\begin{aligned} \langle \mathbf{v}_1,\mathbf{x}\rangle &= 0 \\ &\vdots \\ \langle \mathbf{v}_k,\mathbf{x}\rangle &= 0 \end{aligned}$$

   Since S is linearly independent, there will be k nonzero equations in the equivalent echelon form system. Hence $\dim(W^\perp)$ = number of free variables $= n - k$, so $\dim(W) + \dim(W^\perp) = n$.
   c) If $\mathbf{x} \in W \cap W^\perp$, then $\langle \mathbf{x},\mathbf{x}\rangle = 0$. Hence $\mathbf{x} = \mathbf{0}$, so $W \cap W^\perp = \{\mathbf{0}\}$.
   d) This follows from b) and c). Since $\dim(W) + \dim(W^\perp) = n$ and $\dim(W + W^\perp) = \dim(W) + \dim(W^\perp) - \dim(W \cap W^\perp)$, we have $\dim(W + W^\perp) = n$, so $W + W^\perp = \mathbf{R}^n$.

10. a) Let $\mathbf{v}_1 = (x_1,x_2,x_3)$ and $\mathbf{v}_2 = (y_1,y_2,y_3)$.
Then $\mathbf{v}_1 \times \mathbf{v}_2 = (x_2y_3 - x_3y_2, x_3y_1 - x_1y_3, x_1y_2 - x_2y_1)$, so
$\langle \mathbf{v}_1,\mathbf{v}_1 \times \mathbf{v}_2\rangle = x_1(x_2y_3 - x_3y_2) + x_2(x_3y_1 - x_1y_3) + x_3(x_1y_2 - x_2y_1) = 0$,
and
$\langle \mathbf{v}_2,\mathbf{v}_1 \times \mathbf{v}_2\rangle = y_1(x_2y_3 - x_3y_2) + y_2(x_3y_1 - x_1y_3) + x_3(x_1y_2 - x_2y_1) = 0$.
Hence, by the linearity of the inner product in the first variable, every vector $\mathbf{v} \in V = \text{Span}\{\mathbf{v}_1,\mathbf{v}_2\}$ also satisfies $\langle \mathbf{v},\mathbf{v}_1 \times \mathbf{v}_2\rangle = 0$, so $\mathbf{v}_1 \times \mathbf{v}_2$ is a normal vector to V.

b) Let $\mathbf{v}_l = (v_{l1}, \ldots, v_{ln})$ and let $\mathbf{N} = (D_1, \ldots, D_n)$. Then for each l, $1 \leq l \leq n$, $\langle \mathbf{v}_l,\mathbf{N}\rangle = v_{l1}D_1 + \cdots + v_{ln}D_n$. By the definition of D_i, this is

$$\det \begin{bmatrix} v_{l1} & \cdots & v_{ln} \\ v_{11} & \cdots & v_{1n} \\ & \vdots & \\ v_{n-1,1} & \cdots & v_{n-1,n} \end{bmatrix} = 0,$$

since the matrix has two repeated rows. Hence, by linearity, $\mathbf{N}$ is orthogonal to every vector in V.

11. a) $\mathbf{N} = (1,2,1) \times (0,2,2) = (2,-2,2)$.

13. a) By Proposition (4.4.5b), if $\mathbf{v} \in V$, then $P_W(\mathbf{v}) \in W$, so $P_W^2(\mathbf{v}) = P_W(P_W(\mathbf{v})) = P_W(\mathbf{v})$. Hence $P_W^2 = P_W$.
b) $E_1 = W$.
c) $E_0 = W^\perp$.
d) See Exercise 9 of Section 4.2.

16. Let α be an orthonormal basis for $W \subset \mathbf{R}^n$. We can always find an orthonormal basis α' for $W^\perp$ as well. We claim $\alpha \cup \alpha'$ is an orthonormal basis for $\mathbf{R}^n$. If $\mathbf{x} \in \alpha$ and $\mathbf{x}' \in \alpha'$, then $\langle \mathbf{x},\mathbf{x}'\rangle = 0$, since $\mathbf{x} \in W$ and $\mathbf{x}' \in W^\perp$. Since the $\mathbf{x}$ in α are orthogonal to each other and have length 1, and similarly for the $\mathbf{x}'$ in α', $\alpha \cup \alpha'$ is orthonormal. Since $\mathbf{R}^n = W \oplus W^\perp$, $\alpha \cup \alpha'$ is an orthonormal basis of $\mathbf{R}^n$.

19. a) Let $\alpha = \{\mathbf{x}_1, \cdots, \mathbf{x}_n\}$ be an orthonormal basis for $\mathbf{R}^n$. Let $a_i = l(\mathbf{x}_i)$ and let $\mathbf{y} = a_1\mathbf{x}_1 + \cdots + a_n\mathbf{x}_n$. *Claim:* For all $\mathbf{x} \in \mathbf{R}^n$, $l(\mathbf{x}) = \langle \mathbf{x},\mathbf{y}\rangle$. Write $\mathbf{x} = c_1\mathbf{x}_1 + \cdots + c_n\mathbf{x}_n$. Since α is orthonormal, $\langle \mathbf{x},\mathbf{y}\rangle = \langle c_1\mathbf{x}_1 + \cdots + c_n\mathbf{x}_n, a_1\mathbf{y}_1 + \cdots + a_n\mathbf{y}_n\rangle = c_1a_1 + \cdots + c_na_n$. On the other hand, $l(\mathbf{x}) = l(c_1\mathbf{x}_1 + \cdots + c_n\mathbf{x}_n) = c_1l(\mathbf{x}_1) + \cdots + c_nl(\mathbf{x}_n) = c_1a_1 + \cdots + c_na_n$. $\mathbf{y}$ is unique since if we have $\langle \mathbf{x},\mathbf{y}\rangle = \langle \mathbf{x},\mathbf{y}'\rangle$ for all $\mathbf{x} \in \mathbf{R}^n$, then $\langle \mathbf{x},\mathbf{y} - \mathbf{y}'\rangle = 0$ for all $\mathbf{x}$, so $\mathbf{y} - \mathbf{y}' = \mathbf{0}$.

Chapter 4, Section 5

1. a) If $A^t = B$ and $B^t = B$, $(A + B)^t = A^t + B^t = A + B$, so $A + B$ is symmetric.

2. a) $\lambda = (5 \pm 3\sqrt{5})/2$. A basis for E_λ is $\{(1,(-1 \mp \sqrt{5})/2)\}$.

c) $\lambda = 2, 2 \pm \sqrt{2}$. $E_2 = \text{Span}\{(-1,0,1)\}$. $E_{2\pm\sqrt{2}} = \text{Span}\{(\pm\sqrt{2}/2, 1, \pm\sqrt{2}/2)\}$.

3. a) No.

7. b) The rotation matrix $R = \begin{bmatrix} \cos(\theta) & -\sin(\theta) \\ \sin(\theta) & \cos(\theta) \end{bmatrix}$, $\theta \neq k\pi$, k an integer, is such a matrix: $RR^t = \begin{bmatrix} 1 & 0 \\ 0 & 1 \end{bmatrix} = R^tR$.

c) If A is normal, then

$$\begin{aligned} \|A\mathbf{x}\|^2 &= \langle A\mathbf{x}, A\mathbf{x}\rangle = \langle \mathbf{x}, A^tA\mathbf{x}\rangle, \text{ by Exercise 4,} \\ &= \langle \mathbf{x}, AA^t\mathbf{x}\rangle, \text{ since } A \text{ is normal,} \\ &= \langle A^t\mathbf{x}, A^t\mathbf{x}\rangle, \text{ by Exercise 4 again,} \\ &= \|A^t\mathbf{x}\|^2. \end{aligned}$$

Hence $\|A\mathbf{x}\| = \|A^t\mathbf{x}\|$.

e) If λ is an eigenvalue of A, then $(A - \lambda I)\mathbf{x} = \mathbf{0}$ for some $\mathbf{x} \neq \mathbf{0}$. By part d, $(A - \lambda I)$ is also normal, so by part c, $0 = \|(A - \lambda I)\mathbf{x}\| = \|(A - \lambda I)^t\mathbf{x}\| = \|(A^t - \lambda I)\mathbf{x}\|$. Hence $(A^t - \lambda I)\mathbf{x} = \mathbf{0}$, so $\mathbf{x}$ is also an eigenvector of A^t with eigenvalue λ.

8. a) Let $\mathbf{x} = (x_1,x_2)$, $\mathbf{y} = (y_1,y_2) \in \mathbf{R}^2$

$$\begin{aligned} \langle T(\mathbf{x}),\mathbf{y}\rangle &= [x_1 + x_2 \quad 2x_1 + x_2] \begin{bmatrix} 2 & 1 \\ 1 & 1 \end{bmatrix} \begin{bmatrix} y_1 \\ y_2 \end{bmatrix} \\ &= [x_1 + x_2 \quad 2x_1 + x_2] \begin{bmatrix} 2y_1 + y_2 \\ y_1 + y_2 \end{bmatrix} \\ &= 4x_1y_1 + 3x_1y_2 + 3x_2y_1 + 2x_2y_2 \end{aligned}$$

$$\begin{aligned} \langle \mathbf{x},T(\mathbf{y})\rangle &= [x_1 \quad x_2] \begin{bmatrix} 2 & 1 \\ 1 & 1 \end{bmatrix} \begin{bmatrix} y_1 + y_2 \\ 2y_1 + y_2 \end{bmatrix} \\ &= [x_1 \quad x_2] \begin{bmatrix} 4y_1 + 3y_2 \\ 3y_1 + 2y_2 \end{bmatrix} \\ &= 4x_1y_1 + 3x_1y_2 + 3x_2y_1 + 2x_2y_2 \end{aligned}$$

c) Yes. $\langle T(A),B\rangle = \langle A^t,B\rangle = \mathrm{Tr}((A^t)^tB) = \mathrm{Tr}(AB) = \mathrm{Tr}((AB)^t) = \mathrm{Tr}(B^tA^t) = \mathrm{Tr}(A^tB^t) = \langle A,B^t\rangle = \langle A,T(B)\rangle$.

10. a) $x_1^2/a^2 + x_2^2/b^2 = [x_1 \quad x_2] \begin{bmatrix} 1/a^2 & 0 \\ 0 & 1/b^2 \end{bmatrix} \begin{bmatrix} x_1 \\ x_2 \end{bmatrix}$

Chapter 4, Section 6

1. a) $\lambda = -1,5$. $E_{-1} = \mathrm{Span}(\{(-\sqrt{2}/2, \sqrt{2}/2)\})$ and $E_5 = \mathrm{Span}(\{(\sqrt{2}/2, \sqrt{2}/2)\})$. $\alpha = \{(-\sqrt{2}/2, \sqrt{2}/2), (\sqrt{2}/2, \sqrt{2}/2)\}$ is an orthonormal basis for $\mathbf{R}^2$.

$$Q = \begin{bmatrix} -\sqrt{2}/2 & \sqrt{2}/2 \\ \sqrt{2}/2 & \sqrt{2}/2 \end{bmatrix}$$

$$A = (-1)P_{E_{-1}} + 5P_{E_5} = (-1)\begin{bmatrix} 1/2 & -1/2 \\ -1/2 & 1/2 \end{bmatrix} + 5\begin{bmatrix} 1/2 & 1/2 \\ 1/2 & 1/2 \end{bmatrix}$$

c) $\lambda = 3, -3, -3$. $E_3 = \text{Span}(\{1/\sqrt{3}(1,1,1)\})$ and $E_{-3} = \text{Span}(\{1/\sqrt{2}(-1,1,0), 1/\sqrt{2}(-1,0,1)\})$. $\alpha = \{1/\sqrt{3}(1,1,1), 1/\sqrt{2}(-1,1,0), 1/\sqrt{6}(-1,-1,2)\}$ is an orthonormal basis for $\mathbf{R}^3$.

$$Q = \begin{bmatrix} 1/\sqrt{3} & -1/\sqrt{2} & -1/\sqrt{6} \\ 1/\sqrt{3} & 1/\sqrt{2} & -1/\sqrt{6} \\ 1/\sqrt{3} & 0 & 2/\sqrt{6} \end{bmatrix}$$

$$A = 3P_{E_3} + (-3)P_{E_{-3}} = 3\begin{bmatrix} 1/3 & 1/3 & 1/3 \\ 1/3 & 1/3 & 1/3 \\ 1/3 & 1/3 & 1/3 \end{bmatrix} + (-3)\begin{bmatrix} 2/3 & -1/3 & -1/3 \\ -1/3 & 2/3 & -1/3 \\ -1/3 & -1/3 & 2/3 \end{bmatrix}$$

e) $\lambda = 0, 0, (7 \pm \sqrt{73})/2$, $E_0 = \text{Span}(\{1/\sqrt{2}(-1,0,1,0), 1/\sqrt{6}(1,-2,1,0)\})$, $E_{(7\pm\sqrt{73})/2} = \text{Span}(\{(146 \mp 10\sqrt{73})^{-1/2}(4,4,4,-5 \pm \sqrt{73})\})$, and $\alpha = \{1/\sqrt{2}(-1,0,1,0), 1/\sqrt{6}(1,-2,1,0), (146 \mp 10\sqrt{73})^{-1/2}(4,4,4,-5 \pm \sqrt{73})\}$ is an orthonormal basis for $\mathbf{R}^4$.

$$Q = \begin{bmatrix} -1/\sqrt{2} & 1/\sqrt{6} & 4a & 4b \\ 0 & -2/\sqrt{6} & 4a & 4b \\ 1/\sqrt{2} & 1/\sqrt{6} & 4a & 4b \\ 0 & 0 & -(5-\sqrt{73})a & -(5+\sqrt{73})b \end{bmatrix}$$

where $a = (146 - 10\sqrt{73})^{-1/2}$, $b = (146 + 10\sqrt{73})^{-1/2}$.

$$A = 0 \cdot P_{E_0} + (7 \pm \sqrt{73})/2\, P_{E_{(7\pm\sqrt{73})/2}} = 0\begin{bmatrix} 2/3 & 1/6 & -5/6 & 0 \\ 1/6 & 1/6 & -1/3 & 0 \\ -5/6 & -1/3 & 7/6 & 0 \\ 0 & 0 & 0 & 0 \end{bmatrix}$$

$$+ (7 + \sqrt{73})/2 \cdot (146 - 10\sqrt{73})^{-1}\begin{bmatrix} 16 & 16 & 16 & 4(5-\sqrt{73}) \\ 16 & 16 & 16 & 4(5-\sqrt{73}) \\ 16 & 16 & 16 & 4(5-\sqrt{73}) \\ 4(5-\sqrt{73}) & 4(5-\sqrt{73}) & 4(5-\sqrt{73}) & (5-\sqrt{73})^2 \end{bmatrix}$$

$$+ (7 - \sqrt{73})/2 \cdot (146 + 10\sqrt{73})^{-1}\begin{bmatrix} 16 & 16 & 16 & 4(5+\sqrt{73}) \\ 16 & 16 & 16 & 4(5+\sqrt{73}) \\ 16 & 16 & 16 & 4(5+\sqrt{73}) \\ 4(5+\sqrt{73}) & 4(5+\sqrt{73}) & 4(5+\sqrt{73}) & (5+\sqrt{73})^2 \end{bmatrix}$$

3. If $\lambda_1, \cdots, \lambda_k$ are the distinct eigenvalues of A, $E_{\lambda_1} + \cdots + E_{\lambda_k} = \mathbf{R}^n$. Thus each $\mathbf{x} \in \mathbf{R}^n$ can be written uniquely as $\mathbf{x} = \mathbf{x}_1 + \cdots + \mathbf{x}_k$, where $\mathbf{x}_i \in E_{\lambda_i}$. We claim $\mathbf{x}_i = P_{E_{\lambda_i}}(\mathbf{x})$. By Exercise 2, $(E_{\lambda_i})^\perp = \sum_{j \neq i} E_{\lambda_j}$. Hence $\mathbf{x} = \mathbf{x}_i + (\mathbf{x}_1 + \cdots + \mathbf{x}_{i-1} + \mathbf{x}_{i+1} + \cdots + \mathbf{x}_k)$ gives a decomposition of $\mathbf{x}$ as the sum of a vector in E_{λ_i} and one in $E_{\lambda_i}{}^\perp$. Hence $\mathbf{x}_i = P_{E_{\lambda_i}}(\mathbf{x})$ by the definition of orthogonal projections. Since for all $\mathbf{x} \in \mathbf{R}^n$, $\mathbf{x} = P_{E_{\lambda_1}}(\mathbf{x}) + \cdots + P_{E_{\lambda_k}}(\mathbf{x})$, $I = P_{E_{\lambda_1}} + \cdots + P_{E_{\lambda_k}}$.

4. a) Let $\alpha = \{\mathbf{x}_1, \ldots, \mathbf{x}_n\}$ be an orthonormal basis for $\mathbf{R}^n$ and let Q be the matrix whose columns are the vectors $\mathbf{x}_i$. In Q^tQ, the i,j entry is $\langle \mathbf{x}_i, \mathbf{x}_j \rangle = 0$ if $i \neq j$ and $\langle \mathbf{x}_i, \mathbf{x}_j \rangle = 1$ if $i = j$. Hence $Q^tQ = I$. Similarly, $QQ^t = I$, so $Q^t = Q^{-1}$.

b) If $Q^t = Q^{-1}$, then $Q^tQ = I$. Thus, if we let $\mathbf{x}_i$ be the ith column of Q, then $\langle \mathbf{x}_i, \mathbf{x}_j \rangle = 0$ if $i \neq j$ and $\langle \mathbf{x}_i, \mathbf{x}_j \rangle = 1$ if $i = j$. Hence $\alpha = \{\mathbf{x}_1, \ldots, \mathbf{x}_n\}$ is an orthonormal basis of $\mathbf{R}^n$.

e) Since $Q^tQ = I$, $\det(Q^tQ) = \det(Q^t)\det(Q) = [\det(Q)]^2 = \det(I) = 1$. Hence $\det(Q) = \pm 1$.

5. a) By Proposition 4.5.2a, if Q is orthogonal, $\langle Q\mathbf{x}, Q\mathbf{y} \rangle = \langle \mathbf{x}, Q^tQ\mathbf{y} \rangle = \langle \mathbf{x},\mathbf{y} \rangle$, since $Q^tQ = I$. Conversely, if $\langle Q\mathbf{x}, Q\mathbf{y} \rangle = \langle \mathbf{x},\mathbf{y} \rangle$ for all $\mathbf{x},\mathbf{y} \in \mathbf{R}^n$, letting $\mathbf{x} = \mathbf{e}_i$ and $\mathbf{y} = \mathbf{e}_j$, we see that the ijth entry of Q^tQ is 1 if $i = j$ and 0 otherwise. Hence $Q^tQ = I$, so $Q^t = Q^{-1}$ and Q is orthogonal.

6. a) We have $A^t = A$ and $B = Q^tA\,Q$ for some orthogonal matrix Q. Hence

$$B^t = (Q^tA\,Q)^t = Q^tA^t(Q^t)^t = Q^tA\,Q = B.$$

Hence B is symmetric.

7. By the spectral theorem, there is an orthonormal basis α of $\mathbf{R}^n$ consisting of eigenvectors of A. Let Q be the change of basis matrix from this basis to the standard basis (i.e., the columns of Q are the vectors in α written in standard coordinates). Then $Q^{-1}AQ = D$, a diagonal matrix (the diagonal entries are the eigenvalues of A). Furthermore, since α is orthonormal, by Exercise 4, Q is orthogonal: $Q^t = Q^{-1}$. Hence A is orthogonally similar to D.

10. $\langle \mathbf{x},\mathbf{y} \rangle = \mathbf{x}^tA\mathbf{y}$ is linear in each variable. Further, it is symmetric since $\langle \mathbf{y},\mathbf{x} \rangle = \mathbf{y}^tA\mathbf{x} = (\mathbf{x}^tA\mathbf{y})^t = \langle \mathbf{x},\mathbf{y} \rangle$ using the fact that $A^t = A$. Hence it remains to show that $\langle \mathbf{x},\mathbf{x} \rangle > 0$ if $\mathbf{x} \neq \mathbf{0}$ if and only if all the eigenvalues of A are *positive*. By the spectral theorem (or Exercise 7), A is orthogonally similar to a diagonal matrix D whose diagonal entries are the eigenvalues of A: $Q^tAQ = D$. Hence $\langle \mathbf{x},\mathbf{y} \rangle = (\mathbf{x}^tQ^t)D(Q\mathbf{y}) = (Q\mathbf{x})^tD(Q\mathbf{y})$.

Hence, if $Q\mathbf{x} = \begin{bmatrix} \bar{x}_1 \\ \vdots \\ \bar{x}_n \end{bmatrix}$ and $D = \begin{bmatrix} \lambda_1 & & 0 \\ & \ddots & \\ 0 & & \lambda_n \end{bmatrix}$, $\langle \mathbf{x},\mathbf{x} \rangle = \lambda_1\bar{x}_1^2 + \cdots + \lambda_n\bar{x}_n^2$. If all $\lambda_i > 0$, then $\langle \mathbf{x},\mathbf{x} \rangle > 0$ for all $\mathbf{x} \neq \mathbf{0}$. Conversely, if $\lambda_1\bar{x}_1^2 + \cdots + \lambda_n\bar{x}_n^2 > 0$ for all $\mathbf{x} \neq \mathbf{0}$, then all $\lambda_i = 0$. (If not, say $\lambda_i \leq 0$, then taking

$$\begin{bmatrix} \bar{x}_1 \\ \vdots \\ \bar{x}_n \end{bmatrix} = \begin{bmatrix} 0 \\ \vdots \\ 1 \\ \vdots \\ 0 \end{bmatrix} \leftarrow i\text{th, we have } \lambda_1\bar{x}_1^2 + \cdots + \lambda_n\bar{x}_n^2 = \lambda_i \leq 0.)$$

11. a) $(Q^tQ)^t = Q^t(Q^t)^t = Q^tQ$, so Q^tQ is symmetric. Furthermore, by Exercise 10, since $\mathbf{x}^t(Q^tQ)\mathbf{x} = (Q\mathbf{x})^t(Q\mathbf{x}) = \langle Q\mathbf{x},Q\mathbf{x} \rangle > 0$ for all $\mathbf{x} \neq \mathbf{0}$ (since Q is invertible, $Q\mathbf{x} = \mathbf{0}$ implies $\mathbf{x} = \mathbf{0}$), we have that all the eigenvalues of Q^tQ are positive. Hence Q^tQ is positive-definite.

b) If A is invertible and symmetric, then A^{-1} is symmetric, since for all invertible matrices B, $(B^{-1})^t = (B^t)^{-1}$. Thus $(A^{-1})^t = (A^t)^{-1} = A^{-1}$. If A is positive-definite, each eigenvalue λ of A is positive, $\lambda > 0$. But $\lambda^{-1} > 0$ is an eigenvalue of A^{-1}. Thus, A^{-1} is positive-definite. The converse follows by interchanging A and A^{-1}.

c) If A is symmetric, then A^k is symmetric, since $(A^k)^t = (A^t)^k = A^k$. If A is positive-definite and λ is any eigenvalue of A, then $\lambda > 0$. But then $\lambda^k > 0$ is an eigenvalue for A^k, and these are the only eigenvalues of A^k. Thus A^k is also positive-definite.

14. a) $5x_1^2 + 2x_1x_2 + x_2^2 = 1$. The associated symmetric matrix is $M = \begin{bmatrix} 5 & 1 \\ 1 & 1 \end{bmatrix}$. The eigenvalues of M are $\lambda = 3 \pm \sqrt{5}$. Both are positive, so the conic section is an ellipse. $E_{3\pm\sqrt{5}} = \text{Span}\{(1, -2 \pm \sqrt{5})\}$. If Q is the matrix of normalized eigenvectors and $\mathbf{x}' = Q^{-1}\mathbf{x}$, then $(3 + \sqrt{5})x_1'^2 + (3 - \sqrt{5})x_2'^2 = 1$. Thus $(x_1')^2/a^2 + (x_2')^2/b^2 = 1$, where $a = ((3 - \sqrt{5})/4)^{1/2} \approx .44$ and $b = ((3 + \sqrt{5})/4)^{1/2} \approx 1.14$.

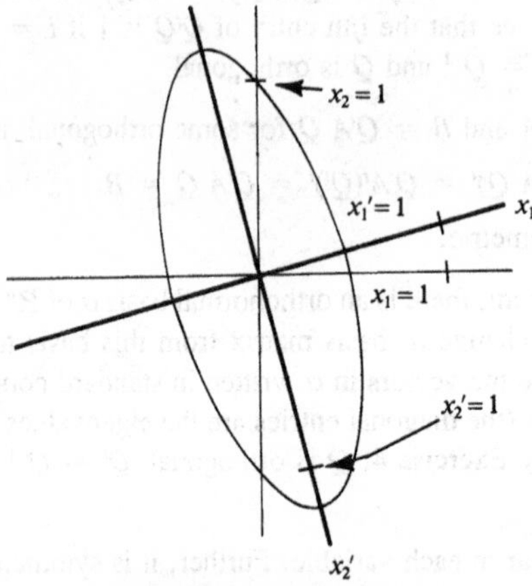

Figure for Exercise 14a.

c) The associated symmetric matrix is $M = \begin{bmatrix} 1 & -3/2 \\ -3/2 & -1 \end{bmatrix}$. The eigenvalues are $\lambda = \pm\sqrt{13}/2$. $E_{\pm\sqrt{13}/2} = \text{Span}(\{1, (2 \mp \sqrt{13})/3\})$. If Q is the matrix of normalized eigenvectors and $\mathbf{x}' = Q^{-1}\mathbf{x}$, then $(x_1')^2/(2/\sqrt{13}) - (x_2')^2/(2/\sqrt{13}) = 1$. The conic section is a hyperbola with asymptotes $x_1' = \pm x_2'$.

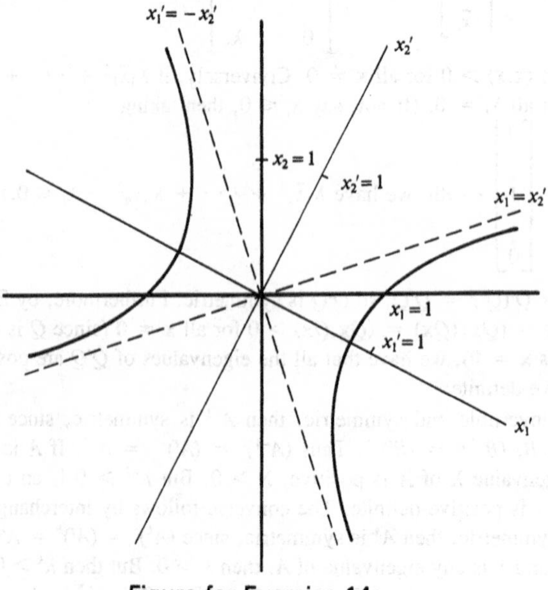

Figure for Exercise 14c.

15. a) The associated symmetric matrix is $M = \begin{bmatrix} -1 & 2 & 2 \\ 2 & -1 & 2 \\ 2 & 2 & -1 \end{bmatrix}$. The eigenvalues of M are $\lambda = -3, -3, 3$. In x' coordinates the conic section is $3(x_1')^2 - 3(x_2')^2 - 3(x_3')^2 = 1$, a hyperboloid of two sheets.

17. a) $\langle A\mathbf{x},\mathbf{x}\rangle = \langle\lambda\mathbf{x},\mathbf{x}\rangle = \lambda\langle\mathbf{x},\mathbf{x}\rangle = \lambda$.

c) By the spectral theorem, there is an orthonormal basis $\alpha = \{\mathbf{x}_1, \cdots, \mathbf{x}_n\}$ of $\mathbf{R}^n$ consisting of eigenvalues of A. Let λ_1 be the largest eigenvalue of A in absolute value, and assume $A\mathbf{x}_1 = \lambda_1\mathbf{x}_1$. If $\mathbf{x} = c_1\mathbf{x}_1 + \cdots + c_n\mathbf{x}_n$ is any unit vector, then

$$\begin{aligned} \| A\mathbf{x} \|^2 = \langle A\mathbf{x},A\mathbf{x}\rangle &= \langle A(c_1\mathbf{x}_1 + \cdots + c_n\mathbf{x}_n), A(c_1\mathbf{x}_1 + \cdots + c_n\mathbf{x}_n)\rangle \\ &= \langle c_1\lambda_1\mathbf{x}_1 + \cdots + c_n\lambda_n\mathbf{x}_n, c_1\lambda_1\mathbf{x}_1 + \cdots + c_n\lambda_n\mathbf{x}_n\rangle \\ &= c_1^2\lambda_1^2 + \cdots + c_n^2\lambda_n^2 \qquad \text{(Since } \alpha \text{ is orthonormal).} \end{aligned}$$

Since $|\lambda_1| \geq |\lambda_j|$ all $2 \leq j \leq n$, we have

$$\begin{aligned} c_1^2\lambda_1^2 + \cdots + c_n\lambda_n^2 &\leq c_1^2\lambda_1^2 + c_2^2\lambda_1^2 + \cdots + c_n^2\lambda_1^2 \\ &= (c_1^2 + \cdots + c_n^2)\lambda_1^2. \end{aligned}$$

Since $\mathbf{x}$ was a unit vector, $c_1^2 + \cdots + c_n^2 = \langle\mathbf{x},\mathbf{x}\rangle = 1$. Hence $\| A\mathbf{x} \|^2 \leq (\lambda_1)^2$, so $\| A\mathbf{x} \| \leq |\lambda_1|$. If we take $\mathbf{x} = \mathbf{x}_1$, then $\| A\mathbf{x}_1 \| = |\lambda_1|$. Hence $|\lambda_1|$ is the maximum value of $\| A\mathbf{x} \|$ as $\mathbf{x}$ ranges over all unit vectors.

Chapter 4, Supplementary Exercises

1. a) True. If $\det(A-\lambda I) = 0$, then $\text{Ker}(A-\lambda I) \neq \{\mathbf{0}\}$, so there is some $\mathbf{x} \neq \mathbf{0}$ with $A\mathbf{x} = \lambda\mathbf{x}$.

b) False. A counterexample is $\begin{bmatrix} 1 & 0 \\ 0 & 0 \end{bmatrix}$.

c) False. Matrices such as $\begin{bmatrix} \cos(\theta) & \sin(\theta) \\ \sin(\theta) & -\cos(\theta) \end{bmatrix}$ are also orthogonal.

d) True. If $\mathbf{x} \in E_\lambda \cap E_\mu$, then $T(\mathbf{x}) = \lambda\mathbf{x} = \mu\mathbf{x}$. Since $\lambda \neq \mu$, this implies $\mathbf{x} = \mathbf{0}$.

e) True. $T(e^{5x}) = 2e^{5x}$ and the eigenvalue is $\lambda = 2$.

f) False. A counterexample is $A = \begin{bmatrix} 3 & 0 \\ 0 & 3 \end{bmatrix}$.

g) True. If $\mathbf{x} \in W \cap W^\perp$, then $\langle\mathbf{x},\mathbf{x}\rangle = 0$, so $\mathbf{x} = \mathbf{0}$.

h) False. $A = \begin{bmatrix} 0 & 1 \\ -1 & 0 \end{bmatrix}$ is skew-symmetric, but has no real eigenvalues.

i) True. $E_\lambda \subset E_\mu^\perp$ since if $A\mathbf{x} = \lambda\mathbf{x}$ and $A\mathbf{y} = \mu\mathbf{y}$, then $\langle A\mathbf{x},\mathbf{y}\rangle = \lambda\langle\mathbf{x},\mathbf{y}\rangle$, but also $\langle A\mathbf{x},\mathbf{y}\rangle = \langle\mathbf{x},A\mathbf{y}\rangle = \mu\langle\mathbf{x},\mathbf{y}\rangle$. Hence $\langle\mathbf{x},\mathbf{y}\rangle = 0$. Moreover, by the spectral theorem, A is diagonalizable, so if A has only two distinct eigenvalues, $\mathbf{R}^n = E_\lambda \oplus E_\mu$. Hence $\dim(E_\lambda) = n - \dim(E_\mu) = \dim(E_\mu^\perp)$ so $E_\lambda = E_\mu^\perp$.

j) False. An orthogonal set can contain $\mathbf{0}$, hence can be linearly dependent.

4. a) Yes. $D = \begin{bmatrix} (7+\sqrt{41})/2 & 0 \\ 0 & (7-\sqrt{41})/2 \end{bmatrix}$, $Q = \begin{bmatrix} 1 & 1 \\ (5-\sqrt{41})/4 & (5+\sqrt{41})/4 \end{bmatrix}$.

c) No. The eigenvalues are $\lambda = -1, -1, -1$, but $\dim E_{-1} = 1 \neq 3 = m_{-1}$.

5. a) $\lambda = 1 \pm \sqrt{29}$. $E_{1\pm\sqrt{29}} = \text{Span}(\{(1,(-1 \pm \sqrt{29})/5)\})$.
$A = (1 + \sqrt{29})\, P_{E_{1+\sqrt{29}}} + (1-\sqrt{29})\, P_{E_{1-\sqrt{29}}}$.

c) $\lambda = 1,1,4$. $E_1 = \text{Span}(\{1/\sqrt{2}(-1,1,0),\ \sqrt{2/3}(-1/2,\ -1/2,\ 1)\})$ and $E_4 = \text{Span}(\{1/\sqrt{3}(1,1,1)\})$. $A = 1\cdot P_{E_1} + 4\cdot P_{E_4}$.

6. c) If A has an eigenvalue $\lambda \in \mathbf{R}$, then T has λ as an eigenvalue of multiplicity 2. If (x_1,x_2) is an eigenvector of A with eigenvalue λ, then $\begin{bmatrix} x_1 & 0 \\ x_2 & 0 \end{bmatrix}$, and $\begin{bmatrix} 0 & x_1 \\ 0 & x_2 \end{bmatrix} \in M_{2\times 2}(\mathbf{R})$ are eigenvectors of T.

9. e) $\langle Tf,g\rangle = \int_{-1}^{1} f(-t)g(t)dt$. Change variables letting $u = -t$, so $t = -u$. Then $\langle Tf,g\rangle = \int_{1}^{-1} f(u)g(-u)(-du) = -\int_{1}^{-1} f(u)g(-u)du = \int_{-1}^{1} f(u)g(-u)du$ $= \langle f,Tg\rangle$.

f) This follows from the same sort of computation as in the finite dimensional case: if $T(f) = f$ and $T(g) = -g$, then $\langle f,g\rangle = \langle T(f),g\rangle = \langle f,T(g)\rangle = -\langle f,g\rangle$. Hence $\langle f,g\rangle = 0$.

12. a) Let $A = \begin{bmatrix} a & b \\ b & c \end{bmatrix}$. By the spectral theorem we know there is a rotated coordinate system (x_1',x_2') in the (x_1,x_2) plane such that $(1/2)\ (ax_1^2 + 2bx_1x_2 + cx_2^2) = (1/2)(\lambda_1(x_1')^2 + \lambda_2(x_2')^2)$ where λ_1 and λ_2 are the eigenvalues of A.

i) The graph $x_3 = (1/2)(\lambda_1(x_1')^2 + \lambda_2(x_2')^2)$ has an isolated minimum at (0,0,0) if and only if $\lambda_1,\lambda_2 > 0$ or A is positive-definite.
ii) The graph has an isolated maximum if and only if $\lambda_1,\lambda_2 < 0$.
iii) The graph has a saddle point if and only if $\lambda_1\cdot\lambda_2 < 0$, so the eigenvalues have opposite signs, or A is indefinite.

c) See Exercise 16 of Section 4.6—the idea is the same.

13. $\rightarrow$: Already done.

$\leftarrow$: If $P^2 = P$ and P is symmetric. Let $W = \text{Im}(P)$. Then since $P^2 = P$, if $\mathbf{x} \in W$, $\mathbf{x} = P(\mathbf{y})$ for some $\mathbf{y} \in \mathbf{R}^n$, so $P(\mathbf{x}) = P^2(\mathbf{y}) = P(\mathbf{y}) = \mathbf{x}$. Moreover if $\mathbf{z} \in W^\perp$, $\langle P(\mathbf{z}),P(\mathbf{z})\rangle = \langle \mathbf{z},P^2(\mathbf{z})\rangle = \langle \mathbf{z},P(\mathbf{z})\rangle$. Since $P(\mathbf{z}) \in W$, this equals $\mathbf{0}$. Hence $P(\mathbf{z}) = \mathbf{0}$, so $W^\perp \subseteq \text{Ker}(P)$. On the other hand, if $\mathbf{z} \in \text{Ker}(P)$, then $\langle \mathbf{z},\mathbf{w}\rangle = \langle \mathbf{z},P(\mathbf{w})\rangle = \langle P(\mathbf{z}),\mathbf{w}\rangle = \langle \mathbf{0},\mathbf{w}\rangle = 0$. Hence $\mathbf{z} \in W^\perp$, so $\text{Ker}(P) \subseteq W^\perp$, and hence $\text{Ker}(P) = W^\perp$. Since $\mathbf{R}^n = W \oplus W^\perp$ and $P|_W = I_W$, and $P|_{W^\perp} = 0_{W^\perp}$, it follows that P is the orthogonal projection on W.

15. a) If P_1P_2 is an orthogonal projection, then P_1P_2 is symmetric, so $P_1P_2 = (P_1P_2)^t = P_2^tP_1^t = P_2P_1$, since P_1 and P_2 are symmetric. Conversely, suppose $P_1P_2 = P_2P_1$. By Exercise 13 it suffices to show P_1P_2 is symmetric and $(P_1P_2)^2 = P_1P_2$. The second condition is immediate: $(P_1P_2)^2 = P_1P_2\cdot P_1P_2 = P_1(P_2P_1)P_2 = P_1(P_1P_2)P_2 = P_1^2P_2^2 = P_1P_2$, since $P_i^2 = P_i$. To show that P_1P_2 is symmetric, let $\mathbf{x},\mathbf{y} \in \mathbf{R}^n$, then $\langle P_1P_2\,\mathbf{x},\mathbf{y}\rangle = \langle P_2\mathbf{x},P_1\mathbf{y}\rangle = \langle \mathbf{x},P_2P_1\mathbf{y}\rangle$, since the P_i are each symmetric. However, since $P_1P_2 = P_2P_1$, this is also equal to $\langle \mathbf{x},P_1P_2\mathbf{y}\rangle$. Hence P_1P_2 is symmetric.

17. Let $P = A(A^tA)^{-1}A^t$. $P^2 = A(A^tA)^{-1}A^tA(A^tA)^{-1}A^t = A(A^tA)^{-1}A^t = P$. Furthermore, $P^t = (A(A^tA)^{-1}A^t)^t = A((A^tA)^{-1})^tA^t = A(A^tA)^{-1}A^t = P$. Hence, by Exercise 13, P is the

matrix of some orthogonal projection P_V with respect to the standard basis. We claim that $V = W = \text{Span}\{\text{columns of } A\}$. First, if $\mathbf{x} \in W^{\perp}$, then $\langle \mathbf{x}_i, \mathbf{x} \rangle = 0$ for every row $\mathbf{x}_i$ of A^t. Hence $A^t\mathbf{x} = \mathbf{0}$, so $P\mathbf{x} = A(A^tA)^{-1}A^t\mathbf{x} = \mathbf{0}$, so $W^{\perp} \subset \text{Ker}(P)$. Next, $PA = A(A^tA)^{-1}(A^tA) = A$, so every column of A is in $\text{Im}(A)$, and hence $W \subset \text{Im}(P)$. Since $W \oplus W^{\perp} = \mathbf{R}^n$ and $\text{Ker}(P) \oplus \text{Im}(P) = \mathbf{R}^n$, this implies $\text{Ker}(P) = W^{\perp}$ and $\text{Im}(P) = W$, so P is the orthogonal projection on W.

19. a) Let Q and B be as defined previously. Let α be the standard basis for $\mathbf{R}^n$. By Exercise 15 of Section 4.3, $[B]_\alpha$ is symmetric. By the spectral theorem, $[B]_\alpha$ is orthogonally similar to a diagonal matrix D whose diagonal entries are the eigenvalues $c_1, \ldots, c_n$ of $[B]_\alpha$. If P is the orthogonal matrix whose columns consist of orthonormal eigenvectors of $[B]_\alpha$, then $[B]_\alpha = P^tDP$ by Exercise 17 of Section 4.3. Let $\mathbf{y} = P\mathbf{x}$. Then $Q(\mathbf{x}) = B(\mathbf{x},\mathbf{x}) = \mathbf{x}^t(P^tDP)\mathbf{x} = (\mathbf{x}^tP^t)D(P\mathbf{x}) = \mathbf{y}^tD\mathbf{y} = \sum_{i=1}^{n} c_iy_i^2$.

b) If we order the eigenvalues c_i of $[B]_\alpha$ so that $c_1, \ldots, c_p > 0$ and $c_{p+1}, \ldots, c_{p+q} < 0$, $c_j = 0$, $j > p - q$, then if

$$D' = \begin{bmatrix} 1/\sqrt{c_1} & & & & & & \\ & \ddots & & & & & \\ & & 1/\sqrt{c_p} & & & & \\ & & & \ddots & & & \\ & & & & -1/\sqrt{c_{p+1}} & & \\ & & & & & \ddots & \\ & & & & & & -1/\sqrt{c_{p+q}} \\ & & & & & & & 0 \end{bmatrix}$$

and $J_{p,q} = \begin{bmatrix} I_{p\times p} & & \\ & -I_{q\times q} & \\ & & 0 \end{bmatrix}$, then $D = (D')^tJ_{p,q}D'$. Letting $\mathbf{z} = D\mathbf{y}' = D'P\mathbf{x}$, we see that

$$Q(\mathbf{x}) = \mathbf{z}^tJ_{p,q}\mathbf{z} = z_1^2 + \cdots + z_p^2 - z_{p+1}^2 - \cdots - z_{p+q}^2.$$

c) The number of positive, negative, and zero terms depends only on the number of positive, negative, and zero eigenvalues of the matrix $[B]_\alpha$. However, the eigenvalues of $[B]_\alpha$ are independent of the choice of orthonormal basis α for $\mathbf{R}^n$ as a consequence of Exercise 17 of Section 3; similar matrices have the same eigenvalues.

20. a) (2,1).

c) (1,0).

Chapter 5, Section 1

1. a) $9 + (2 + \sqrt{7})i$

c) $\sqrt{290}$

e) $(3/34) - (5/34)\,i$

3. Let $z = a + bi$ and $w = c + di$.

a) $\bar{z} + \bar{w} = (a - bi) + (c - di) = (a + c) - (b + d)i = \overline{z + w}$.

c) $\bar{z}/\bar{w} = (a - bi)/(c - di) = \overline{((ac+bd) + (ad-bc)i)/(c^2+d^2)} = ((ac+bd) - (ad-bc)i)/(c^2+d^2) = \overline{(a + bi)/(c + di)} = \overline{(z/w)}$

4. a) If $z = a + 0i$ is real, then $\bar{z} = a - 0i = z$. Conversely, if $z = a + bi$ satisfies $\bar{z} = z$, then $a + bi = a - bi$, so $b = -b$, or $b = 0$.

5. a) Let $p(x) = a_n x^n + \cdots + a_1 x + a_0$, $a_i \in \mathbf{R}$. If $z \in \mathbf{C}$ satisfies $p(z) = a_n z^n + \cdots + a_1 z + a_0 = 0$, then $\overline{p(z)} = \overline{a_n z^n + \cdots + a_1 z + a_0} = \bar{a}_n \bar{z}^n + \cdots + \bar{a}_1 \bar{z} + \bar{a}_0 = 0$. Since $a_i \in \mathbf{R}$, $\bar{a}_i = a_i$. Hence $p(\bar{z}) = a_n \bar{z}^n + \cdots + a_0 = 0$.

6. a) $z = (-2 \pm \sqrt{4-8})/2 = -1 \pm i$

d) $z^3 = -(\sqrt{3}/2) - (1/2)\, i = \cos(7\pi/6) + i\sin(7\pi/6)$, so $z = \cos(7\pi/18) + i\sin(7\pi/18)$, $\cos(19\pi/18) + i\sin(19\pi/18)$, $\cos(31\pi/18) + i\sin(31\pi/18)$.

7. (i) $0' = 0 + 0' = 0$

(ii) $[(-x) + x] + (-x)' = 0 + (-x)' = (-x)'$. But $[(-x) + x] + (-x)' = (-x) + [x + (-x)'] = -x + 0 = -x$. Thus $-x = (-x)'$.

8. b) We show this set is closed under addition, multiplication, and the taking of multiplicative inverses. Let $z = a + bi$, with $a = (p/q)$ and $b = (r/s)$, and $w = c + di$, with $c = (t/u)$ and $d = (v/x)$. Then $z + w = ((p/q) + (t/u)) + ((r/s) + (v/x))i$ also has rational coefficients. Similarly, $zw = ((p/q)(t/u) - (r/s)(v/x)) + ((p/q)(v/x) + (t/u)(r/s))i$ has rational coefficients. If $z \neq 0$, $z^{-1} = a/(a^2 + b^2) - (b/(a^2 + b^2))\, i$ has rational coefficients. Hence $\{a + bi | a, b \text{ rational}\}$ is a field—the axioms for fields hold here since they do for all complex numbers.

9. If $[m]_p = [m']_p = r$, then $m = ap + r$ and $m' = a'p + r$ for some a, a'. Hence $m - m' = (a - a')p$ is divisible by p. Conversely, if $m - m' = qp$, then m and m' have the same remainder on division by p, so $[m]_p = [m']_p$.

10. a) If $[m]_p = [m']_p$ and $[n]_p = [n']_p$, then by Exercise 9, $m - m' = qp$ and $n - n' = q'p$. Hence $(m - m') + (n - n') = (m + n) - (m' + n') = (q + q')p$. Hence, by Exercise 9, $[m + n]_p = [m' + n']_p$.

13. a) $[m]_p[n]_p = [mn]_p$. If $[m]_p \neq [0]_p$ and $[n]_p \neq [0]_p$, then $m = pq + r$, with $1 \leq r < p$, and $n = r'$ with $1 \leq r' < p$. Then $mn = pqr' + rr'$, but rr' is not divisible by p, so $[mn]_p \neq [0]_p$.

b) If $n' = r''$ and $[m]_p[n]_p = [m]_p[n']_p$, then $(pqr' + rr') - (pqr'' + rr'') = p(qr - qr'') + r(r' - r'')$ is divisible by p, so $r(r' - r'')$ is divisible by p. If $1 \leq r, r', r'' \leq p - 1$, this can only be true if $r' = r''$.

c) By b) the mapping $M: F_p \setminus \{[0]_p\} \to F_p \setminus \{[0]_p\}$ defined by $M([n]_p) = [mn]_p = [m]_p[n]_p$ is injective. Since $F_p \setminus \{[0]_p\}$ is finite, it is also surjective, so there is some $[n]_p$ such that $[m]_p[n]_p = [1]_p$.

14. a) $x = [2]_3$ is a double root: $(x + 1)^2 = x^2 + 2x + 1 = x^2 - x - 2$.

c) $x = [1], [2]$.

Chapter 5, Section 2

2. a) If $\mathbf{v} = (v_1, \ldots, v_n)$ and $\mathbf{w} = (w_1, \ldots, w_n)$ both solve the equation, then for all $c \in F$, $c\mathbf{v} + \mathbf{w} = (cv_1 + w_1, \ldots, cv_n + w_n)$ does also:

$$\begin{aligned} a_1(cv_1+w_1) + \cdots + a_n(cv_n+w_n) &= c(a_1v_1 + \cdots + a_nv_n)+(a_1w_1+\cdots+a_nw_n) \\ &= c\cdot 0 + 0 \\ &= 0 \end{aligned}$$

Hence W is a subspace of V.

3. a) If $\mathbf{a} = (a_1, \ldots, a_n) \in F^n$, then $\mathbf{a} = a_1\mathbf{e}_1 + \cdots + a_n\mathbf{e}_n$, thus $\mathbf{a} \in$ Span$\{\mathbf{e}_1, \ldots, \mathbf{e}_n\}$. Furthermore, if $a_1\mathbf{e}_1 + \cdots + a_n\mathbf{e}_n = (a_1, \ldots, a_n) = (0, \ldots, 0)$, then $a_1 = \cdots = a_n = 0$. Hence $\{\mathbf{e}_1, \ldots, \mathbf{e}_n\}$ is linearly independent, hence a basis for F^n.

4. a) Linearly dependent. The dimension of the span is 2.
c) Linearly dependent. The dimension of the span is 3.

5. a) $\dim(\mathrm{Ker}(T)) = \dim(\mathrm{Im}(T)) = 1$ and $\mathrm{Ker}(T) = \mathrm{Span}(\{(1/2 - (1/2)i, 1)\})$, $\mathrm{Im}(T) = \mathrm{Span}(\{(1/2 - (1/2)i, 1)\})$.

6. a) It follows from the definitions that $\hat{\mathbf{C}}^n$ is closed under addition and scalar multiplication. Since addition is defined as it is in $\mathbf{C}^n$, the addition in $\hat{\mathbf{C}}^n$ satisfies 1) through 4) of Definition (5.2.1). In addition, since $\mathbf{R} \subset \mathbf{C}$, that $\mathbf{C}^n$ satisfies 5) through 8) of (5.2.1) implies that $\hat{\mathbf{C}}^n$ satisfies 5) through 8) of (5.2.1).

10. a) $\lambda = 0,2$. $E_0 = \mathrm{Span}(\{(-i, 1)\})$, $E_2 = \mathrm{Span}(\{(i, 1)\})$.
c) $\lambda = i, i, -i$. $E_i = \mathrm{Span}(\{(1, 0, 0)\})$, $E_{-i} = \mathrm{Span}(\{(i, 1, 0)\})$.

11. a) $\lambda = 1$ with multiplicity 2. $E_1 = \mathrm{Span}(\{(1, 0)\})$. A is not diagonalizable.
c) $\lambda = 0, 1$. $E_0 = \mathrm{Span}(\{(1, 1)\})$, $E_1 = \mathrm{Span}(\{(1, 2)\})$. A is diagonalizable over F_3.

12. a) $\lambda = (2\cos(\theta) \pm \sqrt{4\cos^2(\theta) - 4})/2 = \cos(\theta) \pm i\cdot\sin(\theta)$.
$E_{\cos(\theta) \pm i\sin(\theta)} = \mathrm{Span}(\{(1, \mp i)\})$.
b) $\lambda = \pm i\sqrt{a}$. $E_{\pm i\sqrt{a}} = \mathrm{Span}(\{(\pm i\sqrt{a}, 1)\})$.

14. a) $A^{-1} = \begin{bmatrix} 1 & 1 \\ 0 & 1 \end{bmatrix}$ **c)** $A^{-1} = \begin{bmatrix} 0 & 2 & 2 \\ 1 & 2 & 0 \\ 1 & 1 & 1 \end{bmatrix}$

15. b) If $(p/q)1 + (r/s)\sqrt{2} = 0$, and $r/s \neq 0$, then $\sqrt{2} = -(p/q)(s/r) \in \mathbf{Q}$, the rational numbers. This is a contradiction, since $\sqrt{2}$ is an irrational number. Hence $\{1, \sqrt{2}\}$ is linearly independent over $\mathbf{Q}$.

Chapter 5, Section 3

1. a) $\langle(1-i, 1), (-1, 1+i)\rangle = 0$. Yes.
c) Yes.

4. a) If $\mathbf{u} = (u_1, \ldots, u_n)$, $\mathbf{v} = (v_1, \ldots, v_n)$, $\mathbf{w} = (w_1, \ldots, w_n) \in \mathbf{C}^n$ and $a, b \in \mathbf{C}$, then $\langle a\mathbf{u} + b\mathbf{v}, \mathbf{w}\rangle = (au_1 + bv_1)\overline{w}_1 + \cdots + (au_n + bv_n)\overline{w}_n = a(u_1\overline{w}_1 + \cdots + u_n\overline{w}_n) + b(v_1\overline{w}_1 + \cdots + v_n\overline{w}_n) = a\langle\mathbf{u},\mathbf{w}\rangle + b\langle\mathbf{v},\mathbf{w}\rangle$.
b) $\overline{\langle\mathbf{v},\mathbf{u}\rangle} = \overline{v_1\overline{u}_1 + \cdots + v_n\overline{u}_n} = \overline{v}_1u_1 + \cdots + \overline{v}_nu_n = u_1\overline{v}_1 + \cdots + u_n\overline{v}_n = \langle\mathbf{u},\mathbf{v}\rangle$
c) $\langle\mathbf{u},\mathbf{u}\rangle = u_1\overline{u}_1 + \cdots + u_n\overline{u}_n = |u_1|^2 + \cdots + |u_n|^2$.
Hence $\langle\mathbf{u},\mathbf{u}\rangle$ is real and ≥ 0 for all $\mathbf{u} \in \mathbf{C}^n$. If $\langle\mathbf{u},\mathbf{u}\rangle = 0$, then $|u_i| = 0$ for all i, so $\mathbf{u} = \mathbf{0}$.

6. a) $\lambda = (1 \pm \sqrt{5})/2$. $E_{(1\pm\sqrt{5})/2} = \mathrm{Span}(\{(1, i(1 \mp \sqrt{5})/2)\})$.

8. a) If $\langle T(\mathbf{v}), T(\mathbf{w})\rangle = \langle \mathbf{v},\mathbf{w}\rangle$ for all $\mathbf{v},\mathbf{w} \in V$, then $\langle \mathbf{v}, T^*T(\mathbf{w})\rangle = \langle \mathbf{v},\mathbf{w}\rangle$ for all $\mathbf{v},\mathbf{w}$, so $T^*T = I$. Similarly, $TT^* = I$, so T is invertible and $T^* = T^{-1}$.

9. a) First, we claim that if A is unitary and λ is an eigenvalue of A, then $\overline{\lambda}$ is an eigenvalue of $A^* = \overline{A}^t$. Let $\mathbf{x}$ be an eigenvector of A with eigenvalue λ, so that $\|(A - \lambda I)\mathbf{x}\|^2 = 0$. Then $\|(A^* - \overline{\lambda} I)\mathbf{x}\|^2 = \langle (A^* - \overline{\lambda} I)\mathbf{x}, (A^* - \overline{\lambda} I)\mathbf{x}\rangle = \langle \mathbf{x}, (A - \lambda I)(A^* - \overline{\lambda} I)\mathbf{x}\rangle$. Since $A^* = A^{-1}$, A and A^* commute, so this equals $\langle \mathbf{x}, (A^* - \overline{\lambda} I)(A - \lambda I)\mathbf{x}\rangle = \langle (A - \lambda I)\mathbf{x}, (A - \lambda I)\mathbf{x}\rangle = \|(A - \lambda I)\mathbf{x}\|^2 = 0$. Hence $(A^* - \overline{\lambda} I)\mathbf{x} = \mathbf{0}$, so $\overline{\lambda}$ is an eigenvalue of A^*. However, since A is unitary, $\mathbf{x} = A^*A\mathbf{x}$. Hence $|\lambda|^2 = 1$, so $|\lambda| = 1$.

Supplementary Exercises

1. a) False. Counterexample: $1 + i$.
   b) False. Counterexample: $(1+i) + (1-i) = 2$.
   c) False. Counterexample: $ix^2 + 2ix + i = 0$.
   d) False. If $c \in \mathbf{C}$, $\overline{T}(c\mathbf{x}) = \overline{T(c\mathbf{x})} = \overline{cT(\mathbf{x})} = \overline{c}\,\overline{T(\mathbf{x})} = \overline{c}\,\overline{T}(\mathbf{x})$, not $c\overline{T}(\mathbf{x})$.
   e) True. If $ab = 0$ and $a \neq 0$, then a has a multiplicative inverse a^{-1}, so $a^{-1}ab = 1b = b = a^{-1}0 = 0$. So $b = 0$.
   f) True. If $\mathbf{v} = (v_1, \ldots, v_n)$ $v_i \in \mathbf{C}$, then $\overline{\mathbf{v}} = (\overline{v}_1, \ldots, \overline{v}_n)$, so if $\overline{\mathbf{v}} = \mathbf{v}$, $\overline{v}_i = v_i$ for all i, and hence $v_i \in \mathbf{R}$ for all i.
   g) False. $\langle \mathbf{w},\mathbf{v}\rangle = \overline{\langle \mathbf{v},\mathbf{w}\rangle}$, which is different if $\langle \mathbf{v},\mathbf{w}\rangle \notin \mathbf{R}$.
   h) True. Rotation matrices are normal, hence diagonalizable by the spectral theorem for normal operators. See Exercise 11 of Section 5.3.
   i) False. $a_{ji} = \overline{a}_{ij}$.

2. a) $1 + i = \sqrt{2}(\cos(\pi/4) + i\cdot\sin(\pi/4))$, so
$$(1 + i)^5 = (\sqrt{2})^5(\cos(5\pi/4) + i\cdot\sin(5\pi/4)) = 4(-1-i) = -4-4i$$
   c) $2 + 2i = 2^{3/2}(\cos(\pi/4) + i\cdot\sin(\pi/4))$, so
$$\sqrt{2+2i} = 2^{3/4}(\cos(\pi/8) + i\cdot\sin(\pi/8)) \text{ or } 2^{3/4}(\cos(9\pi/8) + i\cdot\sin(9\pi/8))$$

3. a) $z^3 = i = \cos(\pi/2) + i\cdot\sin(\pi/2)$, so
$$z = \cos(\pi/6) + i\cdot\sin(\pi/6) = \sqrt{3}/2 + (1/2)i, \text{ or}$$
$$\cos(5\pi/6) + i\cdot\sin(5\pi/6) = -\sqrt{3}/2 + (1/2)i, \text{ or}$$
$$\cos(9\pi/6) + i\cdot\sin(9\pi/6) = -i$$
   c) $z = [(-1 \pm \sqrt{5})/2]i$

5. We show that $F(z)$ is closed under addition and multiplication. The axioms for a field can be verified by direct calculation. Let $p(z),q(z),r(z),s(z)$ be polynomials in z over F.

Addition: $\dfrac{p(z)}{q(z)} + \dfrac{r(z)}{s(z)} = \dfrac{p(z)s(z) + r(z)q(z)}{q(z)s(z)}$,
which is a quotient of polynomials.

Multiplication: $\dfrac{p(z)}{q(z)}\cdot\dfrac{r(z)}{s(z)} = \dfrac{p(z)r(z)}{q(z)s(z)}$,
which is a quotient of polynomials.

6. b) $S(c\mathbf{z} + \mathbf{w}) = cS(\mathbf{z}) + S(\mathbf{w})$ if $c \in \mathbf{R}$. S is complex linear if and only if $SJ(\mathbf{z}) = S(i\mathbf{z}) = iS(\mathbf{z}) = JS(\mathbf{z})$ for all $\mathbf{z} \in \mathbf{C}^n$.

8. a) If $\mathbf{v} = (v_1, \ldots, v_n)$ and $\mathbf{w} = (w_1, \ldots, w_n) \in \mathbf{C}^n$, then
$\mathrm{Re}\langle \mathbf{v},\mathbf{w}\rangle = \mathrm{Re}(v_1\overline{w}_1 + \cdots + v_n\overline{w}_n)$
$= \mathrm{Re}(v_1)\mathrm{Re}(w_1) + \mathrm{Im}(v_1)\mathrm{Im}(w_1) + \cdots + \mathrm{Re}(v_n)\mathrm{Re}(w_n) + \mathrm{Im}(v_n)\mathrm{Im}(w_n)$
This is exactly the standard inner product on $\hat{\mathbf{C}}^n$ viewed as $\mathbf{R}^{2n}$.

Chapter 6, Section 1

1. a) $\begin{bmatrix} 1 & 3 & 1 \\ 0 & 2 & -1 \\ 1 & 0 & 0 \end{bmatrix}\begin{bmatrix} 1 \\ -1 \\ 1 \end{bmatrix} = \begin{bmatrix} -1 \\ -3 \\ 1 \end{bmatrix} \notin W.$ No.

c) $T(\mathbf{e}_1) = i\mathbf{e}_1 \in W$ Yes.
$T(\mathbf{e}_3) = 2\mathbf{e}_3 \in W$
$T(\mathbf{e}_4) = \mathbf{e}_3 + 2\mathbf{e}_4 \in W$

e) $Tr(AX - XA) = Tr(AX) - Tr(XA) = 0$ for all X, so $T(X) = AX - XA \in W$. Yes.

2. If $\mathbf{x} \in \mathrm{Ker}(T)$, $T(\mathbf{x}) = \mathbf{0} \in \mathrm{Ker}(T)$. Hence $\mathrm{Ker}(T)$ is invariant. If $\mathbf{x} \in \mathrm{Im}(T)$, $T(\mathbf{x}) \in \mathrm{Im}(T)$ by definition. Hence, $\mathrm{Im}(T)$ is invariant.

4. a) $\{\mathbf{0}\}$, $\mathrm{Span}(\{\mathbf{e}_2\})$, $\mathrm{Span}(\{\mathbf{e}_1, \mathbf{e}_3\})$, $\mathbf{R}^3$.

6. a) This is true since if $\beta_1 = \{\mathbf{x}_1, \ldots, \mathbf{x}_n\}$ and $\beta_2 = \{\mathbf{y}_1, \ldots, \mathbf{y}_m\}$, then $T(\mathbf{x}_i) \in \mathrm{Span}(\beta_1)$ and $T(\mathbf{y}_j) \in \mathrm{Span}(\beta_2)$.

b) If $V = W_1 \oplus \cdots \oplus W_k$, and β_i is a basis for W_i, then $\beta = \beta_1 \cup \cdots \cup \beta_k$ is a basis for V. If $T: V \to V$ is any mapping such that $T(W_i) \subseteq W_i$ for all i, $1 < i < k$, then

$$[T]_\beta^\beta = \begin{bmatrix} A_1 & & & 0 \\ & \cdot & & \\ & & \cdot & \\ & & & \cdot \\ 0 & & & A_k \end{bmatrix}, \text{ where } A_i = [T|_{W_i}]_{\beta_i}^{\beta_i}.$$

8. a) Let $p(t) = a_m t^m + \cdots + a_0$. Then

$$\begin{aligned} p(Q^{-1}AQ) &= a_m(Q^{-1}AQ)^m + \cdots + a_1Q^{-1}AQ + a_0I \\ &= a_mQ^{-1}A^mQ + \cdots + a_1Q^{-1}AQ + a_0Q^{-1}IQ \\ &= Q^{-1}(a_mA^m + \cdots + a_1A + a_0I)Q \\ &= Q^{-1}p(A)Q. \end{aligned}$$

10. A defines a linear mapping $T: \mathbf{C}^n \to \mathbf{C}^n$, thus since $\mathbf{C}$ is algebraically closed, all the eigenvalues of A lie in $\mathbf{C}$ and by Theorem (6.1.12), if $p(t) = \det(T - tI)$, then $p(T)$ is the zero mapping and $p(A) = 0$. However, the characteristic polynomial of A is the same whether we view A as a real matrix or a complex matrix. Hence $p(A) = 0$.

11. a) The characteristic polynomial of A is $t^2 - 9t - 4$, so $A^{-1} = (1/4)(A - 9I)$
$= \begin{bmatrix} -2 & 1 \\ 3/4 & -1/4 \end{bmatrix}.$

c) The characteristic polynomial of A is $-t^3 + 2t^2 + 22t + 23$, so $A^{-1} = -(1/23)(-A^2 + 2A + 22I)$. Thus

$$A^{-1} = \begin{bmatrix} -1/23 & -2/23 & 4/23 \\ 5/23 & -13/23 & 3/23 \\ 2/23 & 27/23 & -8/23 \end{bmatrix}.$$

13. As a vector space over F, $\dim(M_{n\times n}(F)) = n^2$. Hence, if we consider the set $S = \{A^k | k = 0, \ldots, n^2\}$, we see that S must be linearly dependent. Hence there exists some polynomial

$$p(t) = a_{n^2}t^{n^2} + a_{n^2-1}t^{n^2-1} + \cdots + a_0 \text{ such that}$$
$$p(A) = a_{n^2}A^{n^2} + a_{n^2-1}A^{n^2-1} + \cdots + a_0 I = 0.$$

Chapter 6, Section 2

1. a) $N^2 = 0$.
   c) $N^5 = 0$.

2. a) $N\mathbf{x} \neq 0$ and $N^2\mathbf{x} = 0$, so length $= 2$. Cycle $= \{(3,3,3,3),(1,2,3,1)\}$.
   c) $N^2\mathbf{x} \neq 0$ and $N^3\mathbf{x} = 0$, so length $= 3$. Cycle $= \{(0,0,0,8i),(0,2i,0,4-i),(-i,1,0,2-i)\}$.
   e) $N^2(p) \neq 0$ and $N^3(p) = 0$, so length $= 3$. Cycle $= \{4,\ 3x + y,\ x^2 + xy\}$.

3. a) If $N^k(\mathbf{x}) = \mathbf{0}$ but $N^{k-1}(\mathbf{x}) \neq \mathbf{0}$, then $N(N^{k-1}(\mathbf{x})) = N^k(\mathbf{x}) = \mathbf{0} = 0\cdot N^{k-1}(\mathbf{x})$, so $N^{k-1}(\mathbf{x})$ is an eigenvector of N with eigenvalue 0.
   b) If $\mathbf{y} \in C(\mathbf{x})$, then $\mathbf{y} = a_{k-1}N^{k-1}(\mathbf{x}) + \cdots + a_1N(\mathbf{x}) + a_0\mathbf{x}$ for some $a_i \in F$. Hence $N(\mathbf{y}) = a_{k-2}N^{k-1}(\mathbf{x}) + \cdots + a_1N^2(\mathbf{x}) + a_0N(\mathbf{x}) \in C(\mathbf{x})$. Hence $C(\mathbf{x})$ is invariant under N.

5. a) For each k, $\text{Ker}(N^{k-1}) \subset \text{Ker}(N^k)$, since if $N^{k-1}(\mathbf{x}) = \mathbf{0}$, then $N^k(\mathbf{x}) = N(N^{k-1}(\mathbf{x})) = N(\mathbf{0}) = \mathbf{0}$. $\{\mathbf{0}\} \subset \text{Ker}(N)$ for every linear mapping N, and $\text{Ker}(N_l) = V$ by assumption.

7. a) Canonical form $= \begin{bmatrix} 0 & 1 & & \\ 0 & 0 & & \\ & & 0 & \\ & & & 0 \end{bmatrix}$

   One canonical basis $= \{(1,1,1,1),(1,0,0,0)\}\cup\{(3,1,0,0)\}\cup\{(0,0,1,3)\}$

   c) Canonical form $= \begin{bmatrix} 0 & 1 & 0 & 0 & 0 \\ 0 & 0 & 1 & 0 & 0 \\ 0 & 0 & 0 & 1 & 0 \\ 0 & 0 & 0 & 0 & 1 \\ 0 & 0 & 0 & 0 & 0 \end{bmatrix}$

   One canonical basis $= \{1944,\ 648-648x,\ 24-144x+108x^2,\ 12x^2-12x^3,\ x^4\}$

   e) Canonical form $= \begin{bmatrix} 0 & 1 & 0 & & & \\ 0 & 0 & 1 & & & \\ 0 & 0 & 0 & & & \\ & & & 0 & 1 & 0 \\ & & & 0 & 0 & 1 \\ & & & 0 & 0 & 0 \end{bmatrix}$

One canonical basis $= \{24, 12x^2, x^4\} \cup \{120x, 20x^3, x^5\}$.

g) Canonical form $= \begin{bmatrix} 0 & 1 & 0 & \\ 0 & 0 & 1 & \\ 0 & 0 & 0 & \\ & & & 0 \end{bmatrix}$

One canonical basis $= \{(2-3i,0,0,0),(4,2+i,1,0),(0,0,0,1)\} \cup \{(0,1/2,1,0)\}$.

9. For $n = 3$ there are 3 possible tableaux:

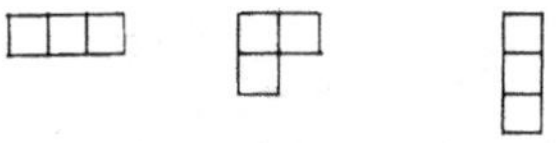

For $n = 4$ there are 5 possible tableaux:

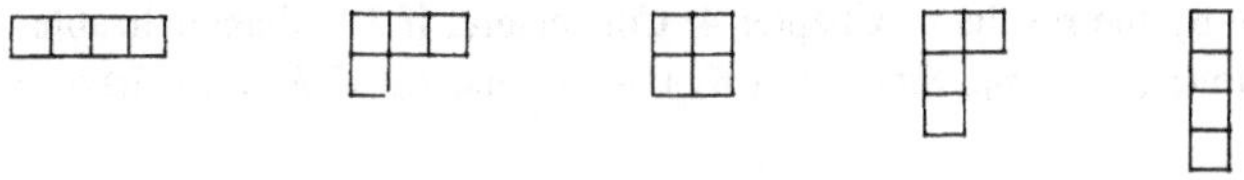

For $n = 5$ there are 7 possible tableaux:

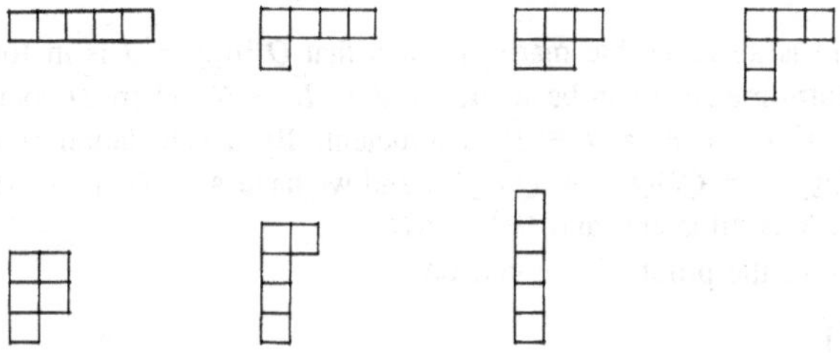

In general, by looking at the number of boxes in the rows (or columns) of the tableaux, it follows that the number of possible tableaux for nilpotent $N:F^n \to F^n$ is the same as the number of partitions of n: The number of ways that n can be written as a sum of positive integers $k \leq n$. There is no easily stated formula for this number.

13. a) If N_1 and N_2 are similar, then we know we can regard them as the matrices of one linear mapping $T:F^n \to F^n$ with respect to two different bases. Since the canonical form is determined by the cycle tableaux, or equivalently by the integers $\dim(\text{Ker}(T^j))$ for $j = 1, \ldots, l$, the canonical forms will be the same.

b) By Exercise 12, there exist invertible matrices P,Q such that $P^{-1}N_1P = C = Q^{-1}N_2Q$, where C is the canonical form. Hence $N_1 = PQ^{-1}N_2QP^{-1} = (QP^{-1})^{-1}N_2(QP^{-1})$, so N_1 and N_2 are similar.

15. a) Let $\alpha = \{\mathbf{x}_1, \ldots, \mathbf{x}_n\}$ be a basis for V, and for each i, $1 \leq i \leq n$ let $k = k_i$ be the smallest integer such that $T^k(\mathbf{x}_i) = \mathbf{0}$. (That is, $T^{k_i}(\mathbf{x}_i) = \mathbf{0}$, but $T^{k_i-1}(\mathbf{x}_i) \neq \mathbf{0}$.) Let $m = \max_{1 \leq i \leq n} k_i$.

Chapter 6, Section 3

1. a) $\lambda = 1$ is an eigenvalue of multiplicity $m = 3$ of T. $(T - I)^3$ is defined by the matrix

$$\begin{bmatrix} 0 & 0 & 0 & 20 \\ 0 & 0 & 0 & 4 \\ 0 & 0 & 0 & 0 \\ 0 & 0 & 0 & 8 \end{bmatrix}, \text{ for which } \begin{bmatrix} 0 & 0 & 0 & 20 \\ 0 & 0 & 0 & 4 \\ 0 & 0 & 0 & 0 \\ 0 & 0 & 0 & 8 \end{bmatrix} \begin{bmatrix} 3 \\ 1 \\ 0 \\ 0 \end{bmatrix} = \begin{bmatrix} 0 \\ 0 \\ 0 \\ 0 \end{bmatrix}$$

Hence $\mathbf{x} \in \text{Ker}((T - I)^3) = K_1$.

c) $\lambda = 0$ is an eigenvalue of multiplicity $m = 5$ of T, and $(T - 0I)^5 = T^5 = 0$. Therefore $p \in \text{Ker}(T^5) = K_0$. (In fact, $T^3(p) = 0$.)

2. a) $K_1 = \text{Span}\{\mathbf{e}_1, \mathbf{e}_2, \mathbf{e}_3\}$.

c) $K_0 = P_4(\mathbf{R})$.

5. a) Let $\lambda_1, \ldots, \lambda_k$ be the distinct eigenvalues and let the multiplicity of λ_i be m_i. Since $E_{\lambda_i} \subset K_{\lambda_i}$ for all i, if $E_{\lambda_i} = K_{\lambda_i}$, then $\dim(E_{\lambda_i}) = m_i$ for each i, so T is diagonalizable by the results of Chapter 4. Conversely, if T is diagonalizable, then for each i, $\dim(E_{\lambda_i}) = m_i$. Since $\dim(K_{\lambda_i}) = m_i$ and $E_{\lambda_i} \subset K_{\lambda_i}$, we have $E_{\lambda_i} = K_{\lambda_i}$ for all i.

b) If T is diagonalizable, then by part a, $K_\lambda = E_\lambda$ for each eigenvalue of T. Hence $(T - \lambda I)|_{K_\lambda} = 0$, so $T|_{K_\lambda} = \lambda I$, and the canonical form of T is its diagonal form.

7. a) By Exercise 3, there is an invertible matrix Q such that $Q^{-1}AQ = J$ is in Jordan form. Any Jordan form matrix J can be written as $J = \overline{D} + \overline{N}$, where $\overline{D}$ contains the diagonal terms of J and $\overline{N} = J - \overline{D}$ is nilpotent. By a calculation $\overline{N}\overline{D} = \overline{D}\overline{N}$. Hence $A = QJQ^{-1} = Q\overline{D}Q^{-1} + Q\overline{N}Q^{-1}$, and we have $A = D + N$, where D is diagonalizable, N is nilpotent, and $DN = ND$.

b) The idea is the same as the proof of Exercise 6b.

10. a) If $B = \begin{bmatrix} \lambda & 1 & & 0 \\ & \ddots & \ddots & \\ & & \ddots & 1 \\ 0 & & & \lambda \end{bmatrix}$ $(\lambda \neq 0)$ is a $k \times k$ block, then

$$B^{-1} = \begin{bmatrix} \lambda^{-1} & -\lambda^{-2} & \lambda^{-3} & \cdots & (-1)^{k+1}\lambda^{-k} \\ & \ddots & \ddots & \ddots & \vdots \\ & & \ddots & \ddots & \lambda^{-3} \\ & & & \ddots & -\lambda^{-2} \\ 0 & & & & \lambda^{-1} \end{bmatrix}$$

b) By part a, B^{-1} has one eigenvalue λ^{-1} with multiplicity n and $B^{-1} - \lambda^{-1}I$ has rank $k - 1$. Hence $\dim(\text{Ker}(B^{-1} - \lambda^{-1}I)) = 1$, so there is only one row in the cycle tableau for B^{-1}. Therefore the Jordan form for B^{-1} is

$$\begin{bmatrix} \lambda^{-1} & 1 & & \\ & \ddots & \ddots & \\ & & \ddots & 1 \\ 0 & & & \lambda^{-1} \end{bmatrix} \text{ — one } k \times k \text{ block.}$$

c) Follows from b).

11. a) $\sum_{j=1}^{l_i} n_{ij} = \dim(K_{\lambda_i}) = m_i$, so

$$\sum_{i=1}^{k} \left[\sum_{i=1}^{l_i} n_{ij} \right] = \sum_{i=1}^{k} \dim(K_{\lambda_i}) = n = \dim(V).$$

Chapter 6, Section 4

2\. a) $(2 - i)$: 3:

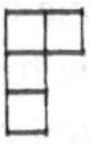
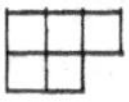

Canonical form:
$$\begin{bmatrix} 2-i & 1 & & & & & & & \\ 0 & 2-i & & & & & & & \\ & & 2-i & & & & & & \\ & & & 2-i & & & & & \\ & & & & 3 & 1 & 0 & & \\ & & & & 0 & 3 & 1 & & \\ & & & & 0 & 0 & 3 & & \\ & & & & & & & 3 & 1 \\ & & & & & & & 0 & 3 \end{bmatrix}$$

b) There are five possibilities for the Jordan blocks for $\lambda = 2 - i$ and seven possibilities for $\lambda = 3$. (See Exercise 9 of Section 6.2) Hence there are $5 \cdot 7 = 35$ possibilities in all.

4\. a) $\lambda = 1,3$. $\lambda = 1$ has multiplicity three and $\dim(E_1) = 1$. Thus the cycle tableau for $\lambda = 1$ is 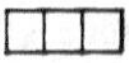. A basis for E_1 is $\{(1,0,0,0)\}$. $\{(1,0,0,0), (-1,1/2,0,0), (0,0,-1/2,0)\}$ is a cycle of length three for $\lambda = 1$. $\lambda = 3$ has multiplicity one and $\dim(E_3) = 1$. The cycle tableau for $\lambda = 3$ is . A basis for E_3 is $\{(5/2,1/2,0,1)\}$. A canonical basis is given by the union of these cycles, $\{(1,0,0,0), (-1,1/2,0,0),(0,0,-1/2,0)\} \cup \{(5/2,1/2,0,1)\}$.

The Jordan form of T is $\begin{bmatrix} 1 & 1 & 0 & \\ 0 & 1 & 1 & \\ 0 & 0 & 1 & \\ & & & 3 \end{bmatrix}$

c) $\lambda = 0$ has multiplicity five and $\dim(E_0) = 2$. Since $\dim(\text{Ker}(T^3)) = 5$ and $\dim(\text{Ker}(T^2)) = 4$, we conclude the cycle tableau for $\lambda = 0$ is 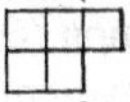. Then 24 is a (conveniently chosen) eigenvector for $\lambda = 0$ that is in the image of T^2 but not T^3. The corresponding cycle is $\{24, 12x^2, x^4\}$. $6x$ is a (conveniently chosen) eigenvector for $\lambda = 0$ that is in the image of T but not of T^2. The corresponding cycle is $\{6x, x^3\}$. Together these cycles form a canonical basis, $\{24, 12x^2, x^4\} \cup \{6x, x^3\}$. The Jordan canonical form is

$$\begin{bmatrix} 0 & 1 & 0 & & \\ 0 & 0 & 1 & & \\ 0 & 0 & 0 & & \\ & & & 0 & 1 \\ & & & 0 & 0 \end{bmatrix}$$

5\. a) $\lambda = 1$ has multiplicity four and $\dim(E_1) = 1$. Thus the cycle tableau for $\lambda = 1$ is 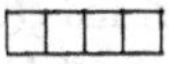 and the Jordan canonical form is

$$\begin{bmatrix} 1 & 1 & 0 & 0 \\ 0 & 1 & 1 & 0 \\ 0 & 0 & 1 & 1 \\ 0 & 0 & 0 & 1 \end{bmatrix}$$

A canonical basis is given by the cycle $\{(-111,-37,0,-37),(21,81,0,44),(5,0,0,7),(0,0,1,0)\}$.

c) $\lambda = 2, -2 \pm i$. $\lambda = 2$ has multiplicity three and $\dim(E_2) = 1$. The cycle tableau of $\lambda = 2$ is $\square\square\square$. A basis for E_2 is $\{e^{4x}\}$. $\{e^{4x}, xe^{4x}, (1/2)\, x^2 e^{4x}\}$ is a cycle for $\lambda = 2$ of length three. $\lambda = -2 \pm i$ has multiplicity one, $\dim(E_{-2\pm i}) = 1$ and a basis for $E_{-2\pm i}$ is $\{\pm i \cdot \sin(x) + \cos(x)\}$. A canonical basis for T is $\{i \cdot \sin(x) + \cos(x)\} \cup \{-i \cdot \sin(x) + \cos(x)\} \cup \{e^{4x}, xe^{4x}, 1/\, 2x^2 e^{4x}\}$.

The Jordan form is $\begin{bmatrix} -2+i & & & & \\ & -2-i & & & \\ & & 2 & 1 & 0 \\ & & 0 & 2 & 1 \\ & & 0 & 0 & 2 \end{bmatrix}$

Chapter 6, Section 5

1. b) Let B be an $l \times l$ Jordan block with diagonal entries λ. Then $(B - \lambda I)^l = 0$, but $(B - \lambda I)^{l-1} \neq 0$. Hence if l_i is the size of the largest Jordan block for the eigenvalue λ_i, then $(B - \lambda_i I)^{l_i} = 0$ for every block B with diagonal entries λ_i, but there is some block for λ_i with $(B - \lambda_i I)^{l_i - 1} \neq 0$. Hence, since $g(t) = (t - \lambda_1)^{l_1} \cdots (t - \lambda_k)^{l_k}$ satisfies $g(B) = 0$ for all blocks in the Jordan form of T, $g(T) = 0$. Since no lower powers of $(t - \lambda_i)$ will do, by part a, this $g(t)$ must be the minimal polynomial of T.

c) $(T - \lambda_i I)^{l_i}|_{K_{\lambda_i}} = 0$ since every block in the Jordan form of T with diagonal entries λ_i satisfies $(B - \lambda_i I)^{l_i} = 0$. Hence $K_{\lambda_i} = \mathrm{Ker}((T - \lambda_i I)^{l_i})$.

2. a) By Exercise 5 of Section 6.3, T is diagonalizable if and only if $E_{\lambda_i} = \mathrm{Ker}\,(T - \lambda_i I) = K_{\lambda_i}$ for all i. By part c of Proposition (6.5.9), this is true if and only if the factor $(t - \lambda_i)$ appears only to the first power in the minimal polynomial for all i (that is, $l_i = 1$ for all i).

b) By part b of Proposition (6.5.9) there is one Jordan block of size m_i for each λ_i if and only if the minimal polynomial of T is $(t - \lambda_1)^{m_1} \cdots (t - \lambda_k)^{m_k}$.

3. a) $(t - 1)^3(t - 3)$

c) t^3

6. a) This is clear for I_1 and I_2. If $q(t)$ and $p(t) \in I_1 \cap I_2$, then $p(t) = p_1(t)f(t)$ for some $p_1(t)$, and $p(t) = p_2(t)g(t)$ for some $p_2(t)$. Similarly, $q(t) = q_1(t)\, f(t)$ and $q(t) = q_2(t)g(t)$ for some $q_1(t)$ and $q_2(t)$. Hence $cp(t) + q(t) = [cp_1(t) + q_1(t)]\, f(t) = [cp_2(t) + q_2(t)]g(t)$, so $cp(t) + q(t) \in I_1 \cap I_2$, and consequently $I_1 \cap I_2$ is a vector subspace of $P(F)$.

Next, if $p(t) \in I_1 \cap I_2$, and $r(t) \in P(F)$, then $p(t)r(t) = [\, p_1(t)r(t)]f(t) \in I_1$ and $p(t)r(t) = [\, p_2(t)r(t)]q(t) \in I_2$, so $p(t)r(t) \in I_1 \cap I_2$. Hence Proposition (6.5.6) applies to $I_1 \cap I_2$.

b) $m(t)$ divides every polynomial which is divisible by both $f(t)$ and $g(t)$, is monic, and is divisible by $f(t)$ and $g(t)$ itself. Hence $m(t)$ is the generator of $I_1 \cap I_2$.

c) This follows from part b of Proposition (6.5.9) and part b of this exercise.

8. Let $s(t) \in P(F)$. We proceed by induction on degree of $s(t)$. If $\deg(p(t)) > \deg(s(t))$, then $q(t) = 0$ and $r(t) = s(t)$. Now assume the claim has been proved for all polynomials $s(t)$ of degree $k \geq \deg(p(t))$ and consider $s(t) = a_{k+1}t^{k+1} + \hat{s}(t)$, where $\hat{s}(t)$ has degree

$\leqslant k$. Write $p(t) = b_m t^m + \cdots + b_0$. We have that $s(t) - p(t) \cdot (a_{k+1}/b_m)\, t^{k+1-m}$ has degree $\leqslant k$. Hence, by induction, there are unique polynomials $\hat{q}, r$ such that $s(t) - p(t) \cdot (a_{k+1}/b_m)\, t^{k+1-m} = p(t)\hat{q}(t) + r(t)$, where $\deg(r(t)) < \deg(p(t))$. Hence $s(t) = p(t)\,[\hat{q}(t) + (a_{k+1}/b_m)\, t^{k+1-m}] + r(t)$, where $q(t) = \hat{q}(t) + (a_{k+1}/b_m)\, t^{k+1-m}$.

11. First, $\dim(W) \leqslant$ the degree of the minimal polynomial, since if the minimal polynomial is $t^m + a_{m-1} t^{m-1} + \cdots + a_0$, the equation $A^m + a_{m-1} A^{m-1} + \cdots + a_0 I = 0$ shows $A^m \in \text{Span}\{A^{m-1}, A^{m-2}, \ldots, A, I\}$. Hence $A^{m+k} \in \text{Span}\{A^{m-1}, A^{m-2}, \ldots, A, I\}$ for all $k \geqslant 0$ as well, so $\dim(W) \leqslant m$. Next, if $\dim(W)$ were less than m, there would be some linear combination $a_{m-1} A^{m-1} + \cdots + a_1 A + a_0 I = 0$, where $a_i \neq 0$ for some i. This gives a polynomial $p(t) = a_{m-1} t^{m-1} + \cdots + a_0$ in I_A which is $\neq 0$, and has degree strictly less than the degree of the minimal polynomial. But this is a contradiction by the definition of the minimal polynomial. Hence $\dim(W) = m$.

Supplementary Exercises

1. a) False. Counterexample: $V = \mathbf{R}^2$ and $T = R_\theta$, rotation mapping with $\theta \neq k\pi$.

   b) True. $W = \text{Span}\{N^{k-1}(\mathbf{x}), \ldots, N(\mathbf{x}), \mathbf{x}\}$ for some $\mathbf{x}$. For each i, $N(N^i(\mathbf{x})) = N^{i+1}(\mathbf{x}) \in W$, so W is invariant under N.

   c) False. Counterexample: $\begin{bmatrix} 0 & 1 \\ -1 & 0 \end{bmatrix}$.

   d) True. The cycle tableau is 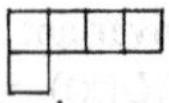 or 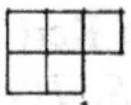.
   In either case, any canonical basis contains a cycle of length 3.

   e) True. Since $E_\lambda \subseteq K_\lambda$ for each λ.

   f) True. By the Cayley-Hamilton theorem $A^n = 0$, so A is nilpotent.

   g) True. Let $A^k = 0$, $B^l = 0$, then $(cA + B)^{k+l+1} = 0$, because when we expand $(cA + B)^{k+l+1}$, using $AB = BA$, every term contains A^k or B^l.

   h) False. $\begin{bmatrix} 0 & 1 \\ -1 & 0 \end{bmatrix}$ has Jordan form $\begin{bmatrix} i & 0 \\ 0 & -i \end{bmatrix}$.

   i) False. It is always infinite-dimensional, since it contains $t^n \det(A - tI)$ for all $n \geqslant 1$.

   j) True. Since $A = Q^{-1}cIQ$ for some invertible Q, $A = cQ^{-1}IQ = cI$.

   k) False. Counterexample: $A = \begin{bmatrix} 2 & 1 \\ 0 & 2 \end{bmatrix}$, $B = \begin{bmatrix} 2 & 1 & \\ 0 & 2 & \\ & & 2 \end{bmatrix}$ both have minimal polynomial $(t - 2)^2$.

2. a) The eigenvalues are $\lambda = 4 \pm \sqrt{5}$ and the matrix is triangularizable over $F = \mathbf{R}$ and $F = \mathbf{C}$.

   c) The only eigenvalue is $\lambda = -2$ with multiplicity four. The matrix is triangularizable over $F = \mathbf{Q}$, $\mathbf{R}$, and $\mathbf{C}$.

4. a) $K_0 = \text{Span}\{\mathbf{e}_1, \mathbf{e}_2\}$.

   c) $K_i = E_i = \text{Span}\{-\sin(x) + i\cos(x)\}$

5. a) $\begin{bmatrix} 0 & 1 & & \\ 0 & 0 & & \\ & & \sqrt{3} & \\ & & & -\sqrt{3} \end{bmatrix}$

c) $\begin{bmatrix} i & & & & & \\ & -i & & & & \\ & & 2i & & & \\ & & & -2i & & \\ & & & & 3i & \\ & & & & & -3i \end{bmatrix}$

6. a) $t^2(t - \sqrt{3})(t + \sqrt{3}) = t^2(t^2 - 3)$.
c) $(t - i)(t + i)(t - i2)(t + i2)(t - i3)(t + i3) = (t^2 + 1)(t^2 + 4)(t^2 + 9)$.

7. a) $\text{Span}\{\begin{bmatrix} 1 \\ i \end{bmatrix}\}$ is such a subspace, since $\begin{bmatrix} 1 & 1 \\ -1 & 1 \end{bmatrix}\begin{bmatrix} 1 \\ i \end{bmatrix} = (1 + i)\begin{bmatrix} 1 \\ i \end{bmatrix}$, and $\begin{bmatrix} 4 & -4 \\ 4 & 4 \end{bmatrix}\begin{bmatrix} 1 \\ i \end{bmatrix} = (4 - 4i)\begin{bmatrix} 1 \\ i \end{bmatrix}$.

b) If $T(\mathbf{x}) \in W$ for all $T \in S$, then for any $c_1T_1 + \cdots + c_nT_n \in \text{Span}(S)$ we have for all $\mathbf{x} \in W$, $(c_1T_1 + \cdots + c_nT_n)(\mathbf{x}) = c_1T_1(\mathbf{x}) + \cdots + c_nT_n(\mathbf{x}) \in W$. Hence W is invariant under all $T \in \text{Span}(S)$.

c) Let λ be any eigenvalue of T. Since $TU = UT$, $E_\lambda(T)$ is invariant under U. Hence $U|_{E_\lambda(T)}$ has some eigenvalue $\mu \in \mathbf{C}$ with an eigenvector $\mathbf{x}$, such that $U(\mathbf{x}) = \mu\mathbf{x}$. Since $T(\mathbf{x}) = \lambda\mathbf{x}$ by definition, we have that $W = \text{Span}\{\mathbf{x}\}$ is invariant under T and U.

9. If $U \in C(S)$ and $\mathbf{x} \in \text{Ker}(U)$, then for any $T \in S$, $U(T(\mathbf{x})) = TU(\mathbf{x}) = T(\mathbf{0}) = \mathbf{0}$. Hence $T(\mathbf{x}) \in \text{Ker}(U)$, so $\text{Ker}(U)$ is invariant under S. Similarly, if $\mathbf{y} = U(\mathbf{x})$ is in $\text{Im}(U)$, then for all $T \in S$, $T(\mathbf{y}) = T(U(\mathbf{x})) = U(T(\mathbf{x})) \in \text{Im}(U)$. Hence, $\text{Im}(U)$ is invariant under S as well. Since S is irreducible, we have either $\text{Ker}(U) = V$ and $\text{Im}(U) = \{\mathbf{0}\}$, so $U = 0$, or $\text{Ker}(U) = \{\mathbf{0}\}$, and $\text{Im}(U) = V$. In the first case $U = 0$; in the second U is invertible, since U is injective and surjective. Hence every $U \in C(S)$ is either 0 or invertible.

12. a) After one month

$SD_1 = (.8)SD_0 + (.1)H_0$
$H_1 = (.2)SD_0 + (.9)H_0$, or

$$\begin{bmatrix} SD_1 \\ H_1 \end{bmatrix} = \begin{bmatrix} .8 & .1 \\ .2 & .9 \end{bmatrix}\begin{bmatrix} SD_0 \\ H_0 \end{bmatrix}.$$

Hence, after n months (since the proportions remain the same in each month), by induction,

$$\begin{bmatrix} SD_n \\ H_n \end{bmatrix} = \begin{bmatrix} .8 & .1 \\ .2 & .9 \end{bmatrix}\begin{bmatrix} SD_{n-1} \\ H_{n-2} \end{bmatrix} = \begin{bmatrix} .8 & .1 \\ .2 & .9 \end{bmatrix}^2\begin{bmatrix} SD_{n-2} \\ H_{n-2} \end{bmatrix} = \cdots$$
$$= \begin{bmatrix} .8 & .1 \\ .2 & .9 \end{bmatrix}^n\begin{bmatrix} SD_0 \\ H_0 \end{bmatrix}$$

b) The Jordan form of $T = \begin{bmatrix} .8 & .1 \\ .2 & .9 \end{bmatrix}$ is $J = \begin{bmatrix} 1 & 0 \\ 0 & .7 \end{bmatrix}$ and $J = \begin{bmatrix} 1/2 & 1 \\ 1 & -1 \end{bmatrix}^{-1} \cdot \begin{bmatrix} .8 & .1 \\ .2 & .9 \end{bmatrix}\begin{bmatrix} 1/2 & 1 \\ 1 & -1 \end{bmatrix}$, where $Q = \begin{bmatrix} 1/2 & 1 \\ 1 & -1 \end{bmatrix}$ is an appropriate change of basis matrix. Since $.7 < 1$, $\lim_{k\to\infty} J^k = \begin{bmatrix} 1 & 0 \\ 0 & 0 \end{bmatrix}$.

c) Hence $L = \lim_{k\to\infty} T^k = Q\begin{bmatrix} 1 & 0 \\ 0 & 0 \end{bmatrix} Q^{-1} = \begin{bmatrix} 1/3 & 1/3 \\ 2/3 & 2/3 \end{bmatrix}$

If $\begin{bmatrix} SD_\infty \\ H_\infty \end{bmatrix} = \begin{bmatrix} 1/3 & 1/3 \\ 2/3 & 2/3 \end{bmatrix} \begin{bmatrix} SD_0 \\ H_0 \end{bmatrix}$, then $SD_\infty/(SD_\infty + H_\infty) = 1/3$, independent of SD_0 and H_0. Similarly, $H_\infty/(SD_\infty + H_\infty) = 2/3$.

Chapter 7, Section 2

1. a) $\mathbf{u}(t) = e^t\mathbf{u}_0$. The solution curves are rays radiating from the origin.

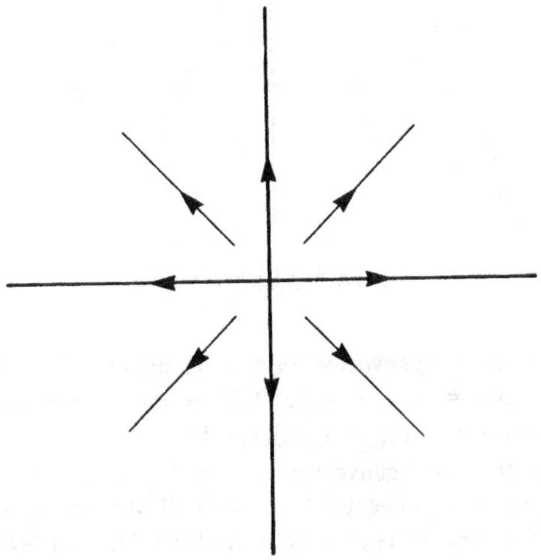

Figure for Exercise 1a.

c) $\mathbf{u}(t) = (u_1(0), e^t u_2(0))$. The solution curves are vertical rays.

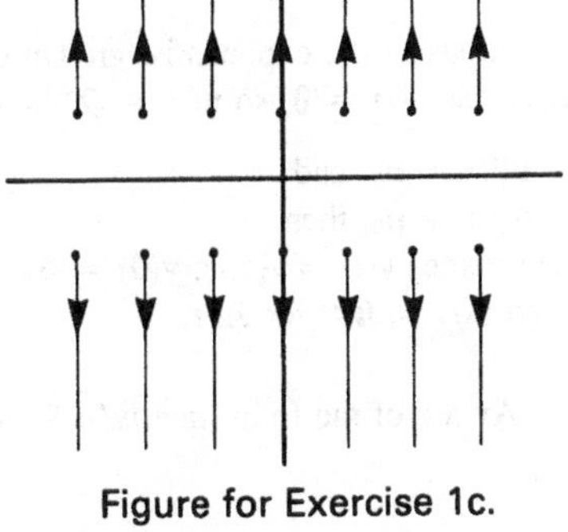

Figure for Exercise 1c.

e) $\mathbf{u}(t) = (e^{-t}u_1(0), e^{-2t}u_2(0))$. The solution curves are parabolas opening up and down with vertices at the origin.

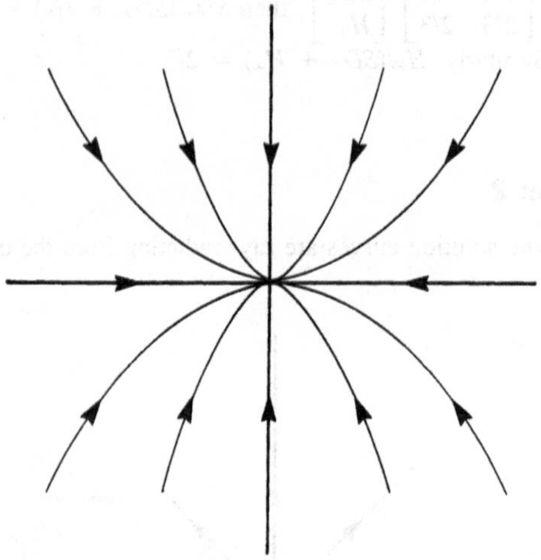

Figure for Exercise 1e.

2. a) $\lambda = 1,3$. $(1,-1)$ is an eigenvector for $\lambda = 1$, and $(1,1)$ is an eigenvector for $\lambda = 3$. If $\mathbf{u}(0) = (x_1,x_2) = ((x_1 + x_2)/2)(1,1) + ((x_1 - x_2)/2)(1,-1)$, then $\mathbf{u}(t) = e^{3t}((x_1 + x_2)/2)(1,1) + e^t((x_1 - x_2)/2)(1,-1)$.

c) $\lambda = 1,2,3$. $(1,0,0)$ is an eigenvector for $\lambda = 1$, $(1,1,0)$ is an eigenvector for $\lambda = 2$, and $(1/2,1,1)$ is an eigenvector for $\lambda = 3$. If $\mathbf{u}(0) = (x_1,x_2,x_3) = (x_1 - x_2 - (1/2)x_3)\,(1,0,0) + (x_2 - x_3)(1,1,0) + x_3(1/2,1,1)$, then $\mathbf{u}(t) = e^t(\,x_1 - x_2 - (1/2)x_3,0,0) + e^{2t}(x_2 - x_3, x_2 - x_3,0) + e^{3t}\,((1/2)x_3,x_3,x_3)$.

3. a) Straight line solutions along Span$\{(1,1)\}$ and Span$\{(1,-1)\}$.

c) Straight line solutions along Span$\{(1,0,0)\}$, Span$\{(1,1,0)\}$, and Span$\{(1/2,1,1)\}$.

5. a) By Exercise 4, if Q is the change of basis matrix to the diagonal form Λ for A, then $\mathbf{v} = Q^{-1}\mathbf{u}$ solves $\mathbf{v}' = \Lambda\mathbf{v}$. If $\mathbf{u}(0) = \mathbf{0}$, $\mathbf{v}(0) = Q^{-1}\mathbf{0} = \mathbf{0}$ as well. The diagonal system has the form

$$v_i'(t) = \lambda_i v_i(t) \qquad v_i(0) = 0$$

By the form of the solutions of the exponential growth equation, the only solution is $v_i(t) = 0$ for all t. Hence $\mathbf{v}(t) = \mathbf{0}$, so $\mathbf{u}(t) = Q^{-1}\mathbf{0} = \mathbf{0}$ as well.

b) If $\mathbf{u}'(t) = A\mathbf{u}(t)$, $\quad \mathbf{u}(0) = \mathbf{u}_0$, and
$\hat{\mathbf{u}}'(t) = A\hat{\mathbf{u}}(t)$, $\quad \hat{\mathbf{u}}(0) = \mathbf{u}_0$, then
$\mathbf{v}(t) = \mathbf{u}(t) - \hat{\mathbf{u}}(t)$ satisfies $\mathbf{v}(t) = A\mathbf{v}(t)$, $\mathbf{v}(0) = \mathbf{0}$.
By part a, $\mathbf{v}(t) = \mathbf{0}$, so $\mathbf{u}(t) = \hat{\mathbf{u}}(t)$ for all t.

6. a) The solutions to $\mathbf{v}' = \Lambda\mathbf{v}$ are of the form (ae^{3t},be^t). Solutions are graphed on page 413.

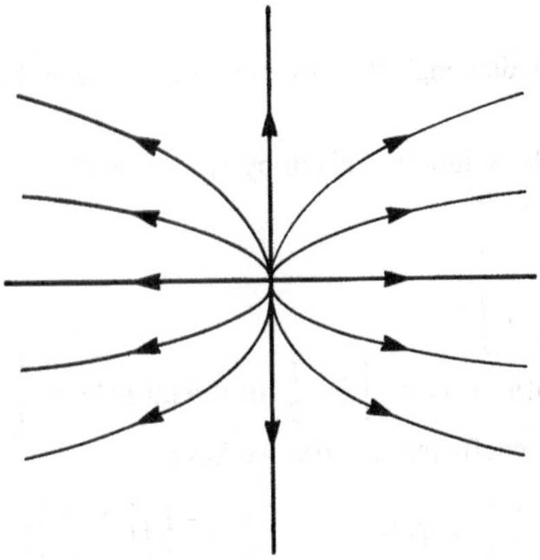

Solutions to $\mathbf{v}' = \Lambda\mathbf{v}$

Figure 1 for Exercise 6a.

We can take $Q = \begin{bmatrix} \sqrt{2}/2 & -\sqrt{2}/2 \\ \sqrt{2}/2 & \sqrt{2}/2 \end{bmatrix}$. Thus, to obtain the solution curves of $\mathbf{u}' = A\mathbf{u}$, we can apply the mapping $Q = R_{\pi/4}$:

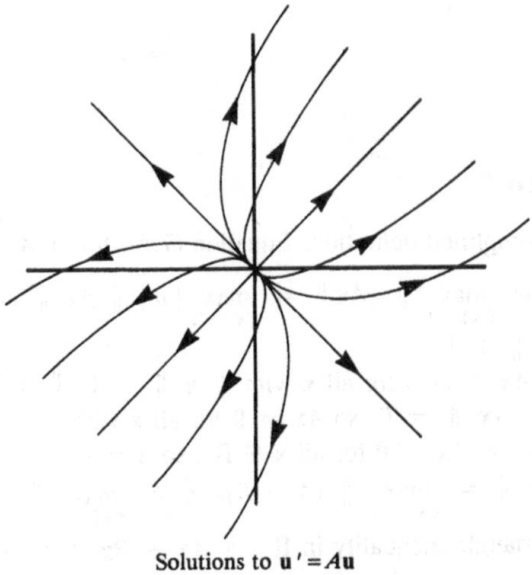

Solutions to $\mathbf{u}' = A\mathbf{u}$

Figure 2 for Exercise 6a.

7. a) Since $\mathbf{u}_0 \in E_+$, we have $\mathbf{u}_0 = \mathbf{v}_1 + \cdots + \mathbf{v}_k$ where $\mathbf{v}_i \in E_{\lambda_i}$ and $\lambda_i > 0$ for all i. Hence, by Proposition (7.2.3), the solution of $\mathbf{u}'(t) = A\mathbf{u}(t)$ with $\mathbf{u}(0) = \mathbf{u}_0$ is $\mathbf{u}(t) = e^{\lambda_1 t}\mathbf{v}_1 + \cdots + e^{\lambda_k t}\,\mathbf{v}_k$. Hence $\lim\limits_{t\to\infty} \| \mathbf{u}(t) \| = \infty$, and $\lim\limits_{t\to-\infty} \mathbf{u}(t) = \mathbf{0}$.

8. If $u(n)$ solves $u(n + 1) = au(n)$ and $u(1) = b$, then $u(2) = a \cdot u(1) = ba$. If $u(k) = ba^{k-1}$, then $u(k + 1) = au\,(k) = a \cdot ba^{k-1} = ba^k$. Hence $u(n) = ba^{n-1}$ for all $n \geqslant 1$ by induction.

10. If $A = \begin{bmatrix} \lambda_1 & & 0 \\ & \ddots & \\ 0 & & \lambda_p \end{bmatrix}$ is diagonal, then we have $u_i(n+1) = \lambda_i u_i(n)$, and $u_i(1) = b_i$, so that by Exercise 8, the solution is given by $u_i(n) = b_i\lambda_i^{n-1}$.

Hence $\mathbf{u}(n) = \begin{bmatrix} b_1\lambda_1^{n-1} \\ \vdots \\ b_p\lambda_p^{n-1} \end{bmatrix}$.

12. a) In matrix form, $\mathbf{u}(n+1) = \begin{bmatrix} 2 & 1 \\ 1 & 2 \end{bmatrix}\mathbf{u}(n)$ with $\mathbf{u}(1) = \begin{bmatrix} 3 \\ 4 \end{bmatrix}$.

Diagonalizing the coefficient matrix, we have

$$\Lambda = \begin{bmatrix} 3 & 0 \\ 0 & 1 \end{bmatrix} = QAQ^{-1} = \begin{bmatrix} 1 & -1 \\ 1 & 1 \end{bmatrix}\begin{bmatrix} 2 & 1 \\ 1 & 2 \end{bmatrix}\begin{bmatrix} 1/2 & 1/2 \\ -1/2 & 1/2 \end{bmatrix}$$

The solution of $\mathbf{v}(n+1) = \Lambda\mathbf{v}(n)$ is $\mathbf{v} = (a3^{n-1}, b)$, $a = v_1(1)$ $b = v_2(1)$, by Exercise 10. Then, by Exercise 11, $\mathbf{u} = Q\mathbf{v} = \begin{bmatrix} 1 & -1 \\ 1 & 1 \end{bmatrix}\begin{bmatrix} a3^{n-1} \\ b \end{bmatrix}$

$= \begin{bmatrix} a3^{n-1} - b \\ a3^{n-1} + b \end{bmatrix}$ satisfies $\mathbf{u}(n+1) = A\mathbf{u}(n)$. When $n = 1$ we obtain

$\begin{bmatrix} a - b \\ a + b \end{bmatrix} = \begin{bmatrix} 3 \\ 4 \end{bmatrix}$, so $a = 7/2$, $b = 1/2$ and $\mathbf{u}(n) = \begin{bmatrix} (7/2)3^{n-1} - 1/2 \\ (7/2)3^{n-1} + 1/2 \end{bmatrix}$.

Chapter 7, Section 3

1. We will use the simplified definition, Equation (7.4), for $\| A \|$.

i) $\| aA \| = \max_{\|\mathbf{x}\|=1} \| aA\mathbf{x} \| = \max_{\|\mathbf{x}\|=1} |a| \, \| A\mathbf{x} \| = |a| \max_{\|\mathbf{x}\|=1} \| A\mathbf{x} \|$
$= |a| \cdot \| A \|$.

ii) Since $\| A\mathbf{x} \| \geqslant 0$ for all $\mathbf{x}$ with $\| \mathbf{x} \| = 1$, $\| A \| \geqslant 0$. If $\| A \| = 0$, then $\| A\mathbf{x} \| = 0$, so $A\mathbf{x} = \mathbf{0}$ for all $\mathbf{x}$ with $\| \mathbf{x} \| = 1$. By linearity, this implies as $A\mathbf{x} = \mathbf{0}$ for all $\mathbf{x} \in \mathbf{R}^n$, so $A = 0$.

iii) $\| A + B \| = \max_{\|\mathbf{x}\|=1} \| (A + B)\mathbf{x} \| = \max_{\|\mathbf{x}\|=1} \| A\mathbf{x} + B\mathbf{x} \|$. For each $\mathbf{x}$ by the triangle inequality in $\mathbf{R}^n$, $\| A\mathbf{x} + B\mathbf{x} \| \leqslant \| A\mathbf{x} \| + \| B\mathbf{x} \|$. Hence

$$\begin{aligned} \max_{\|\mathbf{x}\|=1} \| A\mathbf{x} + B\mathbf{x} \| &\leqslant \max_{\|\mathbf{x}\|=1} [\, \| A\mathbf{x} \| + \| B\mathbf{x} \| \,] \\ &\leqslant \max_{\|\mathbf{x}\|=1} \| A\mathbf{x} \| + \max_{\|\mathbf{x}\|=1} \| B\mathbf{x} \| \\ &= \| A \| + \| B \|. \end{aligned}$$

Hence $\| A + B \| \leqslant \| A \| + \| B \|$ as claimed.

iv) For each $\mathbf{x}$ with $\| \mathbf{x} \|$ with $\| \mathbf{x} \| = 1$, $\| AB\mathbf{x} \| \leqslant \| A \| \cdot \| B\mathbf{x} \| \leqslant \| A \| \cdot \| B \| \cdot \| \mathbf{x} \| = \| A \| \cdot \| B \|$. Hence $\max_{\|\mathbf{x}\|=1} \| AB\mathbf{x} \| \leqslant \| A \| \cdot \| B \|$, so $\| AB \| \leqslant \| A \| \cdot \| B \|$.

2. By Exercise 8 of Section 6.1, for any polynomial $p(t)$, if $A = QBQ^{-1}$, then $p(A) = Qp(B)\,Q^{-1}$. Hence, if we let $p(t) = 1 + t + t^2/2! + \cdots + t^k/k!$, the kth partial sum of the Taylor series for e^x, then

$$\left[I + A + \frac{A^2}{2!} + \cdots + \frac{A^k}{k!}\right] = Q\left[I + B + \frac{B^2}{2!} + \cdots + \frac{B^k}{k!}\right]Q^{-1}.$$

Letting $k \to \infty$, since the series on both sides converge, we have

$$e^A = Qe^BQ^{-I}$$

3. a) By Proposition (7.3.5), for any matrix A, $\| A \|$ is the square root of the largest eigenvalue of A^tA. If A is symmetric, then $A^tA = A^2$. If λ is any eigenvalue of A, then λ^2 is an eigenvalue of A^2, so $\| A \| = \sqrt{\lambda^2} = |\lambda|$, where λ is the eigenvalue of A which is the largest in absolute value.

5. b) $e^{ib} = \sum_{m=0}^{\infty} (1/m!)(ib)^m$. Since i^m is real if $m = 2k$ is even, $i^{2k} = (-1)^k$, and purely imaginary if $m = 2k + 1$ is odd, $i^{2k+1} = (-1)^k i$, we have

$$e^{ib} = \sum_{k=0}^{\infty} \frac{(-1)^k}{2k!} b^{2k} + i \sum_{k=0}^{\infty} \frac{(-1)^k}{(2k+1)!} b^{2k+1}$$

The first term is just the Taylor series expansion of $\cos(b)$, while the second is the Taylor series expansion of $i \cdot \sin(b)$. Since these series converge to $\cos(b)$, $i \cdot \sin(b)$ respectively for all $b \in \mathbf{R}$, we have

$$e^{ib} = \cos(b) + i \cdot \sin(b).$$

6. a) By the computations following Lemma (7.3.13),

$\mathbf{u}(t) = e^{At}\mathbf{u}(0) = \begin{bmatrix} e^{\lambda t} & te^{\lambda t} \\ 0 & e^{\lambda t} \end{bmatrix}\mathbf{u}(0)$. Hence, for $\mathbf{u}(0) = (1,0)$, we get

$\mathbf{u}(t) = \begin{bmatrix} e^{\lambda t} \\ 0 \end{bmatrix}$, a straight-line solution. For $\mathbf{u}(0) = (0,1)$, we get

$\mathbf{u}(t) = \begin{bmatrix} te^{\lambda t} \\ e^{\lambda t} \end{bmatrix}$ which lies on the graph of $x_1 = (x_2/\lambda)\log x_2$, $\lambda > 0$:

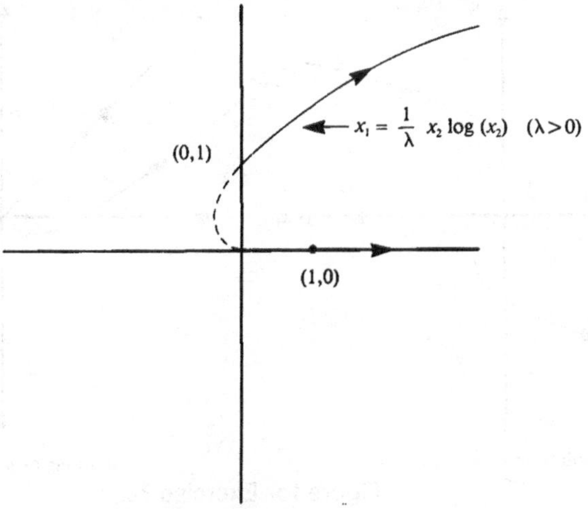

Figure for Exercise 6a.

7. a) $A = \begin{bmatrix} 3 & 1 \\ -1 & 1 \end{bmatrix}$ has eigenvalues $\lambda = 2,2$ and canonical basis $\{(-1,1), (-1,0)\}$. The change of basis matrix $Q = \begin{bmatrix} -1 & -1 \\ 1 & 0 \end{bmatrix}$ and Jordan form $J = \begin{bmatrix} 2 & 1 \\ 0 & 2 \end{bmatrix}$. Thus $A = \begin{bmatrix} -1 & -1 \\ 1 & 0 \end{bmatrix}\begin{bmatrix} 2 & 1 \\ 0 & 2 \end{bmatrix}\begin{bmatrix} -1 & -1 \\ 1 & 0 \end{bmatrix}^{-1}$.

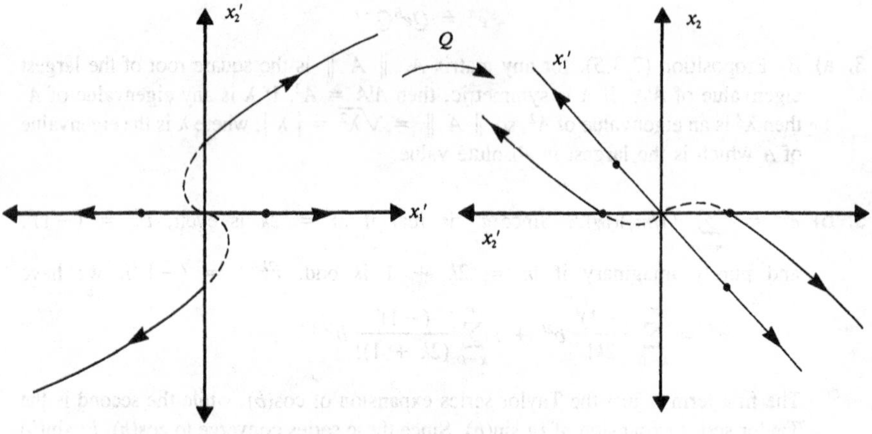

Solutions of $\mathbf{v}' = J\mathbf{v}$ Solutions of $\mathbf{u}' = A\mathbf{u}, \mathbf{u} = Q\mathbf{v}$

Figure for Exercise 7a.

c) $A = \begin{bmatrix} 2 & 1 \\ -1 & 0 \end{bmatrix}$ has eigenvalues $\lambda = 1,1$. The canonical form is $\begin{bmatrix} 1 & 1 \\ 0 & 1 \end{bmatrix}$ and a change of basis matrix is $Q = \begin{bmatrix} 1 & 1 \\ -1 & 0 \end{bmatrix}$. Thus

$$A = \begin{bmatrix} 2 & 1 \\ -1 & 0 \end{bmatrix} = \begin{bmatrix} 1 & 1 \\ -1 & 0 \end{bmatrix}\begin{bmatrix} 1 & 1 \\ 0 & 1 \end{bmatrix}\begin{bmatrix} 1 & 1 \\ -1 & 0 \end{bmatrix}^{-1}.$$

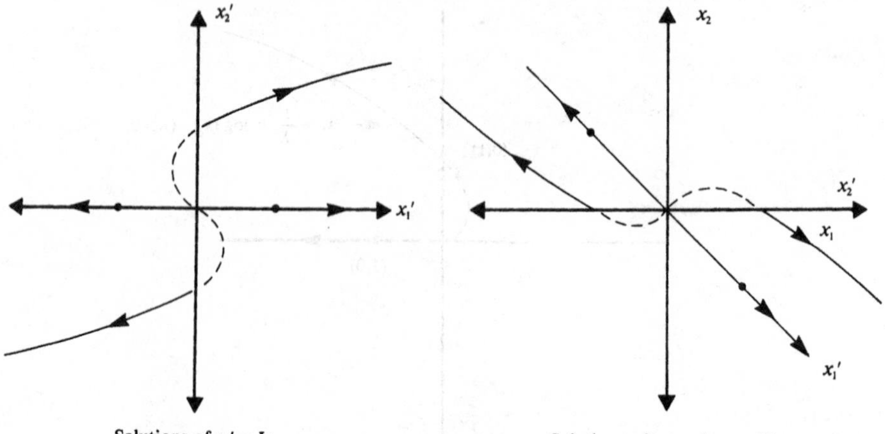

Solutions of $\mathbf{v}' = J\mathbf{v}$ Solutions of $\mathbf{u}' = A\mathbf{u}, \mathbf{u} = Q\mathbf{v}$

Figure for Exercise 7c.

8. a) The Jordan form of A is $J = \begin{bmatrix} 1 & 1 & 0 \\ & 1 & 1 \\ & & 1 \end{bmatrix}$ and the change of basis

matrix is $Q = \begin{bmatrix} 1 & 1 & 0 \\ 0 & -1 & 0 \\ 0 & 0 & 1 \end{bmatrix}$. $A = QJQ^{-1}$, so

$$e^{At} = Qe^{Jt}Q^{-1} = Q\begin{bmatrix} e^t & te^t & (t^2/2)e^t \\ & e^t & te^t \\ & & e^t \end{bmatrix}Q^{-1}$$

$$= \begin{bmatrix} e^t & -te^t & ((t^2/2) + t)e^t \\ 0 & e^t & -te^t \\ 0 & 0 & e^t \end{bmatrix}$$

10. a) $e^{At} = \begin{bmatrix} \cos(bt) & \sin(bt) \\ -\sin(bt) & \cos(bt) \end{bmatrix}$

b) $e^{At} = \begin{bmatrix} e^{at}\cos(bt) & e^{at}\sin(bt) \\ -e^{at}\sin(bt) & e^{at}\cos(bt) \end{bmatrix}$

11. a) $\mathbf{u}(t) = \begin{bmatrix} \cos(bt) \\ -\sin(bt) \end{bmatrix}$. Solution curves are arcs of circles, centered at origin, traversed clockwise if $b > 0$, and counterclockwise if $b < 0$.

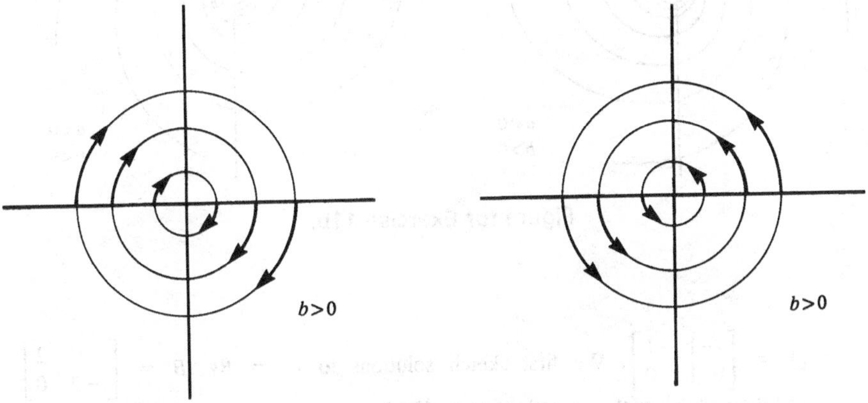

Figure for Exercise 11a.

b) The solutions are spirals (see p. 418).

13. We have $B\mathbf{z} = (a + ib)\mathbf{z}$ and $B\bar{\mathbf{z}} = (a - ib)\,\bar{\mathbf{z}}$. Hence, if $\mathbf{x} = \text{Re}(\mathbf{z}) = (\mathbf{z} + \bar{\mathbf{z}})/2$, and $\mathbf{y} = \text{Im}(\mathbf{z}) = (\mathbf{z} - \bar{\mathbf{z}})/2i$, $B\mathbf{x} = (1/2)(B\mathbf{z} + B\bar{\mathbf{z}}) = (1/2) \cdot [(a + ib)\mathbf{z} + (a - ib)\bar{\mathbf{z}}] = a \cdot (1/2)(\mathbf{z} + \bar{\mathbf{z}}) + (-b) \cdot (1/2i)(\mathbf{z} - \bar{\mathbf{z}}) = a\mathbf{x} + (-b)\mathbf{y}$. Similarly, $B\mathbf{y} = b\mathbf{x} + a\mathbf{y}$, so the matrix of B with respect to the basis $\{\mathbf{x},\mathbf{y}\}$ for $\mathbf{R}^2$ is $Q^{-1}BQ = \begin{bmatrix} a & b \\ -b & a \end{bmatrix}$.

14. a) The eigenvalues of A are $\lambda = \pm 2i$ and corresponding eigenvectors are $(-1 - i, 1)$ and $(-1 + i, 1)$. The change of basis matrix Q (constructed as in Exercise 13) is

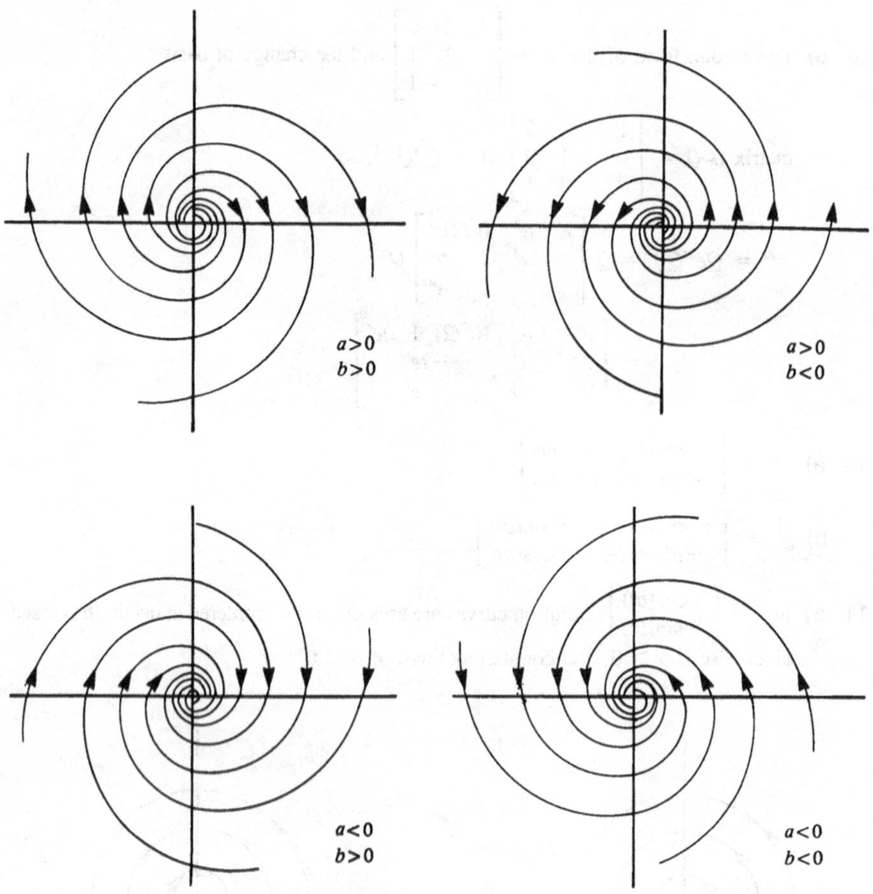

Figure for Exercise 11b.

$Q = \begin{bmatrix} -1 & -1 \\ 1 & 0 \end{bmatrix}$. We first sketch solutions to $\mathbf{v}' = B\mathbf{v}$, $B = \begin{bmatrix} 0 & 2 \\ -2 & 0 \end{bmatrix}$ and then sketch $\mathbf{u}(t) = Q\mathbf{v}(t)$. (See p. 419.)

16. Since $\mathbf{u}(n + 1) = B\mathbf{u}(n) = B^2\mathbf{u}(n - 1) = \ldots = B^n\mathbf{u}(1)$, we have $\mathbf{u}(n + 1) = B^n\mathbf{u}(1)$. If $\lambda = 0$, then B is nilpotent with $B^p = 0$, so, $\mathbf{u}(n) = \mathbf{0}$ for all $n \geqslant p + 1$, and the formula is valid here. For $\lambda \neq 0$, then we have

$$B^n = \begin{bmatrix} \lambda^n & \left[\begin{smallmatrix} n \\ 1 \end{smallmatrix}\right]\lambda^{n-1} & \cdots & \left[\begin{smallmatrix} n \\ p-1 \end{smallmatrix}\right]\lambda^{n-p+1} \\ & \ddots & \ddots & \vdots \\ & & & \left[\begin{smallmatrix} n \\ 1 \end{smallmatrix}\right]\lambda^{n-1} \\ & & & \lambda^n \end{bmatrix}$$

where $\begin{bmatrix} k \\ j \end{bmatrix} = k!/[(k - j)! \cdot j!]$.

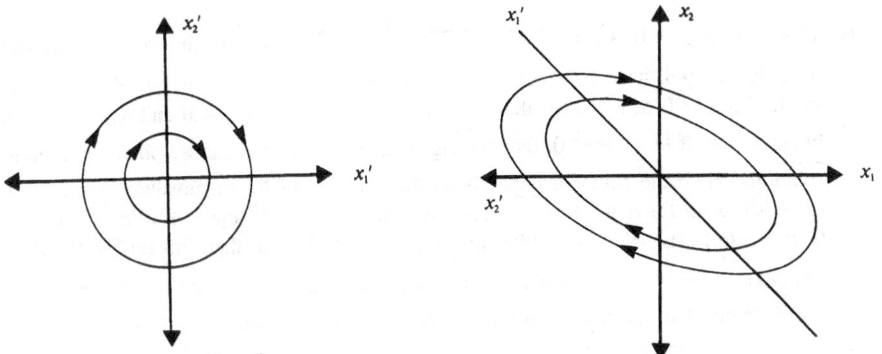

Solutions of $\mathbf{v}' = \begin{bmatrix} 0 & 2 \\ -2 & 0 \end{bmatrix} \mathbf{v}$ Solutions of $\mathbf{u}' = A\mathbf{u}$

Figure for Exercise 14a.

18. a) $A = \begin{bmatrix} 1 & 1 \\ -1 & -1 \end{bmatrix}$ is nilpotent with $A^2 = 0$. Hence $\mathbf{u}(1) = \begin{bmatrix} 1 \\ 0 \end{bmatrix}$, $\mathbf{u}(2) = \begin{bmatrix} 1 \\ -1 \end{bmatrix}$, and $\mathbf{u}(n) = \begin{bmatrix} 0 \\ 0 \end{bmatrix}$ for all $n \geqslant 3$.

Chapter 7, Section 4

1. The proof is the same as the proof of Exercise 18 of Section 4.1. This matrix is the transpose of the companion matrix of the polynomial.

2. $u(t) = (-13/8) + (13/8)e^{2t} - (13/4)te^{2t} + (9/4)\,t^2e^{2t}$.

3. $\alpha_i = \{e^{\lambda_i t}, te^{\lambda_i t}, (t^2/2!)e^{\lambda_i t}, \ldots, (t^{m_i-1}/(m_i - 1)!)\,e^{\lambda_i t}\}$ is a cycle for the mapping $N = d/dt - \lambda_i I$. Hence, by the argument of Proposition (6.2.3c), α is linearly independent, and the same is true of the scalar multiples $\{e^{\lambda_i t}, te^{\lambda_i t}, t^2e^{\lambda_i t}, \ldots, t^{m_i-1}\,e^{\lambda_i t}\}$. Furthermore, by the argument given in Proposition (6.2.4), $\alpha_1 \cup \cdots \cup \alpha_k$ is linearly independent, since the functions $\{e^{\lambda_1 t}, \ldots, e^{\lambda_k t}\}$ form a linearly independent set. (To see this, note that if $c_1e^{\lambda_1 t} + \cdots + c_ke^{\lambda_k t} = 0$, then setting $t = 0, 1, \ldots, k - 1$ we obtain k linear equations

$$\begin{aligned}
c_1 + \cdots + \quad c_k &\geqslant 0 \\
e^{\lambda_1}c_1 + \cdots + \quad e^{\lambda_k}c_k &= 0 \\
(e^{\lambda_1})^2c_1 + \cdots + \quad (e^{\lambda_k})^2c_k &= 0 \\
&\vdots \\
(e^{\lambda_1})^{k-1}c_1 + \cdots + (e^{\lambda_k})^{k-1}c_k &= 0
\end{aligned}$$

Since the λ_i are distinct, so are the e^{λ_i}, hence this Vandermonde system has only the trivial solution $c_1 = \ldots = c_k = 0$. (See Supplementary Exercise 9 in Chapter 3.)

5. a) $u(t) = te^{-2t}$.

c) $u(t) = (1/3)e^t + (2/3)e^{-t/2}\cos(\sqrt{3/2}\,t)$

(or $= (1/3)e^t + (1/3)e^{t(-1+i\sqrt{3})/2} + (1/3)e^{t(-1-i\sqrt{3})/2})$

6. $\lambda^2 + b\lambda + c = 0$, $\lambda_{\pm} = \dfrac{-b \pm \sqrt{b^2 - 4c}}{2}$. If $b^2 - 4c \geq 0$, the roots are real and the solutions will have the form $u(t) = c_1 e^{\lambda_+ t} + c_2 e^{\lambda_- t}$, or $(c_1 + c_2 t)e^{\lambda t}$ if $b^2 - 4c = 0$. If $\lambda_+ > 0$ or $\lambda_- > 0$, then $\lim_{t\to\infty} |u(t)| = +\infty$. If $\lambda_+ < 0$ and $\lambda_- < 0$, then $\lim_{t\to\infty} u(t) = 0$. If $b^2 - 4c < 0$, then the eigenvalues are not real and we obtain oscillating solutions. Here the behavior depends on the real part of the eigenvalues. We have $\lambda = -b/2 \pm id$ for $d = (\sqrt{b^2 - 4c})/2$. Hence $u(t) = e^{-bt/2}(c_1 e^{idt} + c_2 e^{-idt})$. If $b < 0$, then $\lim_{t\to\infty} |u(t)| = +\infty$, while $\lim_{t\to-\infty} u(t) = 0$. If $b > 0$, then $\lim_{t\to\infty} u(t) = 0$, while $\lim_{t\to-\infty} |u(t)| = +\infty$. We obtain solutions bounded for all $t \in \mathbf{R}$ only when $b = 0$, and $c > 0$ (other than the trivial case $u^{(2)} + bu^{(1)} + cu = 0$, $u(0) = u^{(1)}(0) = 0$).

7. a) $x(t) = \sqrt{m/k}\,\sin(\sqrt{k/m}\,t)$.

b) Let $\alpha = \sqrt{1 - 8k}$. Then $x(t) = (2/\alpha)\,e^{t(-1+\alpha)/4} - (2/\alpha)\,e^{t(-1-\alpha)/4}$. (If $8k > 1$, so $\alpha = \sqrt{1 - 8k}$ is purely imaginary, this can be expressed as $(4/\omega)e^{-t/4}\sin(\omega t/4)$, where $\omega = (1/i)\alpha$.)

8. b) If $v(t)e^{-at}$ solves $u^{(1)}(t) + au(t) = w(t)$, then $v'(t)e^{-at} - av(t)e^{-at} + av(t)e^{-at} = w(t)$, so $v'(t) = e^{at}w(t)$. Hence $v(t) = \int e^{at}w(t)dt$. So $e^{-at}\int e^{at}w(t)dt$ solves the inhomogeneous equation.

9. a) If $(D - \lambda_n)v = w$ and $(D - \lambda_{n-1}) \cdots (D - \lambda_1)u = v$, then substituting for v in the first equation $(D - \lambda_n) \cdots (D - \lambda_1)\,u = p(D)u = w$, so u solves the equation. Conversely, if $p(D)u = (D - \lambda_n) \cdots (D - \lambda_1)u = w$, let $v = (D - \lambda_{n-1}) \cdots (D - \lambda_1)u$. Then $(D - \lambda_n)v = p(D)u = w$, and $(D - \lambda_{n-1}) \cdots (D - \lambda_1)u = v$. Hence solving the original equation is the same as solving $(D - \lambda_n)v = w$ and $(D - \lambda_{n-1}) \cdots (D - \lambda_1)u = v$.

12. a) If $u^{(1)}(t) + p(t)u(t) = 0$, then $u^{(1)}(t) = -p(t)u(t)$, so $\dfrac{u'(t)}{u(t)} = -p(t)$. Integrating both sides yields $\log|u(t)| = -\int p(t)dt + \hat{c}$, where $\hat{c}$ is some constant of integration. Hence $u(t) = e^{-\int p(t)dt + \hat{c}} = ce^{-\int p(t)dt}$, where $c = \pm e^{\hat{c}}$, is the general solution of (7.18).

b) Multiply both sides of (7.17) by $e^{\int p(t)dt}$. We have

$$e^{\int p(t)\,dt}[u'(t) + p(t)u(t)] = e^{\int p(t)dt}q(t)$$

The left hand side is just $[e^{\int p(t)dt} \cdot u(t)]'$, so integrating, we obtain

$$e^{\int p(t)dt}u(t) = \int e^{\int p(t)dt}q(t)dt, \text{ or}$$
$$u(t) = e^{-\int p(t)dt} \cdot \int e^{\int p(t)dt}q(t)dt.$$

The steps in the above derivation can all be reversed, hence u_p is a particular solution.

c) This follows from part d of Exercise 11.

13. a) $p(t) = q(t) = t$, so $u(t) = ce^{-\int tdt} + e^{-\int tdt} \cdot \int e^{\int tdt}\,tdt = ce^{-t^2/2} + e^{-t^2/2}\int e^{t^2/2}tdt = ce^{-t^2/2} + e^{-t^2/2} \cdot e^{t^2/2} = ce^{-t^2/2} + 1$. From the initial condition $u(0) = 2$, we have that $c = 1$. Hence $u(t) = e^{-t^2/2} + 1$.

14. Let $u_1(n) = u(n + 1)$
$u_2(n) = u(n + 2)$
$\vdots$
$u_{k-1}(n) = u(n + k - 1)$,
then the equation (7.19) is equivalent to the following system for

$$\mathbf{u} = (u, u_1, \ldots, u_{k-1}): \mathbf{u}(n + 1) = \begin{bmatrix} 0 & 1 & & & \\ 0 & 0 & 1 & & \\ 0 & & & \ddots & \\ \vdots & & & 0 & 1 \\ -a_0 & \cdots & & & -a_{k-1} \end{bmatrix} \mathbf{u}(n).$$ This is true

since $u_i(n + 1) = u_{i+1}(n)$ for all $i \leq k - 1$, and by (7.19) $u_{k-1}(n + 1) = u(n + k) = -a_{k-1} u_{k-1}(n) - \cdots - a_0u(n)$.

Chapter 7, Section 5

2. a) $\langle cf + g,h \rangle = \int_0^\pi (cf + g)(x)\overline{h(x)}dx$
$= \int_0^\pi (cf(x)\overline{h(x)} + g(x)\overline{h(x)})dx$
$= c\int_0^\pi f(x)\overline{h(x)}dx + \int_0^\pi g(x)\overline{h(x)}dx$
$= c \langle f,h \rangle + \langle g,h \rangle$

b) $\langle h,g \rangle = \int_0^\pi h(x)\overline{g(x)}dx$
$= \int_0^\pi \overline{g(x)\overline{h(x)}}dx$
$= \overline{\int_0^\pi g(x)\overline{h(x)}dx} = \overline{\langle g,h \rangle}$.

c) $\langle f,f \rangle = \int_0^\pi f(x)\overline{f(x)}dx$
$= \int_0^\pi |f(x)|^2dx$.

Since $|f(x)|^2 \geq 0$ for all $x \in [0,\pi]$, we see $\langle f,f \rangle \geq 0$. Moreover, if $\langle f,f \rangle = 0$, then $\int_0^\pi |f(x)|^2dx = 0$. Since $|f|^2$ is continuous, and $|f|^2 \geq 0$, this implies $|f(x)|^2 = 0$ for all x, so $f(x) = 0$ for all x. Hence the inner product $\langle , \rangle$ is positive definite.

4. The equation can be rewritten as $u^{(2)}(t) + ku(t) = 0$. The roots of $\lambda^2 + k = 0$ are $\lambda = \pm i\sqrt{k}$. By Theorem (7.4.5), this equation has solutions $u(t) = c_1e^{i\sqrt{k}t} + c_2e^{-i\sqrt{k}t}$ for some c_1,c_2.

6. The roots of $\lambda^2 + m^2c^2 = 0$ are $\lambda = \pm mci$. Hence, by Theorem (7.4.5), the solutions of $h''(t) + m^2c^2 h(t) = 0$ are $h(t) = c_1e^{mcit} + c_2e^{-mcit}$. If $h(0) = 1$ and $h'(0) = 0$, we see

$$c_1 + c_2 = 1$$
$$(mci)c_1 - (mci)c_2 = 0.$$

Hence $c_1 = c_2 = 1/2$, so $h(t) = (1/2)(e^{mcit} + e^{-mcit}) = \cos(mct)$.

9. a) For any $f,g \in V$, $\langle T(f),g\rangle$

$$= \int_0^{\pi} xf(x)\overline{g(x)}dx = \int_0^{\pi} f(x)\overline{x\cdot g(x)}dx \quad (x \in [0,\pi] \text{ is real})$$
$$= \langle f,T(g)\rangle$$

Hence T is self-adjoint or Hermitian.

b) T has no eigenvectors or eigenvalues, since if $T(f) = \lambda f$, then $xf(x) = \lambda f(x)$ for all x. If $f \neq 0$, then this would say $x = \lambda$, which is a contradiction (the equation is supposed to hold for all $x \in [0,\pi]$).

11. a) The Fourier coefficients are $((2/\pi)^{3/2}/m^2)\sin(m\pi/2)$.

b) $u(x,t) = \sum_{m=0}^{\infty} ((2/\pi)^{3/2}/m^2)g_m(x) \cdot h_m(t)$.

Supplementary Exercises

1. a) False. The solution curve could be a circle ($A = \begin{bmatrix} 0 & b \\ -b & 0 \end{bmatrix}$).

b) False. Example: $A = \begin{bmatrix} 0 & 1 \\ 0 & 0 \end{bmatrix}$, $\mathbf{u}(0) = (0,1)$. The solution is $\mathbf{u} = \begin{bmatrix} t \\ 1 \end{bmatrix}$, which is a horizontal line. $(0,1)$ is not an eigenvector of A, though.

c) True. If $\mathbf{u}(0) = c_1\mathbf{v}_1 + c_2\mathbf{v}_2$ with $\mathbf{v}_1 \in E_\lambda$ and $\mathbf{v}_2 \in E_\mu$, then $\mathbf{u}(t) = c_1e^{\lambda t}\mathbf{v}_1 + c_2e^{\mu t}\mathbf{v}_2 \in E_\lambda + E_\mu$.

d) False. It is true if $AB = BA$, but not in general.

e) True. Since $(At)(-At) = (-At)(At)$, $e^{-At}e^{At} = e^0 = I$, so e^{At} is invertible, and $(e^{At})^{-1} = e^{-At}$.

f) True. $e^{At} = \sum_{m=0}^{\infty} (1/m!)(At)^m$. Since all the entries of the terms $(1/m!)(At)^m$ are real, e^{At} also has real entries.

g) True. Let $a + bi$ and $c + di$ be the roots. Since they are in $\mathbf{C}\setminus\mathbf{R}$, $b \neq 0$ and $d \neq 0$. The only solutions are $c_1e^{(a+bi)t} + c_2e^{(c+di)t}$ by Theorem (7.4.5).

h) False. This is not necessarily true if f and g do not vanish at the endpoints of the interval.

i) False. $f(t)$ can be zero for some t. For example, $f(x) = \sqrt{2/\pi} \cdot \sin(mx)$, which is zero at $x = \pi/m$.

2. a) For $\mathbf{u}(0) = \begin{bmatrix} a \\ b \end{bmatrix}$ the solution is

$$\mathbf{u}(t) = \begin{bmatrix} e^t(a\cos(t) - b\sin(t)) \\ e^t(a\sin(t) + b\cos(t)) \end{bmatrix}$$

c)
$$e^{At} = \begin{bmatrix} (1/2)(e^{3t} + e^t) & (1/2)(e^{3t} - e^t) & 0 & 0 \\ (1/2)(e^{3t} - e^t) & (1/2)(e^{3t} + e^t) & 0 & 0 \\ 0 & 0 & e^t\cos(t) & -e^t\sin(t) \\ 0 & 0 & e^t\sin(t) & e^t\cos(t) \end{bmatrix}$$

$$\text{If } \mathbf{u}(0) = \begin{bmatrix} a \\ b \\ c \\ d \end{bmatrix}, \text{ then } \mathbf{u}(t) = e^{At}\begin{bmatrix} a \\ b \\ c \\ d \end{bmatrix}$$

$$= \begin{bmatrix} \left[\dfrac{a+b}{2}\right]e^{3t} + \left[\dfrac{a-b}{2}\right]e^{t} \\ \left[\dfrac{a+b}{2}\right]e^{3t} + \left[\dfrac{-a+b}{2}\right]e^{t} \\ (c \cdot \cos(t) \quad - d \cdot \sin(t))e^{t} \\ (c \cdot \sin(t) \quad + d \cdot \cos(t))e^{t} \end{bmatrix}$$

3. a) $u(t) = \sin(\sqrt{a}t)$

c) $u(t) = (1/2a)e^{at} - (1/2a)\, e^{-at}$.

5. a) If all $\mu_i > 0$, then if $Q^{-1}AQ = \begin{bmatrix} \mu_1 & & \\ & \ddots & \\ & & \mu_n \end{bmatrix}$

$$Q^{-1}AQ = \begin{bmatrix} e^{\log(\mu_1)} & 0 \\ & \ddots \\ 0 & e^{\log(\mu_n)} \end{bmatrix} = e^{\bar{B}}, \text{ where } \bar{B} = \begin{bmatrix} \log(\mu_1) & 0 \\ & \ddots \\ 0 & \log(\mu_n) \end{bmatrix}.$$

Hence $A = Qe^{\bar{B}}Q^{-1} = e^{(Q\bar{B}Q^{-1})}$. Conversely, if $e^B = A$, all the eigenvalues of A are positive by Exercise 4a, Thus $A = e^B$ for some B if and only if all the eigenvalues of A are strictly positive.

7. a) When we compute the derivative using the product rule, every term contains at least one factor of $t^2 - 1$. Hence $\dfrac{d^k}{dt^k}(t^2 - 1)^n\Big|_{t=\pm 1} = 0$ for $k = 1, \ldots, n-1$.

b) We prove this by induction on k. $\langle q_n(t), t^k\rangle = \displaystyle\int_{-1}^{1} \frac{d^n}{dt^n}(t^2-1)^n \cdot t^k dt$

For $k = 1$: $\langle q_n(t), t\rangle = \displaystyle\int_{-1}^{1}\left[\frac{d^n}{dt^n}(t^2 - 1)^n\right] t\, dt.$

Integrate by parts with $u = t$, $dv = \dfrac{d^n}{dt^n}[(t^2 - 1)^n]$ to obtain

$$= t \cdot \frac{d^{n-1}}{dt^{n-1}}(t^2 - 1)^n \Bigg|_{-1}^{1} - \int_{-1}^{1} \frac{d^{n-1}}{dt^{n-1}}[(t^2 - 1)^n]dt$$

$$= 0 - \frac{d^{n-2}}{dt^{n-2}}(t^2 - 1)^n \Bigg|_{-1}^{1}$$

$= 0$, using part a.

Now, assume true the result is true for k, and consider

$\langle q_n(t), t^{k+1}\rangle = \displaystyle\int_{-1}^{1}\frac{d^n}{dt^n}[(t^2 - 1)^n]t^{k+1}\, dt$, where $k + 1 \leq n$.

Integrating by parts again:

$$= t^{k+1}\frac{d^{n-1}}{dt^{n-1}}[(t^2 - 1)^n]\Big|_{-1}^{1} - (k+1)\int_{-1}^{1} t^k \cdot \frac{d^{n-1}}{dt^{n-1}}[(t^2 - 1)^n]dt$$
$$= 0$$

by part a and induction.

c) $q_n(t)$ is a polynomial of degree n. Assume without loss of generality that $n > m$. Then $q_m(t) \in \text{Span}\{1, t, \ldots, t^m\}$. By part b, we have $\langle q_n(t), q_m(t)\rangle = 0$, since $\langle q_n(t), t^k\rangle = 0$, for all $k \leq n - 1$.

8. c) $\| q_n(t) \| = \sqrt{\frac{(n!)^2}{(2n+1)}2^{2n+1}} = n!\,2^n\sqrt{\frac{2}{2n+1}}$ and $L_n(t) = \frac{1}{n!2^n}q_n(t)$.

9. a) $L_0(t) = 1$
$L_1(t) = t$
$L_2(t) = (3/2)t^2 - 1/2$
$L_3(t) = (5/2)t^3 - (3/2)t$
$L_4(t) = (35/8)t^4 - (15/4)t^2 + 3/8.$

10. a) $t^2 + t + 1 = (2/3)L_2 + L_1 + (4/3)L_0.$

c) $a_n = \begin{cases} 0 & \text{if } n = 2k+1, \\ \binom{2k}{k-1}/((2k)(2k-1)2^{2k-1}) & \text{if } n = 2k. \end{cases}$

12. Let $v(t) = (t^2 - 1)^n$. Then differentiating, we have $v' = 2nt(t^2 - 1)^{n-1}$, so $(1 - t^2)v' - 2ntv = 0$. Differentiating again, $(1 - t^2)v'' - 2tv' + 2ntv' + 2nv = 0$, so $(1 - t^2)v'' + (2n - 2)tv' + 2nv = 0$. Continuing in this way, after k differentiations we will have $(1 - t^2)v^{(k)} + (2n - 2(k-1))tv^{(k-1)} + 2[n + (n-1) + \cdots + (n-k-2)]v^{(k-2)} = 0$. When $k = n + 2$ we obtain

$$\begin{aligned} 0 &= (1 - t^2)v^{(n+2)} - 2tv^{(n+1)} + 2[n + (n-1) + \cdots + 2 + 1]v^{(n)} \\ &= (1 - t^2)v^{(n+2)} - 2tv^{(n+1)} + n(n+1)v^{(n)}. \end{aligned}$$

Since $v^{(n)}(t) = q_n(t) = \frac{d^n}{dt^n}[(t^2 - 1)^n]$, this shows that $u = q_n(t)$ satisfies the Legendre equation:

$$0 = (1 - t^2)u^{(2)} - 2tu^{(1)} + n(n+1)u.$$

The Legendre polynomial $L_n(t)$ is also a solution, since it is a constant multiple of $q_n(t)$.

Appendix 1, Solutions

1. a)

p	q	$p \to q$	$\leftrightarrow$	$(\sim p \vee q)$
T	T	T	T	T
T	F	F	T	F
F	T	T	T	T
F	F	T	T	T

c)

p	q	r	$p\to(q\vee r)$	$\leftrightarrow$	$(p\wedge\sim q)\to r$
T	T	T	T	T	T
T	T	F	T	T	T
T	F	T	T	T	T
T	F	F	F	T	F
F	T	T	T	T	T
F	T	F	T	T	T
F	F	T	T	T	T
F	F	F	T	T	T

2. a) p is "Jim Rice bats .300"; q is "the Red Sox will win the pennant." Converse: If the Red Sox win the pennant, then Jim Rice bats .300. Contrapositive: If the Red Sox do not win the pennant, then Jim Rice will not bat .300.
   c) p is "the function f is odd"; q is "$f(0) = 0$." Converse: If $f(0) = 0$, then f is odd. Contrapositive: If $f(0) \neq 0$, then f is not odd.

3. a) For all $x \in \mathbf{R}, f(x) \neq 0$.
   c) For all $x \in \mathbf{R}$, there exists $y \in \mathbf{R}$ such that $x + y \leqslant 0$.

4. $A \cup B = \{1, 3, 4, 7, 8, 9, 10, 12\}$, $A \cap B = \{4, 9\}$, $A \setminus B = \{1, 7, 10\}$, $B \setminus A = \{3, 8, 12\}$

7. $A \cap (B \cup C)$

$$
\begin{aligned}
&= \{x|\ x \in A \text{ and } x \in B \cup C\}\\
&= \{x|\ x \in A \text{ and } (x \in B \text{ or } x \in C)\}\\
&= \{x|\ (x \in A \text{ and } x \in B) \text{ or } (x \in A \text{ and } x \in C)\}\\
&= \{x|\ x \in A \cap B \text{ or } x \in A \cap C\}\\
&= (A \cap B) \cup (A \cap C)
\end{aligned}
$$

The second part is similar.

8. a) Yes.
   c) Yes.

9. a) Neither.
   c) f is surjective, but not injective.

10. a) $f^{-1}([1,\infty)) = \{x \in \mathbf{R}|\ |x| \geqslant 1\} = (-\infty, -1] \cup [1,\infty)$
    c) $f^{-1}([1,\infty)) = \{x \mid -\sqrt{3} \leqslant x \leqslant 0\} \cup \{x \mid x \geqslant \sqrt{3}\}$

11. 1a) $F(C \cup D) = \{y \in B|\ y = F(x) \text{ for some } x \in C \cup D\}$
    $= \{y \in B|\ (y = F(x) \text{ for some } x \in C) \text{ or } (y = F(x) \text{ for some } x \in D)\}$
    $= \{y \in B|\ y = F(x) \text{ for some } x \in C\} \cup \{y \in B|y = F(x) \text{ for some } x \in D\}$
    $= F(C) \cup F(D)$
    1b) is similar.
    3) $F^{-1}(B \setminus E) = \{x|\ F(x) \in B \setminus E\}$
    $= \{x|\ (F(x) \in B) \text{ and } (F(x) \notin E)\}$
    $= \{x|\ x \in F^{-1}(B) \text{ and } x \notin F^{-1}(E)\}$
    $= F^{-1}(B) \setminus F^{-1}(E)$
    5) $C = \{x \in A|x \in C\} \subseteq \{x \in A|\ F(x) \in F(C)\} = F^{-1}(F(C))$

14. The following chart gives the composition $f \circ g$.

$f \backslash g$	x	$1/x$	$1 - x$	$(x - 1)/x$	$x/(x - 1)$	$1/(1 - x)$
x	x	$1/x$	$1 - x$	$(x - 1)/x$	$x/(x - 1)$	$1/(1 - x)$
$1/x$	$1/x$	x	$1/(1 - x)$	$x/(x - 1)$	$(x - 1)/x$	$1 - x$
$1 - x$	$1 - x$	$(x - 1)/x$	x	$1/x$	$1/(1 - x)$	$x/(x - 1)$
$(x - 1)/x$	$(x - 1)/x$	$1 - x$	$x/(x - 1)$	$1/(1 - x)$	$1/x$	x
$x/(x - 1)$	$x/(x - 1)$	$1/(1 - x)$	$(x - 1)/x$	$1 - x$	x	$1/x$
$1/(1 - x)$	$1/(1 - x)$	$x/(x - 1)$	$1/x$	x	$1 - x$	$(x - 1)/x$

15. a) We prove the contrapositive. If $[s] \cap [s'] \neq \phi$, then there is a $t \in S$ such that (s, t) and $(s', t) \in R$. By symmetry, $(t, s') \in R$. Hence, by transitivity, $(s, s') \in R$.

b) Let $t \in [s]$. Then $(s, t) \in R$, so by symmetry, $(t, s) \in R$ as well. By transitivity, since $(s, s') \in R$, $(t, s') \in R$ as well. Hence $t \in [s']$, so that $[s] \subseteq [s']$. The opposite inclusion is proved in the same way, reversing the roles of s and s'. Hence $[s] = [s']$.

c) Each $s \in S$ is an element of its own equivalence class, $[s]$, by the reflexive property. Hence $S = \bigcup_{s \in S} [s]$. We must show that if two equivalence classes have a non-empty intersection, then they are in fact equal. If $[s] \cap [s'] \neq \phi$, then by the contrapositive of part a, $(s, s') \in R$. Hence, by part b, $[s] = [s']$.

d) $[m] = \{m' \in \mathbf{Z} | m' = m + nk \text{ for some integer } k\}$

Appendix 2, Solutions

1. (i) $p(1)$ is true since $1^2 = (1/6) \cdot 1 \cdot (1 + 1) \cdot (2 \cdot 1 + 1)$.

(ii) Assume that $\sum_{m=1}^{k} m^2 = (1/6)k(k + 1)(2k + 1)$. Then

$$\begin{aligned}\sum_{m=1}^{k+1} m^2 &= (k + 1)^2 + \sum_{m=1}^{k} m^2 \\ &= (k + 1)^2 + (1/6)k(k + 1)(2k + 1) \text{ by the induction hypothesis} \\ &= (1/6)(k + 1)(2k^2 + 7k + 6) \\ &= (1/6)(k + 1)(k + 2)(2k + 3),\end{aligned}$$

as claimed.

3. a) By definition, $\begin{bmatrix} n \\ k \end{bmatrix} = n!/[k! \cdot (n - k)!]$. Hence

$$\begin{bmatrix} n \\ j \end{bmatrix} + \begin{bmatrix} n \\ j + 1 \end{bmatrix} = n!/[j! \cdot (n - j)!] + n!/[(j + 1)! \cdot (n - j - 1)!]$$

Putting these two terms over a common denominator, we have

$$\begin{aligned}&= n! \cdot [(j + 1) + (n - j)]/[(n - j)! \cdot (j + 1)!] \\ &= (n + 1)!/([(n + 1) - (j + 1)]! \cdot (j + 1)!) \\ &= \begin{bmatrix} n + 1 \\ j + 1 \end{bmatrix}\end{aligned}$$

b) (i) $p(1)$ is the product rule for derivatives:

$$(fg)'(x) = f(x)g'(x) + f'(x)g(x)$$

(ii) Assume that $p(k)$ is true:

$$(fg)^{(k)}(x) = \sum_{j=0}^{k} \begin{bmatrix} k \\ j \end{bmatrix} f^{(k-j)}(x)g^{(j)}(x).$$

Then

$$\begin{aligned}(fg)^{(k+1)}(x) &= ((fg)^{(k)})'(x) \\ &= [\sum_{j=0}^{k} \begin{bmatrix} k \\ j \end{bmatrix} f^{(k-j)}(x)g^{(j)}(x)]' \text{ (by the induction hypothesis)} \\ &= \sum_{j=0}^{k} \begin{bmatrix} k \\ j \end{bmatrix} [f^{(k-j)}(x)g^{(j+1)}(x) + f^{(k-j+1)}(x)g^{(j)}(x)] \\ &\text{(by the sum and product rules for derivatives)} \\ &= \sum_{j=0}^{k+1} \left[\begin{bmatrix} k \\ j-1 \end{bmatrix} + \begin{bmatrix} k \\ j \end{bmatrix} \right] f^{(k+1-j)}(x)g^{(j)}(x) \\ &= \sum_{j=0}^{k+1} \begin{bmatrix} k+1 \\ j \end{bmatrix} f^{(k+1-j)}(x)g^{(j)}(x) \text{ (by part a),}\end{aligned}$$

which is $p(k + 1)$.

5. From Exercise 4, each natural number n can be written as a product of prime numbers, $n = p_1 \cdot \ldots \cdot p_k$. Let $n = q_1 \cdot \ldots \cdot q_l$ be a second such factorization. We must show that $k = l$ and the p_i and the q_j are the same, up to the order of the factors. Let $p(n)$ be this statement. We will proceed by induction on n, using the modified form (A2.5).

(i) $p(2)$ is true, since 2 is prime itself.

(ii) Assume that $p(2)$ through $p(n - 1)$ are true, and consider $p(n)$. If $n = p_1 \cdot \ldots \cdot p_k = q_1 \cdot \ldots \cdot q_l$, where the p_i and the q_j are prime, then since p_1 divides n, p_1 must divide one of the q_j. By rearranging the q's, we may assume that $j = 1$. Since q_1 is also prime, we must have that $p_1 = q_1$. Canceling that factor, we are left with the equality $p_2 \cdot \ldots \cdot p_k = q_2 \cdot \ldots \cdot q_l$. Since $p_1 = q_1 \geqslant 2$, both sides of this equation are strictly less than n, so our induction hypothesis applies to these two factorizations. Hence $k = l$, and the p_i and q_j are the same, up to order.

7. Proceed by induction on n.

(i) The first case is $p(0)$. The only subset of the empty set is the empty set itself, so $P(\phi) = \{\phi\}$, which has $2^0 = 1$ element.

(ii) Assume that $p(n)$ is true; that is, for all sets X with n elements, $P(X)$ has 2^n elements. Let X be a set with $n + 1$ elements, $X = \{x_1, \ldots, x_{n+1}\}$. Let $X' = \{x_1, \ldots, x_n\}$. The subsets of X can be broken into two groups: those which contain x_{n+1} and those which do not. Those which do not are subsets of X'; those which do all have the form $A \cup \{x_{n+1}\}$, where A is a subset of X'. Hence

$$P(X) = P(X') \cup \{A \cup \{x_{n+1}\} | A \in P(X')\}$$

By induction, there are 2^n of each of the two kinds of elements of $P(X)$, hence $P(X)$ contains $2^n + 2^n = 2^{n+1}$ elements.

9. a) To show that $a_{n+1} = a_n + a_{n-1}$, we express $a_n + a_{n-1}$ as a sum of quotients with denominator 2^{n+1}:

$$a_n + a_{n-1} = [(1 + \sqrt{5})/2]^n + [(1 - \sqrt{5})/2]^n + [(1 + \sqrt{5})/2]^{n-1} + [(1 - \sqrt{5})/2]^{n-1}$$
$$= [(1 + \sqrt{5})/2]^{n-1}[(1 + \sqrt{5})/2 + 1] + [(1 - \sqrt{5})/2]^{n-1}[(1 - \sqrt{5})/2 + 1]$$
$$= [(1 + \sqrt{5})/2]^{n+1} + [(1 - \sqrt{5})/2]^{n+1}$$
$$= a_{n+1}.$$

b) Proceed by induction.
  (i) $p(2)$ is the statement that a_2 is an integer. This follows since $a_0 = 2$, $a_1 = 1$, so $a_2 = 1 + 2 = 3$.
  (ii) $p(n)$ is the statement that a_n is an integer. Since $a_n = a_{n-1} + a_{n-2}$, this follows immediately from the induction hypothesis.

Index

H

I

J

K

L

M

A 8
B 9
C 0
D 1
E 2
F 3
G 4
H 5
I 6
J 7

A CATALOG OF SELECTED

DOVER BOOKS

IN SCIENCE AND MATHEMATICS

Mathematics-Bestsellers

Mathematics–Logic and Problem Solving

PERPLEXING PUZZLES AND TANTALIZING TEASERS, Martin Gardner. Ninety-three riddles, mazes, illusions, tricky questions, word and picture puzzles, and other challenges offer hours of entertainment for youngsters. Filled with rib-tickling drawings. Solutions. 224pp. 5 3/8 x 8 1/2. 0-486-25637-5

MY BEST MATHEMATICAL AND LOGIC PUZZLES, Martin Gardner. The noted expert selects 70 of his favorite "short" puzzles. Includes The Returning Explorer, The Mutilated Chessboard, Scrambled Box Tops, and dozens more. Complete solutions included. 96pp. 5 3/8 x 8 1/2. 0-486-28152-3

THE LADY OR THE TIGER?: and Other Logic Puzzles, Raymond M. Smullyan. Created by a renowned puzzle master, these whimsically themed challenges involve paradoxes about probability, time, and change; metapuzzles; and self-referentiality. Nineteen chapters advance in difficulty from relatively simple to highly complex. 1982 edition. 240pp. 5 3/8 x 8 1/2. 0-486-47027-X

SATAN, CANTOR AND INFINITY: Mind-Boggling Puzzles, Raymond M. Smullyan. A renowned mathematician tells stories of knights and knaves in an entertaining look at the logical precepts behind infinity, probability, time, and change. Requires a strong background in mathematics. Complete solutions. 288pp. 5 3/8 x 8 1/2. 0-486-47036-9

THE RED BOOK OF MATHEMATICAL PROBLEMS, Kenneth S. Williams and Kenneth Hardy. Handy compilation of 100 practice problems, hints and solutions indispensable for students preparing for the William Lowell Putnam and other mathematical competitions. Preface to the First Edition. Sources. 1988 edition. 192pp. 5 3/8 x 8 1/2. 0-486-69415-1

KING ARTHUR IN SEARCH OF HIS DOG AND OTHER CURIOUS PUZZLES, Raymond M. Smullyan. This fanciful, original collection for readers of all ages features arithmetic puzzles, logic problems related to crime detection, and logic and arithmetic puzzles involving King Arthur and his Dogs of the Round Table. 160pp. 5 3/8 x 8 1/2. 0-486-47435-6

UNDECIDABLE THEORIES: Studies in Logic and the Foundation of Mathematics, Alfred Tarski in collaboration with Andrzej Mostowski and Raphael M. Robinson. This well-known book by the famed logician consists of three treatises: "A General Method in Proofs of Undecidability," "Undecidability and Essential Undecidability in Mathematics," and "Undecidability of the Elementary Theory of Groups." 1953 edition. 112pp. 5 3/8 x 8 1/2. 0-486-47703-7

LOGIC FOR MATHEMATICIANS, J. Barkley Rosser. Examination of essential topics and theorems assumes no background in logic. "Undoubtedly a major addition to the literature of mathematical logic." – *Bulletin of the American Mathematical Society.* 1978 edition. 592pp. 6 1/8 x 9 1/4. 0-486-46898-4

INTRODUCTION TO PROOF IN ABSTRACT MATHEMATICS, Andrew Wohlgemuth. This undergraduate text teaches students what constitutes an acceptable proof, and it develops their ability to do proofs of routine problems as well as those requiring creative insights. 1990 edition. 384pp. 6 1/2 x 9 1/4. 0-486-47854-8

FIRST COURSE IN MATHEMATICAL LOGIC, Patrick Suppes and Shirley Hill. Rigorous introduction is simple enough in presentation and context for wide range of students. Symbolizing sentences; logical inference; truth and validity; truth tables; terms, predicates, universal quantifiers; universal specification and laws of identity; more. 288pp. 5 3/8 x 8 1/2. 0-486-42259-3

Mathematics-Algebra and Calculus

VECTOR CALCULUS, Peter Baxandall and Hans Liebeck. This introductory text offers a rigorous, comprehensive treatment. Classical theorems of vector calculus are amply illustrated with figures, worked examples, physical applications, and exercises with hints and answers. 1986 edition. 560pp. 5 3/8 x 8 1/2. 0-486-46620-5

ADVANCED CALCULUS: An Introduction to Classical Analysis, Louis Brand. A course in analysis that focuses on the functions of a real variable, this text introduces the basic concepts in their simplest setting and illustrates its teachings with numerous examples, theorems, and proofs. 1955 edition. 592pp. 5 3/8 x 8 1/2. 0-486-44548-8

ADVANCED CALCULUS, Avner Friedman. Intended for students who have already completed a one-year course in elementary calculus, this two-part treatment advances from functions of one variable to those of several variables. Solutions. 1971 edition. 432pp. 5 3/8 x 8 1/2. 0-486-45795-8

METHODS OF MATHEMATICS APPLIED TO CALCULUS, PROBABILITY, AND STATISTICS, Richard W. Hamming. This 4-part treatment begins with algebra and analytic geometry and proceeds to an exploration of the calculus of algebraic functions and transcendental functions and applications. 1985 edition. Includes 310 figures and 18 tables. 880pp. 6 1/2 x 9 1/4. 0-486-43945-3

BASIC ALGEBRA I: Second Edition, Nathan Jacobson. A classic text and standard reference for a generation, this volume covers all undergraduate algebra topics, including groups, rings, modules, Galois theory, polynomials, linear algebra, and associative algebra. 1985 edition. 528pp. 6 1/8 x 9 1/4. 0-486-47189-6

BASIC ALGEBRA II: Second Edition, Nathan Jacobson. This classic text and standard reference comprises all subjects of a first-year graduate-level course, including in-depth coverage of groups and polynomials and extensive use of categories and functors. 1989 edition. 704pp. 6 1/8 x 9 1/4. 0-486-47187-X

CALCULUS: An Intuitive and Physical Approach (Second Edition), Morris Kline. Application-oriented introduction relates the subject as closely as possible to science with explorations of the derivative; differentiation and integration of the powers of x; theorems on differentiation, antidifferentiation; the chain rule; trigonometric functions; more. Examples. 1967 edition. 960pp. 6 1/2 x 9 1/4. 0-486-40453-6

ABSTRACT ALGEBRA AND SOLUTION BY RADICALS, John E. Maxfield and Margaret W. Maxfield. Accessible advanced undergraduate-level text starts with groups, rings, fields, and polynomials and advances to Galois theory, radicals and roots of unity, and solution by radicals. Numerous examples, illustrations, exercises, appendixes. 1971 edition. 224pp. 6 1/8 x 9 1/4. 0-486-47723-1

AN INTRODUCTION TO THE THEORY OF LINEAR SPACES, Georgi E. Shilov. Translated by Richard A. Silverman. Introductory treatment offers a clear exposition of algebra, geometry, and analysis as parts of an integrated whole rather than separate subjects. Numerous examples illustrate many different fields, and problems include hints or answers. 1961 edition. 320pp. 5 3/8 x 8 1/2. 0-486-63070-6

LINEAR ALGEBRA, Georgi E. Shilov. Covers determinants, linear spaces, systems of linear equations, linear functions of a vector argument, coordinate transformations, the canonical form of the matrix of a linear operator, bilinear and quadratic forms, and more. 387pp. 5 3/8 x 8 1/2. 0-486-63518-X

Mathematics–Probability and Statistics

BASIC PROBABILITY THEORY, Robert B. Ash. This text emphasizes the probabilistic way of thinking, rather than measure-theoretic concepts. Geared toward advanced undergraduates and graduate students, it features solutions to some of the problems. 1970 edition. 352pp. 5 3/8 x 8 1/2. 0-486-46628-0

PRINCIPLES OF STATISTICS, M. G. Bulmer. Concise description of classical statistics, from basic dice probabilities to modern regression analysis. Equal stress on theory and applications. Moderate difficulty; only basic calculus required. Includes problems with answers. 252pp. 5 5/8 x 8 1/4. 0-486-63760-3

OUTLINE OF BASIC STATISTICS: Dictionary and Formulas, John E. Freund and Frank J. Williams. Handy guide includes a 70-page outline of essential statistical formulas covering grouped and ungrouped data, finite populations, probability, and more, plus over 1,000 clear, concise definitions of statistical terms. 1966 edition. 208pp. 5 3/8 x 8 1/2. 0-486-47769-X

GOOD THINKING: The Foundations of Probability and Its Applications, Irving J. Good. This in-depth treatment of probability theory by a famous British statistician explores Keynesian principles and surveys such topics as Bayesian rationality, corroboration, hypothesis testing, and mathematical tools for induction and simplicity. 1983 edition. 352pp. 5 3/8 x 8 1/2. 0-486-47438-0

INTRODUCTION TO PROBABILITY THEORY WITH CONTEMPORARY APPLICATIONS, Lester L. Helms. Extensive discussions and clear examples, written in plain language, expose students to the rules and methods of probability. Exercises foster problem-solving skills, and all problems feature step-by-step solutions. 1997 edition. 368pp. 6 1/2 x 9 1/4. 0-486-47418-6

CHANCE, LUCK, AND STATISTICS, Horace C. Levinson. In simple, non-technical language, this volume explores the fundamentals governing chance and applies them to sports, government, and business. "Clear and lively ... remarkably accurate." – *Scientific Monthly*. 384pp. 5 3/8 x 8 1/2. 0-486-41997-5

FIFTY CHALLENGING PROBLEMS IN PROBABILITY WITH SOLUTIONS, Frederick Mosteller. Remarkable puzzlers, graded in difficulty, illustrate elementary and advanced aspects of probability. These problems were selected for originality, general interest, or because they demonstrate valuable techniques. Also includes detailed solutions. 88pp. 5 3/8 x 8 1/2. 0-486-65355-2

EXPERIMENTAL STATISTICS, Mary Gibbons Natrella. A handbook for those seeking engineering information and quantitative data for designing, developing, constructing, and testing equipment. Covers the planning of experiments, the analyzing of extreme-value data; and more. 1966 edition. Index. Includes 52 figures and 76 tables. 560pp. 8 3/8 x 11. 0-486-43937-2

STOCHASTIC MODELING: Analysis and Simulation, Barry L. Nelson. Coherent introduction to techniques also offers a guide to the mathematical, numerical, and simulation tools of systems analysis. Includes formulation of models, analysis, and interpretation of results. 1995 edition. 336pp. 6 1/8 x 9 1/4. 0-486-47770-3

INTRODUCTION TO BIOSTATISTICS: Second Edition, Robert R. Sokal and F. James Rohlf. Suitable for undergraduates with a minimal background in mathematics, this introduction ranges from descriptive statistics to fundamental distributions and the testing of hypotheses. Includes numerous worked-out problems and examples. 1987 edition. 384pp. 6 1/8 x 9 1/4. 0-486-46961-1

Mathematics-Geometry and Topology

PROBLEMS AND SOLUTIONS IN EUCLIDEAN GEOMETRY, M. N. Aref and William Wernick. Based on classical principles, this book is intended for a second course in Euclidean geometry and can be used as a refresher. More than 200 problems include hints and solutions. 1968 edition. 272pp. 5 3/8 x 8 1/2. 0-486-47720-7

TOPOLOGY OF 3-MANIFOLDS AND RELATED TOPICS, Edited by M. K. Fort, Jr. With a New Introduction by Daniel Silver. Summaries and full reports from a 1961 conference discuss decompositions and subsets of 3-space; n-manifolds; knot theory; the Poincaré conjecture; and periodic maps and isotopies. Familiarity with algebraic topology required. 1962 edition. 272pp. 6 1/8 x 9 1/4. 0-486-47753-3

POINT SET TOPOLOGY, Steven A. Gaal. Suitable for a complete course in topology, this text also functions as a self-contained treatment for independent study. Additional enrichment materials make it equally valuable as a reference. 1964 edition. 336pp. 5 3/8 x 8 1/2. 0-486-47222-1

INVITATION TO GEOMETRY, Z. A. Melzak. Intended for students of many different backgrounds with only a modest knowledge of mathematics, this text features self-contained chapters that can be adapted to several types of geometry courses. 1983 edition. 240pp. 5 3/8 x 8 1/2. 0-486-46626-4

TOPOLOGY AND GEOMETRY FOR PHYSICISTS, Charles Nash and Siddhartha Sen. Written by physicists for physics students, this text assumes no detailed background in topology or geometry. Topics include differential forms, homotopy, homology, cohomology, fiber bundles, connection and covariant derivatives, and Morse theory. 1983 edition. 320pp. 5 3/8 x 8 1/2. 0-486-47852-1

BEYOND GEOMETRY: Classic Papers from Riemann to Einstein, Edited with an Introduction and Notes by Peter Pesic. This is the only English-language collection of these 8 accessible essays. They trace seminal ideas about the foundations of geometry that led to Einstein's general theory of relativity. 224pp. 6 1/8 x 9 1/4. 0-486-45350-2

GEOMETRY FROM EUCLID TO KNOTS, Saul Stahl. This text provides a historical perspective on plane geometry and covers non-neutral Euclidean geometry, circles and regular polygons, projective geometry, symmetries, inversions, informal topology, and more. Includes 1,000 practice problems. Solutions available. 2003 edition. 480pp. 6 1/8 x 9 1/4. 0-486-47459-3

TOPOLOGICAL VECTOR SPACES, DISTRIBUTIONS AND KERNELS, François Trèves. Extending beyond the boundaries of Hilbert and Banach space theory, this text focuses on key aspects of functional analysis, particularly in regard to solving partial differential equations. 1967 edition. 592pp. 5 3/8 x 8 1/2. 0-486-45352-9

INTRODUCTION TO PROJECTIVE GEOMETRY, C. R. Wylie, Jr. This introductory volume offers strong reinforcement for its teachings, with detailed examples and numerous theorems, proofs, and exercises, plus complete answers to all odd-numbered end-of-chapter problems. 1970 edition. 576pp. 6 1/8 x 9 1/4. 0-486-46895-X

FOUNDATIONS OF GEOMETRY, C. R. Wylie, Jr. Geared toward students preparing to teach high school mathematics, this text explores the principles of Euclidean and non-Euclidean geometry and covers both generalities and specifics of the axiomatic method. 1964 edition. 352pp. 6 x 9. 0-486-47214-0

Mathematics–History

THE WORKS OF ARCHIMEDES, Archimedes. Translated by Sir Thomas Heath. Complete works of ancient geometer feature such topics as the famous problems of the ratio of the areas of a cylinder and an inscribed sphere; the properties of conoids, spheroids, and spirals; more. 326pp. 5 3/8 x 8 1/2. 0-486-42084-1

THE HISTORICAL ROOTS OF ELEMENTARY MATHEMATICS, Lucas N. H. Bunt, Phillip S. Jones, and Jack D. Bedient. Exciting, hands-on approach to understanding fundamental underpinnings of modern arithmetic, algebra, geometry and number systems examines their origins in early Egyptian, Babylonian, and Greek sources. 336pp. 5 3/8 x 8 1/2. 0-486-25563-8

THE THIRTEEN BOOKS OF EUCLID'S ELEMENTS, Euclid. Contains complete English text of all 13 books of the Elements plus critical apparatus analyzing each definition, postulate, and proposition in great detail. Covers textual and linguistic matters; mathematical analyses of Euclid's ideas; classical, medieval, Renaissance and modern commentators; refutations, supports, extrapolations, reinterpretations and historical notes. 995 figures. Total of 1,425pp. All books 5 3/8 x 8 1/2.
Vol. I: 443pp. 0-486-60088-2
Vol. II: 464pp. 0-486-60089-0
Vol. III: 546pp. 0-486-60090-4

A HISTORY OF GREEK MATHEMATICS, Sir Thomas Heath. This authoritative two-volume set that covers the essentials of mathematics and features every landmark innovation and every important figure, including Euclid, Apollonius, and others. 5 3/8 x 8 1/2.
Vol. I: 461pp. 0-486-24073-8
Vol. II: 597pp. 0-486-24074-6

A MANUAL OF GREEK MATHEMATICS, Sir Thomas L. Heath. This concise but thorough history encompasses the enduring contributions of the ancient Greek mathematicians whose works form the basis of most modern mathematics. Discusses Pythagorean arithmetic, Plato, Euclid, more. 1931 edition. 576pp. 5 3/8 x 8 1/2. 0-486-43231-9

CHINESE MATHEMATICS IN THE THIRTEENTH CENTURY, Ulrich Libbrecht. An exploration of the 13th-century mathematician Ch'in, this fascinating book combines what is known of the mathematician's life with a history of his only extant work, the Shu-shu chiu-chang. 1973 edition. 592pp. 5 3/8 x 8 1/2. 0-486-44619-0

PHILOSOPHY OF MATHEMATICS AND DEDUCTIVE STRUCTURE IN EUCLID'S ELEMENTS, Ian Mueller. This text provides an understanding of the classical Greek conception of mathematics as expressed in Euclid's Elements. It focuses on philosophical, foundational, and logical questions and features helpful appendixes. 400pp. 6 1/2 x 9 1/4. 0-486-45300-6

BEYOND GEOMETRY: Classic Papers from Riemann to Einstein, Edited with an Introduction and Notes by Peter Pesic. This is the only English-language collection of these 8 accessible essays. They trace seminal ideas about the foundations of geometry that led to Einstein's general theory of relativity. 224pp. 6 1/8 x 9 1/4. 0-486-45350-2

HISTORY OF MATHEMATICS, David E. Smith. Two-volume history – from Egyptian papyri and medieval maps to modern graphs and diagrams. Non-technical chronological survey with thousands of biographical notes, critical evaluations, and contemporary opinions on over 1,100 mathematicians. 5 3/8 x 8 1/2.
Vol. I: 618pp. 0-486-20429-4
Vol. II: 736pp. 0-486-20430-8

Engineering

FUNDAMENTALS OF ASTRODYNAMICS, Roger R. Bate, Donald D. Mueller, and Jerry E. White. Teaching text developed by U.S. Air Force Academy develops the basic two-body and n-body equations of motion; orbit determination; classical orbital elements, coordinate transformations; differential correction; more. 1971 edition. 455pp. 5 3/8 x 8 1/2. 0-486-60061-0

INTRODUCTION TO CONTINUUM MECHANICS FOR ENGINEERS: Revised Edition, Ray M. Bowen. This self-contained text introduces classical continuum models within a modern framework. Its numerous exercises illustrate the governing principles, linearizations, and other approximations that constitute classical continuum models. 2007 edition. 320pp. 6 1/8 x 9 1/4. 0-486-47460-7

ENGINEERING MECHANICS FOR STRUCTURES, Louis L. Bucciarelli. This text explores the mechanics of solids and statics as well as the strength of materials and elasticity theory. Its many design exercises encourage creative initiative and systems thinking. 2009 edition. 320pp. 6 1/8 x 9 1/4. 0-486-46855-0

FEEDBACK CONTROL THEORY, John C. Doyle, Bruce A. Francis and Allen R. Tannenbaum. This excellent introduction to feedback control system design offers a theoretical approach that captures the essential issues and can be applied to a wide range of practical problems. 1992 edition. 224pp. 6 1/2 x 9 1/4. 0-486-46933-6

THE FORCES OF MATTER, Michael Faraday. These lectures by a famous inventor offer an easy-to-understand introduction to the interactions of the universe's physical forces. Six essays explore gravitation, cohesion, chemical affinity, heat, magnetism, and electricity. 1993 edition. 96pp. 5 3/8 x 8 1/2. 0-486-47482-8

DYNAMICS, Lawrence E. Goodman and William H. Warner. Beginning engineering text introduces calculus of vectors, particle motion, dynamics of particle systems and plane rigid bodies, technical applications in plane motions, and more. Exercises and answers in every chapter. 619pp. 5 3/8 x 8 1/2. 0-486-42006-X

ADAPTIVE FILTERING PREDICTION AND CONTROL, Graham C. Goodwin and Kwai Sang Sin. This unified survey focuses on linear discrete-time systems and explores natural extensions to nonlinear systems. It emphasizes discrete-time systems, summarizing theoretical and practical aspects of a large class of adaptive algorithms. 1984 edition. 560pp. 6 1/2 x 9 1/4. 0-486-46932-8

INDUCTANCE CALCULATIONS, Frederick W. Grover. This authoritative reference enables the design of virtually every type of inductor. It features a single simple formula for each type of inductor, together with tables containing essential numerical factors. 1946 edition. 304pp. 5 3/8 x 8 1/2. 0-486-47440-2

THERMODYNAMICS: Foundations and Applications, Elias P. Gyftopoulos and Gian Paolo Beretta. Designed by two MIT professors, this authoritative text discusses basic concepts and applications in detail, emphasizing generality, definitions, and logical consistency. More than 300 solved problems cover realistic energy systems and processes. 800pp. 6 1/8 x 9 1/4. 0-486-43932-1

THE FINITE ELEMENT METHOD: Linear Static and Dynamic Finite Element Analysis, Thomas J. R. Hughes. Text for students without in-depth mathematical training, this text includes a comprehensive presentation and analysis of algorithms of time-dependent phenomena plus beam, plate, and shell theories. Solution guide available upon request. 672pp. 6 1/2 x 9 1/4. 0-486-41181-8